FLORA UNVEILED

Flora Unveiled

THE DISCOVERY AND DENIAL OF SEX IN PLANTS

Lincoln Taiz and Lee Taiz

OXFORD
UNIVERSITY PRESS

Oxford University Press is a department of the University of Oxford. It furthers the University's objective of excellence in research, scholarship, and education by publishing worldwide. Oxford is a registered trade mark of Oxford University Press in the UK and certain other countries.

Published in the United States of America by Oxford University Press
198 Madison Avenue, New York, NY 10016, United States of America.

CIP data is on file at the Library of Congress
ISBN 978–0–19–049026–3

9 8 7 6 5 4 3 2 1

Printed by Sheridan Books, Inc., United States of America

Contents

Contents

Preface

IN THIS BOOK, we attempt to answer two basic questions: why did it take so long to discover sex in plants, and why, after its proposal and experimental confirmation in the late seventeenth century, did the debate continue for another 150 years?

The answers to these questions have long roots, extending deep into human history, and some of the same cultural factors that delayed the discovery of sex in plants also prolonged the debate afterward. Some of these cultural factors are so entrenched that they are still with us today. Consequently, we found that significant progress in understanding plant sex did not always depend on advances in technology. For example, the crucial insight that led to the full resolution of the question of plant sex and to the taxonomic unification of the plant kingdom in the mid-nineteenth century was achieved by the severely near-sighted Wilhelm Hofmeister by examining his hand-cut fresh sections using a relatively crude light microscope of a type that was widely available at the time. Clearly, conceptual factors were barriers that stymied progress.

In the absence of scientific data, cultural concepts and assumptions fill the gaps in our knowledge, and these concepts and assumptions often remain durable long after those gaps have been unequivocally filled. They are embedded in custom, economics, religion, and even language, and they can make it difficult to uncover or accept new insights. Even when new scientific theories have been thoroughly documented, older paradigms persist in obscuring fact, propping up and "proving" the truth of groundless convictions. Invidious fantasies concerning race are an example of this kind of fossilized perception.

Any discussion of sex in humans, excepting the most clinical, almost inevitably leads to discussions of gender, and ideas about gender often take precedence over what is scientifically demonstrable about biological sex per se. Throughout human history, much of what has been taken to be attributes of sex are actually assumptions about gender—the subjective qualities we attribute to males or females. The association of plants with women and the entanglement of ideas about plants with ideas about gender roles appears early in human history. This association goes beyond mere analogy based on the resemblance of the gravid female to the swelling fruit that contains a seed that can reproduce its own kind. Social and economic roles come into play as well. The association of plants, especially flowers, with women is a thread that can be found weaving through vast periods of human history, influencing the way in which plants were, and are, perceived. It is an important thread that we try to follow.

Thus, simply focusing on the chain of successful theories and experiments that led to our current understanding of sex in plants would be inadequate to answering our questions. Failed theories and experimental artifacts were not mere curiosities or impediments to progress, as they most certainly were from a strictly scientific perspective. Examining them became crucial to the understanding of process, to answering the question "Why?"

Our approach to the history of the discovery of sex in plants is both "whiggish" and contextual. On the one hand, we consider the problem over a sweeping timespan, from the Paleolithic to the middle of the nineteenth century. On the other hand, the book is divided into historical periods in specific geographic regions that were significant in terms of the development of science. We restrict ourselves to Western civilization because the endpoint of the narrative is the full elucidation of the sexual cycle of plants and the subsequent consolidation of the plant kingdom as a taxonomic unit, which occurred in Europe during the modern period. For the time being at least, we leave aside parallel developments in Asia and the New World that did not lead to the scientific understanding of plant sex.

The challenge to discover what human beings thought about the lives of plants in different eras of human history, and how these perceptions advanced or hindered the understanding of sex in plants, was formidable. We realized at the outset that, in order to do justice to our subject, we needed to go far beyond conventional research. We needed to inhabit the landscapes where the significant events of our history had actually occurred and engage first-hand with the places where important *dramatis personae* had actually lived and worked. As a result, research for this book became both a pilgrimage and an adventure. Over many years, it took us through some of the world's most significant archaeological sites, libraries, and museums, and its most fascinating historical gardens. We believe that our journeys impart a resonance to this history that could not have been achieved otherwise. But although we were able to explore a wealth of relevant locales, some of the places that we wished to encounter were inaccessible because of war or political turmoil. Like a great many others, we have shed tears as irrecoverable archaeological sites, museums, and monuments of the past have been looted or destroyed in recent years, some of these crimes committed in senseless wars, some in the name of religion. Contemplating the disasters that have ensured that countless treasures can never again be seen in the history of the world has increased our boundless gratitude to the people and institutions that have rescued so much irreplaceable material from destruction.

Acknowledgments

OUR MOST VALUABLE resource during the writing of *Flora Unveiled* has been the many scholars, librarians, curators, students, and friends we have consulted and interacted with over the two decades that it has taken us to complete our research. *Flora Unveiled* is a broad synthesis that required considerable input from multiple disciplines outside our own areas of expertise in plant biology, including archaeology, archaeobotany, linguistics, history, literature, history of science, gender studies, and philosophy. Among those who provided significant guidance are Elizabeth A. Augspach, Ofer Bar-Yosef, Paul Bahn, Alix Cooper, Oliver Dickenson, Mark Elvin, Yosef Garfinkel, Diane Gifford-Gonzalez, Jaako Hämeen-Anttila, Wolfgang Heimpel, Ian Hodder, Jules Janick, Helmut W. Klug, John Lynch, Michael Marder, Joan Marler, Naomi Miller, Gary Myles, Olga Sofer, Catherine Preece, Karen Reeds, Sophia Rhizopoulou, Ann Macy Roth, Gonzalo Rubio, Lawrence Guy Straus, and Peter M. Warren. The generous assistance of these scholars, as well as many others, is gratefully acknowledged.

In addition, we would like to thank Michael Leapman for his kind hospitality during one of our visits to London and for allowing us access to his personal collection of papers on Thomas Fairchild. We also wish to thank our long-time friend and colleague, Heven Sze, at the University of Maryland, for a critical reading of preliminary versions of the chapters on Alternation of Generations.

We are especially indebted to Gildas Hamel, our colleague at the University of California, Santa Cruz, for providing translations of many key Latin passages from the Medieval, Renaissance, and Early Modern periods, and to Mark Elvin for sending us preliminary drafts of his forthcoming comprehensive volume of English translations and interpretations of the Latin works by Camerarius and his contemporaries. Finally, we thank Georges

Métailié for stimulating discussions and for making available his unpublished manuscript on ancient Chinese knowledge of sex in plants.

All of the preceding individuals deserve much of the credit for whatever we got right, while the mistakes are purely our own.

We are also grateful to many others who offered encouragement at various times during the long gestation period of the manuscript, among them Judith Aissen, Barry and Emma Jean Bowman, Ben Leeds Carson, James Clifford, Jennifer Colby, Joanie DeNeffe, David Federman, Shimon and Amira Gepstein, Jean Langenheim, Linda Locatelli, Kerrie McCaffrey, Wendy Miller, Angus Murphy, Wendy Peer, Gary Silberstein, Bernardo Taiz, Sherry Walker, Edmond and Beverly Weiss, and Joel and Nicole Yellin.

Finally, we thank Elizabeth Morales for her excellent illustrations in Chapters 17 and 18, and our editor Jeremy Lewis at Oxford University Press, whose faith in a book on sex in plants has brought it to fruition.

FLORA UNVEILED

There is much to be said for failure. It is more interesting than success.
—MAX BEERBOHM (1872–1956)

1

The Quandary Over Plant Sex

ANYONE WHO HAS ever cultivated a garden or tended an orchard and watched with eager anticipation as their crops progressed through the stages of plant development is aware of the organic relationship between flowers and fruits. Think of your favorite fruits—apples, watermelons, papayas, cherries, peaches, tomatoes, peppers, avocados, walnuts, or one of the hundreds of others sold in marketplaces throughout the world. Whether destined for the fruit bowl, the salad bowl, or the cereal bowl, all fruits contain seeds and grow from small, vase-shaped structures called *carpels* (or *pistils*) located at the centers of flowers.[1]

To the casual observer, the transformation of these tiny, green carpels into their respective fruits occurs seamlessly, as if fruit production were simply a continuation of vegetative growth. But it is not quite so simple. A series of marvelous, complex, and more or less random events must occur before fruits and seeds can be formed. If any of the steps in this elaborate succession of events fails, barrenness results, and flowers wither and drop uselessly to the ground. When this happens in an agricultural context, the consequences can be catastrophic. It's an occurrence farmers and growers know only too well, the threat of which can rob them of sleep at the start of each growing season.

Apart from their agricultural importance, flowers also appeal to our senses. What could be more fragrant than a rose, more alluring than an orchid, more exquisite than Queen Anne's lace, more brilliant than a sunlit field of golden poppies? Because of their attractiveness, fragrance, and transitory existence, flowers are associated with youthful beauty, especially female beauty. As Samantha George and others have pointed out, "Flowers are traditionally emblematic of the female sex in literary texts."[2] Flowers give rise to fruits and swell with seed. The analogy to female pregnancy is inescapable, as in the biblical phrase "fruit of the womb."[3] Yet, at the same time, fruit formation appears to be quite different. Unlike female pregnancy, the pistil of a rose blossom may expand into a seed-filled rose hip apparently without carnality of any sort.

Among early Christians, flowers inspired a sense of piety and spiritual immanence, embodying the union of the earthly and the divine. From the Middle Ages onward, flowers have served as symbols of the core mystery in Christianity: the Virgin who gives birth miraculously to the divine child. Throughout the Middle Ages and the Renaissance, the lily appears almost as regularly as the angel Gabriel in paintings of the Annunciation. In Dante's *Paradiso*, a mystic white rose becomes a symbol of the poet's pure and unconsummated love for Beatrice, the paragon of female beauty and virtue. In Dante's rose, the divine light of the sun becomes material, and earthly love and divine love become one.

Gabriel and Dante notwithstanding, the reputations of lilies and roses for virginity are entirely undeserved. "Virgin birth" can and does occur in some flowers (it is called "parthenogenesis"), but it is relatively rare. Nearly always, sex is involved, although not of the carnal sort associated with animals. *Pollination*, not copulation, is the vehicle for sexual union in plants.

The essential elements of pollination are quite simple. Pollen grains the size of dust particles are released by the stamens of a flower and find their way—usually with the help of wind, insects, or other animal agents—to the upper surface of the pistil of another flower. There they germinate, forming a thread-like tube that grows down through the neck of the pistil and emerges in the hollow base called the *ovary*. The pollen tube next empties its cargo of two sperm cells into a future seed, called the *ovule*. The sperm cells enter the embryo sac inside the ovule, and one of the sperm cells fuses with the egg, a process called *fertilization*. *Voila!* The first cell of the new embryonic plant is born. This, in broad strokes, is what happens during sexual encounters of the floral kind.

As familiar as this basic description of plant sex may seem, the biological role of pollen in plant reproduction is still widely misunderstood. This point was first driven home to us when we came across a newspaper article about a survey given to visitors to the new Pollinarium Exhibit at the National Zoo in Washington, DC. The purpose of the exhibit was to educate the public about the function of flowers in sexual reproduction and the role of insects in bringing about pollination. Before the exhibit opened, 100 visitors to the zoo were interviewed to determine what they knew about pollination. When asked to define pollen in the pre-exhibit interview, about 70% of those surveyed failed to connect pollen with sexual reproduction in plants. After the exhibit had opened, the same survey was repeated with a different set of 100 visitors. The post-exhibit interviews were conducted just outside the exit door of the Pollinarium.[4] Once again, 70% of those surveyed failed to connect pollen with sexual reproduction in plants. In other words, there was zero improvement in visitors' understanding of the sexual role of pollen *after viewing the pollinarium exhibit*.

Intrigued by the results of the poll, we decided to visit the Pollinarium to evaluate its effectiveness first-hand. The exhibit was housed in a small tropical greenhouse attached to the Invertebrate Building. We arrived on a cold rainy day in December. Upon entering the muggy warmth of the greenhouse, we were immediately surrounded by hundreds of fluttering, iridescent butterflies, flitting delicately from flower to flower and from shoulder to shoulder, oblivious of their human observers. It was as if we had walked straight into a tropical rainforest, vibrant with beating life. A large color poster made it abundantly clear what the exhibit was about. It read:

PLANT SEX?
Birds do it, bees do it, and even flowers and trees do it!

No punches pulled here. The body of the poster included a clearly drawn life cycle of a strawberry plant and a picture of a butterfly in the act of pollinating a flower. The process of pollination was clearly and simply explained. It seemed to us that the message that pollen is involved in plant sexual reproduction came across loud and clear. But the clarity of the message was not reflected in the Pollinarium survey. Taken at face value, the survey suggests that, however technologically sophisticated the average person may be when it comes to using the latest digital device, he or she still has a very limited understanding of the fundamental biological principle of the sexual role of pollen in plant reproduction. A disheartening number of people in the survey understood the function of flowers much as the ancients did, as beautiful, colorful, fragrant adornments, whereas pollen was perceived in negative terms as the cause of allergies.[5]

We wondered, is pollination—that crucial step in the sexual reproduction of all seed-producing plants—such an arcane, esoteric phenomenon that most people have either never heard of it or find too difficult to comprehend? Thinking about this question, we began to suspect that the problem was not so much one of education as a subjective bias that most people have, and perhaps have always had, about plants. If people tend to view plants as inherently *asexual*, if they think of sexuality as somehow incompatible with the very nature of plants, it would take far more than a casual stroll through the Pollinarium Exhibit to challenge their preconceptions.

The Pollinarium survey was much too small to be statistically significant and may not accurately reflect the average zoo-goer's true understanding of how plants reproduce. On the other hand, there is historical precedent for the apparent ignorance about plant sexuality reflected in the survey. Historians of science have long marveled at the extraordinarily long delay before sex in plants was discovered at the end of the seventeenth century. This delay is all the more remarkable when one considers that sex in animals was probably discovered around 14,000 years ago, when dogs were first domesticated and bred. Thus, a span of more than 13,000 years separates the discovery of sex in animals and plants, even though humans have depended on plants and agriculture for their survival for at least 10,000 years. Equally astonishing, after the new sexual theory was first proposed by the British physician Nehemiah Grew in 1684, it was summarily rejected by some of the leading botanical lights of the day, and it continued to be challenged on philosophical, moral, and religious grounds for another 150 years until the middle of the nineteenth century! Perhaps the apparent resistance to the idea of sex in plants suggested by the Pollinarium survey has its roots in something deep within the human psyche, an ancient prejudice rooted in "common sense" going back to prehistoric times.

A well-known example of how an entrenched bias based on common sense can block scientific discovery is the delayed acceptance of the heliocentric model of the solar system. Aristarchus of Samos, a Pythagorean astronomer who lived in the third century BCE was the first to postulate that the Earth revolved in a circular orbit around the sun, but the theory failed to take hold because it violated common sense. Instead, the geocentric Ptolemaic system prevailed, along with its famous epicycles. It wasn't until the fifteenth century that Copernicus proposed the modern heliocentric model. However, Copernicus's model lacked predictive power because it was based on perfectly circular planetary orbits, an idea dating back to the Greeks. Another 200 years were to pass before Johann Kepler validated the heliocentric model by demonstrating that the orbits of the planets around the sun were actually elliptical.

Why did it take 2,000 years for the acceptance of the heliocentric model of the solar system? Clearly, common sense combined with theological teachings posed an impenetrable epistemological barrier that prevented even the greatest thinkers of the ancient world from formulating the correct heliocentric model. It seems likely that a similar conceptual obstacle hindered the discovery of plant sexuality and delayed the theory's universal acceptance long after it was proposed. We can visualize the relationship between science and culture in the following simple diagram:

Culture ↔ Belief ↔ Perception ← Reality

What we call "science" is actually a subset of our overall belief system, in the sense that we *believe* that it is based on a correct interpretation of external reality. However, as illustrated in the diagram, our beliefs are constantly being modified by two major influences: culture (including religion) and perception. We learn to view the world through the lens provided to us by our culture. However, our beliefs are capable of being modified by our perceptions, which are derived from sensory data obtained directly or indirectly from external reality.[6] When our belief system becomes altered, it feeds back into our culture, so the relationship between culture and belief is a dynamic one. At the same time, our perceptions can also be influenced by our beliefs. Hence, the saying "seeing is believing" has its converse: "believing is seeing."

What cultural biases might have stood in the way of the discovery of sex in plants? Before addressing this question, we first need to distinguish between two related but distinct terms: sex and gender. *Sex* refers to the biologically determined characteristics of males and females; *gender* refers to those culturally defined qualities traditionally associated with either of the two sexes. Despite their somewhat arbitrary nature, ideas about gender can become such familiar items of our mental furniture that they acquire the patina of common sense.

The association of gender and color provides a striking example. As everyone *knows*, the proper colors for dressing infants and toddlers are pink for girls and blue for boys. However, the gender color code was once the reverse of the present one. Prior to World War I, American infants and toddlers of both sexes typically wore white frocks. There was nothing inappropriate about dressing little boys in essentially the same outfits as little girls, and no one worried that little boys would become effeminate as a result. According to Dr. Jo Paoletti of the University of Maryland, who specializes in the history of textiles and apparel, all that changed "shortly after the turn of the century . . . when psychologists suggested that gender identification was influenced by nurture as well as nature."[7] Instead of white frocks, boys now wore trousers and girls wore dresses, and the sexes were further distinguished by color—pink versus blue. However, the gender/color associations were the opposite of what they are today: boys wore pink and girls wore blue. An article published in the trade journal *Infant's Department* (later renamed *Earnshaw's*) provided the following rationale for the colors:

> There has been a great diversity of opinion on the subject, but the generally accepted rule is pink for the boy and blue for the girl. The reason is that pink being a more decided and stronger color is more suitable for the boy; while blue, which is more delicate and dainty is prettier for the girl.[8]

Between 1900 and 1940, pink and blue gradually switched sexes, but even as late as 1939 the editors of *Parent's Magazine* were still arguing in favor of the original color coding:

> There seems to be more reasons for choosing blue for girls, than the customary pink . . . red symbolizes zeal and courage, while blue is symbolic of faith and constancy . . . all these points lead to blue for girls.[9]

By the end of World War II, the matter was no longer being debated, and the baby boomers became the first generation to regard the association of pink for girls and blue for boys as the norm.

The practice of sex-typing by color illustrates that no matter how arbitrary gender associations may be, they can quickly become accepted as common sense. Note that the character traits associated with blue versus pink—"zeal and courage" for pink (male), "faith and constancy" for blue (female)—reflect even more ancient gender biases. Such deeply ingrained prejudices, like those applied to race, potentially can provide the rationale and create the conditions for social and economic inequality.

Gender stereotypes have also had an impact on the fields of biology and medicine. Aristotle has the dubious distinction of being regarded as the first to argue on quasi-scientific grounds that women are physiologically inferior to men. The argument was initially based on the concepts, inherited from Empedocles, of the four elements (earth, air, fire, and water) and the four primary qualities (heat, cold, wetness, and dryness) that characterize them. In ranking the sexes, heat seems to have been the decisive factor. According to Aristotle, "in man the male is much superior to the female in natural heat . . . females are weaker and colder in nature, and we must look upon the female character as being a sort of natural deficiency."[10] This "deficiency" is evident during procreation:

> The male and the female differ from each other in the possession of an ability and in the lack of an ability. The male is able to concoct, formulate and to ejaculate the sperm which contains the origin of the form [of the newborn]—I do not mean here the material element out of which it is born resembling its parent but the initiating formative principle whether it acts within itself or within another. The female, on the other hand, is that which receives the seed but is unable to formulate or to ejaculate it.[11]

Unable to produce and ejaculate seed herself, the female is relegated to a passive role as a mere incubator of the man's seed. Although in other passages Aristotle allows that the female contributes gross matter to the growing embryo, he maintains that the male semen contributes the all-important soul, which, like the hands of a carpenter, directs the formation of the developing fetus.

Earlier Greek writings on sex differences, specifically those of the Hippocratic School, recognized the same differences between the sexes as Aristotle did but did not attach hierarchical values to these differences.[12] Aristotle went further than his predecessors in emphasizing the superiority of males over females. He believed he was applying reason and logic to the question of the nature of the sexes, but in fact his conclusions were strongly influenced by the tradition of social inequality that prevailed in the Athens of his day.

In his book *Making Sex: Body and Gender from the Greeks to Freud*, Thomas Laqueur discusses the role played by gender in the development of scientific theories about sexual differences in humans.[13] According to Laqueur, from the time of Aristotle to the seventeenth century in Europe, the male reproductive system was considered the fundamental and primary pathway for sexual development, whereas the female reproductive system was regarded as a secondary, less robust version of the male's. Aristotle's ideas were endorsed by Galen, the second-century Greek physician from Asia Minor. Like Aristotle, Galen saw the female reproductive organs as essentially an internalized version of the male sexual organs:

> Think too, please, of . . . the uterus turned outward and projecting. Would not the testes [ovaries] then necessarily be inside it? Would it not contain them like the scrotum? Would not the neck [the cervix and vagina], hitherto concealed inside the perineum but now pendant, be made into the male member?[14]

In other words, the female reproductive system was equivalent to a male reproductive system that had simply failed to protrude. Galen attributed this failure to the female's lack of heat. Thus, the female reproductive system is simply a colder, and therefore defective, version of the male's. Laqueur refers to this explanation as the "one-sex model." The idea of the female as a defective male reinforced the Greek patriarchal social structure, providing a "scientific" justification for depriving women of their full rights of citizenship.

According to Laqueur, this one-sex model held sway in Europe throughout the Middle Ages. However, as early as the sixteenth century, anatomists began to reject the Galenic view that the female reproductive system was merely an inverted form of the male's. After the founding of the first medical schools in Italy in the thirteenth century, human dissection was no longer the sole province of illiterate barber-surgeons conducting postmortems, as it had been in medieval times. Dissections were now overseen by medical professors who wrote textbooks based on their observations for their students. But anatomy had not yet entered the scientific stage, and no one was yet prepared to challenge the authority of Aristotle and Galen.

Some time in the sixteenth century, medical professors began performing the dissections themselves, which soon led them into uncharted territory. Novel anatomical structures were discovered that were not mentioned in the classical texts, and the cumulative effect of these discoveries began to undermine the authority of the Greek authors. Gradually, a scientifically accurate view of the complexity and functionality of human sexual differences began to emerge.[15]

Despite isolated advances, however, the one-sex model remained the dominant paradigm in medical textbooks of the Renaissance and Reformation periods.[16] In 1543, for example, the Flemish anatomist Andrea Vesalius published an illustration of a vagina and uterus that bears a striking resemblance to an internal penis (Figure 1.1).[17]

According to Laqueur, the two-sex model for animals was "invented" some time in the eighteenth century during the Enlightenment period. Laqueur uses the term "invented" to indicate that the recognition of the distinct nature of the female reproductive system was prompted as much by cultural and political factors as by scientific progress. During the Enlightenment, with its emphasis on equality, reason, and science, many archaic notions left over from the classical and medieval periods were tossed out, including the belief that the female reproductive system was a defective version of the male's. It had already become clear during the seventeenth century that there were too many differences between the male and female reproductive systems to be accounted for by the one-sex model. In the

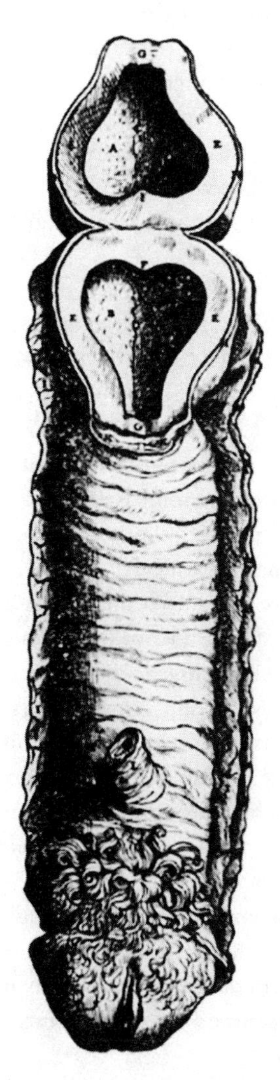

FIGURE 1.1 Vesalius' rendering of Galen's conception of the female reproductive organs as an internal penis.
From *De humani corporis fabrica* (Basel, 1543, plate 60). *Source*: Boston Medical Library in the Francis A. Countway Library of Medicine.

new way of thinking, the woman's reproductive system was simply different from the man's, not inferior to it. The two-sex model was born. Comparisons of male and female reproductive systems were soon augmented by comparative studies of the skeletons, musculature, and nervous systems of men and women. It became fashionable among anatomists to catalog the many physical differences between the sexes in order to differentiate them functionally.

Why the sudden rush to validate the two-sex model? Once again, gender politics played an important role. As discussed by Londa Schiebinger in her book *The Mind Has No Sex?*, the lofty ideal of human equality in the abstract, so prevalent during the eighteenth-century Enlightenment period, posed something of a dilemma for men who perceived women's sociopolitical aspirations as a threat.[18] If women were indeed equal to men, men could no longer rule at home, nor could they exclude women from positions of power and responsibility in business or government. Science again came to the rescue of those who sought to justify the traditional patriarchal arrangement on the basis of "natural law." A burst of new anatomical studies were published demonstrating that the physical differences between males and females were not limited to the sex organs, but included the skeleton and the nervous system as well. A man, by virtue of his distinct physiology and anatomy, was destined to lead and dominate, whereas a woman's anatomy and physiology determined her role as wife and mother. The two-sex model may have indeed represented scientific progress, but, according to Schiebinger, it owed much of its popularity in the eighteenth century to the arguments it seemingly provided to opponents of equal rights for women.

Based on the foregoing, we can say that from the Greeks onward, scientific progress on the nature of human sexuality has been influenced by gender ideology. *Did attitudes toward gender also influence the history of ideas about sexuality in plants?* Could scientific progress on the sexual role of pollen have been hindered by a reciprocal gender bias even older than the Greek gender bias regarding animal sexuality? In this book, we will argue that a one-sex model for plants predated the two-sex model by thousands of years—but instead of being gendered male, plants were gendered female. This distinction between animals and plants based on gender may, in fact, have had its origin in the oldest human societies, which often manifested a division of labor between men as hunters and women as gatherers.

To our knowledge, the only historian of botany ever to have recognized the existence of the ancient one-sex model of plants is Edward Lee Greene, the founder of the Botany Department at the University of California, Berkeley. In his book *Landmarks of Botanical History*, written around 1936 and published posthumously in 1983, Greene concluded that "it should be clear that men long ago held that plants are not asexual, but unisexual and feminine," and he expressed surprise that this "ancient doctrine" had been overlooked by other historians of botany.[19] Greene's important insight, which he noted in passing in his chapter on the Renaissance botanist Jean Ruel (discussed in Chapter 11), has languished in obscurity ever since.

To uncover the deep history of the one-sex model of plants, we must travel back through time to the last Ice Age, around 40,000 years ago, to an archaeological period known as the Upper Paleolithic. Here, we are confronted with an even more fundamental question than sex in plants: When did modern humans discover sex in themselves? The answer, it turns out, is not intuitively obvious.

NOTES

1. Collectively, all the female reproductive units of a flower are referred to as the *gynoecium* (based on the Greek terms for "woman" and "house"). The basic unit of the gynoecium is the *carpel* (from the Greek word for "fruit"). *Pistil* (from the Latin word for pestle) is another term for carpel, but there is a subtle difference: a pistil can refer either to an individual carpel (*simple pistil*) or to multiple carpels that are united into a single structure (*compound pistil*). When multiple individual carpels (simple pistils) are present, they are referred to collectively as the gynoecium. In this book, we will use the terms "carpels" and "pistils" interchangeably when referring generally to the female reproductive structures of flowers.

2. George, S. (2007), *Botany, Sexuality and Women's Writing*. Manchester University Press, p. 2.

3. In Genesis 30:2, when Rachel complains to her husband Jacob that he has not made her pregnant, he replies, "Am I in the place of God, who has withheld from you the fruit of the womb?"

4. The survey ran from April 1995 to June 1996.

5. To a certain extent, the allergy idea may have been suggested by the exhibit itself. We noticed that the very first item that greeted our eyes upon entering the Pollinarium greenhouse (besides the cloud of butterflies) was a blow-up of a photograph of a pollen grain in all its spiky glory, along with a description of its role in causing hay fever. Perhaps the hay fever sign should have been placed closer to the back door as an afterthought, where it belongs! TV commercials have promoted the fear of pollen so effectively (some stations even present "pollen and mold spore forecasts" as part of the weather report) that the average person today has been conditioned to regard pollen as an agent of disease rather than one of the keys to human survival.

6. Most of us would agree that, at the macro level, external reality cannot be influenced by our perceptions, hence the single arrow from "reality" to "perception." At the nano level of quantum mechanics, however, perception strongly influences external reality (the essence of Heisenberg's "Uncertainty Principle") and we would have to insert a double arrow between "perception" and "reality." Fortunately, we are dealing with pollination biology, not quantum mechanics, so a single arrow will suffice!

7. Salmans, Sandra, "When an it is labeled a he or a she," *New York Times*, November 16, 1989, p. C1.

8. "Pink or blue?," *Infant's Department* (June 1918):161, cited by Paoletti, 1997.

9. "What color for your baby?," *Parents'* 14, no. 3 (March 1939): 98.

10. Aristotle, *De Generatione Animalium*, Book IV, 6. Trans. William Ogle.

11. Aristotle, On the Generation of Animals, 765b9–16.

12. Cadden, Joan (1995), *Meanings of Sex Differences in the Middle Ages*, Medicine, Science, and Culture. Cambridge University Press.

13. Laqueur, T. (1990), *Making Sex: Body and Gender From the Greeks to Freud*. Harvard University Press.

14. Cited by Laqueur, p. 26.

15. Stolberg, Michael (2003), A woman to her bones: The anatomy of sexual difference in the sixteenth and early seventeenth centuries. *Isis* 94:274–299; Ian Maclean (1980), *The Renaissance Notion of Woman*. Cambridge University Press, pp. 28–46.

16. Laqueur, Thomas W. (2003), Sex in the flesh, *Isis* 94:300–306; Londa Schiebinger (2003), Skelettestreit, *Isis* 94:307–313.

17. Laqueur, *Making Sex*.

18. Schiebinger, L. (1991), *The Mind Has No Sex?: Women in the Origins of Modern Science*. Harvard University Press.

19. Greene, E. L. (1983), *Landmarks of Botanical History*, F. N. Egerton, ed. Stanford University Press, part ii, chapter 16, p. 651.

And the eyes of them both were opened, and they knew that they were naked.

—GENESIS 3:7

2

The Discovery of Sex

BEFORE PEOPLE COULD discover sex in plants, they first had to discover it in themselves. Not the sexual act itself, of course, which humans and all living creatures are programmed to perform more or less spontaneously. It is also self-evident that women give birth. The question we are posing is one of comprehension: When did humans learn the role of men in childbirth? This question raises a second, even more basic question: Is it even possible to answer the first question?

The short answer to the second question is, no, we can never know when humans first understood the cause of childbirth. However, we can make an educated guess when the earliest archaeological evidence for such knowledge appears. We can say with absolute certainty, for example, that the discovery of the contribution of the male to childbirth predates the historic period, which begins with the invention of writing in ancient Mesopotamia around 3500 BCE.* From the earliest cuneiform tablets onward, poets have celebrated love's passion as well as love's outcome: the birth of a child. To uncover the first glimmerings of sexual knowledge, we must travel further back in time to the remarkable florescence of human self-expression and creativity that began during the last Ice Age—a period known as the Upper Paleolithic, between 40,000 and 11,000 years ago (see Table 2.1). Although modern humans—*Homo sapiens*—had spread from Africa to Europe, Asia, and Australia by this time, most of our understanding of this period is derived from two centuries of research on the artistically rich sites of Western Europe. We will therefore focus our discussion on this geographical area as an exemplar of Upper Paleolithic culture, with the caveat that significant regional differences are known to exist.[1]

* Following modern convention in archaeology, we will use BCE/CE ("Before the Common Era" and "Common Era") in place of the traditional BC/AD.

TABLE 2.1

Archaeological periods of Western Europe[a]	
Lower Paleolithic[b] (*Homo habilis* and *Homo erectus*)	c. 2,500,000–200,000 BP
Middle Paleolithic (Mousterian—period of Neantherthals)	c. 350,000–45/35,000 BP
Châtelperronian[c]	c. 37,000–35,000 BP
Upper Paleolithic (Cro-Magnons)	
Aurignacian	c. 35,000–29,000 BP
Gravettian	c. 29,000–22,000 BP
Solutrean	c. 22,000–17,000 BP
Magdalenian	c. 18,000–11,000 BP
Epi-Paleolithic in North Africa & Levant	c. 20,000–11,500 BP
Mesolithic in Europe	c. 11,500–8,000 BP
Early Neolithic	c. 11,500–6,500 BP (9,500–4500 BCE)
Late Neolithic or Chalcolithic (From The Greek *chalkos* meaning copper.)	c. 4,500–3,000 BCE
The Bronze Age	c. 3,000–1,000 BCE
The Iron Age	c. 1,200 BC–100 BCE

[a] The dates given are very approximate. The ranges within periods, and the overlapping between consecutive periods, reflect variability in cultural development between Europe and the Near East. Dates reported in years "before present" (BP = "years ago") are based only on uncalibrated radiocarbon measurements. Calibration is needed to obtain a calendar date because the level of atmospheric ^{14}C has not been constant over time and must be corroborated by independent methods, such as the annual growth rings of trees. Dates reported in calendar years (BCE) are based on calibrated radiocarbon measurements.

[b] The names of the periods are associated with specific stages of cultural development. These periods overlap because the progression from one stage to another varies with location. The Upper Paleolithic has been further subdivided into periods corresponding to different cultures of the Franco-Cantabrian school of Paleolithic art, which flourished in Cantabria, Spain (exemplified by Altamira cave) and in the Dordogne region of France—Lascaux cave being the most famous example of the latter. Each of the names is based on a specific archaeological site in France.

[c] The Châtelperronian, which temporally overlaps with the Upper Paleolithic, represents a Mousterian (Neanderthal) culture.

The term "Ice Age" in Europe conjures up visions of subzero temperatures and bleak landscapes of glaciers and snow—hardly the conditions likely to foster a thriving, prosperous human community. Yet, despite the hardships, the last Ice Age was a period of unprecedented abundance for early European *Homo sapiens*. The solution to the apparent paradox lies in our flawed image of the Ice Age. Even at the peak of glaciation, only the northernmost regions of Europe—England and Scandinavia—and mountainous regions, such as the Alps and Pyrenees, were actually glaciated. The nonglaciated regions of central and southern Europe were characterized by low-lying vegetation, with fringes of forest along the coasts.[2] In fact, the summer temperatures in the nonglaciated regions were only slightly lower than they are today. On the other hand, winter temperatures were significantly lower, with the result that the average temperature in central and southern Europe was about 18°F (10°C) lower than modern levels. Nevertheless, these conditions, which are similar to those of present-day southern Alaska or northern Scandinavia, supported a rich fauna of large herbivores, such as reindeer, horses, and bison, which provided an ample food supply for early modern humans. So, despite the often harsh conditions that prevailed in Western Europe, the abundance of wild game in southern refugia ensured the survival of modern humans and even provided them with a measure of prosperity and leisure.[3]

Perhaps as a result of their material success, Western Europeans in the Upper Paleolithic experienced an artistic explosion unlike anything that had gone before. Such an outpouring of creative expression must have reflected a quantum leap in overall cultural complexity, including new developments in art (painting, sculpture, music, dance, story-telling), religion (shamanism, totemism, etc.), proto-science (natural history, herbalism, calendrical record keeping, celestial reckoning), and technology (tools and textiles). It is within the context of this surprisingly sophisticated intellectual ferment that we begin our enquiry into the discovery of sex.

BARRIERS TO UNDERSTANDING SEX IN HUMANS DURING THE ICE AGE

It has been argued that, because hunter-gatherers of the Upper Paleolithic typically consisted of small nomadic groups of about twenty-five individuals, their survival as a group depended on restricting their population size. Did the need to control population size necessitate some form of birth control? If so, the survival of these hunter-gatherers might have depended on their discovering the role of the male in childbirth, as well as techniques to prevent or counteract pregnancy. John M. Riddle reviewed the evidence that effective herbal contraceptives and abortifacients were employed by women as early as the Bronze Age in Egypt and concluded that some of the treatments were scientifically plausible.[4] For example, in Libya, from about 600 BCE to the first century CE, the city of Cyrene grew wealthy exporting vast quantities of the herb *Silphium*, a now-extinct member of the parsley family, largely because of its reputation as an effective contraceptive and abortifacient. The herb was even featured on Cyrenean coins. Timothy Taylor has speculated that similar herbal contraceptives and abortifacients may have originated thousands of years earlier, in the Upper Paleolithic.[5] However, there is no evidence that effective herbal methods of population control were known during the Ice Age, and controlling population size may not even have been necessary under Ice Age conditions. The lifestyle of hunter-gatherers may have been sufficiently stressful to reduce the number of pregnancies without recourse to artificial measures.

One way in which population growth might have been controlled unconsciously is by prolonging the period of breastfeeding beyond infancy. According to Luigi Luca and Francesco Cavalli-Sforza:

> The hunter-gatherers of earlier times were presumably like those of today, who have an average of five children, one about every four years. A four-year gap means that the parents can always carry the youngest child on their backs or in their arms, while the older ones are already able to walk at a reasonable pace. Longer gaps also mean that

children can be breastfed until the age of three, and this in turn lowers the probability of another pregnancy. An average of five children per woman keeps the population substantially stable, because more than half will generally die at an early age.[6]

Breast-feeding can potentially inhibit pregnancy because stimulation of the nipple triggers the secretion of prolactin, a hormone that inhibits ovulation. But don't try this at home! The effectiveness of this method of birth control depends heavily on the frequency and duration of the nursing episodes. Hence, breast-feeding as a method of birth control only works well in cultures in which infants are carried by their mothers more or less continuously and are nursed whenever they wish. For this reason, women in hunter-gatherer or agricultural societies have had greater success using lactation as a method of birth control than have women in industrial societies. If Paleolithic mothers nursed their babies for several years, as seems likely, the number of pregnancies resulting from sexual intercourse would have been reduced.

There are, however, still gaps in our understanding of the contraceptive effect of lactation, such as the relative importance of the intensity of breast-feeding versus maternal nutrition. Valleggia and Ellison attempted to disentangle these two factors in studies of well-nourished Toba women in Formosa, Argentina.[7] Here, they found that the correlation between lactation-induced amenorrhea was not correlated with either nursing intensity or maternal nutrition alone, but also could be at least partially explained by the ratio of food energy intake to energy expenditure in individual women—which varied widely. Like the variable time between conception and the awareness of quickening, this much wider variation could obscure the identification of cause and effect concerning pregnancy.

Changes in menstrual and hormonal function characteristic of nonlactationally induced amenorrhea have been reported in very thin, dieting, and exercising Western women, including athletes. Similar hormonal changes have been found in rural farming villages in Nepal, where Catherine Panter-Brick and colleagues showed that fecundity was correlated with the women's energy balance—that is, the balance between caloric intake versus energy expenditure as work.[8] They found that during the monsoon season when food was scarce, the level of progesterone in the women's bodies decreased, and so did their fertility. In winter, energy expenditure decreased, progesterone levels increased, and the fecundity of the women increased. Peter Ellison has argued that the hypothalamus of the brain tracks short-term changes in energy balance, not just total energy reserves, tending to restrict conception at times that are inappropriate to sustain the energy drain of pregnancy and lactation.[9]

In a 2004 study in rural Poland, Jasienska and Ellison found that ovarian function was suppressed when women engaged in hard physical labor, even when their caloric intake kept up with the amount of energy expended. That is, ovarian function was suppressed by hard work alone even when the total energy balance remained unchanged.[10]

Jasienska and Ellison have attempted to apply their findings about lactational amenorrhoea in southern Poland to Paleolithic hunter-gatherers.[11] They speculated that because the Paleolithic diet was relatively poor in calories, periods of high energy expenditure by our human ancestors typically were not balanced by commensurate increases in caloric intake, thus leading to a negative energy balance. Under such conditions, ovulation would be suppressed and female fertility would decline. Alternatively, even when the energy expenditure *was* compensated by high caloric intake, high energy expenditure alone might interfere with the body's ability to allocate food energy to reproduction. According to the authors, this is because "a woman's ability to meet the metabolic cost of pregnancy and lactation depends on her ability to down-regulate her own metabolic requirements." When this ability is affected by high workload, a temporary suppression of ovulation may be an adaptive mechanism, preventing pregnancy when the body is under physical stress.

Because wild food resources are seldom uniformly available at all times and places, hunter-gatherer societies tend to develop regular patterns of group movement, aggregation, and dispersal that are synchronized with the annual fluctuations of food abundance, variety, and availability. Inevitably, this results in periods of abundance interspersed with periods of hunger. From the standpoint of women's fertility, the combination of prolonged nursing and high energy expenditure (during harvesting or migration) may have combined to suppress ovulation, thus lowering the birthrate at these times. The birthrate would thus increase during periods of plenty when the group settled in one place for a time and decrease in lean times when the group migrated in search of resources. Instead of being associated with sexual intercourse, cycles in the birth rate would have correlated with periods of abundance and leisure, especially for women.

Another factor that may have obscured the causal relationship between copulation and childbirth is prepuberty sexual activity. In studies of the Dobe Ju of Botswana (1963–76), Richard Lee found that sex education of children began at a young age.[12] Children slept under the same blanket with their parents and were thus exposed to their parents having sexual intercourse. Sexual play was considered a normal part of childhood, and nearly all boys and girls had some sexual experience, including intercourse, by age 15. Of course, pregnancy only resulted if both partners had passed through puberty.

A similar attitude toward children's sexual activity by the Trobriand Islanders was observed in the early twentieth century by Bronislaw Malinowski.[13] Not surprisingly, given the cultural attitudes of the period, many of his interpretations have had to be revised. Nevertheless, Malinowski's accounts are still the best record we have of Trobriand Islander beliefs and customs at something close to the pre-contact state.

In a paper published in 1916 entitled *Baloma; the Spirits of the Dead in the Trobriand Islands*, Malinowski reported that sexual activity among the Trobriand Islanders began well before puberty and that both sexes were expected to have many lovers before marriage:

> The sexual freedom of unmarried girls is complete. They begin intercourse with the other sex very early, at the age of six to eight years. They change their lovers as often as they please, until they feel inclined to marry. Then a girl settles down to a protracted and, more or less, exclusive intrigue with one man, who, after a time, usually becomes her husband.[14]

Under such an arrangement, instances are bound to occur when an unmarried girl becomes pregnant, but when Malinowski inquired of his informants "who was the father of an illegitimate child," he received the circular answer that there was no father because the girl was not married. When he pursued the point and asked who was the biological father, his question drew a blank. In Malinowski's day, at least, Trobriand Islanders had no concept of, or perhaps no interest in, biological paternity. Upon further questioning, the informant explained that "it is the *baloma* who gave her this child."

Malinowski reported that the Trobriand Islanders' concept of human reproduction was based on the idea of reincarnation. The *baloma* was thought of as a person's reflection or shadow image; only people had *balomas*. Each newly born child was a reincarnated *baloma*. Malinowski concluded that Trobriand Islanders did not recognize any material contribution of the father to the newborn. Malinowski further speculated that the failure to understand the father's role was the basis for the practice of matrilineal descent among the different clans. However, in his book *Magic, Science and Religion and Other Essays*, Malinowski also pointed out that the islanders did have "a vague idea as to some *nexus* between sexual connection and pregnancy."[15] They understood, for example, that a woman who is a virgin typically didn't become pregnant. However, they believed that the function of sexual intercourse was purely mechanical, to "open up" or "pierce" the vagina, making it "easier for a spirit child to enter."

Trobriander accounts of procreation obtained by later anthropologists have revealed a surprising diversity of opinion about the role of the father. According to Montague, some informants report that during intercourse the penis pounds on the cervix, which prevents the loss of menstrual blood that participates in the formation of the fetus.[16] According to another version (possibly influenced by outside contact), the semen acts a coagulant forming a clot in the menstrual blood into which the spirit-child may enter. Thus the *baloma* reincarnation model for childbirth has proved flexible enough to allow for some mechanical or chemical role of the father without contradicting the core assumption that it is the *baloma* and not the biological father that is essential for pregnancy to occur. Beliefs similar to those of the Trobriand Islanders at the turn of the twentieth century may also have been held by some groups in the Ice Age.

In summary, then, many factors could have combined to obscure the connection between sexual intercourse and childbirth during the Upper Paleolithic, and it is therefore quite possible that the role of copulation in pregnancy was not understood for some time during the early hunter-gatherer stage of human social evolution.

COULD ICE AGE PEOPLE COUNT TO NINE?

In his book *A Prehistory of Sex*, Timothy Taylor quotes the British scientist James Biment as saying, "Stone Age man probably didn't associate sex with something that came along nine months later . . . I doubt he could count up to nine." If Biment is correct and Upper Paleolithic people couldn't count to nine, tracing childbirth to its cause nine months earlier would have been a daunting task. However, even if they could count to nine, did they know how long a month was? In other words, did they have calendars? Let's consider counting first and calendars second.

As their cave paintings make clear, European early modern humans were able to create stunning works of art, equal in skill to those of modern artists.[17] Nor is there any reason to believe that the innate arithmetical abilities of Ice Age humans were any less impressive than their artistic abilities. It is only the slow, incremental way in which mathematics progresses that obscures the numerical achievements of our Paleolithic forebears. Had he been born in the Upper Paleolithic instead of the seventeenth century, Sir Isaac Newton couldn't possibly have invented calculus without the "shoulders of giants" to stand on. But it seems quite likely that he could have counted to nine had he any reason to do so. In fact, there is plausible archaeological evidence for numeracy (the numerical equivalent of "literacy") as early as 30,000 years ago based on a group of Upper Paleolithic artifacts inscribed with repetitive and regular lines, spots, or other markings that suggest counting.

The most arresting of the possible depictions of an Upper Paleolithic calendar has been found in the Dordogne region of France near the famous Lascaux cave. Carved into limestone blocks in a rock shelter in Laussel, France, are a group of bas-reliefs depicting three women, animals, and a sexually ambiguous figure, all of which date to the Gravettian period, around 25,000–20,000 years ago. The most famous of these reliefs is the "Lady of Laussel" or "Lady of the Horn" (Figure 2.1A,B). The relief depicts a nude, woman with her left hand resting against her belly. Her right arm is bent at the elbow, and she holds in her hand a crescent-shaped bison horn inscribed with thirteen parallel lines. The sculptor intentionally emphasized the belly—and by implication, fertility—by carving the figure on a convex surface, so that the arc of the body reaches its apogee precisely at the abdomen, where her hand is resting. However, the abdomen itself is flat, and the navel does not protrude as it often does during pregnancy, suggesting that the woman is in

FIGURE 2.1 Two views of Lady of Laussel, Gravettian, Dordogne, France.
Original in the Musée d'Aquitaine à Bordeaux. Photos courtesy of Don Hitchcock and donsmaps.com.

her child-bearing years, but not actually pregnant. Alexander Marshack postulated that the horn represents the crescent moon, and the thirteen lines represent the number of crescent moons that make up a lunar year—a kind of lunar calendar.[18] The connection between the lunar calendar, the female menstrual cycle, and the fertility implied by the overall image is inescapable.

THE VENUS FIGURINES: ICONS OF FEMALE SEXUALITY?

The most famous of the portable anthropomorphic images of the Upper Paleolithic are small statuettes referred to as "Venus figurines." The epithet "Venus" was first applied to these objects in 1864, in connection with a headless, armless, and footless ivory statuette of a nude female discovered in the Dordogne region by French archaeologist Paul de Vibraye (Figure 2.2A). Vibraye named his statuette "Vénus impudique" or "immodest Venus," a play on "Venus pudica" ("modest Venus"), which refers to a genre of classical sculpture depicting the Venus modestly covering her breasts and pubic area with her hands. The risqué term "Vénus impudique" implies that the statuette represents a love goddess who, in contrast to her more refined classical counterpart, openly flaunts her sexuality. However, though the figure is clearly female, it is probably that of a young girl rather than a sex object. The "Vénus impudique" figurine dates to the Magdalenian period around 16,000–12,500 years ago.

The term "Venus figurine" is usually reserved for a collection of mostly female figurines that were produced during the Aurignacian and Gravettian periods between 29,000 and 21,000 years ago (see Table 2.1).[19] The first of these were discovered in a cave near Grimaldi, Italy, between 1883 and 1895. At last count, well over 200 of these mostly palm-sized statuettes have been unearthed from Paleolithic sites all over Europe—from the Pyrenees in

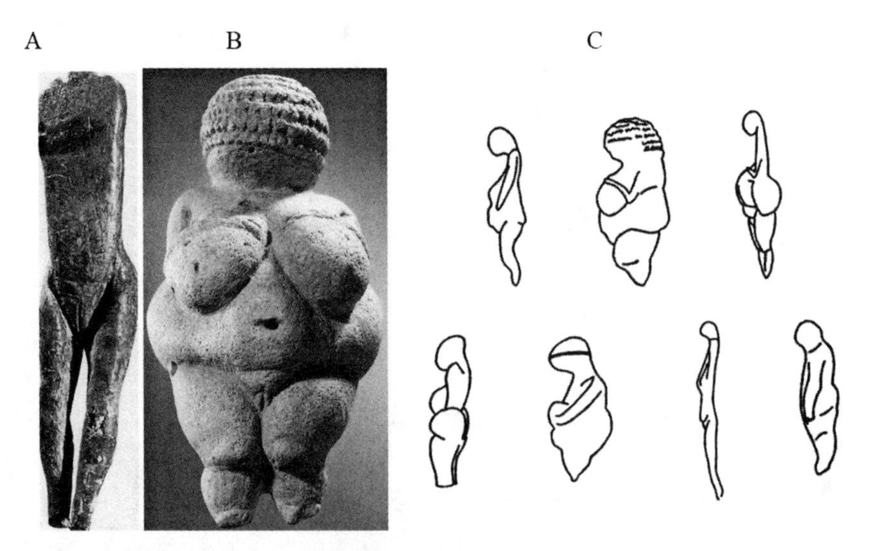

FIGURE 2.2 Examples of "Venus" figurines. A.Venus impudique. B. Venus of Willendorf. C. Profiles of Venus figurines from various European locations.
A & B. Photos courtesy of Don Hitchcock and donsmaps.com. C. From Smith, N. W. (1992), *An Analysis of Ice Age Art: Its Psychology and Belief System*. Peter Lang, p. 175.

southern France to the Don River in Russia, with scattered occurrences in Siberia. Most of them were carved in ivory, although other materials such as antlers, clay, limestone, and steatite (soapstone) were also used. Although some of the best known examples, such as the Venus of Willendorf (Figure 2.2B), discovered in 1908 in Austria, depict corpulent, possibly pregnant women with exaggerated breasts, hips, and buttocks, there is, in fact, a great variety of shapes among the figurines found at different locations. According to one subjective survey, about 23% represented slim young girls, 38% mature but not pregnant women, 17% pregnant women, and 22% elderly women.[20] Another study found that "only 39% of these figurines could possibly represent pregnancy."[21] Thus, the "Venus figurines" as a whole seem to depict various stages of female development (Figure 2.2C).

What was the function of these female figurines, and what, if anything, do they tell us about what the Ice Age people knew about sex? Because we know so little of the cultures that produced them, there is a wide divergence of opinion about the meaning of the Venus figurines, as there is for all of Upper Paleolithic art. Indeed, the various interpretations of the Venus figurines often seem to shed more light on the archaeologist than on the figurines themselves.

A male bias is apparent in some of the early interpretations of the Venus figurines, beginning with the sexually charged epithet "Vénus impudique." When the first of the Gravettian group of obese figurines was discovered at Grimaldi and later at Willendorf, much attention was focused on their large breasts and buttocks. Many male archeologists of the late nineteenth and early twentieth century immediately assumed that they represented Stone Age erotica produced by men.[22] However, this theory ignores the fact that roughly half of the figurines depict young girls or old women.

Another idea that has been around for a long time is that the figurines represent sympathetic fertility magic or perhaps good luck charms to ward off evil spirits during pregnancy and childbirth. However, studies have shown that only 17% of the figurines appear to be pregnant, and none of them is shown in the process of childbirth, so this cannot be the whole story.

In 2009, a sensational headless Venus figurine, carved from mammoth ivory, was discovered in a deposit at Hohle Fels Cave in Swabian Jura of southwestern Germany.[23] Dating to 35,000 years ago in the early Aurignacian period, it is the oldest of the known Venus figurines (Figure 2.3). It is also the most highly sexualized. The extreme exaggeration of the breasts, belly, and genitals, together with the decorative surface engraving, suggest that's its purpose is symbolic rather than erotic. In the words of archaeologist Nicholas J. Conard:

> There can be no doubt that the depiction of oversized breasts, accentuated buttocks and genitalia results from the deliberate exaggeration of the sexual features of the figurine. . . . Although there is a long history of debate over the meaning of Palaeolithic Venuses, their clearly depicted sexual attributes suggest that they are a direct or indirect expression of fertility.[24]

"Fertility" in this context could refer to human fertility, abundance in a general sense, or both. Thus far, no comparable male figurine has been found, suggesting either that the male role in reproduction was not yet fully recognized, or it was accorded a lesser symbolic value in the prevailing belief system.

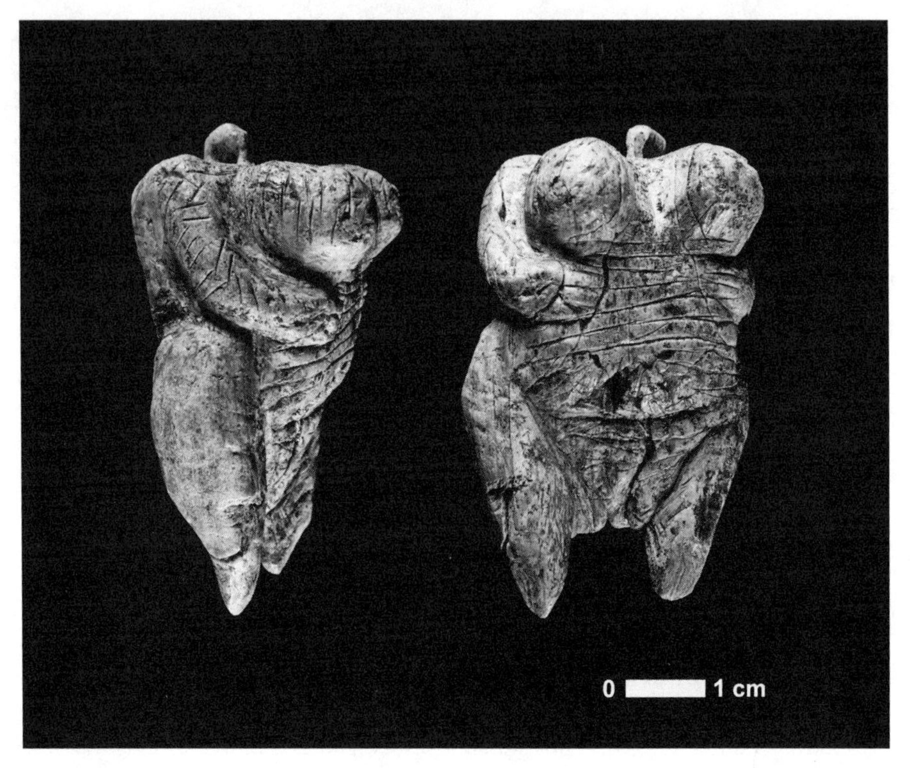

FIGURE 2.3 The Venus of Hohle Fels.
From Conard, N. J. (2009), Female figurine from the basal Aurignacian of Hohle Fels cave in Southwestern Germany. *Nature* 459:248–252, figure 1.

SHUTTLE DIPLOMACY: VENUS AT THE LOOM

Archeologists had long noted that Venus figurines may be either nude or partially clothed. Partially clothed examples are adorned with incised caps, snoods (a type of hair-net), belts, string skirts, bandeaux,[25] fillets or hair-bands, and jewelry (Figure 2.4A–D). The incised parallel lines on the Venus of Hohle Fels, for example, could represent a ceremonial apron. Archaeologists had long assumed that such articles of clothing were made of animal furs and hides, but Olga Soffer and her colleagues demonstrated that they were actually made of plant-based textiles. Because no actual woven material from the Ice Age has survived, this important Paleolithic industry was completely unknown to archeologists until Soffer's important discovery. Small clay fragments bearing crude impressions of rope, textiles, and basketry provided the clues that led to the breakthrough. Some of the clay fragments had been fired, some unfired, and the impressions seemed to have been produced accidentally. Only thirty-six of them, each less than 2 cm in diameter, have now been found at Paleolithic sites throughout Europe, from France to Russia.[26]

Fortunately, there was sufficient detail preserved in the clay impressions to enable Soffer and her team to determine that the textiles were made of plant fibers rather than wool or hair. They all dated to around 29,000 to 23,000 years ago, during the Aurignacian and Gravettian periods. Their appearance in the Aurignacian and Gravettian periods

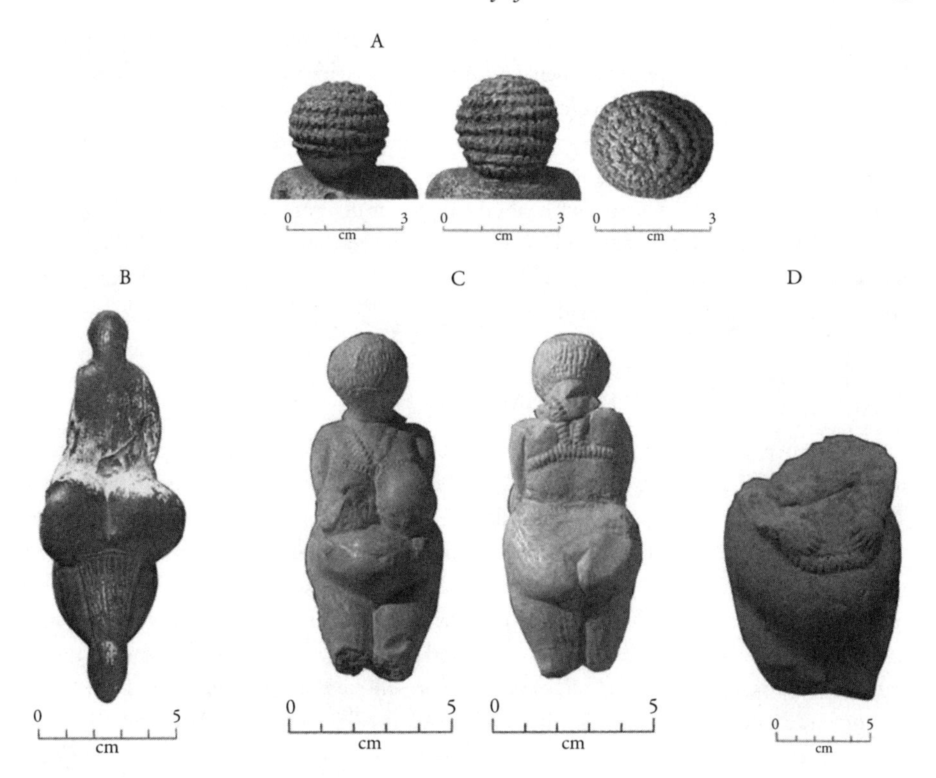

FIGURE 2.4 Headgear and clothing made from textiles as seen in Venus figurines. A. Venus of Willendorf. B. Low skirt from Venus of Lespugue, posterior view. C. Front and back views of Venus of Kostenki, showing bandeaux attached to straps. D. Fabric belt around waist of a possibly pregnant female.
From Soffer, O., J. M. Adovasio, and D. C. Hyland (2000), The "Venus" figurines. Textiles, basketry, gender, and status in the upper Paleolithic. *Current Anthropology* 41:511–537.

(see Table 2.1) coincides with the occurrence of tools and implements, including eyed needles, associated with sewing, weaving, and net-making. Some needles were large enough to have been used for sewing hides and leather, but others are quite small and thus were probably used for stitching and embroidering cloth.[27]

The discovery of representations of plant-based textiles on some of the Venus figurines has profound implications for gender and the status of women in Paleolithic society. An impressive body of ethnographic data on hunter-gatherer societies supports the notion that women are closely associated with the harvesting and processing of plant materials, as well as with the transformation of processed plant material into more complex products, such as textiles and baskets. The labor-intensive nature of the textiles and their surprisingly high quality suggest that wearing such clothing was a mark of prestige and possibly part of a sacred ritual. Textiles evidently became an important commodity in the new economy of the Gravettian period.

In addition to cloth used for clothing and ritual occasions, evidence for plant fiber-based nets for hunting small animals has also been found at the same sites. Mary Stiner and colleagues have documented the critical importance of small game and fish to the economy and survival of Paleolithic societies.[28] Since it is likely that small game hunting and fishing in the Paleolithic were carried out by women, children, and the elderly, the evidence further supports the view that the roles of women in Paleolithic societies were important and diverse.

Prior to Soffer's discovery, the high quality of the textiles produced during the Upper Paleolithic was not appreciated. Given the apparent division of labor between men and women in the Upper Paleolithic, with women taking primary responsibility for the gathering and processing of plants for food and fiber, women were most likely the inventors of the new technology. Consistent with this idea, the textile revolution coincides with the sudden appearance of the Venus figurines all over Europe.

SEXUAL SYMBOLISM IN PARIETAL ART

In the absence of written records, Upper Paleolithic wall art has the greatest potential to tell us what Ice Age people knew, or did not know, about the role of the male in procreation.[29] A hypothetical triptych that would unequivocally demonstrate that people understood the causal relationship between copulation and pregnancy would be a sequence of illustrations depicting copulation, pregnancy, and childbirth. Unfortunately, no such set of images has ever been found in all the Paleolithic caves of Europe. Indeed, humans were rarely depicted in parietal art. Most of the painted images are of large animals, such as bison, aurochs, horses, and deer. When humans are represented, they are usually female. Images of women can be naturalistic, as in the "Lady of Laussel" (see Figure 2.1) or symbolic, such as the schematic vulvas inscribed or painted on walls or stone blocks at many European Upper Paleolithic sites, examples of which are shown in Figure 2.5. Consistent with the

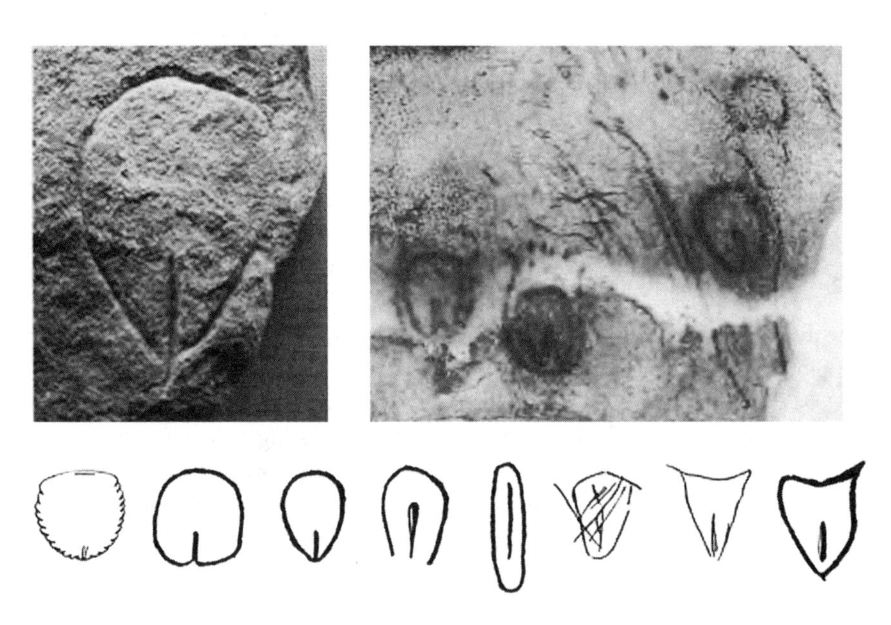

FIGURE 2.5 Upper Paleolithic images of vulvas.
(Top) From Angulo Cuesta, J., and M. Garcia Diez (2005), *Sexo en Piedra. Sexualidad, Reproducción y Erotismo en Época Paleolítica*; (bottom) from Marshack, A. (1991), *The Roots of Civilization: The Cognitive Beginnings of Man's First Art, Symbol and Notation*. McGraw-Hill.

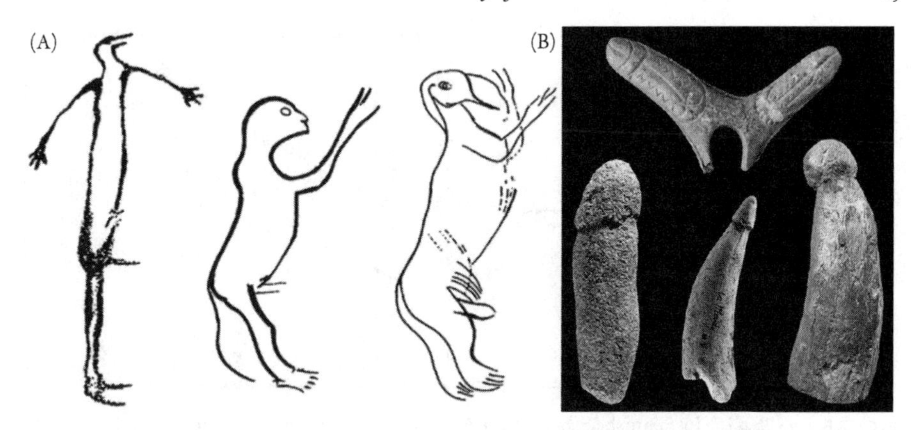

FIGURE 2.6 Ithyphalic male figures (A) and phallic symbols (B).
From Angulo Cuesta, J., and M. Garcia Diez (2005), *Sexo en Piedra. Sexualidad, Reproducción y Erotismo en Época Paleolítica.*

interpretation of these signs as pubic triangles or vulvas, M. R. González Morales and L. G. Straus have recently identified the possible representation of a woman, including a pubic triangle, engraved on a Magdalenian-age cave wall. The engraving appears to have marked the site of a human female burial.[30]

According to Angulo Cuesta and Garcia Diez, "only a few figures explicitly show clearly masculine sexual organs in the Paleolithic art now known in Western Europe." Some examples of ithyphallic (the Greek word *ithys* means "straight") human-like figures are shown in Figure 2.6A. In addition, phallic imagery is apparent in a few portable artifacts, as shown in Figure 2.6B. However, in contrast to the female images, many of which date to the Aurignacian and Gravettian periods (contemporary with the "Venus figurines"), the majority of the male figures and phallic images found at Les Combarelles, Saint-Cirq, Altamira, La Madeleine, and other Late Paleolithic sites, date to the Magdalenian period—16,000 to 8,500 years ago.

Possible representations of human sexual activity, including copulation and childbirth, occur later in the archaeological record.[31] The view that copulation was indeed represented in Upper Paleolithic cave art has received support in recent years. For example, Figure 2.7A from Les Combarelles illustrates what has been interpreted as a "pre-coital scene," while Figure 2.7B, from Los Casares, is thought to represent face-to-face coitus. Angulo Cuesta and Garcia Diez have speculated that the "couple" on the left in Figure 2.7B may illustrate the only example of penetration, although the site of the penetration is anatomically problematical. As in the case of the male images, the putative copulation depictions do not begin until the Magdelanian period.

The next stage after copulation is pregnancy. Although plausible depictions of pregnancy can be found among the "Venus figurines," there are no clear examples from parietal art. In contrast, the image found engraved on the face of a reindeer bone, showing what appears to be a male reindeer standing in front of—and over—a supine pregnant woman, seems clear, even if the significance of the image is obscure (Figure 2.8). This portable art object was found at a Middle Magdalenian site in the Dordogne. The arms of the woman are raised, suggesting that she and the male animal may be part of a mythic ritual concerned with

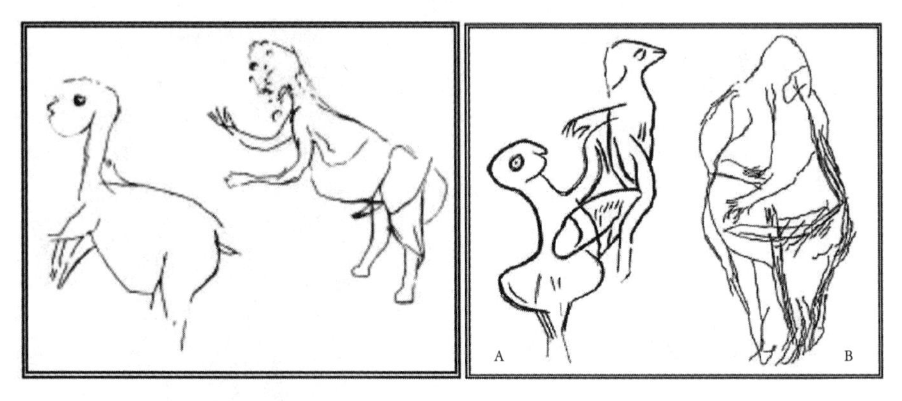

FIGURE 2.7 Tracing by Spanish archaeologist Aguilo Cabré of an image scratched on the wall of the cave of Los Casares.
From Angulo Cuesta, J., and M. Garcia Diez (2005), *Sexo en Piedra. Sexualidad, Reproducción y Erotismo en Época Paleolítica.*

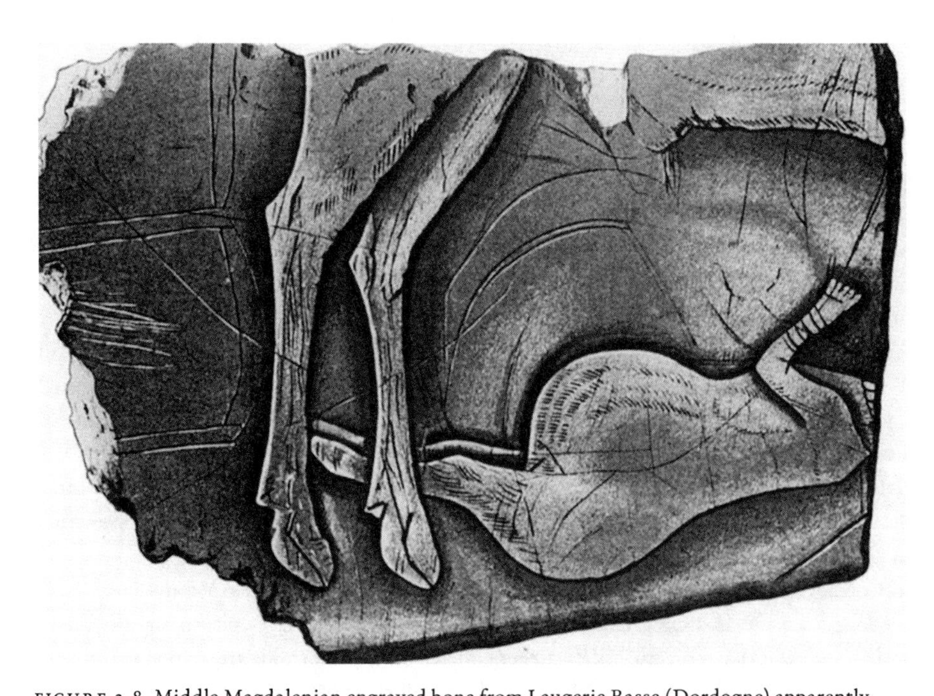

FIGURE 2.8 Middle Magdalenian engraved bone from Laugerie Basse (Dordogne) apparently showing the hindquarters of a male reindeer over a pregnant woman.
From Marshack, A. (1991), *The Roots of Civilization: The Cognitive Beginnings of Man's First Art, Symbol and Notation.* McGraw-Hill, figure 189, p. 320. Original from Piette, E. (1907), *L'Art Pendant L'Age du Renne*, Masson et Cie.

reproduction.[32] Assuming the reindeer is male, the association with the pregnant woman may imply some understanding of the role of the male in childbirth.

If copulation images are rare and difficult to interpret, representations of childbirth are even more so. Alexander Marshack described one stone with a possible birth scene inscribed on it, but in the absence of any other plausible scenes of childbirth from the Upper Paleolithic it remains doubtful at best.[33]

THE MAGDALENIAN TRANSITION IN ICONOGRAPHY: A SEXUAL REVOLUTION?

During the Aurignacian and the Gravettian periods, cave paintings and portable art objects focused mainly on animals, especially large ones that formed an important element in the diet. Humans were rarely represented, and when they were it was usually women rather than men. Although *some* of the Venus figurines appear to be pregnant, and vulva symbols are widespread, there are very few comparable male images or phallic symbols dating to this early period. Nor have any depictions of either copulation or childbirth surfaced from the Aurignacian and the Gravettian periods. From this we can infer that although reproduction was an important artistic element, it was regarded primarily as a female function.

Beginning with the Magdalenian period, there is a significant increase in the frequency of male iconography, as well as the first plausible depictions of copulation and childbirth. Angulo Cuesta and Garcia Diez have cited this artistic shift during the Magdalenian as evidence for a sexual revolution in which eroticism, as opposed to mere reproduction, was emphasized:

> The earliest images related almost exclusively to reproduction, whereas others express a more actual view of sexual relations and sexuality, tied not only to reproduction but also to enjoyment, pleasure and sexual attraction.[34]

The authors point to the increase in male iconography and possible representations of copulation, bestiality, and masturbation during the Magdalenian as signs that sexual attitudes and practices had become more like those modern society:

> From the Magdalenian period (16,000–10,000 BC) onward, the artistic evidence provides expressive and narrative images of sex as reproduction, pleasure and probably play. They undoubtedly reflect a varied sex life. Sensual love and sexual appetite are innate to humanity. It could be said that their sexual practices were, at least since this time, similar to our own.[35]

If the sexual repertoire of people did indeed expand during the Magdalenian, why at this particular time? Humans had tens, if not hundreds, of thousands of years prior to the Magdalenian to explore their sexuality. The notion that a range of previously unknown sexual practices was adopted as late as the Magdalenian seems rather improbable. Another possibility is that the practices themselves didn't change, but that social attitudes and artistic styles changed, allowing artists to represent sexuality more freely and playfully. Yet the earlier images of female nudity, pregnancy, and vulva symbols dating to the Aurignacian and the Gravettian don't seem particularly modest. Presumably, puritanical attitudes toward sex developed much later in human history.

Alternatively, the Magdalenian expansion of sexual iconography to include more phallic and copulation images might have been prompted by a new awareness of the essential, as opposed to the *facilitative*, role of sexual intercourse in procreation. As outlined at the beginning of the chapter, physiological, developmental, and behavioral factors could have delayed the discovery, given that there is precedent for a lack of understanding about the male role in procreation among recent hunter-gatherer societies. Perhaps by the Magdalenian period in Europe, beginning around 18,000 years ago, the mystery was solved. No longer was the male perceived as a mere facilitator of childbirth, but as an equal partner. This is pure speculation, but it provides a reasonable upper limit for the time of discovery of the role of the male in childbirth. A more conservative estimate of when the discovery actually took place would be anytime between 100,000 and 18,000 years ago!

THE VULVA AND THE PLANT

Upper Paleolithic artists rarely depicted plants, and when they did so the results were barely recognizable as botanical in origin, especially when compared with the meticulous renderings of animals.[36] Alexander Marshack was the first to point out that plant images were mainly employed as seasonal indicators, which may explain their simplified, pictographic style.

Marshack noted that images of plant-like symbols on cave walls are occasionally juxtaposed with symbols of vulvas. Figure 2.9 shows a striking association of three possible vulva signs with what could be a flower. The three bell-shaped symbols were painted in red, while the "plant" was painted black. This panel was originally interpreted by French archaeologist André Leroi-Gourhan as a combination of three female symbols and one male symbol. However, if the interpretation of the tufted object as a flower is correct, it would provide a striking illustration of the association of plants and women during the Upper Paleolithic.

FIGURE 2.9 Painting possibly representing three vulvas and a flower, from the early Magdalenian site of El Castillo cave, northwestern Spain.
From Bahn, P. G., and J. Vertut (1997), *Journey Through the Ice Age*. University of California Press.

NOTES

1. Variations from the Western European model during the Upper Paleolithic are particularly evident in the Mediterranean area, Eastern China, South East Asia, and Australia. For example, according to Ofer Bar-Yosef (personal communication), "prehistoric foragers in the Mediterranean areas lived better, ate more plants, and did not hunt a lot of animals" compared to their Western European counterparts.

2. Mellars, P. (2001), The Upper Paleolithic revolution, in Barry Cunliffe, ed., *Oxford Illustrated History of Prehistoric Europe*. Oxford University Press, pp. 42–78.

3. There was considerable climatic and environmental diversity in Europe within the Upper Paleolithic time frame (i.e., the Last Glacial Maximum [LGM] between the Gravettian and Magdalenian periods). For example, during the LGM, when humans had to abandon the North and survived in refugia in the south, there was polar desert in parts of Northern Europe, while other areas were relatively mild with surviving trees and abundant game (Lawrence Guy Straus, personal communication).

4. Riddle, J. M. (1992), *Contraception and Abortion from the Ancient World to the Renaissance*. Harvard University Press.

5. Taylor, T. L. (1996), *The Prehistory of Sex: Four Million Years of Human Sexual Culture*. Bantam Press.

6. Cavalli-Sforza, L. L. and F. Cavalli-Sforza (1995), *The Great Human Diasporas: The History of Diversity and Evolution*. Addison-Wesley, p. 134.

7. Valeggia, C., and P. Ellison (2004), Lactational amenorrhoea in well-nourished Toba women of Formosa, Argentina. *Journal of Biosocial Science* 36:573–595.

8. Panter-Brick, C., D. S. Lotstein, and P. T. Ellison (1993), Seasonality of reproductive function and weight loss in rural Nepali women. *Human Reproduction* 8:684–690; Panter-Brick, C. (1996), Proximate determinants of birth seasonality and conception failure in Nepal. *Population Studies* 50:203–220.

9. Jasienska, G. and P. T. Ellison (2004), Energetic factors and seasonal changes in ovarian function in women from rural Poland. *American Journal of Human Biology* 16:563–580.

10. Ibid.

11. Ibid.

12. Lee, R. B. (2003), *The Dobe Ju/'hoansi*, 3rd ed. Thomson Learning/Wadsworth.

13. Malinowski, B. (1929). *The Sexual Life of Savages in North-Western Melanesia: An Ethnographic Account of Courtship, Marriage and Family Life Among the Natives of the Trobriand Islands, British New Guinea*. Readers League.

14. Malinowski, B. (1916), Baloma: The spirits of the dead in the Trobriand Islands. *Journal of the Royal Anthropological Institute of Great Britain and Ireland* 46:353–430.

15. Malinowski, B. (1948), *Magic, Science and Religion and Other Essays*. The Free Press.

16. Montague, S. (1971), Trobriand kinship and the virgin birth controversy. *Man, New Series* 6:353–368.

17. According to a widely published anecdote, Picasso, after viewing the cave paintings of either Lascaux or Altamira, exclaimed "We have invented nothing!" However, archaeologist Paul Bahn extensively researched the quote and was unable to verify it. Unfortunately, the oft-cited quotation appears to be apocryphal. The closest authentic quote from Picasso on the quality of Paleolithic art that Bahn was able to track down was the statement, "Primitive sculpture

has never been surpassed," which he made to his secretary. See Bahn, P. G. (2005), A lot of bull? Pablo Picasso and Ice Age cave art, in Homenaje a Jesús Altuna, *Antropologia-Arkeologia* 57:219–225.

18. Marshack, A. (1991), *The Roots of Civilization: The Cognitive Beginnings of Man's First Art, Symbol And Notation*. McGraw-Hill.

19. Dobres, M.- A. (1992*a*), Re-considering Venus figurines: A feminist-inspired re- analysis, in *Ancient Images, Ancient Thought: The Archaeology of Ideology—Proceedings of the 23rd Annual Chacmool Conference*, A. Sean Goldsmith, Sandra Garvie, David Selin, and Jeannette Smith, eds., Archaeological Association of the University of Calgary, pp. 245–262; Dobres, M. -A. (1992*b*) Representations of Palaeolithic visual imagery: simulacra and their alternatives. *Kroeber Anthropological Society Papers* 73–74:1–25.

20. Rice, P. C. (1981), Prehistoric Venuses: Symbols of motherhood or womanhood? *Journal of Anthropological Research* 37: 402–414.

21. Nelson, S. M. (1997), *Gender in Archaeology: Analyzing Power and Prestige*. Alta Mira Press.

22. Nowell, A., and M. L. Chang (2014), Science, the media, and interpretations of Upper Paleolithic figurines. *American Anthropologist* 116:562–577.

23. Conard, N. J. (2009), Female figurine from the basal Aurignacian of Hohle Fels cave in Southwestern Germany. *Nature* 459:248–252.

24. Ibid.

25. A bandeau (pl. bandeaux) is a garment made from a strip of cloth, usually referring to a band around the breasts.

26. Soffer, O., J. M. Adovasio, and D. C. Hyland (2000), The "Venus" figurines. Textiles, basketry, gender, and status in the upper Paleolithic. *Current Anthropology* 41:511–537.

27. Ibid.

28. Stiner, M. C., N. D. Munro, and T. A. Surovell (2000), The tortoise and the hare: Small-game use, the broad spectrum revolution, and Paleolithic demography. *Current Anthropology* 41:39–73.

29. Angulo Cuesta, J., and M. Garcia Diez (2005), *Sexo en Piedra. Sexualidad, Reproducción y Erotismo en Época Paleolítica*. Angulo Cuesta, J., and M. Garcia Diez (2006), Diversity and meaning of Paleolithic phallic male representations in Western Europe. *Actas Urológicas Españolas* 30:254–267; Delluc, G., and B. Delluc (2006), *Le Sexe Au Temps Des Cro-Magnons*. Édition Pilote 24; Duhard, J. -P. (1996), *Réalisme de l'image masculine paléolithique*. Millon, J., and Duhard, J. -P. (1993), *Réalisme de l'image féminine paléolithique*. Jérôme Millon.

30. González Morales, M. R., and L. G. Straus (2015), Magdalenian-age graphic activity associated with the El Mirón Cave human burial. *Journal of Archaeological Science* 60:125–133; Straus, L. G., M. R. González Morales, J. M. Carretero, and A. B. Marín-Arroyo (2015), "The Red Lady of El Mirón": Lower Magdalenian life and death in Oldest Dryas Cantabrian Spain: an overview. *Journal of Archaeological Science* 60:134–137.

31. Angulo Cuesta and Garcia Diez, 2005, 2006; Delluc, 2006; Duhard, 1993, 1994.

32. Marshack, *The Roots of Civilization*.

33. Marshack, A. (1975), Exploring the mind of Ice Age man. *National Geographic* 147:62–89.

34. Angulo Cuesta, J., and M. Garcia Diez, Diversity and meaning, pp. 254–267.

35. Ibid.

36. Tyldesley, J. A., and P. G. Bahn (1983), Use of plants in the European Paleolithic: A review of the evidence. *Quaternary Science Reviews* 2:53–81.

3

Crop Domestication and Gender

"IN 1928, THE British School in Jerusalem undertook the excavation of the Cave of Shukba, discovered by Père Alexis Mallon, and I spent two months on the site. . . ." So begins British archaeologist Dorothy Garrod's 1932 account of her revolutionary discovery, at the Wady en-Natuf in the Judaean hills, of a new type of hunter-gatherer society that showed signs of early *sedentism*—living in permanent settlements.[1] The surface layer (layer A) of the Shukba cave contained numerous artifacts ranging from the Early Bronze Age to the Byzantine, with the majority of material dating to the Bronze Age. Just below the surface layer was a substratum (layer B) with an abundance of small flint tools called "microliths," which were characteristic of the Late Upper Paleolithic or "Mesolithic" in the case of European sites. In addition, there were larger flint knives with a "peculiar polish produced by cutting corn [wheat] or grass." These flint blades, apparently used to harvest wild grains, were the first archaeological evidence that the art of gathering had taken the first step toward cultivation.

After two months, Garrod and her colleagues closed down the *Shukba* cave excavation, intending to return the following spring. However, fate intervened in the form of an urgent request from the Department of Antiquities. Garrod explains the sudden change of plans as follows:

> In 1929 the excavation of Shukba was postponed by request of the Department of Antiquities, in order that the British School might undertake the more urgent work of digging out a site that was threatened by quarrying. This was the cave known as the Mugharet el-Wad, near Haifa. It is one of a group of caves lying at the western foot of Mount Carmel, at the point where the Wady el-Mughara (Valley of the Caves) opens on to the coastal plain.[2]

Mount Carmel in northern Israel is actually a mountain range that begins on the Mediterranean coast, where Haifa perches on its northern slope, and stretches southeast about 15 miles toward Megiddo. The name is derived from the Hebrew word *HaKarmel*, which means "fresh-planted garden or vineyard." During the period of the British Mandate following World War I, plans were made to build a large harbor in Haifa using stone quarried from the Carmel range. Massive construction of the port of Haifa did, in fact, begin late in 1928, and much of the stone was quarried from the Wadi el-Mughara, the site of four archaeologically significant caves in Mount Carmel. Fortunately, E. T. Richmond, the Director of Antiquities, ordered preliminary surveys to be conducted of the caves, the results of which clearly demonstrated the presence of abundant artifacts from the Upper Paleolithic. A rescue operation was urgently needed to document and preserve whatever prehistoric artifacts were present.

In April of 1929, Garrod began excavating "The Valley of the Caves" in Mount Carmel. In the largest cave of the group, the Mugharet el-Wad, she and her coworkers struck pay dirt.[3] Although the stratigraphic levels were badly disturbed in most places, they found a site on the right side of the cave where level B, a 50 cm layer of black earth, was still intact, and it contained microlithic implements similar to those Garrod had found at the Shukba cave. The B layer also contained tools—sickle blades along with their bone handles—confirming that these Late Upper Paleolithic gatherers practiced an incipient form of agriculture (Figure 3.1).

Like the blades found at Shukba, the blades at Mugharet el-Wad were glossy, as if they had been polished. This high gloss was shown to be the result of abrasion from cereal grass stalks, which are naturally impregnated with tiny glass particles (silicon dioxide) called *phytoliths*. Microscopic analyses suggested that the sickles were used mainly to harvest wild wheat in addition to other wild species, including brome grass, barley, oats, and legumes.

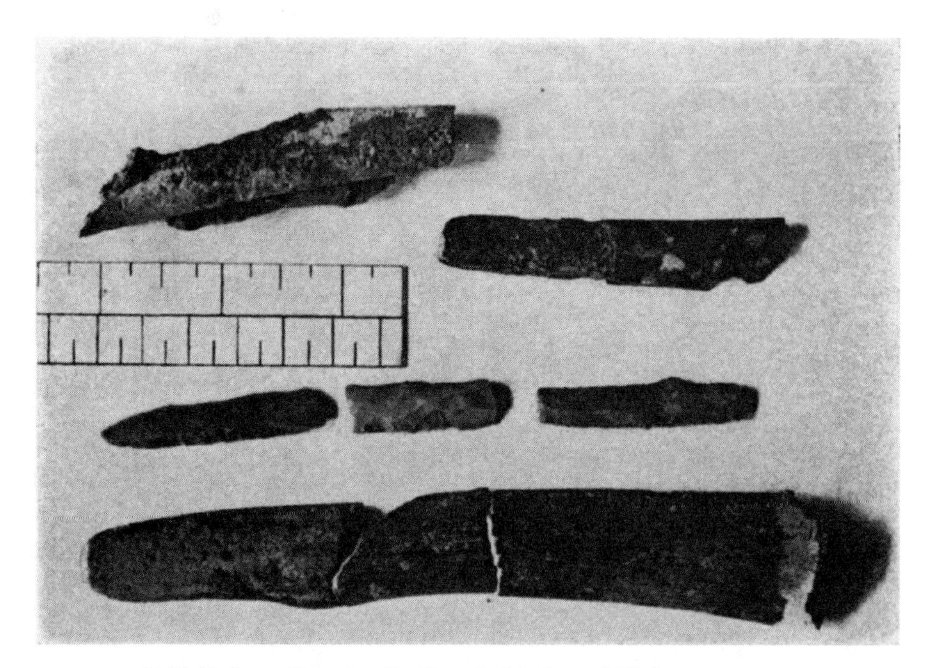

FIGURE 3.1 Sickle blades and bone handles from the Mugharet el-Wad cave. From Garrod, D. A. E. (1932), figure 2.

The particular pattern of striations on the sickle blades further indicated that the wild cereals were harvested while still moist and green and were cut low near the soil level, rather than directly under the seed head.[4] Wild cereals can only be effectively gathered while green because the fully mature seed heads shatter easily under the blows of a sickle, scattering the precious grain. Only after thousands of years of selection would the stalks of ripened cereal heads become tough enough to remain intact during harvesting.

Similar sickle blades were subsequently found at nearby locations, and Garrod named the prehistoric culture to which these people belonged the Natufian culture, after Wadi el-Natuf, where the Shukba cave was located.

THE NATUFIANS: THE FIRST SEDENTARY HUNTER-GATHERERS

The Natufians occupied the Levant[5] and other sites in the Near East from around 14,500 to 11,500 years ago, during the period now called the Epipaleolithic, a transitional phase between the Upper Paleolithic and the Neolithic periods in the Near East (see Table 3.1).

Thousands of years earlier, at the time of the Last Glacial Maximum (LGM) of the Pleistocene Epoch (25,000–19,000 years ago), the climate in the Levant had been cold and dry and the land relatively unproductive except along the coastal regions, which remained forested. The ancestors of the Natufians had been hunter-gatherers who led a nomadic existence, tracking herds of wild antelope, gazelle, and wild cattle, and supplementing their animal diet with a wide variety of wild plant foods.

As the glaciers receded and temperatures rose, so did the amount of rainfall. The period from 14,500 to 11,500 years ago, the pre-Holocene,[6] was especially wet, and the enhanced water availability generated a wealth of plant and animal life. It was under conditions of

TABLE 3.1

Chronological chart of the Late Epipaleolithic and Early Neolithic in the Northern and Southern Levant

<table>
<tr><th>Dates cal BP</th><th colspan="2">Northern Levant</th><th colspan="2">Southern Levant</th></tr>
<tr><td>14,000</td><td colspan="2">Early Natufian?</td><td colspan="2">Early Natufian</td></tr>
<tr><td>13,000</td><td colspan="2">Late Natufian</td><td colspan="2">Late Natufian</td></tr>
<tr><td>12,000</td><td colspan="2">Final Natufian</td><td colspan="2">Final Natufian</td></tr>
<tr><td rowspan="3">11,000</td><td>Khiamian</td><td rowspan="2">PPNA</td><td>Khiamian</td><td rowspan="2">PPNA</td></tr>
<tr><td>Mureybetian</td><td>Sultanian</td></tr>
<tr><td colspan="2">Early PPNB</td><td colspan="2">Early PPNB</td></tr>
<tr><td>10,000</td><td colspan="2">Middle PPNB</td><td colspan="2">Middle PPNB</td></tr>
<tr><td>9,000</td><td colspan="2">Late PPNB</td><td colspan="2">Late PPNB</td></tr>
<tr><td>8,000</td><td colspan="2">Final PPNB</td><td colspan="2">Final PPNB (PPNC)</td></tr>
<tr><td>7,000</td><td colspan="2">Pottery Neolithic</td><td colspan="2">Pottery Neolithic</td></tr>
</table>

Dates are given as "calibrated years ago" (years before the present), in which radiocarbon measurements are calibrated using tree rings. Suggested correlations with climatic events are indicated. The cultural terms are marked with the chronological boundaries (*dashed lines*) based on the currently available calibrated radiocarbon dates. From Bar-Yosef, O. (2014), The origins of sedentism and agriculture in Western Asia, in C. Renfrew and P. Bahn, eds., *The Cambridge World Prehistory*, Cambridge University Press, chapter 3.4: 408–1438.

relative plenty that the Natufian culture arose in what is now modern Israel, then composed of woodlands dominated by oak and pistachio trees and grasslands rich in wild cereals. So abundant were food resources in their homeland that the ancestors of the Natufians began depending less and less on migratory herds, which necessitated a nomadic lifestyle, and more on local herds of gazelle, equids, aurochs, deer, wild boar, wild goats, and small game.[7] Simultaneously, in addition to foraging and processing wild plant foods, they began, either consciously or unconsciously, to manage the habitats where their food sources were growing.[8]

Prior to the discovery of the Natufians, it had been assumed that the transition from migrant hunter-gatherer societies to permanent settlements coincided with the invention of agriculture. But the evidence of the Natufians forced a re-evaluation of this long-held belief. The subsequent discovery of semi-subterranean, dry-stone houses organized into villages showed that the Natufians had begun to live in permanent settlements prior to the planting of the first gardens, when they were still making their living by foraging. This new way of life was made possible in part by the abundance of local wild animal and plant foods, as well as by geographical features that greatly extended the growing season. It was the beginning of sedentism.

Natufian villages were typically located at the border between the mountains and the plains. For example, the Natufian settlement at Jericho was a spring-fed oasis on the western edge of the Jordan Valley, bounded by Mt. Nebo to the east and the Central Mountains to the west. As a consequence of the wide variation in elevation up and down the mountains, the cereals and nut crops growing there ripened and were available for harvesting at different times throughout the year. The earliest harvests occurred at the lower elevations, and the later harvests were at the highest elevations, thus ensuring an abundant and varied diet of wild plant foods for much of the year. At the permanent settlement, storage pits, or in some cases baskets, were filled with provisions, assuring the villagers' survival over the winter months. Each dwelling probably had its own storage capability as well as grinding stones and other implements for processing food. Plant foods formed the basis of the diet and were supplemented with meat from migratory herds and local game, which, like the Natufians, took advantage of the sequential ripening of plant food resources at different elevations.

Temporary satellite camps were established in the mountains where the food could be processed before transport to the main settlement. As the Natufians became increasingly proficient at harvesting, processing, and storing wild plant foods, the populations of the settlements increased. In fact, this new settled forager way of life functioned so well that the Natufians planted no crops for 2,000 years, even though it is highly likely that by this time methods for growing plants from seed were well known.

As to why the Natufians avoided planting crops, the answer is very simple: farming is hard work. Recall that tilling the soil was considered by the writers of the Bible to be God's curse on Adam and Eve for eating the fruit from the Tree of Knowledge in the Garden of Eden:

> [C]ursed is the ground for thy sake;
> in sorrow shalt thou eat of it all the days of thy life;
> thorns also and thistles shall it bring forth to thee;
> and thou shalt eat the herb of the field;
> in the sweat of thy face shalt thou eat bread,
> till thou return unto the ground;
> since from it you were taken:
> for dust you are, and to dust you will return.[9]

Written sometime during the Iron Age, these terrifying lines remind us that early sedentary foragers were blessed with an abundant year-round food supply. They would have little reason to convert to an agricultural lifestyle, exchanging relative leisure for back-breaking work. Even today, some groups living in naturally productive areas avoid agriculture, preferring the less arduous lifestyle of hunting and gathering. The bountiful supplies of wild cereals, legumes, nuts, and fruits were a disincentive for the Natufians to plant crops. No doubt powerful cultural disincentives existed as well.

PARADISE LOST, AGRICULTURE GAINED?

According to a widely held theory, the Natufian lease on Paradise eventually ran out. Foraging conditions began to worsen around 12,500 years ago. The climate became drier, the supply of wild foods decreased, and the Natufian culture came under environmental stress. This stress is thought to be associated with a period of global cooling referred to as the Younger Dryas climatic episode (13,100–11,000 years ago), during which the temperatures in the north approached those of the LGM. (Both the Older Dryas, which occurred before the Natufians, and the Younger Dryas episodes are named after a flower, *Dryas octopetala*, a small arctic-alpine member of the Rose family whose pollen[10] is particularly abundant in the archaeological record during these cooler periods.)

Although the environmental stress theory based on a reduction in rainfall during the Younger Dryas has been challenged, pollen analyses of the Fertile Crescent region have confirmed a relatively rapid decline in forest habitats during this period in the Levant, consistent with a marked reduction of moisture in the spring and early summer. Some archaeologists believe it was this drought that caused a catastrophic decline in the supply of wild foods and helped spur the transition from sedentary foraging to agriculture. Those living in relatively dry areas were forced to seek new locations where water was abundant. Those already living by springs or rivers, such as the inhabitants of Jericho, stayed put, but gradually reduced their foraging lifestyle in favor of the more labor-intensive but less risky agricultural lifestyle.

The agriculturally unfavorable Younger Dryas episode eventually ran its course, and, by around 10,000 years ago, the former pluvial (i.e., rainy) conditions returned to the Levant. Temperatures rose, the annual rainfall increased by nearly 30%, and the growing season was extended, resulting in a major surge in plant biomass. Once again, wild food resources became abundant—emmer and einkorn wheat, barley, almonds, acorns, and pistachio nuts. The end of the Younger Dryas roughly coincides with the appearance of a new culture called the Khiamians, named after the El Khiam terrace in Wadi Khareitoun, near Bethlehem, where a new type of arrowhead, signaling the presence of a new, post-Natufian culture, was found.[11] Dating to between 12,200 and 11,800 years ago, Khiamian culture belongs to a period called the Pre-Pottery Neolithic A (PPNA) because ceramic pottery, usually considered a marker for the Neolithic, was not present at the site.[12]

The second PPNA culture in the Northern Levant, the Mureybetian, is named after an ancient settlement mound or "tell" on the bank of the Euphrates in northern Syria. Both are found primarily in the Northern Levant, although some Khiamian deposits have been found in the Southern Levant as well (Figure 3.2).

The Khiamians have been called the "first farmers" because of the presence of a wide variety of carbonized cereal grains and legume seeds found at the site. However, the term "farming" is defined as the utilization of domesticated plants and/or animals for food and other resources.[13] This raises the question whether the carbonized seeds found at El Khiam were domesticated. As we'll discuss later, one of the diagnostic features of domesticated cereals is the presence of nonshattering rachises. According to George Willcox, the earliest appearance of fully domesticated cereals exhibiting a relatively high percentage of nonshattering ears is thought to have occurred between 10,300 and 10,000 years ago, corresponding to the early Pre-Pottery Neolithic B (PPNB), the stage following PPNA.[14] If so, it may be more accurate to describe Khiamian cultivation practices as "horticulture" or "gardening," which can involve either wild or domesticated species. Since some of the genetic changes accompanying domestication are invisible, such changes would not show up in the archaeological record, so it is possible that the crops grown by the Khiamians may have been partially domesticated.

The Sultanians of the Southern Levant, named after the Ein as-Sulṭān spring near Jericho where the first settlement was found, were contemporaries of the Mureybetians in the north, and, together with the Khiamians, they comprise the PPNA period of the Levant region (see Figure 3.2). Sultanian settlements began appearing along a corridor of land extending from Jericho in the south to Abu Hureyra and Murybet in Syria to the north. Although the Sultanians were similar in some respects to the Khiamians, they were distinct in terms their tools, art works, and means of sustenance. A marked increase in the

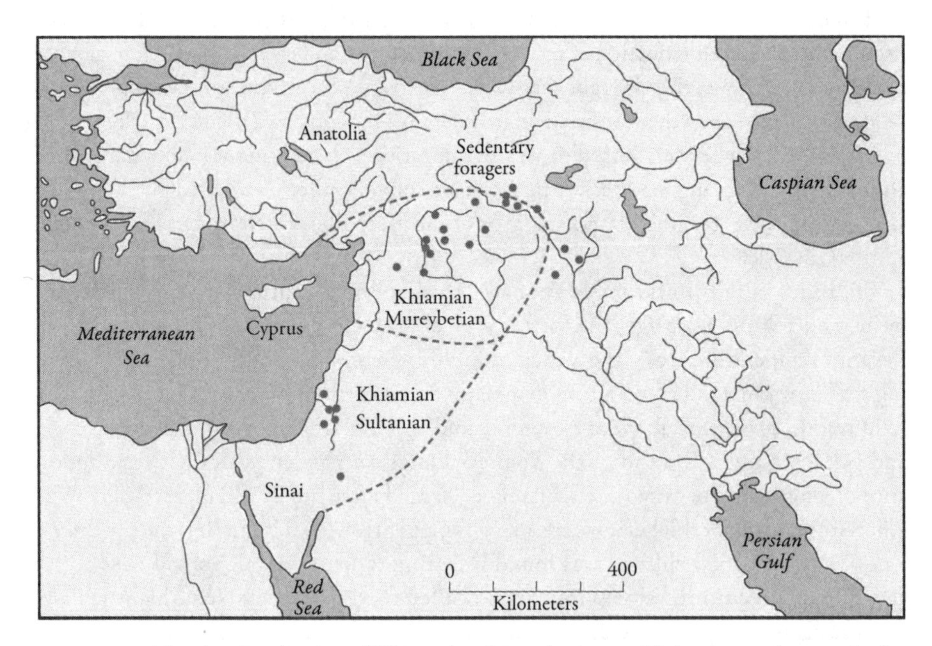

FIGURE 3.2 Map showing the sites of Khimanian, Mureybetian, and Sultanian settlements in the Northern and Southern Levant (indicated by the dashed lines).
Map based on: Bar-Yosef, O. (2014), South Turkish Neolithic: A view from the Southern Levant, in M. Özdoğan, N. Başgelen, and P. Kuniholm, eds., *The Neolithic in Turkey*, Vol. 6. Archaeology and Art Publications, Instanbul, pp. 293–320, figure 2.

presence of cereal pollen grains is observed at Sultanian sites, coupled with an abundance of carbonized seeds, including rye, barley, oats, einkorn wheat, and legumes. In addition, there is a spike in the number of seeds of unwanted weed species, which, as every gardener knows, quickly colonize any break in the ground cover. Consistent with these signs of cultivation, archaeologists have also found sickle blades exhibiting the type of abrasions indicative of harvesting cereal crops. In contrast to the earlier Natufian sickle blades found by Garrod and others, which showed signs of use for a variety of vegetation,[15] the Sultanian sickles appear to have been used exclusively for cereals.

The Sultanians of the south thus appear to be the first true agriculturalists, depending mainly on the cultivation of domesticated plants and animals for their diet.[16] Sultanian communities greatly expanded the area of land under cultivation and diversified the crops they grew. As the practice of crop cultivation increased, so too did the populations of the permanent settlements.

We should not underestimate the profound impact the transition to agriculture had on the lives of Epipaleolithic people. In addition to the increased labor required, the practice of agriculture would have disrupted people's ethical and spiritual ideas about their relationship to their environment, fundamentally changed their concepts of resource ownership and rights of access, and forced a revision of how work was performed and distributed within and between groups.[17] The increase in population led to an increase in social complexity and stratification based on the unequal distribution of the world's first artificially generated wealth. Thousands of years later, the Greek poet Hesiod imagined that the sexual union between the grain goddess Demeter and her consort, Iasion, in a thrice-plowed wheat field resulted in the birth of Plutos, the god of cereal wealth, from which the term "plutocracy" is derived.

For better or for worse, the Agricultural Revolution had begun.

CROP DOMESTICATION IN TIME AND SPACE

All wild species exhibit variability within their populations because of mutations that occur continually in the DNA of their germ cells—sperm and egg. When subjected to natural selection, the best-adapted individuals will tend to have more progeny than those less well-adapted, and their descendants will gradually come to dominate the population. In the course of cultivating wild plant species in the Neolithic period, humans inadvertently accelerated the process of evolution by selecting, either intentionally or unintentionally, certain individuals over others in the population.[18] Although the domestication of animals was achieved by both artificial selection and breeding, the option of breeding was not available to early farmers simply because plant sex had not yet been discovered. Plant domestication, in contrast to animal domestication, was brought about exclusively by artificial selection.

Virtually all of the important crop plants are angiosperms (flowering plants), of which there are about 235,000 species. Of these, only a few hundred have been more or less domesticated. In fact, 80% of the world's annual harvest of plant food is derived from only a dozen plant species: wheat, corn, rice, barley, sorghum, soybean, potato, manioc, sweet potato, sugar cane, sugar beet, and banana.[19]

The dependence of global agriculture on such a small number of species means that agriculture initially was restricted to those few areas where the wild relatives of these major crop species evolved. As noted earlier, the remains of the earliest agricultural settlements are found in the Fertile Crescent, the broad, arching zone of grassland and woodland that begins at the Levant and curves around some 2000 km (1,243 miles) eastward to the Zagros Mountains. However, agriculture arose independently at least six more times in various regions around the world, including South China, North China, Sub-Saharan Africa, the South Central Andes, Central America, and the Eastern United States.

THE FOUNDER CROPS OF THE FERTILE CRESCENT WERE ALL ANNUALS AND SELFERS

All of the seven founder crops of the Fertile Crescent region have two important features that made them easier to domesticate than other wild species: they are all annuals, and they are all self-pollinators, or *selfers*. Selfers are plants that release their pollen from the anthers onto the stigmas of the carpel before the opening of the floral bud ("floret"), a process called *anthesis* (Figure 3.3A–C). As a result, pollination in selfers occurs prior to anthesis. In contrast, nearly all the wild species of plants are "outcrossers"—that is, they require pollen from other plants to produce seed.

Outcrossing is an evolutionary adaptation in most wild species that ensures genetic mixing, thus preventing the accumulation of deleterious mutations from one generation to the next. The build-up of deleterious mutations can lead to the phenomenon of *inbreeding depression*—a common occurrence in cereals. However, the eight founder crops are not obligate selfers. Occasionally, they are pollinated by pollen from nearby plants, so that some genetic mixing occurs, although on a limited scale. Wind serves as the agent for outcrossing among the cereals, whereas insects act as vectors for cross-pollination among the pulses and in flax.

The obvious advantage of annuals over perennials for seed crop domestication is that annuals produce seed in a single growing season; in contrast, perennials typically pass through a juvenile phase of two or more years before reproducing sexually. The fact that annuals die after a single growing season also facilitates selection because the death of the previous generation allows newly selected traits to spread more rapidly in the population.

The advantage of selfing for plant domestication is that it ensures that mutant plants selected for their favorable traits can be propagated by seed more or less indefinitely, even when grown in proximity to "wild type" plants. Their habit of selfing makes them nearly immune to the pollen from the other plants, so the desired trait is seldom diluted in the next generation by outcrossing. Although the property of selfing may have been an advantage for plant domestication, it was a major hindrance to the discovery of sex in plants. First, the process of pollination was completed before anthesis occurred and so was completely concealed within the emerald confines of the unopened floret. Second, there were fewer opportunities for viewing hybridization events, which might have led to the discovery of sex.

SELECTION FOR THE TOUGH RACHIS OF DOMESTICATED CEREALS

The Fertile Crescent region in the Near East is particularly well stocked with wild cereals suitable for domestication. Of the world's fifty-six species of large-seeded grasses, thirty-two grow wild in the Mediterranean area. The wild ancestors of the three core domesticates—einkorn wheat, emmer wheat, and barley—have all been identified there.[20]

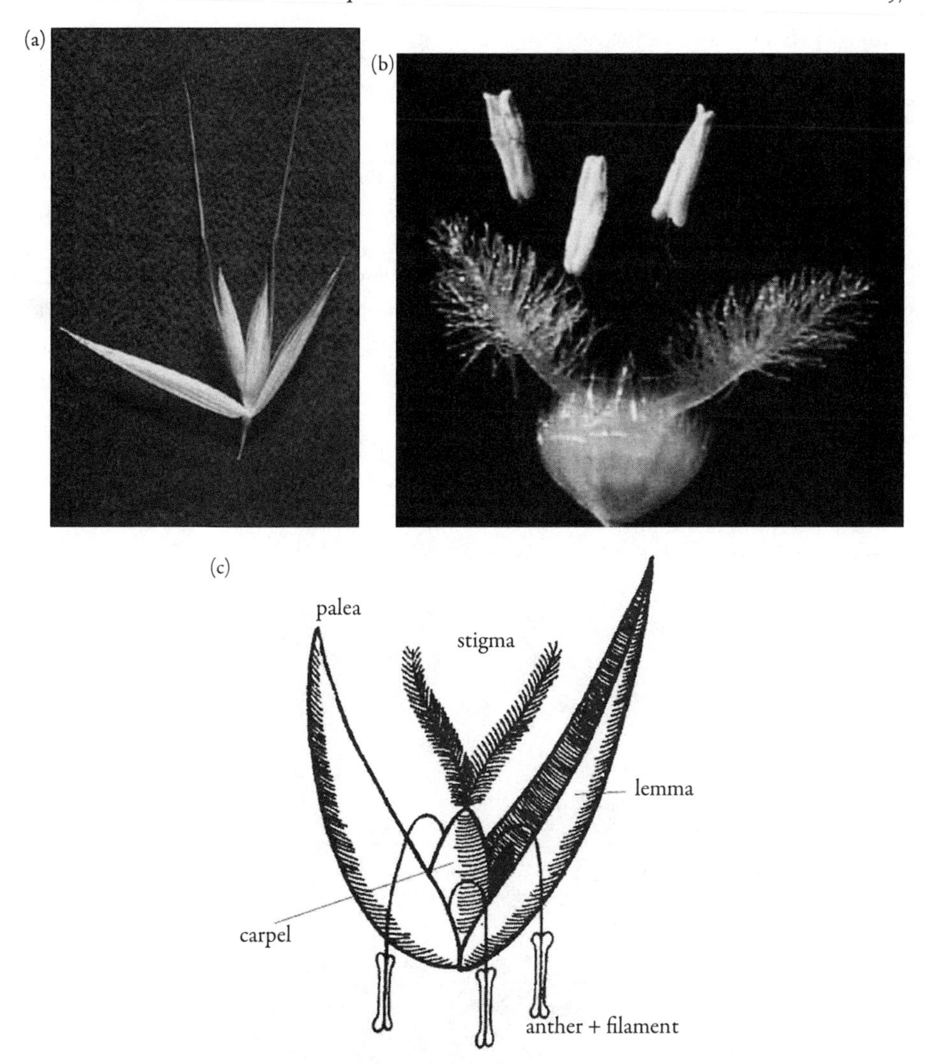

FIGURE 3.3 Three views of the oat flower. A. An oat spikelet contains a pair of florets. B. Oat flower dissected from a single floret. C. Diagram of oat floret. The ovary, topped by two feathery stigmas, is surrounded by three stamens. The reproductive structures are enclosed in two modified scales, the larger lemma with its bristle-like extension, the awn, and the smaller palea (missing in B). During selfing, pollen released from the anthers is deposited on the stigmas of the same flower before the opening of the floret.
B is from: http://www.biology.iastate.edu/Courses/202L/New%20Site%20S05/27AngioReprod/Plantrepro.htm).

Of all the differences between wild and domesticated cereals, two are regarded as most critical for domestication. The first is the loss of seed dormancy—the inability of certain seeds to germinate when planted in moist soil unless given additional environmental stimuli, such as a cold treatment. Because the loss of seed dormancy trait is invisible, it does not show up in the archeological record. Such a change may have begun either during the phase of ecosystem management prior to the adoption of a sedentary lifestyle, or later, in small garden plots planted near permanent settlements during the

PPNA. It is easy to imagine how the loss of seed dormancy might be selected for in small gardens because the first gardeners would have preferentially harvested the rapidly growing crops.

The second critical change that took place during cereal domestication was the development of a tough rachis, the main stem or axis of the cereal head (ear). In wild cereals, the rachis normally falls apart, or disarticulates, into dispersal units beginning at the top of the inflorescence (Figure 3.4A,B). The rachis of a ripe head of wild wheat is thus easily shattered into individual spikelets when touched or blown by the wind; hence, it is referred to as a *brittle rachis*. A brittle rachis is an extremely undesirable trait from a human standpoint because it drastically decreases the efficiency of harvesting. Plants with tough rachises occasionally appear in wild cereal populations, but they are never selected for under natural conditions because they prevent seed dispersal. During harvesting with sickles, however, tough rachises would be artificially selected for by humans.

After cutting the tough-rachised domesticated varieties, the stalks can be transported to another location for threshing. It is easy to distinguish the grains of wild and domestic wheat varieties because the wild varieties have a smooth abscission scar where the spikelet

FIGURE 3.4 Disarticulation of spikelet of wild versus domesticated wheat. A. Photograph of ripe ear of wild wheat shedding a spikelet. B. Comparison of the abscission zones of wild versus domesticated variety. Note the rough surface on the abscission zone of the domesticated wheat variety.

A is from Tenenbaum, David, http://whyfiles.org/shorties/199wheat/; B is from Smith, Bruce (1999), *The Emergence of Agriculture*. Scientific American Library. W.H. Freeman & Co.

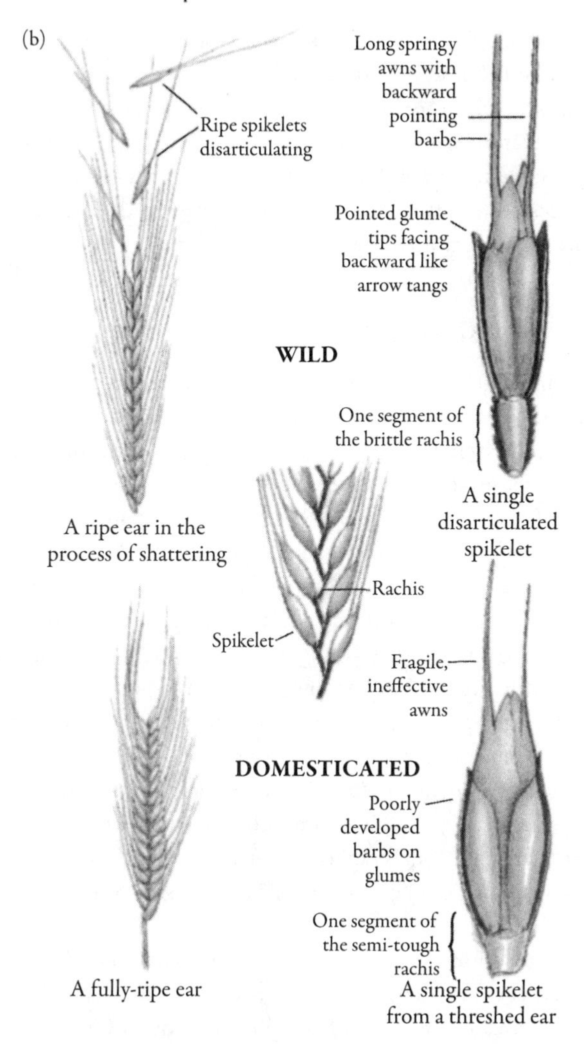

FIGURE 3.4 Continued

has broken cleanly from the brittle rachis, whereas the cultivated grain leaves a rough edge where the spikelet has torn from the tough rachis (Figure 3.4B).

The artificial selection of a tough-rachised wheat variety probably took place under conditions of horticulture during the early Neolithic period around 10,300–10,000 years ago.[21] In other words, farmers began planting the seeds of wild cereals in gardens before they had altered their traits to the domesticated forms by artificial selection. Different methods of harvesting the cultivated grain would have selected for different traits. For example, the easiest harvesting method, beating the ripe spikelets of intact plants into a basket, would tend to select the seeds of brittle-rachised individuals and leave the tough-rachised ears behind. However, if stalks at the nearly ripe, yellow-green stage (when the abscission zones of the spikelets have not yet fully matured) were either harvested by sickle or by uprooting, spikelets with tough-rachises that were able to survive this rough treatment might be preferentially harvested. This method of harvesting could in principle favor the selection of mutations that delayed the abscission process.

To determine how quickly the shift from wild to domesticated forms actually occurred, Tanno and Willcox examined 9,844 charred spikelets of einkorn wheat from four different Neolithic sites in northern Syria and southeastern Turkey.[22] A comparison of the spikelets of various ages indicated that there was a gradual increase in the percentage of the tough-rachised domesticated grain from 10,250 to 8,500 years ago.[23]

AGRICULTURE AND THE REVOLUTION IN ICONOGRAPHY

Any time a society undergoes a change in lifestyle on the scale of the transition from foraging to agriculture—a radically different way of life based on an entirely new technology—we can assume that the psychological and spiritual make-up of the members of that society will undergo a commensurate adaptational adjustment over time. Moreover, when belief systems and cultural institutions begin to alter, they can initiate a positive feedback loop, adding momentum to the impetus for change. Once the validity and value of crop cultivation was fully accepted and integrated into belief systems, the stage was set for further innovations and wider applications of the new technology. Such a positive feedback loop would explain why, at the end of the Younger Dryas period, when conditions once again became favorable for foraging, the Sultanians pressed on with their labor-intensive agricultural lifestyle—clearing, digging, hoeing, planting, and weeding—rather than reverting to the more leisurely existence of hunting and gathering.

Archeologists Jacques Cauvin and Ofer Bar-Yosef have documented a revolution in iconography, evidenced by small art objects found in the settlements, that took place in the Levant during the transition from foraging to crop cultivation. These art objects were often associated with graves, consistent with some sort of ritual significance. The pre-agricultural Natufians used shell, stone, and bone to make a variety of implements, decorations, and small figurines. Nearly all of their figurines represent animals, typically small herbivores such as deer or gazelle, but also dogs, owls, and possibly even baboons (Figure 3.5A). Human forms are rare and consist of small, highly simplified heads with no indication of sex.[24] If we assume that the zoomorphic figurines were not merely aesthetic objects or toys, but also had important symbolic significance for the lives of the individuals in whose graves they were often found, we can infer that they represent elements of the basic mythological system of the Natufians. For example, images of ruminants that the Natufians frequently carved as decorations on sickle handles could represent a totemic symbol associated with a clan or group.[25] The absence of human images at Natufian sites may indicate that they had not yet begun to use humans as abstract symbols. This is somewhat surprising considering the highly sophisticated "Venus" figurines produced during the Aurignacian and Gravettian periods of the Paleolithic. Clearly, there is no direct cultural lineage between the Paleolithic people of Eurasia and the Natufians of the Levant.

The Khiamians, who succeeded the Natufians, introduced various changes in settlement architecture, tools, burial practices, and horticultural practices. In the spiritual realm, they also initiated an important change in the types of symbols used in their portable art. Instead of the zoomorphic figurines of the Natufians, the Khiamians created small

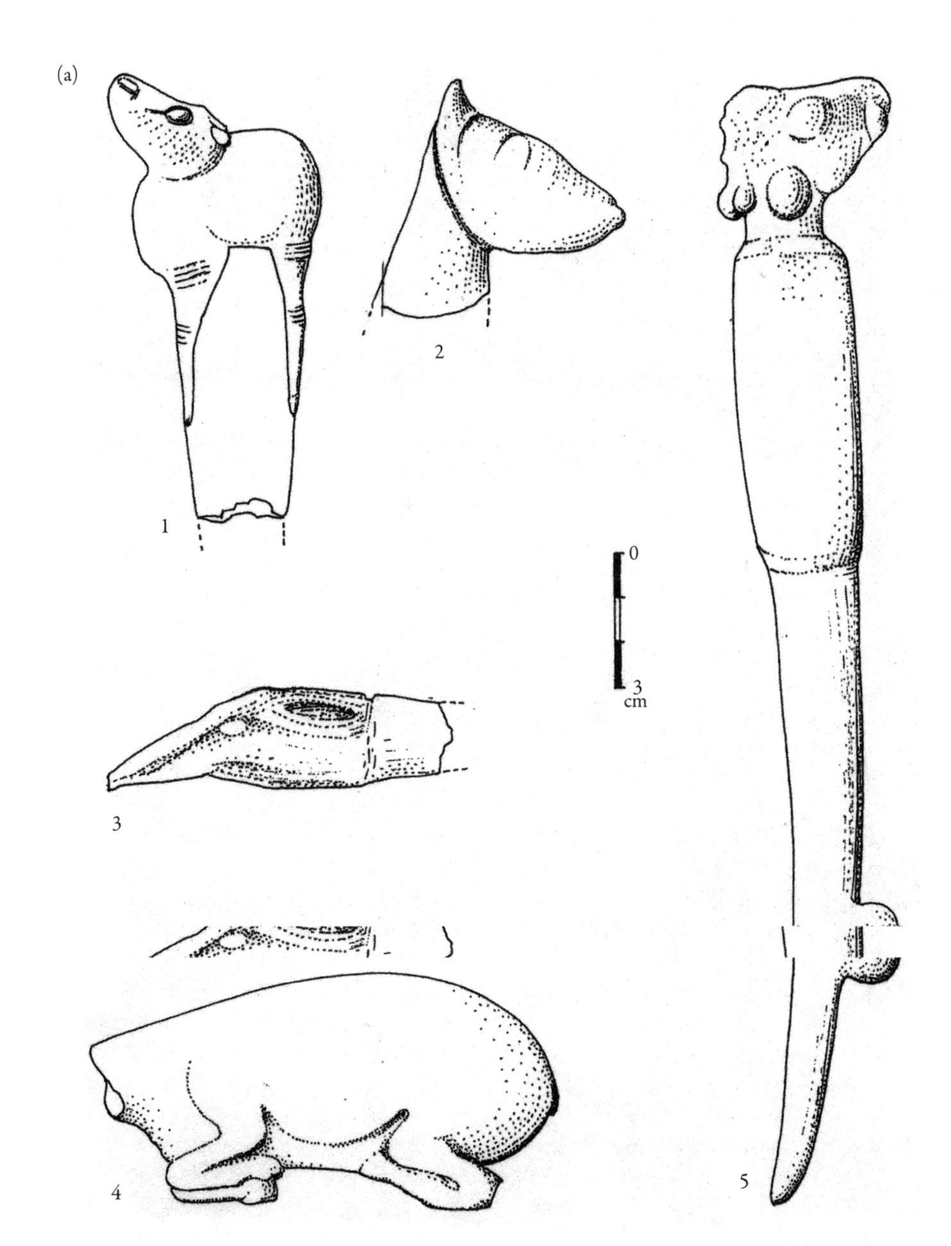

FIGURE 3.5 Changes in representation in Neolithic figurines from the Levant. A. Natufian. B Khiamian; C. Sultanian.
A and B from Cauvin, J. (2000), *The Birth of the Gods and the Origins of Agriculture*, Trans. Trevor Watkins. Cambridge University Press, figures 3 and 6, respectively. C from Bar-Yosef, O. (1998), The Natufian culture in the Levant, threshold to the origins of agriculture. *Evolutionary Anthropology* 5(6):159–177, figure 13.

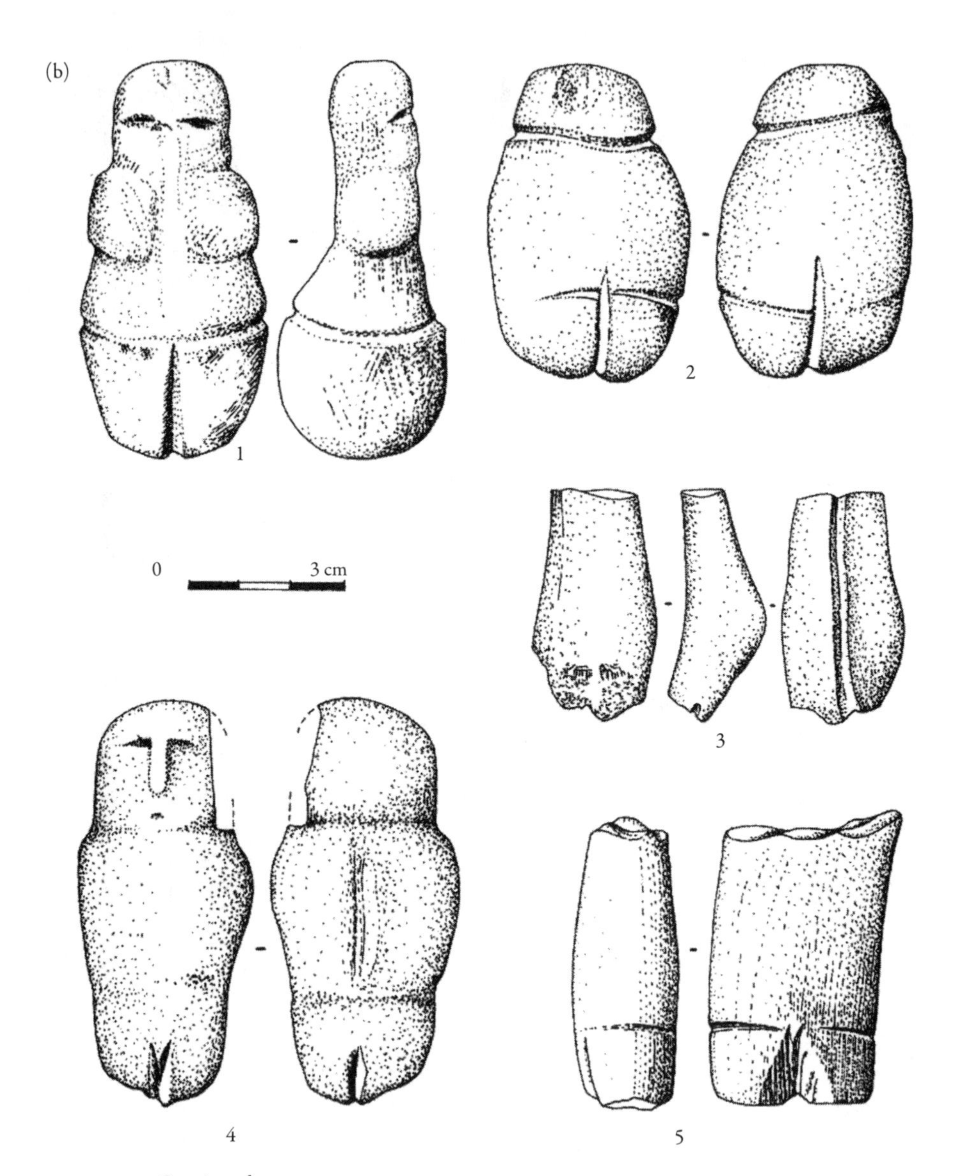

FIGURE 3.5 Continued

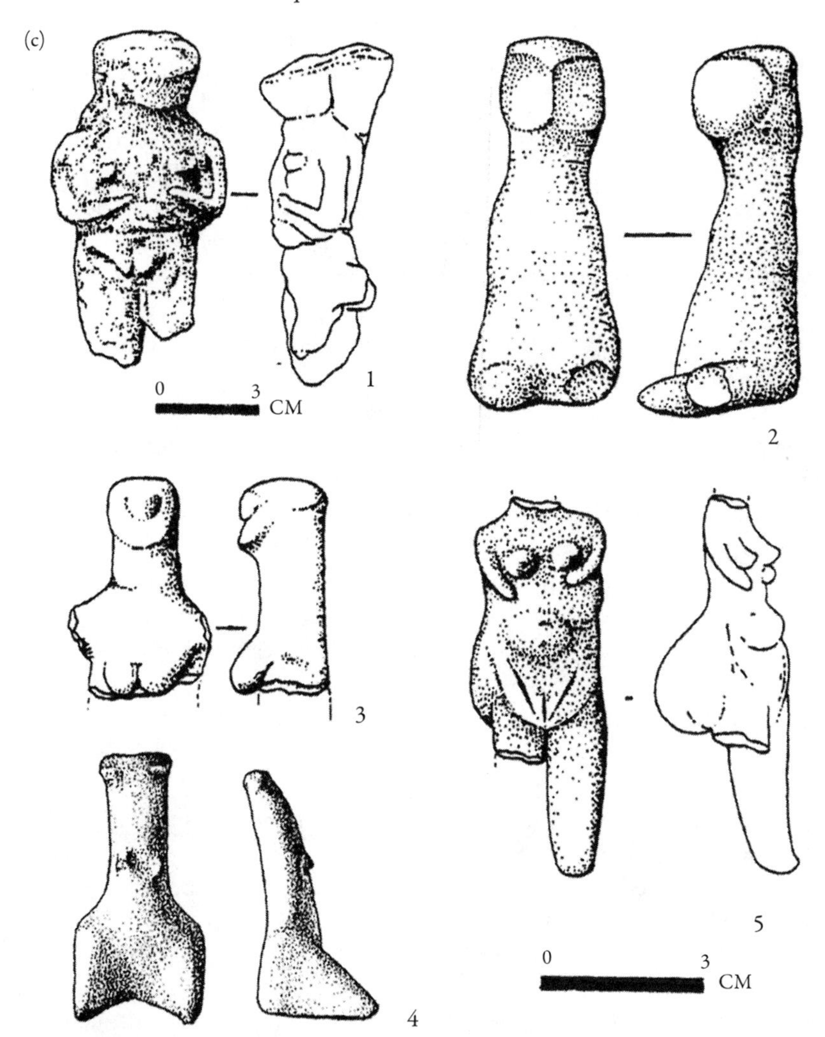

FIGURE 3.5 Continued

limestone statuettes representing female forms (Figure 3.5B). Virtually no animal figurines are known from the Levant during the Khiamian period. What the spiritual or cultural transformation was that gave rise to the change in symbols is unclear, but the emphasis on the creation of these female figurines arises concurrently with the increasing importance of horticulture as an important means of subsistence. According to Bar-Yosef, the change in iconography from animals to women "epitomizes a departure from being equal partners within the natural environment to humans as major players on the scene."[26]

The Khiamians incorporated symbolic objects into the very structures of their houses, suggesting that the houses themselves were taking on a new symbolic significance. Although no animal figurines were found at Khiamian sites, auroch skulls, complete with horns, were buried under houses and were also incorporated into raised clay benches inside the houses. The embedded auroch skulls were often accompanied by the shoulder-blades of aurochs and onagers (wild asses).[27] Because Khiamian villagers hunted mainly herbivores of the steppe regions and rarely hunted aurochs (doubtless because of their ferocity), the use of auroch skulls in their houses cannot be viewed as mere hunting magic. Rather, the incorporation of auroch skulls into Khiamian houses suggests a "spiritual dimension" of some kind.[28]

The Sultanians continued the Khiamian tradition and carried it still further. Most of the figurines, constructed of either baked clay or limestone, are either clearly females or sexually ambiguous (Figure 3.5C). There are no obvious male figurines among them, although one or two limestone phalli have been found dating from this period. Mortars and pestles have also been cited as possible sexual symbols.[29] Assuming women were primarily responsible for food preparation, they would have been the ones using such sexual symbols.

Like the Khiamians, the Sultanians incorporated auroch bones into the walls of their houses. According to Cauvin, there is a transition in iconography from zoomorphic representations to two major symbols, the woman and the bull, just at the time when the transition from foraging to agriculture is taking place.[30] This specification of gender, not evident among the Natufian figurines, is commonly interpreted as a reflection of the emerging role of women in a society increasingly dependent on agriculture.[31]

Another striking example of a Neolithic site with abundant female figurines is the ancient settlement of Sha'ar Hagolan, located in the Jordan Valley near the southern shore of Lake Kinneret (Sea of Galilee) on the north bank of the Yarmuk River.[32] First excavated by archaeologist Moshe Stekelis of the Hebrew University of Jerusalem in 1949–52, the site, which is situated within Kibbutz Sha'ar Hagolan, revealed the remains of a unique Late Neolithic culture (8,400–7,800 years ago) that came to be called the Yarmukian culture.

Over the years, scores of artifacts have been uncovered, and the Sha'ar Hagolan Museum now houses one of the most impressive collections of prehistoric art in Israel. Of the 298 "figurative objects" collected as of 2011, 115 depict seated, steatopygous women with narrow, slitted eyes (Figure 3.6).[33] Although the peculiar slitted eyes, common in figurines throughout the ancient Mediterranean region, are often referred to as "coffee-bean eyes," it seems more likely that they were patterned on cereal grains and are therefore emblematic of agriculture. The original excavators of Sha'ar Hagolan, concluded that distinctive female figurines, which are quite similar in appearance, represented "fertility of the soil and its fruits" and may also have been used as "amulets for fertility, protection against evil, relief in childbearing, and the cult of the dead."[34]

According to Yosef Garfinkel, the recent excavator of Sha'ar Hagolan, the uniformity of design and the contexts in which they were found suggest that the Sha'ar Hagolan female figurines represent a "goddess" worshipped within households by the nuclear or extended family:

> [A]ll depict seated women according to a fixed formula. . . . This is a canonical figure in which certain features always appear, a rigid code that is typical of the religious sphere in contrast to art for its own sake or children's toys. Consequently the female figurines should be seen as representing a Yarmukian goddess who was worshipped by the ancient inhabitants of Sha'ar Hagolan. Since the figurines were found in residential courtyard houses, they must have belonged to a household cult that was conducted by the nuclear or extended family.[35]

FIGURE 3.6 Neolithic "Goddess" figurines from Sha'ar Hagolan (~8000 BP).
From Garfinkel, Y. (2004), *The Goddess of Sha'ar Hagolan*. Israel Exploration Society, Jerusalem.

Freikman and Garfinkel have emphasized that, in contrast to the zoomorphic figurines found at the site, which were schematically and crudely executed, the female figurines "are among the highest achievements of the Neolithic Period in the Near East, being made with great care and with much attention to detail."[36] The authors concluded that "the anthropomorphic items were indeed used as venerated cultic items, representing supernatural powers."

Freikman and Garfinkel's hypothesis doesn't exclude the presence of other additional deities or cultic figures. For example, some ithyphallic male figurines have also been recovered from the site, although male images are extremely rare. It is also possible that other entities were formally worshipped in dedicated shrines or temples by multiple households. So far, no public cult buildings dating to the Neolithic have been found at Sha'ar Hagolan.

In addition to portable objects such as figurines, human images were engraved on stone, painted onto pottery, and painted on walls and floors. Garfinkel has interpreted a vast body of human representation in art from the Neolithic to the Chalcolithic, or Copper Age, as ritual dancing, which he believes played an important role in maintaining social coherence and group identity once people began to live in permanent settlements.[37] Some examples of possible dancing rituals painted onto pottery are shown in Figure 3.7A, and a gallery of female dancers taken from many such examples is shown in Figure 3.7B. While both men and women are depicted, women account for some 75% of the dancing figures during the

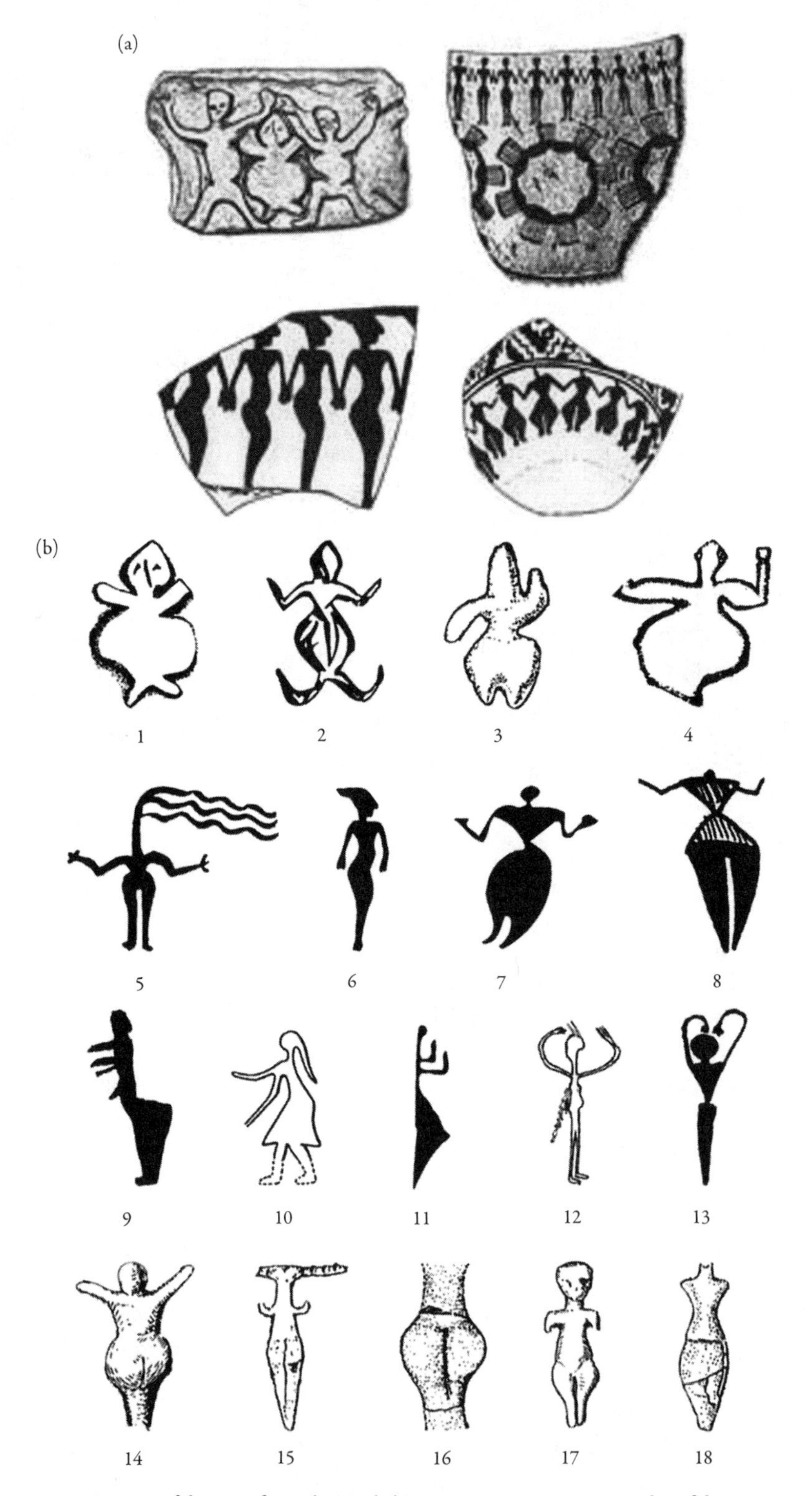

FIGURE 3.7 Images of dancing from the Neolithic Near East. A. Four examples of dance scenes painted on pottery. B. Dancing figures that are clearly gendered female.
From Garfinkel, Y. (2003), The earliest dancing scenes in the Near East. Near Eastern Archaeology 66: 84–95; and Garfinkel, Y. (2003), *Dancing at the Dawn of Agriculture*. University of Texas Press, p. 50.

Neolithic. Thus, once again we find a correlation between the ascendance of a plant-based, female-led industry (agriculture) and an upsurge in female iconography. The predominance of female dancers during this period suggests that women were playing a leading role in the social and spiritual spheres of their communities.[38]

CRYPTIC AGRICULTURAL SYMBOLS AT ÇATALHÜYÜK

Çatalhüyük is located at the southern edge of the Anatolian plateau some 20 miles southeast of the Turkish town of Konya. Çatalhüyük (pronounced CHAH-tahl-HOO-yook) literally means "forked mound," a name suggested by a path that divides into three trails at the foot of the site (Hodder 2006). Çatalhüyük actually consists of two tells,[39] referred to as the East and West Mounds. The tells were originally situated on opposite banks of the Çarşamba River, which flowed northward from the Taurus Mountains and emptied into the Konya Plain, before being converted into an irrigation canal in the early 1900s. The Neolithic East Mound was settled around 7400 BCE, some 2,000 years after the first crops were planted in the region. Thus, the earliest inhabitants of Çatalhüyük were agriculturalists from the start, although it is clear, based on the remains of fauna and flora found at the site, that hunting and gathering still played an important role in their overall diet.

The houses of Çatalhüyük were multilevel structures packed tightly together, like the pueblo-style Native American communities of the Southeastern United States, except that there were no doors or windows. The interiors were accessed through rooftop openings via ladders and stairs. All interior walls were plastered to a smooth, white finish, and in some cases they were decorated with striking polychrome images. Indeed, what sets the houses of Çatalhüyük apart from all other known Neolithic dwellings is the startling impact of their interior artwork, which, in the case of the most elaborate examples, combine enigmatic polychrome murals and monumental relief sculptures of felines and other animals with rows of breast-like protuberances, each containing the skull, teeth, or jawbone of an animal—all set into walls and benches bristling with the plastered skulls and horn cores of aurochs. Little wonder that James Mellaart, the first excavator of the site, initially designated the most complex of these rooms as "shrines" to distinguish them from the less ornately decorated rooms.

The murals of Çatalhüyük have often been compared in their beauty and power to the Upper Paleolithic cave paintings of France and Spain. In both cases, images of wild animals predominate, but an important shift has taken place since the Upper Paleolithic. The wild animals depicted at Çatalhüyük are more isolated from their fellows and more completely dominated by humans than their Franco-Cantabrian counterparts, which are rarely shown with humans. In the dynamic murals at Çatalhüyük, enormous wild animals, such as stags and wild boars, are shown as the targets of relatively diminutive, but highly energetic people who are either hunting or baiting them. Such a contrast in the power relations between animals and humans is in keeping with the shift in emphasis from hunting and gathering to agriculture.

The dearth of images depicting agriculture in the wall art of Çatalhüyük is puzzling. All evidence from the site indicates that domesticated plants and animals constituted the core, year-round diet of the settlement, yet the wall paintings, relief sculptures, and even many figurines and statuettes continued to emphasize wild animals. Not until the art of the Bronze Age do we find explicit images of agricultural activity, usually centered around deities and rulers whose job it was to ensure a bountiful harvest (see Chapter 5). The art of Çatalhüyük thus seems to reflect a state of mind in transition from hunting and gathering to agriculture.

Discrepancies between art and everyday life are not without precedent. Art, like religion, does not always keep pace with technological innovation. The vast majority of people today subscribe to religious narratives and beliefs hundreds, if not thousands, of years old, while at the same time making use of all the latest electronic conveniences. Such may have been the case with the inhabitants of Çatalhüyük, firmly committed to agriculture as the cornerstone of their material prosperity while at the same time adhering to pre-agricultural narratives and images in their artistic and religious lives. Eventually, "modern" ideas and motifs drawn from the contemporary agricultural way of life begin to seep in, until a tipping point is reached when the new agricultural narratives, beliefs, and symbols become dominant, consolidating a cultural shift that, in practice, has already occurred. Here, we focus on five such cryptic agricultural symbols at Çatalhüyük: the sacred fruit tree motif; animal-baiting scenes; clay balls; relief sculptures of leopards; and female figurines associated with wild plants, grain, and cats.

THE SACRED FRUIT TREE MOTIF

In his fascinating book, *The Leopard's Tale*, Ian Hodder observes that plants were rarely represented in the wall paintings of Çatalhüyük and that there is also a scarcity of domestic animals in the murals.[40] From this scarcity, he infers that neither plants nor domesticated animals were considered appropriate subjects for wall paintings. However, there are two remarkable exceptions to the generalization that plants were deemed unworthy of depiction: two large paintings of fruit trees discovered by Mellaart in houses on level VII (~6600 BCE). The more complete example shown in Figure 3.8 consists of a vertical trunk, over 5 feet tall, with roughly symmetrical branches slanting upward, apparently leafless, and accompanied by four ibexes (wild mountain goats) arranged in an elliptical orbit among the lower branches (Figures 3.8A and the tracing in Figure 3.8B). Small, egg-shaped spots, presumably fruits, are scattered throughout, some still attached to their branches, others apparently detached and possibly in the process of falling. The absence of leaves and the apparent shedding of the fruits could be seasonal markers indicating that it is harvest time.

Unlike the chaotic and "noisy" animal-baiting tableaus, humans are nowhere to be seen in the fruit tree painting, which conveys a mood of harmonious serenity. Goats, not people, tend this tree, although they undoubtedly serve human interests rather than their own. The impression that the goats are *apotropaic*, or protective, "tree spirits" is reinforced by their encirclement of the boughs. The theme of abundance is evinced by the many fruits, which, together with the "guardian" goats, lend the tree an abstract, iconographic, dare we say "sacred" quality.

The fruit tree painting stands in direct contrast to the ritualistic wild animal-baiting scenes associated with feasting and masculine derring-do. As pointed out by Mellaart et al., the fruit tree painting at Çatalhüyük is reminiscent of the Mesopotamian sacred tree, which was often depicted with a female deity and protective spirits, including goats.[41] The Bronze Age version of the tree represents the agriculture-based prosperity of the state, with the Goddess Ishtar providing both protection to the state and divine sanction to its rulers.

FIGURE 3.8 Wall painting of stylized tree with fruit and ibexes at Çatalhöyük. A. Actual image on wall of house designated "Shrine E. VII/44" by Mellaart. B. Outline of the main features of the painting, including a possible fig tree with four goats or ibexes circling through the branches. The design is reminiscent of sacred "Tree of Life" images from Bronze Age in Mesopotamia and the Levant. From Mellaart, J., U. Hirsch, and B. Balpinar (1989), *The Goddess from Anatolia*. Ezkenazi.

No such elaborate, urban religious symbolism is evident in the Neolithic "sacred tree" at Çatalhüyük, which lacked the concept of a state. On the other hand, deciphering the significance of the Çatalhüyük fruit tree may not only shed light on Neolithic society; it may also help us to understand the origins of the sacred tree of Mesopotamia. We will have more to say about the sacred tree motif of the Bronze and Iron Ages in Chapters 5 and 6.

As for the identity of the fruit tree, it is possible that it was meant to be generic, symbolizing all fruit or nut trees. Alternatively, the artist may have had a specific tree in mind, perhaps one that was highly esteemed. The marl soil in the area immediately surrounding the settlement was nutrient-poor and subject to seasonal flooding, rendering on-site fruit tree cultivation difficult, if not impossible. As evidenced by the plant remains found in and around the houses, the inhabitants of Çatalhüyük collected a variety of fruits and nuts, presumably from steppe and mountain regions to the south, including apples, pears, plums, acorns, pistachios, almonds, and hackberries.[42] Ian Hodder's team also found fig (*Ficus carica*) remains in the rake-out from an oven.[43] Wild figs were widely distributed throughout the Fertile Crescent region during the Neolithic, including southeastern Anatolia and the Mediterranean coast of southern Anatolia.[44] The fig remains at Çatalhüyük thus could represent either wild or domesticated varieties of *Ficus caria*. Given the lack of easy access to areas suitable for fig cultivation, it seems probable that domesticated figs would have been obtained via trade routes to the Levant. Domesticated figs could have been one of the valuable foodstuffs for which obsidian miners and artisans traded.

In addition to its general appearance, our tentative identification of the "sacred tree" as *Ficus carica* is based on two other criteria: economic value and the association with goats. Figs, especially the domesticated varieties, would have been highly prized for their sweetness, just as dates were esteemed in Mesopotamian societies during the Bronze Age. Like dates, dried figs could have provided a portable, highly nutritious food for long-distance journeys south to the Taurus Mountains. From a supply and demand perspective, domesticated figs would have been valued precisely because they were so rare at the settlement compared with the remains of other fruits and nuts, as the archaeological record suggests. Figs may thus have been the costliest fruit at the settlement.

The close association between figs and goats is well-attested during the historical period. Greeks referred to wild fig trees as "caprifigs" because the less succulent syconia (the fleshy, hollow receptacles containing multiple carpels that comprise the fig "fruit") were considered suitable only for goats. It is possible that that the goat/caprifig artistic motif actually arose in the Levant sometime during the Neolithic, predating the Greeks by several thousands of years. The presence of ibexes or goats among the branches of the tree is by no means fanciful, since goats can and do climb trees. Obsidian traders from Çatalhüyük may have seen artistic images of goats and fig trees during their trading expeditions to the Levant and brought the motif back with them to the settlement for use in decorating the walls of houses.[45]

LARGE ANIMAL-BAITING SCENES: THE SIGNIFICANCE OF THE LONE WOMAN

Individual women are sometimes depicted in the wild animal-baiting scenes found on the wall of Çatalhüyük, but they are usually detached from the main action. For example, in the scene of an outsized red stag being harassed by a raucous band of hunters from a wall

FIGURE 3.9 Transcription of a stag-baiting scene involving slim, bearded men. The more rounded, obese figure on the right (*arrow*) appears to be a female.
From Mellaart, J. (1966), Excavations at Catal Hüyük, 1965: Fourth Preliminary Report *Anatolian Studies* 16:165–191, plate LII, redrawn from original by Raymonde Enderlé Ludovici.

painting on level V, the single figure of a woman stands apart from the group to the right of the painting (Figure 3.9).[46] Her broad hips and heavy legs are emphasized, in contrast to the wiry, supple forms of the men, which twist and flex with double-jointed agility. Her arms seem to be resting on her belly, which could be a reference to an impending feast. If she is pregnant, the hands-on-belly gesture could symbolize ideas of transformation and regeneration. In outline, she resembles the figurines of obese women so ubiquitous at other Neolithic sites—solid, immobile, and iconic. She is defined by her symbolic significance rather than by her actions. A similar figure of a lone woman with her hands on her belly is found below a bull in another hunting scene on level V. [47] In reality, large wild animals comprised a relatively minor part of the diet at Çatalhüyük, which relied mainly on plant foods and meat from domesticated sheep and goats. The bones of aurochs, however, are associated with feasting deposits, thus, the social significance of the animal-baiting scenes may be ceremonial, commemorating special events.

The static, composed, self-referential pose of the woman is reminiscent of the many female figurines found at Neolithic sites throughout Eurasia. We have interpreted this particular tableau as a metaphor for transformation and regeneration, which culminates in the figure of the woman, but this is only one of many possible explanations. It fails to address, for example, the issue of women's work and role in society. The evidence thus far seems to suggest that whereas hunting and otherwise engaging with large wild animals was largely a male activity, the sowing, harvesting, and processing of crops was primarily the responsibility of women. Accordingly, we might expect to find some clue in the rich artistic legacy of Çatalhüyük that connects women to agriculture.

Hodder has observed that "the few paintings that unmistakably depict women appear to show them gathering plants."[48] A possible depiction of a female plant gatherer is present in the stag-baiting scene from level V, which is prominently displayed just inside the entrance of the Anatolian Museum in Ankara. The painting shows the usual array of active male hunters harassing a stag. They are for the most part oriented vertically and their arms are raised in threatening gestures (Figure 3.10A). At the upper right-hand corner is an elongated, horizontal black figure, which Mellaart interpreted as one of the male hunters leaning forward as if sprinting. Upon closer inspection, however, we favor an alternative interpretation: the horizontal figure may actually be a female harvester (Figure 3.10B). Our interpretation is based on two criteria. First, although not entirely clear, the figure seems to have breasts. Second, the large conical object perpendicular to her waist is quite different from the leopard skins typically attached to the waists of the male hunters figures and resembles a basket.

What are lone women doing in a wild animal-baiting scenes dominated by men? Perhaps there is a trivial explanation, that they were added later and have nothing to do with the original tableau. However, if the presence of women in these scenes is intentional, they might be telling us something of central importance. We propose that the women in these paintings have iconic value, beyond their individual personhood. They could, for example, signify agricultural fields, which were located primarily on dry ground several

FIGURE 3.10 Stag-baiting mural from house at level V. A. Stag with vertical male hunters and a possible horizontal female plant harvester. B. Enlarged view of putative female harvester with basket.

miles away from the settlement. The grain fields of Çatalhüyük would have been easy pickings for wild ruminants, such as deer and aurochs. When such invasions occurred, threatening the food supply of the settlement, the men immediately would have been summoned from the village to drive the interlopers away. Perhaps they adorned themselves with leopard skins to instill fear in the formidable ruminants, as well as to access the occult aspects of leopard power. Perhaps the men in these scenes are protecting agricultural fields from the depredations of large, fierce herbivores.[49] If possible, the intruders would have been killed and eaten for good measure. Thus, the agricultural fields might have provided the added dividend of luring large ruminants near the settlement for killing. However, the original inspiration for the animal-baiting scenes may have had an agricultural context.

If the preceding interpretation is correct, the men and women portrayed in these paintings are actually united in a single purpose: protecting the crops. Alternatively, if the act of driving large herbivores away from planted fields became enshrined in myth, the paintings could depict reenactments of that myth. As a bonus, such festive occasions—lubricated by the local brew—would have provided an opportunity for young men to showcase their courage and athleticism before other members of the settlement.

Of course, an agricultural (economic) interpretation of the animal-baiting murals could overlap with a spiritual dimension as well. As David Lewis-Williamson has pointed out, what appears to be "economic behavior" cannot always be disentangled from "symbolic behavior."[50] In many societies, hunting wild animals, or in this case the possible harassment of wild animals, is bound up with important symbolic and spiritual constructs. It is possible that the people portrayed in the murals may be intent on acquiring the supernatural power possessed by the animals they attack. In the stag mural, one of the figures appears to be seizing the tongue of the animal, while others grasp different parts of its body, perhaps accessing its supernatural power as well as gaining prestige for courage in the community. Thus, the protection of fields and the acquisition of food represent only one aspect of meaning. The acquisition of spiritual power and prestige serves the individual and the community on another level.

EMBEDDED SEEDS

The identification of a female harvester in the stag-baiting scene just described is highly speculative, so we must ask ourselves whether there are any other artworks at Çatalhüyük that support the association of women and agriculture, or at least plants. The small female figurine shown in Figure 3.11 appears to provide such evidence. Found by Hodder's group in a midden at the upper levels of the mound, the figurine contains a seed embedded in its back. Based on its position and depth, the seed appears to have been pressed into the soft clay intentionally, as if mimicking the act of planting a seed in soil. However, this seed does not appear to have been derived from any of the known cultivated crops, so the association of the figurine with agriculture is weakened.[51] Still, a variety of wild plants were still being collected at Çatalhüyük for food and other purposes, so the figurine does exemplify a strong connection between women and plants. In this context, the implanted seed may symbolize generation, an attribute commonly associated with women.

FIGURE 3.11 A small clay female figurine, 2.8 cm high, with a seed pressed into its back, found in a midden near the surface layers of the South Area of the East Mound.
From Hodder, I. (2004), Catalhoyuk Research Project.

CLAY BALL IMPRESSIONS OF PLANTS

Many of the ovens at Çatalhüyük contain rich deposits of large, spherical objects called "clay balls." These clay balls were apparently used as "boiling stones" to cook stews, gruels, and porridges in containers prior to the advent of ceramic pottery.[52] Sonya Atalay has carried out a detailed analysis of the clay balls at Çatalhüyük and has classified them according to their shapes, sizes, conditions, and surface elaborations. Of particular interest are the "intentional markings" that appear to be too deeply impressed into the clay to have been accidental (Figure 3.12). Among the various types of intentional clay ball markings are impressions of baskets, mattings, and cordage, all of which are plant-based materials, as well as actual impressions of plants and seeds. Atalay found that about 18% of the clay balls seem to have been intentionally decorated in some way. The two most common forms of decoration were plant and seed impressions (27%) and basket or matting impressions (21%). Thus, approximately half of all the intentional markings on the ball surfaces were either plants or plant-based products. As Atalay points out, the freshly made clay balls may have been set inside baskets or on matted surfaces to dry. Nevertheless, whoever made the balls was probably well aware of these markings and used them for aesthetic reasons or for purposes of identification:

> [T]he people crafting the balls may have also actively used the drying surface, other nearby woven materials, plants, grasses, seeds, their fingers, palms, or other devices, to intentionally elaborate the surface of the balls they made, possibly for decoration or as a sign of ownership or as a sort of maker's mark, or possibly, as in the case of the plant and seed impressions, to demonstrate a symbolic association of the balls with plants.[53]

Since it is likely that women were primarily responsible for cooking, the clay ball impressions provide us with another symbolic link between women and plants.

FIGURE 3.12 Clay ball with leaf impressions.
From Atalay, Sonya (2003), Domesticating clay. Engaging with 'They': The Social Life of Clay Balls from Çatalhöyük, Turkey and Public Archaeology for Indigenous Communities. Ph.D. Dissertation. UC Berkeley Anthropology Department.

THE CATS OF ÇATALHÜYÜK

Like the Neolithic inhabitants of Çatalhüyük, modern villagers along the banks of Dongting Lake in the Hunan province of China are used to periodic flooding. Every year during the flooding season, rodents, especially the eastern field mouse, flee their waterlogged homes and invade nearby rice paddies, causing significant crop losses. In the summer of 2007, however, the crop losses caused by mice were catastrophic. From September to June, a prolonged drought had caused Lake Dongting to recede more than usual, which led to an explosion of the rodent population within the newly exposed lake bed. In mid-June, torrential rains brought devastating floods throughout the region. To relieve pressure on dams, engineers opened the sluice gates, causing a sudden rise in the level of the lake. As a consequence, millions of mice, driven from their lake bed burrows, took up residence in the rice paddies of Binhu. Unfortunately, most of the natural predators of the mice—snakes, owls, weasels, and cats—had been killed off earlier by the profligate use of pesticides, so there was no defense against the rodent onslaught. According to one farmer, the scritching of their little teeth was deafening as the field mice nibbled at the stalks and grains through the night. By morning, the verdant paddies of Binhu, upon which the village depended for its livelihood, were reduced to stubble.[54]

We relate this story because we believe that the flooding of Binhu Village may be telling us something important about agricultural symbolism at Çatalhüyük. As a consequence of its proximity to the Çarsamba river, Çatalhüyük experienced flooding every spring. During these months, much of the surrounding area was submerged, which is the primary reason archaeologists believe that crops must have been planted on permanent dry ground located as far away as 6 kilometers (nearly 4 miles) from the settlement. What was the impact of the annual spring floods on the settlement's rodent population? It seems likely that something similar to what the people of Binhu Village experience may have occurred annually at Çatalhüyük. Displaced rodents, mainly mice, would have swarmed through the settlement and adjacent dry ground, some of which may have been planted with crops. The annual spring floods would only have exacerbated a rodent problem that plagued Çatalhüyük throughout the year. Hodder's team has found clear evidence for severe mouse infestations in the houses, often associated with food storage bins.[55]

Without knowing the precise locations of Çatalhüyük's farmland, it is difficult to assess the possible impact of the spring floods on the influx of rodents. Still, assuming the majority of the crops were planted at remote locations at elevations safely above the flood-zone (but relatively close to the high-water mark), masses of field mice could have been driven into the planted fields in the spring by the encroaching flood water. Any crops planted closer to home, on nearby hills and hummocks, would almost certainly have become infested. What defenses did the farmers of Çatalhüyük have against such rodent plagues? In the fields, small wild carnivores— snakes, owls, weasels, and foxes—would have helped to keep the rodent populations in check.[56] Domesticated dogs might also have been of some use in combating the rodent menace. But the farmers of Çatalhüyük may have had an even more potent weapon in their never-ending war against rodents: domesticated cats. And herein lies the possible connection to the agricultural symbolism at Çatalhüyük.

It was once thought that cats were first domesticated during the Bronze Age in Egypt. That cats were revered and that they attained cultic status in ancient Egypt, an area that was also subject to periodic inundation and that relied heavily on grain crops and the storage of those crops, has been thoroughly documented.[57] However, there is now clear evidence that cat domestication coincided with the early stages of agriculture in the Neolithic, when "grain storage attracted large mice populations" and cats were "encouraged to settle in villages to control the mice."[58] A cat was found buried in association with a human at the archeological site of Shillourokambos in Cyprus, demonstrating that "a close relation had developed by ~9500 years ago."[59] Given the crucial importance of the grain crop to human survival, it would not be surprising to find artworks honoring cats in Neolithic settlements. At Shillourokambos, a stone carving of a head, believed to represent a cat-human hybrid has been found (Figure 3.13). The stone sculpture is contemporary with the cat burial. According to Jean Guilaine, "This representation might belong to a mythology of continental origin transferred to the island with other components of the process of 'neolithization.' But it might also represent a concern linked to an ideology elaborated by the first agro-pastoralists of the island."[60]

At Çatalhüyük, felines are represented by the numerous impressive images of paired "leopards," typically in the form of painted wall reliefs. In Figure 3.14, the "leopard" on the right appears to be pregnant, suggesting a mated couple. Intriguingly, directly below the swollen belly of the female there is a second painting of a "Tree of Life" motif with

FIGURE 3.13 Stone carving of the head of a cat or cat–human hybrid found at the aceramic Neolithic site of Shillourokambos, Cyprus, ~9,500 years ago.
From Guilaine, Jean (2001), Tête sculptée dans le Néolithique pré-céramique de Shillourokambos (Parekklisha, Chypre). *Paléorient* 26:137–142.

ibexes (see Figure 3.14). This pairing of the "Tree of Life" motif with the (possibly pregnant) "leopard" relief sculpture suggests that they may be ideologically linked to abundance or protection.

Today, leopards are largely confined to the rugged Taurus Mountains, but in Neolithic times they may have also roamed the plateaus and grasslands of central Anatolia, so the inhabitants of Çatalhüyük may have encountered them with some frequency. The Anatolian leopard hunted wild ungulates—deer, chamois, mountain goats, and occasionally wild boar—and would only hunt field mice as a last resort. However, mice are not the only threat to crops, and the leopard might have been perceived as offering protection against grazing by deer and other ungulates. Thus, leopards may have had a dual symbolic significance as fierce carnivores whose skins were worn ritualistically by men in hunting and baiting scenes and as protectors of the grain crops.

According to Mellaart, most of the elaborate wall art, including the "leopard" reliefs, begins around level VII, which corresponds to about 8,600 years ago.[61] As noted earlier, a close relationship between people and cats was already in evidence in Cyprus by 9,500 years ago. Thus, by the time the feline reliefs become prominent at Çatalhüyük, Neolithic farmers throughout the Fertile Crescent had been using tamed or domesticated cats to protect

FIGURE 3.14 Relief sculpture of a pair of "leopards," which had been replastered and repainted numerous times. Mellaart noted some sexual dimorphism in the "leopard" reliefs. In this case, the cat on the right seems to be pregnant female. (Note the presence of a second "Tree of Life" with ibexes motif directly below the pregnant female. Parallel rows of fruits, indicating branches, along with their ibexes are indicated by arrowhead.)
From Mellaart, J. (1967), *Çatal Hüyük: A Neolithic Town in Anatolia*. Thames & Hudson; Mellaart, J., U. Hirsch, and B. Balpinar (1989), *The Goddess from Anatolia*. Ezkenazi.

their fields and grain stores against rodents for nearly a thousand years. It seems reasonable to assume that the inhabitants of level VII or earlier must also have established a mutualistic relationship with cats. A few remains of the African wild cat, *Felis silvestris ssp. lybica*, have been found at the site, but so far no joint burials with humans have been reported.

Perhaps because domesticated plants and animals were not considered suitable subjects for art, there are no clear depictions of domesticated cats among the wall reliefs, paintings, and figurines at Çatalhüyük, but we can assume that the first domesticated cats were large and had tabby-like markings with banding patterns similar to those of their ancestor, *F.s. lybica*. In the artworks at Çatalhüyük, a variety of patterns are employed for the coats of the felines. Some suggest Anatolian leopard (*Panthera pardus*) spots, as in the case of the skins carried by the men in the animal-baiting murals and on some of the figurines, but others are quite abstract and could be interpreted as banding patterns consistent with wildcat markings. In the example shown in Figure 3.12, the X's could be interpreted as spots, but the horizontal bars between them could represent banding patterns.

We favor the interpretation that the "leopard" reliefs of Çatalhüyük, on at least one level of meaning, are the Anatolian analogs of the stone head of the cat at Shillourokambos in Cypress and ideologically linked to agriculture via the new cat–human mutualistic relationship. Consistent with such an interpretation, Mellaart found "numerous offerings of grain and crucifer seeds as well as one fine [female] figurine" on the platform set in front of the "leopard" relief on level VIA.[62] The finding of both grains and crucifer seeds on a platform before the leopard relief sculpture, together with a female figurine, is strong evidence that both women and plant foods were associated with felines. Such a finding supports the hypothesis that the relief sculptures of leopards functioned as apotropaic images whose roles included the protection of crops and stored food supplies from other animals, chiefly rodents and ungulates. If the reliefs represent actual leopards, leopards could have been viewed in this context as the wild apotheosis of domestic cats seen as protective spirit animals, which would make them worthy subjects for art. Alternatively, the "leopard" reliefs may depict mythical hybrids of *F. s. lybica* and *P. pardus*—the partially domesticated wildcat (hereafter referred to as "cat") and the Anatolian leopard.

THE "GRAIN BIN GODDESS" OF ÇATALHÜYÜK

If we accept the widely held view that the first agriculturalists of the Neolithic were predominately women, it follows that the first humans to establish the special relationship with cats that ultimately led to their domestication were also women. Women would have viewed cats as their natural allies in their perennial war against rodents, both in the agricultural fields and in the grain storage rooms of their houses. The domestication of cats, like that of dogs, probably involved self-selection for tameness around human middens, where waste from meals was scavenged, as well as in planted fields where cats were attracted to rodent concentrations. The human propensity for adopting baby animals as pets and the selection of those that were easiest to tame, together with the culling of the more resistant animals, no doubt played a role in the domestication process as well.[63] Eventually, a subpopulation evolved that was tame enough to mingle with humans, and people would have been able to introduce the tamed cats into their homes. This would have been a natural thing to do, given the severe mouse infestations in the storage areas. Cats, unlike dogs, are adept at climbing ladders, so they would have been able to come and go freely despite the absence of side doors in the houses of Çatalhüyük.

The most iconic example of an artwork that associates felines and women[64] is the striking clay statue of a seated female that was discovered in a grain bin. Dubbed the "Grain Bin Goddess," it consists of an imposing woman seated on what has been described as a "leopard throne" (Figure 3.15). She is all curves and roundnesses, lending her bulk a pillowy softness. Her ample belly overflows onto her lap like rising dough. Her heavy, well-demarcated breasts are pushed apart, forming an inverted V that nearly contacts the upright V formed by her resting belly. The knees and navel are marked by three inverted half-circles arranged in a triangle, which echoes the angle of the breasts. Another triangle is formed by the (restored) head of the woman and the heads of the two leopards on either side. Whether or not such patterns have symbolic meaning, they contribute to the profound artistic and emotional impact the sculpture evokes even in modern viewers totally outside the frame of reference of the culture that produced it.

Two enigmatic features of the "Grain Bin Goddess" are worth noting. First, there is a mysterious small object between the woman's ankles, which James Mellaart interpreted as the head of a newborn, thus making this a depiction of childbirth. However, Mellaart also remarked that the woman's left foot, though damaged, "rests on what looks like a human cranium" (Figure 3.15A, arrow). Mellaart's conclusion was that the statuette represents a "Great Goddess, Mistress of all life and death." In contrast, Hodder interpreted the object

(a)

(b)

FIGURE 3.15 Different views of the "Grain Bin Goddess" at the Anatolian Museum in Ankara. A–B. Partially restored model of original statue. C. Drawings showing the probable positions of the hands of the woman on the heads of the felines, and the tails of the felines draped over the woman's shoulders.
A–B. Photos by the authors. C is from drawings from Mellaart, J. (1963). Excavations at Catal Hüyük, 1962: Second Preliminary Report *Anatolian Studies* 13:24–103.

(c)

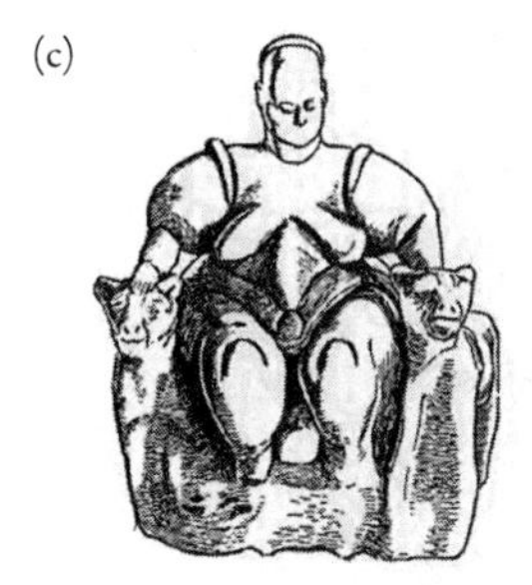

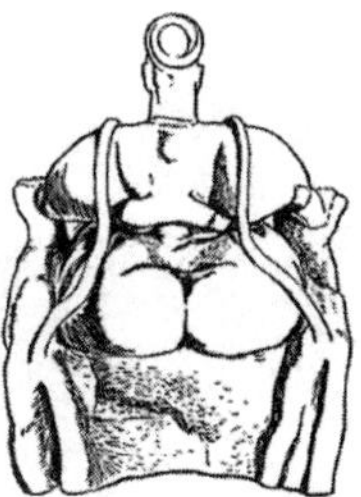

FIGURE 3.15 Continued

between her ankles as a plastered skull representing ancestors, similar to those found in burial contexts.[65] If Hodder's interpretation is correct, it reinforces the association of the figurine with birth and lineage. Perhaps the ambiguity is intentional. Newborn, skull, or both, the overall gestalt of this magnificent figure is one of strength, continuity with the past, and continuation into the future.

Much has been written about the possible religious significance of the "Grain Bin Goddess." Those who side with Mellaart's interpretation cite the leopard throne itself as a clear indication that this is no ordinary human, but a supernatural being of some sort, which would make it a cultic object. Skeptics of Mellaart's "Goddess" interpretation cite the lack of any obvious religious context—an altar or dedicated temple, for example—that would clearly identify the woman as a specific deity. Intriguingly, a figurine thought to represent the Phrygian Mother Goddess Matar (known by the Greeks as Kybele), dating to the Iron Age, resembles the "Grain Bin Goddess" in that she is seated on a throne with her arms supported by two quadrupeds, probably felines (Figure 3.16). Çatalhüyük is situated near the former southeastern border of Phrygia, suggesting the possibility of a continuous cultic or artistic tradition in that region of Anatolia.

Recently, Hodder and his team have made the exciting discovery of an intact 8000-year-old female figurine with a striking resemblance to the Grain Bin Goddess. Rather than being seated on a leopard throne, the figure, which was carved out of marble with "very fine craftsmanship," was depicted lying down with her hands folded under her breasts. The fact that the figurine had been carefully buried beneath a platform in a house alongside a valuable piece of obsidian indicates "some sort of ritual deposition."[66]

For our purposes, the most important aspect of the "Grain Bin Goddess" is not only its cultic significance, but its physical context—inside a grain bin. The simplest explanation of her location there is that the figure was placed in the bin as a talisman to protect the grain stores. Indeed, in the upper levels of both Çatalhüyük and the later Neolithic site of Hacilar, several of the bins for storing grain and legumes contained clay female figurines.[67]

We have also suggested that domesticated or semi-domesticated cats were recruited, probably by women, to protect the grain in the fields and in the storage bins in the homes from the ravages of mice and other rodents. As noted earlier, some of the relief sculptures of leopards found by Mellaart were associated with a deposits of grain together with a female figurine that was, perhaps, a votive offering.

Another significant discovery made by Hodder's team was that of a leopard's claw pendant in the grave of a woman at Çatalhüyük. This unique artifact presumably conferred high status on the wearer, and the fact that it was in the grave of a woman is consistent with

FIGURE 3.16 Statuette of Kybele at the Anatolian Museum in Ankara, Turkey, probably dating to the Iron Age.
Photo by the authors.

a special relationship between felines and women. The woman in whose grave the leopard claw pendant was found was "holding a plastered skull to her chest and face," suggesting a "special social significance." According to Hodder, the "claw and its context evoke in an uncanny way the most iconic image from Çatalhüyük—the woman seated on a seat of felines. . . . Whether she wore the pendant or held the skull wearing the pendant, the central role of powerful images of women in the upper part of the site is reinforced."[68]

Unlike the relationship between men and wild animals in the hunting tableaus, the relationship between women and cats was not one of aggressive dominance. Quite the contrary, the relationship appears affectionate and harmonious, as indicated by the gesture of the leopards' tails, which are draped over the woman's shoulders in a kind of embrace. The protective, "maternal" relationship between women and cats is further reinforced by the figurine of a seated woman with two leopard cubs on her shoulders (Figure 3.17A).

A mutualistic relationship with cats is also suggested in a figurine from the nearby late Neolithic village of Hacilar (Figure 3.17B). Hacilar VI, the level at which the figurine in Figure 3.15B was found, dates from the end of the East Mound sequence at Çatalhüyük to around 7,400 years ago, making it contemporaneous with the Chalcolithic (Copper Age) West Mound. The abundance of plant remains relative to animal remains at that site indicates that agriculture played an even greater role at Hacilar VI than at the East Mound of Çatalhüyük, and the figurines suggest that cats were held in very high esteem.[69]

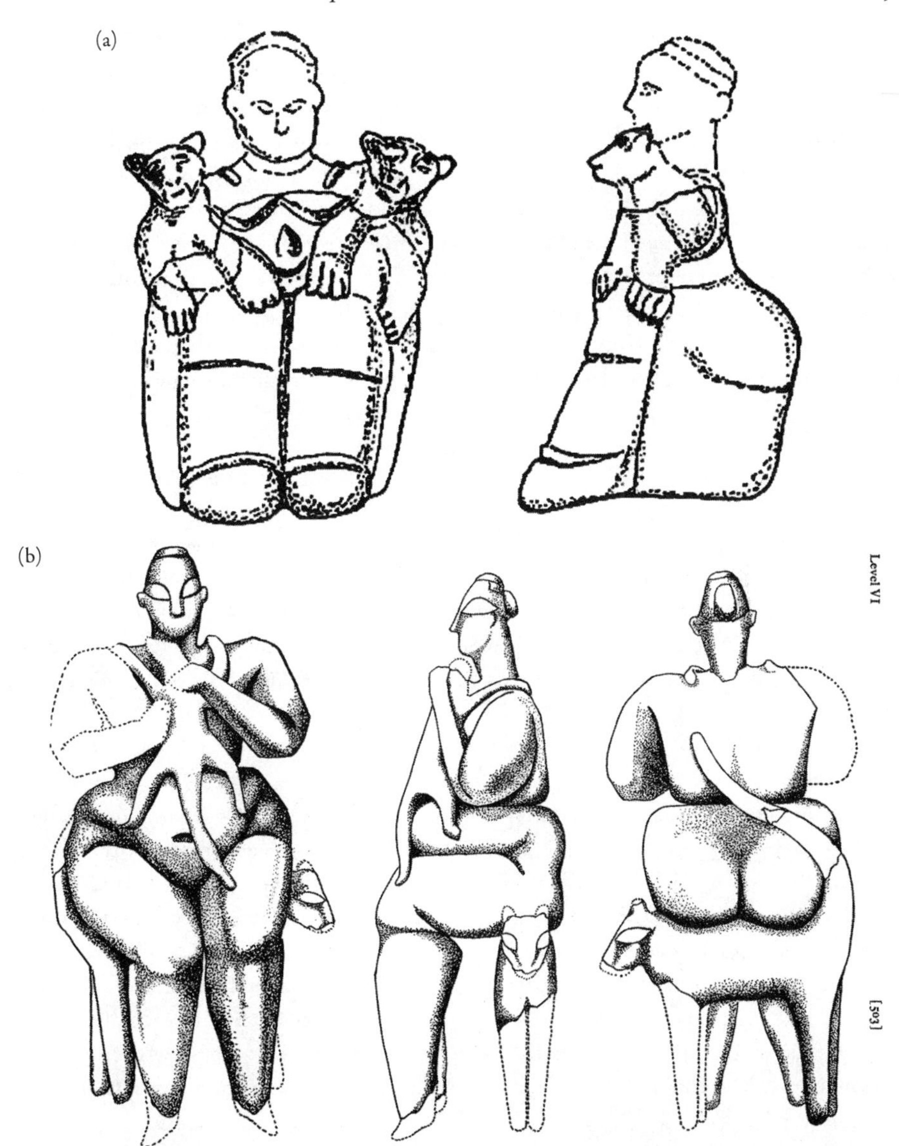

FIGURE 3.17 A. Female with two leopard cubs, from Çatalhüyük. B. Female seated on a feline, holding a kitten, from Hacilar. Discovered by Mellaart in a house in level III.
From Mellaart, J. (1970), Excavations at Hacilar. Edinburgh University Press; Mellaart, J. (1975), The Neolithic of the Near East. Charles Scribner's Sons.

In the female statuette in Figure 3.17B, a woman sits directly on a large feline while holding in her arms what appears to be a cat or kitten, whose spread-eagled embrace implies domestication. As in the case of the "Grain Bin Goddess," the tail of the supporting feline is pressed against the woman's back. Indeed, the pose resembles similar statuettes from Hacilar VI, which show mothers cradling and embracing their infants. Thus, the woman is clearly exhibiting maternal behavior toward the cat, as if it has become a member of the human household.

All this suggests that there may have been two co-equal subcultures at Çatalhüyük, each with its own mythology and practical concerns: a men's subculture based in part on the mythology of the hunting and domination of dangerous animals and a woman's subculture based, among other things, on myths related to growing crops, giving birth and rearing children, preparing food, and grooming cats to combat vermin. We propose that these agricultural roles for the women of Çatalhüyük are variously encrypted in the "sacred tree" paintings, female figurines (including the "Grain Bin Goddess"), the leopard reliefs, and the clay balls with plant impressions used for cooking. If we are correct that the paintings of the "animal-baiting scenes" are actually inspired by the collective need to protect crops from large ruminants, crop protection would have been a common activity shared by both sexes, just as the symbolism of the leopard is likely multilayered, encompassing women's lives as well as well as those of men.

NOTES

1. Boyd, B. (1999), "Twisting the kaleidoscope": Dorothy Garrod and the Natufian Culture, in W. Davies and R. Charles, eds., *Dorothy Garrod and the Progress of the Paleolithic*. Oxbow Books; Price, K. M. (2009), One vision, one faith, one woman: Dorothy Garrod and the crystallisation of prehistory, in R. Hosfield, F. F. Wenban-Smith, and M. Pope, eds., *Great Prehistorians: 150 Years of Palaeolithic Research, 1859–2009*, Special Volume 30 of *Lithics*: 135–155, Lithic Studies Society, London.

2. Garrod, D. A. E. (1932), A New Mesolithic Industry: The Natufian of Palestine. *Journal of the Royal Anthropological Institute of Great Britain and Ireland* 62: 257–269.

3. A former student of Abbe Breuil's, Dorothy Garrod made a point of including women on her excavation teams, against the prevailing attitudes of her male peers. In 1939, she was elected Professor of Archaeology at Cambridge, becoming the first woman to hold a Chair at either Cambridge or Oxford. Her election electrified the female students at Cambridge's women's colleges. The newsletter of Newnham's College declared, "Miss Garrod's election to the Disney Professor has been the outstanding event of the year and has filled us with joy." Nevertheless, Garrod faced discrimination from her male colleagues. For example, she was barred from sitting at High Table in the men's colleges, where academic policy decisions were often discussed. See Smith, P. J. (2000), Dorothy Garrod, first woman professor at Cambridge. *Antiquity* 74:131–136.

4. Unger-Hamilton, R. (1989), The Epi-Palaeolithic of southern Levant and the origins of cultivation. *Current Anthropology* 31:88–103.

5. From the French word *levant* (rising), referring to the point where the sun rises.

6. The Holocene epoch corresponds to the present warm period, which began at the end of the Pleistocene about 11,700 years ago. It is part of the Quaternary period. The name is based on the Greek words meaning "wholely new." The recent phenomenon of accelerated global warming due to human industrial activity is sometimes referred to as the "Anthropocene."

7. Bar-Yosef, O. (2014), The origins of sedentism and agriculture in Western Asia, in C. Renfrew and P. Bahn, eds., *The Cambridge World Prehistory*. Cambridge University Press, chapter 3.4: 408–1438.

8. Management has been defined by Ofer Bar-Yosef as "manipulation and some degree of control of wild species (plants or animals) without cultivation or morphological changes."

Bar-Yosef, O. (2011), The origins of agriculture: New data, new ideas. *Current Anthropology* 52(S4):S163–S174.

9. In *Genesis*, Eve is represented as a gatherer: she picks the fruit. This passage also seems to reflect the well-established idea that women, as the primary gatherers of plant products, were responsible for initiating the "curse" of farming for a living.

10. The outer cell wall of pollen grains is composed of sporopollenin, one of the most inert of all biological polymers. For this reason, pollen is often preserved in soils and sediments and provides useful archeological markers for palynologists.

11. Bar-Yosef, O. (2011), Climatic fluctuations and early farming in West and East Asia. *Current Anthropology* 52(S4):S175–S193.

12. Bar-Yosef, The origins of sedentism and agriculture in Western Asia.

13. Bar-Yosef, The origins of agriculture.

14. Zeder, Melinda A. (2011), The origins of agriculture in the Near East. *Current Anthropology* 52(S4):S221–S235.

15. Unger-Hamilton, The Epi-Palaeolithic of southern Levant and the origins of cultivation.

16. Bar-Yosef, Climatic fluctuations and early farming in West and East Asia.

17. Hillman, G. C., R. Hedges, A. Moore, S. Colledge, and P. Pettitt (2001), New evidence of late glacial cereal cultivation at Abu Hureyra on the Euphrates. *Holocene* 11(4):383–393.

18. New insights gained from the field of evolutionary ecology have now pushed the beginning of crop domestication even further back in time. Even before the planting of the first gardens, foragers were already manipulating and modifying their local environments in ways that increased the abundance of desirable species and reduced the abundance of undesirable species. By doing so, they also may unconsciously have been selecting for certain *invisible*, genetically determined, physiological traits, such as rapid seed germination. Therefore, the true beginnings of plant "domestication" may stretch as far back as the Late Glacial Maximum, around 23,000 BP, even though the first *visible* (morphological) signs of plant domestication don't appear in the archaeological record until the early Pre-Pottery Neolithic B. See Zeder, The origins of agriculture in the Near East.

19. Diamond, J. (1999), *Guns, Germs, and Steel*. W. W. Norton, p. 132.

20. Einkorn wheat (*Triticum monococcum*) is derived from wild einkorn (*T. boeoticum*), which is abundant throughout the Near East. Emmer wheat (*T. turgidum*) evolved from a wild relative, *T. dicoccoides*. See Feuillet, C., P. Langridge, and R. Waugh (2007), Cereal breeding takes a walk on the wild side. *Trends in Genetics* 24:24–32; Salamini, F., H. Özkan, A. Brandolini, R. Schäfer-Pregl, and W. Martin (2002), Genetics and geography of wild cereal domestication in the Near East. *Nature Genetics* 3:429–441.

21. Zeder, The origins of agriculture in the Near East.

22. Tanno, K., and G. Willcox (2006), How fast was wild wheat domesticated? *Science* 311:1886.

23. Recent studies indicate that domestication of barley with tough rachises involved the artificial selection of mutations in two adjacent genes, named *btr1* and *btr2*. The *btr1*-type mutation was selected for in the Southern Levant, whereas the *btr2*-type mutation appears to have occurred subsequently in the Northern Levant. See Pourkheirandish et al. (2015), Evolution of the grain dispersal system in barley. *Cell* 162: 527–539.

24. Cauvin, J. (2000), *The Birth of the Gods and the Origins of Agriculture*, Trans. Trevor Watkins. Cambridge University Press.

25. Bar-Yosef, O. (1998), The Natufian culture in the Levant: Threshold to the origins of agriculture. *Evolutionary Anthropology* 5(6):159–177.

26. Bar-Yosef, The origins of sedentism and agriculture in Western Asia.

27. Cauvin, The Birth of the Gods and the Origins of Agriculture.

28. Ibid.

29. Mithen, Steven, Bill Finlayson, and Ruth Shaffrey (2005), Sexual symbolism in the early Neolithic of the southern Levant: pestles and mortars from WF16. *Documenta Praehistorica* 159:103–110.

30. Cauvin, *The Birth of the Gods and the Origins of Agriculture.*

31. Bar-Yosef, The Natufian culture in the Levant.

32. Garfinkel,Y., D. Ben-Schlomo, and N. Marom (2011), Sha'ar Hagolan: A major Pottery Neolithic settlement and artistic center in the Jordan Valley. Eurasian Prehistory 8:97–143.

33. Ibid.

34. Stekelis, Moshe, cited by Yosef Garfinkel (2004), *The Goddess of Sha'ar Hagolan*. Israel Exploration Society, p. 149.

35. Garfinkel, *The Goddess of Sha'ar Hagolan*, p. 149.

36. Freikman, Michael, and Yosef Garfinkel (2009), The zoomorphic figurines from Sha'ar Hagolan: Hunting magic practices in the Neolithic Near East. *Levant* 41:5– 17.

37. Garfinkel, Y. (2003), *Dancing at the Dawn of Agriculture*. University of Texas Press.

38. A gradual decline in representations of female dancers to about 13% after the Neolithic coincides with the rise of urban centers during the Chalcolithic period, which is associated with a shift to more patriarchal and more fully stratified societies.

39. "Tell" is the Arabic word for an artificial hill built up by successive layers of human habitation. Bellwood, P. (2004), *First Farmers*. Wiley.

40. Hodder, I. (2006), *The Leopard's Tale: Revealing the Mysteries of Çatalhüyük*. Thames & Hudson.

41. Mellaart, J., U. Hirsch, and B. Balpinar (1989), *The Goddess from Anatolia*. Ezkenazi.

42. Mellaart, J. (1967), *Çatal Hüyük: A Neolithic Town in Anatolia*. Thames & Hudson; Atalay, S., and C. A. Hastorf (2006), Food, meals, and daily activities: Food *habitus* at Neolithic Çatalhüyük. *American Antiquity* 71:283–319.

43. *Çatalhüyük* (1999), Archive Report.

44. Zohary, D., M. Hopf, and E. Weiss (2012), *Domestication of Plants in the Old World*, 4th edition. Oxford University Press.

45. According to Hodder, date palm phytoliths recovered at the Çatalhüyük site suggest long-distance trade or exchange with "Mesopotamia or the Levant." Hodder, *The Leopards's Tale*, p. 80.

46. Mellaart, J. (1966), Excavations at Catal Hüyük, 1965: Fourth Preliminary Report Anatolian Studies 16:165–191, plate LII, redrawn from original by Raymonde Enderlé Ludovici.

47. Ibid.

48. Hodder, I. (2004), Men and women at Çatalhüyük. *Scientific American* 290:67–73. This statement appears to refer to recently uncovered wall paintings from the upper levels of the settlement.

49. In Classical mythology, we find the theme of driving out or destroying monstrous animals and birds that threaten crops enshrined in tales of the great heroes. The boar-baiting mural

from Çatalhüyük could function very well as an illustration for the myth of Meleager and his band of heroes who destroy the Calydonian boar. Interestingly, three of the labors of Heracles also involve variations on the same theme: the Erymanthian boar, the Cretan bull, and the birds of Stymphalos. Moreover, animal depredation and the heroic efforts to defend against them are not just myths from the ancient world; they are also completely contemporary. According to the World Wildlife Fund, in India and Africa, substantial numbers of people—and elephants—are killed every year in conflicts over crops and gardens, and, in the United States, struggles against crop-destroying wild animals, particularly wild and feral pigs, result in the loss of millions of dollars every year and sometimes even in human fatalities.

50. Lewis-Williams, David (2004). Constructing a cosmos: Architecture, power and domestication at Çatalhöyük. *Journal of Social Archaeology* 4:28–59.

51. Hodder, *The Leopard's Tale.*

52. Atalay, Sonya (2005), Domesticating clay: The role of clay balls, mini balls, and geometric objects in daily life at Çatalhöyük, in I. Hodder, ed., *Changing Materialities at Çatalhöyük: Reports from the 1995–99 Seasons*, Çatalhöyük Project Volume 5. McDonald Institute Monographs/British Institute of Archaeology at Ankara, chapter 6, p. 139–168; Atalay, Sonya, and Christine Hastorf (2005), Foodways at Çatalhöyük, in I. Hodder, ed., *Çatalhöyük Perspectives: Themes from the 1995–99 Seasons.* Çatalhöyük Project Volume 6. McDonald Institute Monographs/British Institute of Archaeology at Ankara, chapter 8, pp. 109–124.

53. Atalay, Sonya (2003), *Domesticating clay. Engaging with 'They': The Social Life of Clay Balls from Çatalhöyük, Turkey and Public Archaeology for Indigenous Communities.* Ph.D. Dissertation. UC Berkeley Anthropology Department.

54. Fan, Maureen (2007, July 15), Chinese rice farmers battle a plague of munching mice. *Washington Post*, A16.

55. The early spread of the house mouse (*Mus musculus*) is closely associated with the spread of sedentism in the Levant, beginning around 12,000 years ago. Wherever people settled down and began storing plant foods, the house mouse settled in with them. Once crops began to be planted, field mice would have moved in as well. As the complex of plant domesticates spread throughout the Fertile Crescent, the house mouse followed along (Bursot et al., 1993). For example, *M. musculus* makes its first appearance on the island of Cyprus during the Early Preceramic Neolithic, around 10,000 years ago, coincident with appearance of the first crops (Cucchi et al., 2002).

56. According to Lewis-Williams, some of these small carnivores can also be seen as helpers in the context of a symbolic/spiritual configuration. In some rooms at Çatalhüyük are plaster forms that appear to be breasts molded on the walls. Contained within the "breasts" are vulture beaks, fox teeth, and, in one instance, a weasel skull. To Lewis-Williams they suggest "spirit beings" emerging through walls. At Çatalhüyük "the walls were like 'membranes' between components of the cosmos; behind them lay a realm from which spirits and spirit-animals could emerge and be induced to emerge." He writes, "wild animals have spiritual counterparts that inhabit . . . another tier of the shamanistic cosmos and that can become spirit guides or helpers. It is perhaps in terms of these concepts . . . that the beaks, tusks and teeth set in moulded breasts should be seen. It was the mouths of wild creatures that were being associated with breasts. From both breasts and the mouths of wild animals emerged sustaining spiritual power."

Lewis-Williams, David (2004), Constructing a cosmos: Architecture, power and domestication at Catal Hoyuk. *Journal of Social Archaeology* 4:28–59.

57. The reverence for cats in ancient Egypt reached its apogee during the Ptolemaic period, when the cat came to manifest a plethora of apotropaic functions and to become identified with a number of deities. For example, the male cat was identified as a manifestation of the sun god Ra, and the sun god, in the guise of a tomcat, was believed to battle each night with the "typhonic serpent of darkness." The cat also came to represent the goddess Bast or Bastet whose annual cult festival, at its height around 450 BCE, was possibly the largest in Egypt. During the Late Ptolemaic Period (664–630 BCE), cats were held sacred, and killing a cat in Egypt was a considered capital offense (except for religious purposes). See Serpell, James A. (2014), The domestication and history of the cat, in D. C. Turner and P. Bateson, eds., *The Domestic Cat: The Biology of Its Behavior*, 3rd edition. Cambridge University Press, pp. 89–92.

58. Vigne, J.-D., J. Guilaine, K. Debue, L. Haye, and P. Gérard (2004), Early taming of cats in Cyprus. Science 304:259.

59. Ibid.

60. Guilaine, Jean (2001), Tête sculptée dans le Néolithique pré-céramique de Shillourokambos (Parekklisha, Chypre). *Paléorient* 26:137–142.

61. Mellaart, *Çatal Hüyük.*

62. Mellaart, J. (1963), Excavations at Catal Hüyük, 1962: Second Preliminary Report *Anatolian Studies* 13:24–103, at 45.

63. Serpell (2014, pp. 87–88) describes an often repeated process of interaction between wild animals and people that may underlie the early stages of domestication: "In the Amazon region, where hunting and gathering and subsistence horticulture is still practiced by a handful of surviving Amerindian groups, hunters commonly capture young wild animals and take them home where they are then adopted as pets, *usually—although not invariably—by women* . . . . (t)hese animals do not need to serve any functional or economic purpose to be valued by their owners. Rather, they are viewed, cared for and indulged much like adopted children" (emphasis added).

64. Serpell (2014, p. 88) also points out what is probably a long and persistent association of domestic cats with women in ancient Egypt, where, in tomb paintings "from about 1450 BCE onwards, images of cats in domestic settings become increasingly common in Theban tombs. . . . *The cats are usually illustrated sitting, often tethered, under the chairs of the tomb-owners' wives*" (emphasis added).

65. Hodder, *The Leopard's Tale.*

66. Hodder, Personal communication.

67. Atalay and Hastorf, Food, meals, and daily activities; Hamilton, Naomi (1996), Figurines, clay balls, small finds and burials, in I. Hodder, ed., *On the Surface: Çatalhüyük 1993–1995*. British Institute of Archaeology at Ankara/McDonald Institute Monographs, Cambridge, p. 215–263.

68. Hodder, *The Leopard's Tale*, pp. 260–261.

69. Mellaart, J. (1970), *Excavations at Hacilar.* Edinburgh University Press; Mellaart, J. (1975) *The Neolithic of the Near East.* Charles Scribner's Sons.

4

Plant-Female Iconography in Neolithic Europe

IT TOOK ABOUT a thousand years, from about 8000 to 7000 BCE, for the complete agricultural ensemble consisting of the eight founder crops and four livestock animals to permeate all the societies of the Near East. These early Neolithic cultures were economically reliant on three major industries strongly associated with women: textiles, agriculture, and pottery. Livestock were maintained on a small, farmyard scale and were used almost exclusively as a source of meat, although hunting persisted as in the Epipaleolithic period. (This period is referred to as the Mesolithic period when discussing Northern Europe.)

Several lines of archaeological evidence suggest that early Neolithic societies were quite egalitarian in their social organization and relatively peaceful compared to the highly stratified, hierarchical Bronze Age societies that succeeded them. Although obsidian projectile points and other weapons were produced in abundance, these were used primarily for hunting animals rather than for warfare. Perhaps the ready availability of nearby sparsely inhabited lands acted as a safety valve to prevent conflicts as the population densities of individual settlements increased.

Around 6900 BCE, agriculture, accompanied by the new ceramic technology, began to spread westward across the Aegean from Hacilar and other central Anatolian settlements to Greece and Crete.[1] The Franchthi cave in the northeastern region of the Peloponnese,[2] which had been occupied intermittently by hunter-gatherers since 20,000 BCE, provides a microcosm illustrating the rapid changes that occurred in Greece during the period of agricultural expansion. Prior to 7000 BCE, the remains of local wild game (including wild cattle and pigs) and tuna from the Aegean were abundant, whereas wild cereals, lentils, pistachios, and almonds made up the bulk of the plant remains. By 6900 BCE, however, after a period of abandonment, there was an abrupt shift to domesticated food sources, including emmer wheat, barley, sheep, and goats—and simple pottery is found for the first time.[3] At the same period, living quarters expanded to a settlement beyond the cave, indicative of an increase in population.

The rapid transformation observed at this site strongly suggests that agriculture came to mainland Greece through a process of migration and colonization from the Near East, rather than through the indigenous adoption of Near Eastern agricultural practices. The case for migration is even stronger in Cyprus and Crete, which had no Epipaleolithic human inhabitants and no indigenous wild progenitors of the suite of domesticated animals and plants that suddenly appear around 9000 BCE in Cyprus[4] and 7000 BCE in Crete.

The agriculturalists who colonized Greece and Crete in the Neolithic were once believed to have spoken an ancient Anatolian language, unrelated to Indo-European languages such as Greek, Latin, and Sanskrit. More recently, however, the Anatolian language family, which also includes Hittite, Lydian, Lycian, Palaic, and Luwian, has been shown to be an early branch of the Indo-European (IE) language phylogenetic tree.[5] The hypothetical ancestor of all the Indo-European languages, including the Anatolian group, has been reconstructed and termed Proto-Indo-European (PIE). According to one estimate, the Anatolian language family branched off from the main PIE trunk around 6500 BCE.[6]

THE RAPID EAST–WEST MIGRATION OF CROP PLANTS

Once agriculture had taken hold in Greece, Neolithic settlements began to appear in the Balkans to the north, bringing an end to the remnants of the Mesolithic semi-nomadic lifestyle that had persisted there. European Mesolithic sites were typically short-term camps that were used repeatedly for hundreds of years. Rarely, more substantial permanent settlements were built reminiscent of Çatalhüyük in Anatolia and the Natufian settlements in the Levant.

Overall, it took only 3,000 more years for the new farming way of life to spread from Greece to the rest of Europe.[7] It was not just the idea of agriculture that spread, but the domesticates themselves. By the early fourth millennium BCE, virtually all of the eight founder crops in Europe, as well as the four animal domesticates, were descended from ancestors in the Fertile Crescent region. With the exception of the pig, not one of them had been independently domesticated by indigenous Europeans.[8]

As outlined by Jared Diamond, the rapid transmission of the eight founder crops was facilitated by the predominant east–west migration axis, which meant that the plants remained at approximately the same latitude.[9] Since climate and day length, both of which strongly affect plant growth, are closely linked to latitude, the founder crops were, in effect, preadapted for European, especially southern European, climates. As discussed in Chapter 3, none of the eight founder crops could have spread without human intervention. Artificial selection had resulted in the loss of the seed dispersal mechanisms of the wild progenitors of cereals. The heads of ripened grain remained intact because of their tough rachises. Similarly, the ripened pods of domesticated legumes failed to split open as they normally would, due to the loss of the dehiscence mechanism. Instead, humans took over the role of seed dispersal.

DEMIC VERSUS CULTURAL DIFFUSION

There are two contrasting views on the mechanism by which agriculture spread to Europe: by the physical migration of people, sometimes referred to as *demic diffusion*, or by information transfer, or *cultural diffusion*. The demic diffusion model can be traced back to the renowned Australian archaeologist and philologist V. Gordon Childe who first introduced the term "agricultural revolution" in 1935. According to Childe's hypothesis, agriculture led to increases in population in the Near East, and it was this population pressure that led to the migration of people to the less densely populated areas of Europe, using the Balkan Peninsula as a "bridge" to Europe.[10] These Neolithic people brought their agricultural way of life with them.

The demic diffusion model is supported by two main lines of evidence: archaeological and genetic. The archaeological evidence is based on radiocarbon dating of the ages of European Neolithic sites. When these dates are assembled on a map together with dates for the original Neolithic settlements in the Near East, a clear gradient is evident from the oldest sites in the Near East to the most recent sites along the northwest coast and Britain. However, such a gradient would be expected whether agriculture spread by demic or cultural diffusion.

The genetic evidence for demic diffusion is based on a comparable east–west gradient in gene frequencies (i.e., the frequency of occurrence of particular versions of certain genes) among populations of Europeans today. Such a genetic gradient could only be established by the physical migration and mixing of human populations. Accordingly, population geneticist Luigi Luca Cavalli-Sforza and his colleagues, who first discovered the genetic gradient, attributed it to successive waves of expansion of populations out of the Near East that occurred during the Neolithic period.[11] It has been estimated that as little as a 10% increase in dietary intake due to agriculture can cause a 50% rise in population size, which could be a significant driver of migration.[12] However, others have suggested that the east–west genetic gradient actually reflects movements of people that took place much earlier, during the Paleolithic period.

In recent years, greater attention has focused on the role of cultural diffusion, that is, the adoption of farming and animal husbandry by indigenous foraging societies without their being assimilated by an outside group. The importance of cultural diffusion in the spread of agriculture is consistent with recent analyses of mitochondrial DNA. Although an east–west gradient in mitochondrial DNA sequences was confirmed, it could only account, on average, for about 13% of the total variability observed.[13] In agreement with this more modest assessment of the amount of genetic mixing that took place as agriculture spread to Europe, Cavalli-Sforza has pointed out that the east–west genetic gradient he identified only accounted for about 25% of the total genetic variability. In other words, both molecular and genetic studies agree that although demic diffusion probably occurred, the number of Neolithic colonists who spread out from the Near East did not exceed about 20% of the indigenous populations.

Despite the evidence against large-scale migration of Neolithic farmers from the Near East into Europe, colonies of Neolithic migrants were probably crucial to the spread of agriculture. Because writing had not yet been invented, it is difficult to conceive how such a thorough transformation in lifestyle could have been brought about by word of mouth alone, especially if, as Colin Renfrew has suggested, Mesolithic hunter-gatherers initially spoke a variety of non-Indo-European languages and became PIE speakers around the time of the adoption of agriculture.[14] Based on these and other considerations, there is a growing consensus that agriculture, along with PIE, spread throughout Europe during the Neolithic by a patchy, mosaic mechanism, involving both demic and cultural diffusion.[15]

PLANT–FEMALE ASSOCIATIONS IN EUROPEAN NEOLITHIC FIGURINES

The Neolithic period in the Aegean (mainland Greece and surrounding islands) extends from ca. 6800 BCE (Aceramic Period) to ca. 3200 BCE (Final Neolithic). Accompanying the transition to agriculture in the Aegean was the production of a large corpus of anthropomorphic figurines, the vast majority of which (84%) were modeled in clay. Archaeologist Maria Mina has classified the total corpus of the published Aegean figurines into categories according to gender.[16] She found that female and probable female figurines account for about 60% of the total, whereas males and probable males represent less than 3%. Clearly, the genre of anthropomorphic clay figurines from the Aegean Neolithic was dominated by images of women at this time.

In addition to their greater numbers, Mina also reported that female figurines differed stylistically from male figurines. For example, female figurines were frequently shown "with hands resting on the breasts and breast area or with hands below the breasts." Another stylistic difference was the presence or absence of decorations, such as applied paint or inscribed motifs representing "body decoration, attire, and jewelry." About 43% of the figurines were decorated, all of which were either female or asexual. None of the male figurines was decorated. Colored paints were sometimes used to highlight specific anatomical parts of the female figurines: "Red was applied on the breasts and pubic area, black demarcated the abdomen and pubic area, with white being restricted to the pubic area." During the late Neolithic, blue and green were added. In one late Neolithic female figurine, green was used to mark the pubic area, suggesting a possible analogy between women's fertility and "agricultural lushness." According to Mina, such decorative elements generally emphasized reproduction and fertility:

> Overall, the predominance of *Female* figurines implies the strong preoccupation of Neolithic people with women's bodies and, no doubt, physical aspects related to them, such as pregnancy, birth, and menstrual cycle. The close parallelism of the female body and the phenomena of nature cannot have escaped Neolithic people, the physical body acting as a common metaphor to explain the natural and social world.[17]

As noted here, the use of green paint to indicate the pubic region of an Aegean female figurine from the Neolithic is a possible example of the association of women and agriculture. Similar symbolic associations between plants and women appear to have accompanied the spread of agriculture from the Near East to Europe beginning around 7000 BCE. In the Upper Dneister Valley, for example, cereal grains were sometimes pressed into the wet clay of female figurines, leaving a pattern of grain impressions (Figure 4.1). Such terracotta figurines date from the mid fifth millennium BCE.

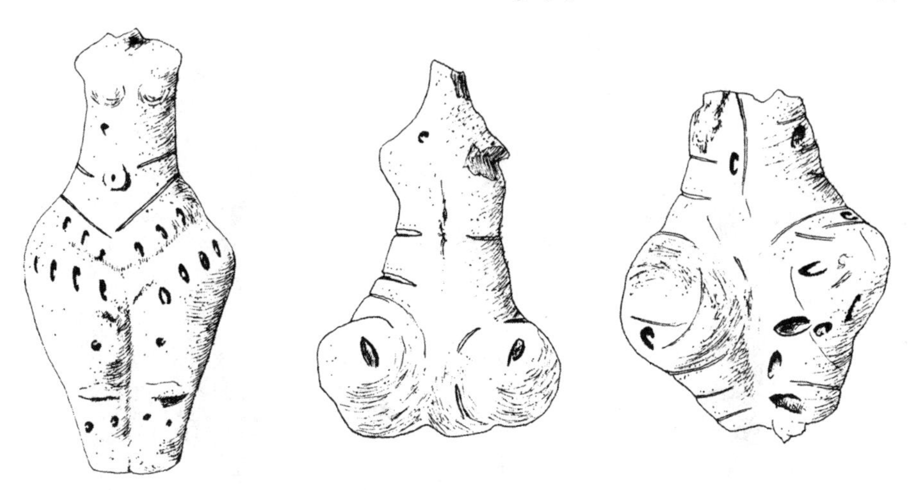

FIGURE 4.1 Female figurines bearing grain impressions, from the Luka-Vrublevetskaya settlement in the upper Dneister Valley, proto-Cucuteni, late fifth millennium BCE.
From Gimbutas, M. (1982), *The Goddesses and Gods of Old Europe: Myths and Cult Images*. University of California Press, p. 204.

Figure 4.2A,B illustrate plant-like designs incised on the abdomens of female figurines. In Figure 4.2A, the putative plant appears to be sprouting from the vulva area, whereas the figurine in Figure 4.2B has what appears to be an inverted plant inscribed on the belly, with the flower bud positioned in region of the vulva.[18]

The "dot and lozenge motif" was proposed as a possible symbolic representation of a seed by archaeologist Marija Gimbutas, with the dot representing the seed and the diamond-shaped border representing the sown field. The symbol is sometimes incised or painted on the belly of female figurines, with the dot also serving as the navel. Figures 4.3A,B show two examples from Yugoslavia, a stylized female torso with a dot and lozenge design on the abdomen and a standing female figurine with four dot and lozenge designs on her belly. Figure 4.3C shows a figurine of a seated pregnant female from Bulgaria with a dot and lozenge design on her swollen belly. If Gimbutas's interpretation is correct, the location of a symbolic seed on the women's belly would be consistent with a symbolic association of seeds with generation, lineage, and agricultural abundance.

As we discussed in Chapter 3, numerous female figurines with "coffee-bean eyes" were found at the Neolithic settlement of Sha'ar Hagolan in the Levant dating to about 6000 BCE. Like coffee beans, cereal grains also have a distinctive *furrow*, or pigmented groove, that runs longitudinally down the ventral side of each grain, and we proposed that the eyes of the Sha'ar Hagolan figurines were intended to represent cereal grains. A similar style of eyes was used in figurines of later European farming settlements. For example, Figure 4.4 illustrates a comparable female figurine, dating to about 5800–5600 BCE, from the Sesklo Neolithic culture located in Thessaly, Greece. According to Alasdair Whittle, there is a strong emphasis among the figurines on female reproduction, and "cereal grain eyes" probably symbolized abundance and fecundity in relation to the growth of crops.[19]

Many of the Neolithic female figurines are vase-shaped with long necks. One of the most striking of the vase-shaped figurines is the "Hamangian type" from Dobruja, in Bulgaria/

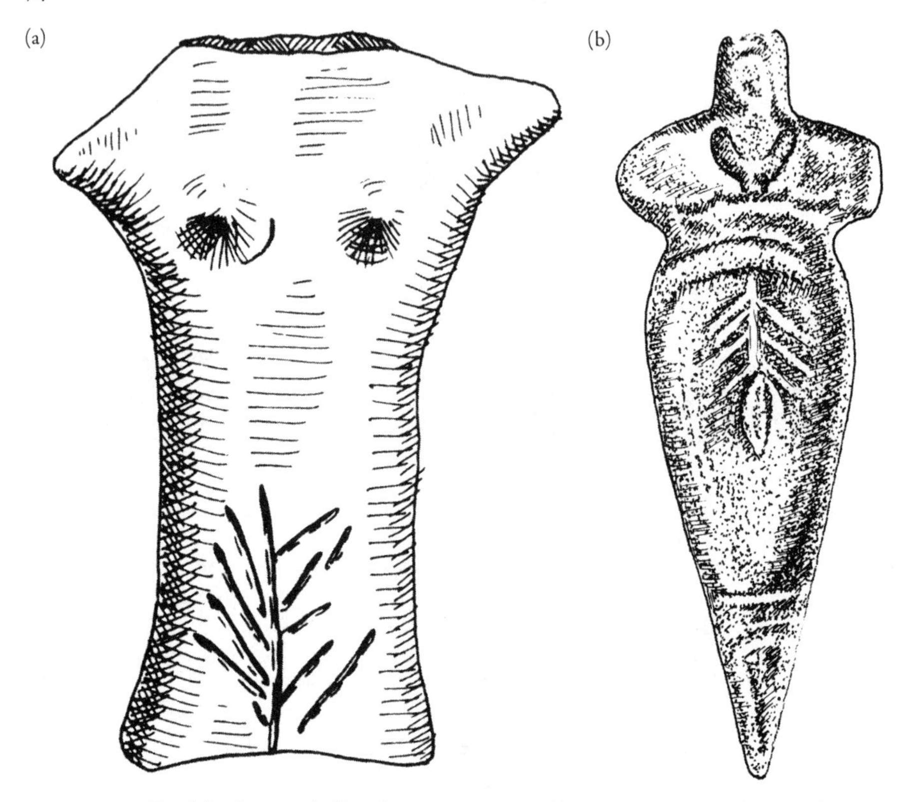

FIGURE 4.2 Possible plant symbolism from European Neolithic. A. Terracotta figurine from Vinča culture, Jela, Serbia, 5200 BCE. B. Bone plate female figurine from Gaban cave, near Trento, N. Italy, ca. 4800 BCE.
From Gimbutas, M. (1991), *The Language of the Goddess*. Harper San Francisco, p. 103.

Romania, dating from 5000 BCE (Figure 4.5A). Another example is from Neolithic Crete, ca. 5500 BCE (Figure 4.5B). The long necks of these figurines have sometimes been interpreted as snake symbols connected to the worship of a snake goddess or as a phallic symbol indicating the figurine's androgynous nature. When viewed through the eyes of plant biologists, however, the elongated neck surmounting a swollen base evokes the style of a flower attached to its ovary, a stem attached to a ripened fruit, or a persistent calyx above a fruit, such as the pomegranate. A rattle shaped like a figurine evokes both pregnancy and fruit because of its pistil-like shape and because of the presence of seed-like stones within the swollen base (Figure 4.5C). The grid pattern around the swollen base of the upper rattle, which was found in a grave along with painted bowls, could represent a textile pattern, further associating this object with female activities.

On the island Malta, some 80 km south of Sicily, a thriving Neolithic society existed from the sixth to the third millennium BCE. A characteristic feature of Maltese Neolithic society was the elaborate construction of megalithic monuments and architecture, including cave sanctuaries, tombs, and "temples." Altogether, about thirty sculptures representing females have been found, ranging from figurines 20 cm in length, to statues 3 meters in length. According to Cristina Biaggi, "Some are nude, others clothed; some seated, some

(b)

(a)

(c)

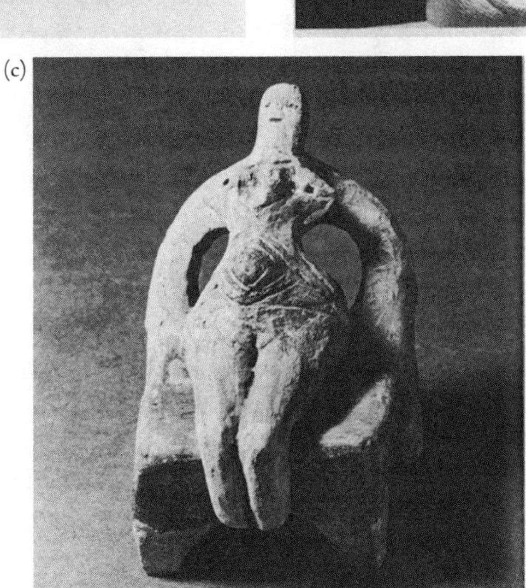

FIGURE 4.3 Dot and lozenge motifs on female figurines. A. From Gladnice near Priština, in Kosovo. ca. 6000 BCE. B. From the Cucuteni culture in northern Moldavia, ca. 4500 BCE. C. From Kolekovats in central Bulgaria, ca. 4500 BCE.
From Gimbutas, M. (1982), *The Goddesses and Gods of Old Europe: Myths and Cult Images*. University of California Press, p. 206.

FIGURE 4.4 Female figurine with "cereal grain" eyes. A. Sesklo Neolithic culture located in Thessaly, Greece, 5800–5600 BCE.
From Gimbutas, M. (1982), *The Goddesses and Gods of Old Europe: Myths and Cult Images*. University of California Press, p. 41.

standing; some without primary sexual characteristics, but all obese—or, at least, very heavy."[20] The monumental proportions of these statues and their contexts within "temple" complexes suggests that they had religious significance.

Of particular interest from the point of view of agriculture are images carved into the base of one of the huge seated females at the Tarxien West Temple, dating to ca. 3000 BCE (Figure 4.6). If the statue were complete, it would be nearly 3 meters tall. At the base there is a frieze consisting of a row of symbols that have been interpreted as cereal grains separated by what could be sheaves of wheat.[21] If this interpretation is correct, the combination of the seated female—surely a supernatural or mythic figure—on a platform decorated with wheat or barley seeds is evidence for a strong connection between the large seated female and agriculture.

Located a short distance northwest of Tarxien, encompassing an area of about 600 square meters, is an underground catacomb-like structure, the Hypogeum of Ħal-Saflieni, that may have served both as a sanctuary and necropolis or a collective tomb. Descending from the Main Hall, one reaches the rectangular Oracle Room, the ceiling of which is decorated with four stylized trees painted with red ochre. Each of the curlicued branches bears a large circular fruit resembling a pomegranate (Figure 4.7A). The pomegranate fruit was an important symbol of fertility and abundance in Crete, and the association was later transmitted to classical Greece. Given the temporal overlap of the Minoan palace period and the megalithic period of Malta, the common use of this symbol is not unexpected. The fact that the tree painting is located underground in a tomb complex strongly evokes the myth

(a)

(b)

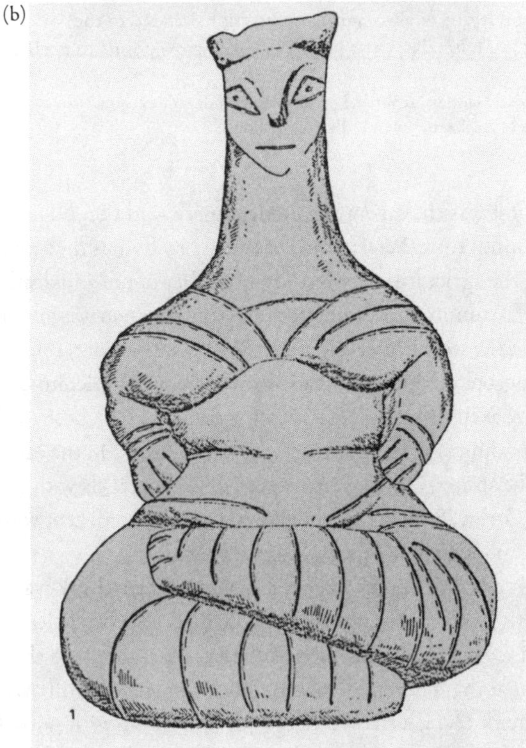

FIGURE 4.5 Pistil- or fruit-shaped Neolithic female figurines. A. "Hamangian type" from Dobruja, in Bulgaria/Romania, dating from 5000 BCE. B. Seated "snake goddess" from Neolithic Crete, ca. 5,500 BCE. C. Rattles in the shape of a pregnant woman from a grave that also included richly painted black-on-red bowls. Vykhvatintsi, Soviet Moldavia, Lake Cucuteni, fourth millennium BCE.
From Gimbutas, M. (1982), *The Goddesses and Gods of Old Europe: Myths and Cult Images*. University of California Press, pp. 202, 204.

FIGURE 4.6 Lower half of a monumental statue of a woman, 2 meters high, located in the vestibule of the Tarxien temple complex in Malta, dating to 3000 BCE. The arrow points to a relief of cereal grains at the base of the statue.
From Biaggi, C. (1994), *Habitations of the Great Goddess*. Knowledge Ideas & Trends; the original stone sculpture is housed in the National Museum of Archaeology in Valletta, Malta.

of Kore/Persephone and Demeter, for it was the eating of a pomegranate seed that doomed Kore to live part of the year underground. Kore/Persephone, of course, embodies the fertility of the soil and, in a broader sense, the agricultural cycle and the duality of life and death.

In one of the pits of the Hypogeum[22] a stunning clay statuette of a sleeping woman was found, one of several similar figurines found on the second level (Figure 4.7B). She has been interpreted as representing regeneration and healing, but it is likely she is also connected with agriculture in some way, especially since it is thought that the third level of the Hypogeum may have been used to store grain.[23] Perhaps she is the embodiment of the dormant grain that "sleeps" in the earth during the winter and "awakens" in the spring. Once again, we see possible parallels with the Kore/Persephone myths of Crete and Greece. Such an interpretation would be consistent with the cereal grain motif at the base of the monumental Tarxien Temple "goddess."

Having begun the transition to agriculture relatively late compared to the Levant, the Neolithic people of Malta were contemporary with the Bronze Age societies of Mesopotamia, and, like them, they incorporated trees such as the pomegranate into their iconography. This is especially evident at the "temple" of Haqar Qim, situated on a hill overlooking the southern coast of the island. Here, a freestanding limestone altar with potted trees carved on its sides was placed within the main entrance to the "temple" (Figure 4.8A).

(a)
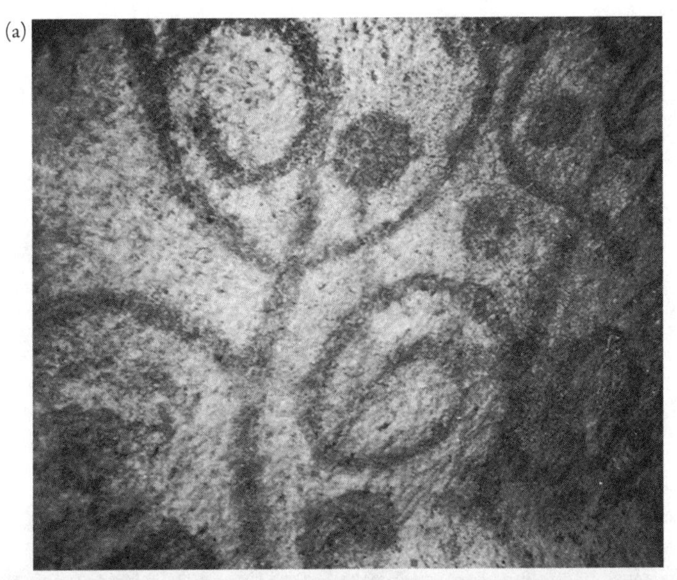

(b)

FIGURE 4.7 Images from the Hypogeum in Malta. A. Pomegranate design on the ochre-painted ceiling of the Oracle Room of the middle level of the Hypogeum, ca. 3000–3600 BCE. B. The terracotta figurine of the "Sleeping Lady," 12 cm in length, was found in one of the pits of the Hypogeum.
From Biaggi, C. (1994), *Habitations of the Great Goddess*. Knowledge Ideas & Trends.

(a)

(b)

FIGURE 4.8 Plant/female images at Hagar Qim. A. Limestone altar with carved trees on the sides at entrance to the temple. B. "Mother Goddess" figurine from Hagar Qim Temple.
From Biaggi, C. (1994), *Habitations of the Great Goddess*. Knowledge Ideas & Trends.

The precise meaning of the trees is not known, but their placement on the altar suggests that they had religious significance. Within the "temple," the famous "Mother Goddess" of Hagar Qim was found (Figure 4.8B). Although no direct connection can be made between the tree altar and the female figurine of Hagar Qim, their presence within the same context provides yet another example of the association between plant and female images.

THE SECONDARY PRODUCTS REVOLUTION OF THE CHALCOLITHIC

The type of farming that initially spread from the Fertile Crescent throughout Europe from 7000 to 4500 BCE was hoe-agriculture or horticulture, although some archaeologists believe that *ards*—primitive plows consisting of forked branches—made their earliest appearance during this interval. Hand-held implements and ards were adequate for certain types of soils with lighter textures, but were not suitable for hard-packed, stony soils. The labor intensiveness of hand hoeing also limited the amount of acreage that could be placed under cultivation, thus limiting the population sizes of the settlements. The central importance of women's labor and expertise within these early agricultural communities may have been key factors promoting egalitarianism between the sexes, as indicated by the equality of their burials and the preponderance of female figurines found at many Neolithic settlements. However, by the Bronze Age in the fourth millennium BCE, all this had changed. The agricultural enterprise had been greatly expanded; populations had vastly increased, leading to urbanization; and the societies that emerged from the first cities were now socially stratified and patriarchal in nature. As we enter the historic period, the identities of deities become clear for the first time. What happened to bring about the radical transformation of agricultural societies that began to take place around 5000 BCE during the Chalcolithic, or Copper Age?

According to Andrew Sherratt's widely accepted model for the sweeping changes to Neolithic society that ultimately gave rise to Bronze Age states, the most important innovations were the plow, the exploitation of animals for traction, and the increased use of animals for milk and wool.[24] Sherratt proposed that the livestock of the Near East (sheep, goat, and cattle) had originally been domesticated for their "primary products" (meat, hide, and bone). "Secondary products," such as milk, wool, and traction, were utilized only on a small scale during the Neolithic. During the Chalcolithic, the use of animal secondary products greatly intensified in the Near East, and the innovations then spread to Europe and Asia.[25] Since it was first published in the early 1980s, Sherratt's overall synthesis has so far survived intense scrutiny, although many questions remain.[26]

Simple ards may have first appeared in the Neolithic, but the widespread use of plows apparently dates to the early fourth millennium BCE in Mesopotamia, and images of ox-drawn plows are already in evidence in Eastern Europe by around 3500 BCE. Wooden plowshares were soon replaced by bronze, greatly increasing their hardness and strength. A "seeding funnel" was invented that sowed seeds deep into the furrow as it was dug. With the invention of wheeled vehicles around 3000 BCE, oxen were also recruited to pull heavy four-wheeled carts, providing increased mobility and transport capacity. Images of ox-drawn carts or plows always show men operating this equipment as well as harvesting the crops, signaling that the long span of time in which women had held primary responsibility for agriculture, a period lasting several thousand years and shaping Neolithic culture in important ways, was over.

The use of animal secondary products such as milk and wool during the Late Neolithic also had profound effects on agriculture and the division of labor along gender lines. Chemical evidence for dairying has been found in pottery in Britain dating from around 4500 BCE onward.[27] By consuming milk and milk byproducts, four to five times more protein could be obtained than by directly consuming the cow for meat, so the introduction of milking greatly increased the food value of cattle.[28] Judging from Bronze Age depictions from the Near East, Egypt, and southeast Europe, milking itself was carried out by men, although the processing of milk into yogurt and cheese was probably a female task.

Another secondary product that became suddenly important in the Chalcolithic was wool. The fleece of sheep is actually made up of three types of fibers: kemp, hair, and wool. Kemps are coarse, thick, and brittle, and are unsuitable for dyeing or spinning into yarn; hairs are intermediate in thickness and stiffness, but can be spun, especially when combined with wool; wool is the finest of the fibers and can be readily spun to make yarn. In addition to its fineness, wool has several other properties that make it superior for textiles. First, wool's surfaces are scaly rather than smooth, as in the case of flax fibers, and the scales act like tiny hooks that allow the fibers to snag each other like Velcro, forming felts. Second, wool fibers tend to be kinky, and the kinkiness creates tiny air pockets in woolen textiles, giving them superior insulating properties. Flax fibers, which are smooth and straight, don't form such air pockets when spun, and flax-based textiles are accordingly poor insulators. Third, nonpigmented wool fibers bind most dyes more avidly than vegetable fibers, allowing for a greater variety of color in woolen textiles.

In wild sheep, the kemp, which forms the outer coat of the fleece, is more abundant than the wool, which forms an insulating inner coat. Wild sheep molt in the spring, and the first wool must have been plucked during the molting season, an activity probably carried out by women. This mixture of kemp, hair, and wool was probably used for making felt at first. However, the potential for spinning the wool was probably recognized relatively quickly, and sheep with higher proportions of wool in their fleece were probably selected over their kempy colleagues. This seems to have happened first in southern Iran around 7000 BCE and is in evidence in the Northern Fertile Crescent and Anatolia by around 4000 BCE. The job of processing wool fibers, spinning them into yarn, and weaving the yarn into textiles was probably the domain of women. The proliferation of clay spindle whorls at many Chalcolithic sites attests to the increase in wool textile production that took place at this time.

Margaret Ehrenberg has argued that, as a consequence of the Secondary Products Revolution, men began assuming formerly female roles as farmers and herders.[29] As women's role in agriculture contracted, their roles in the spinning and weaving of textiles, food preparation, and child-rearing probably intensified. An emphasis on child-rearing could be seen as a response to the perceived need to increase the workforce in an expanding economy. But the new wealth and authority generated by the expanding economies was not being equally distributed. By the Chalcolithic and Early Bronze Age, the relatively egalitarian farming communities had become socially stratified and patriarchal.[30] The first recorded religions of the Eastern Mediterranean region—in Egypt, the Levant, and Mesopotamia—all reflect this patriarchal, hierarchical organization of society. Although polytheistic, a male warrior-deity stands at the apex of all the ancient Near Eastern religions of the Bronze Age.

Despite the transformative nature of the Secondary Products Revolution, one aspect of Neolithic traditions that shows evidence of continuity with those of the highly stratified, patriarchal societies of the Bronze Age is the strong identification of women with plants. In addition, date palm cultivation led to the first intimations of plant sexuality.

NOTES

1. With the exception of the island of Cyprus, which had been settled by pre-ceramic agriculturalists from the Levant around 8500 BCE, the first Neolithic settlers in Europe all possessed ceramic technology. Starting around 7500 BCE, a pre-ceramic version of agriculture also diffused southeast to the Kachi Plain near the Indus Valley in modern-day Pakistan, where Mehrgarh, an extensive Neolithic town five times larger than Çatalhüyük was founded around 7000 BCE.

2. Peninsula at the southern end of the mainland connected via the Isthmus of Corinth.

3. Bellwood, Peter (2005), *First Farmers*. Blackwell.

4. As noted in Chapter 3, the first signs of cat domestication in Cyprus are evident 1,500 years later, around 7500 BCE.

5. Renfrew, Colin (1987), *Archaeology and Language: The Puzzle of Indo-European Origins*. Cambridge University Press; Bellwood, *First Farmers*; Bouckaert, R. et al. (2012), Mapping the origins and expansion of the Indo-European language family. *Science* 337:957–960.

6. Renfrew, Colin (2003), Time depth, convergence theory, and innovation in proto-Indo-European: "Old Europe" as a PIE Linguistic Area. In A. Bammesberger and T. Vennemann, eds., *Languages in Prehistoric Europe*. Universitätsverlag Winter GmBH, pp. 17–48; Bellwood, *First Farmers*.

7. Coward et al. (2008) were able to trace several possible routes for the dispersal of the founder crops by analyzing European archaeobotanical assemblages using phylogenetic techniques.

8. Analysis of mitochondrial DNA has shown that shortly after the introduction of the pig into Europe from the Near East, Europeans breeds began to appear and soon replaced the Near Eastern domesticates. See Larsen, G., et al. (2007), Ancient DNA, pig domestication, and the spread of the Neolithic into Europe. *PNAS* 104:15276–15281.

9. Diamond, J. (1999), *Guns, Germs, and Steel: The Fates of Human Societies*. W.W. Norton & Company, pp. 176–191.

10. The Balkan Peninsula, the area south of the Balkan Mountains, includes Serbia, Montenegro, Croatia, Bosnia Herzegovina, Macedonia, Bulgaria, and Greece.

11. Cavalli-Sforza, L. L., and F. Cavalli-Sforza (1995), *The Great Human Diasporas: The History of Diversity and Evolution*. Translated by S. Thorne, Addison-Wesley.

12. Bellwood, P. (2014), *First Migrants: Ancient Migration in Global Perspective*. John Wiley & Sons.

13. Richards, M. (2003), The Neolithic invasion of Europe. *Annual Review of Anthropology* 32:135–162.

14. Bellwood, *First Farmers*.

15. But see Haak, W. et al. (2015), Massive migration from the steppe was a source for Indo-European languages in Europe. *Nature* 522:207–211, for an alternative theory.

16. Mina, M. (2008), Carving out gender in the prehistoric Aegean: Anthropomorphic figurines of the Neolithic and early Bronze Age. *Journal of Mediterranean Archaeology* 21:213–239.

17. Ibid.

18. From Gimbutas, M. (1991), *The Civilization of the Goddess*. HarperSanFrancisco.

19. Whittle, A. (2001), The first farmers. In Barry Cunliffe, ed., *Oxford Illustrated History of Prehistoric Europe*. Oxford University Press, p. 144.

20. Biaggi, C. (1994), *Habitations of the Great Goddess*. Knowledge Ideas & Trends.

21. Fergusson, Ian G. G. (1986), New views on the Hypogeum and Tarxien, in Bonnano, ed. *Archaeology and Fertility Cult in the Ancient Mediterranean*. B.R. Grüner Publishing, p. 152; cited by Biaggi, *Habitations of the Great Goddess*.

22. Cilia, D. ed. (2004), *Malta Before History*. Miranda Publishers.

23. The Hypogeum was most likely used and progressively elaborated throughout the Temple Period until ca. 2500 BCE when the Temple Period abruptly ended. (Joan Marler, personal communication.)

24. Sherratt, A. G. (1981), Plough and pastoralism: Aspects of the Secondary Products Revolution. In I. Hodder, G. Isaac, and N. Hammond, eds., *Pattern of the Past*. Cambridge University Press, pp. 261–306; Sherratt, A. G. (1983), The Secondary Products Revolution of animals in the Old World. *World Archaeology* 15:90–104; Sherratt, A. G. (1986), Wool, wheels, and ploughmarks: Local developments or outside introductions in Neolithic Europe? *Bulletin of the Institute of Archaeology* 2:31–15; Sherratt, A. G. (1997). *Economy and Society in Prehistoric Europe: Changing Perspectives*, Princeton University Press.

25. Ibid.

26. Greenfield, H. J. (2010), The Secondary Products Revolution: The past, the present and the future. *World Archaeology* 42:1,29–54.

27. Copley, M. S., et al. (2005), Dairying in antiquity. III. Evidence from absorbed lipid residues dating to the British Neolithic. *Journal of Archaeological Science* 32:523–546.

28. Ehrenberg, M. (1990), *Women in Prehistory*. University of Oklahoma Press.

29. Ibid.

30. Ibid.

5

Sacred Trees and Enclosed Gardens

MORE THAN ANY other domesticated tree in the Near East, date palms evoke a sense of majesty, awe, and reverence. At noon, their symmetrically radiating leaves provide cooling shelter from the harsh summer sun. Because of their preference for wet soils—along river banks and in desert oases—their appearance on the horizon signals the presence of water to the thirsty traveler. At night, their tall trunks ascend to the stars, forming a symbolic bridge between heaven and earth. Every part of the tree is useful, seeds, leaves, trunk, roots, and even its sap, from which a fermented drink can be prepared. But the date palm's primary value, for which it is so justly esteemed, lies in the sticky-sweet, nutrient-packed fruits that hang in heavy clusters from its stately crown.

Date palms probably first came under cultivation during the Late Neolithic Period. By the second millennium BCE, Babylonians had not only recognized the phenomenon of *dioecism* (male and female individuals) in date palms, they had established laws governing the practice of artificial pollination.

Arboriculture represents the second major wave of the agricultural revolution, which helped to sustain the growth and wealth of the great urban centers. Many species of cultivated trees came to be regarded as sacred, but of all the sacred trees of the Near East, the date palm was the most celebrated in art and literature. The Sumerian goddess Inanna was closely identified with the date palm harvest, and Inanna's sexuality, perhaps inspired by the dioecism of palms, became synonymous with agricultural abundance. Accordingly, any discussion of what people of the Near East knew about sex in plants during the Bronze Age must begin with the date palm.

ORIGIN OF CULTIVATED DATE PALMS

The natural distribution of wild date palms (*Phoenix dactylifera)* prior to domestication has never been satisfactorily established. This is because domestication occurred so long ago and the cultivation of date palms spread so quickly over so large an area that it is no longer possible for botanists to identify a particular population of wild *P. dactylifera* trees that gave rise to the domesticated forms.[1] However, archeology has provided some clues to the origins of the cultivated trees.

The earliest known remains of *P. dactylifera* seeds were found at a Neolithic site on the island of Dalma off the coast of the United Arab Emirates. Two carbonized date seeds, dating to around 5000 BCE and 4650 BCE, were found at the site, but it's unclear whether they were cultivated, collected from wild trees, or imported from a distant location.[2] A larger cache of carbonized date seeds dating to about 4000 BCE were found in the city of Eridu in Lower Mesopotamia. Eridu is now situated in the desert south of the Euphrates, but during the Bronze Age it was located next to a branch of the Euphrates River. The combination of searing desert heat and abundant water made Eridu an ideal location for date cultivation. The archaeological record suggests that about 300 years later, date orchards spread to the Jordan River in the Levant area.[3] By around 2500 BCE, date palms were being cultivated in the Indus Valley region of the Indian subcontinent, and the practice finally reached Egypt during the Middle Kingdom period around 2040 BCE.

"Mesopotamia" is the Greek name for the region that lies between the Tigris and Euphrates rivers in modern Iraq. During the Bronze Age, Mesopotamia was a vast, teeming wetland larger than the Florida Everglades. So biologically rich and favorable was the area for human habitation that biblical scholars often associated it with the fabled Garden of Eden. It is here in Lower Mesopotamia that the original wild population of *P. datylifera* may have been located. According to an old Arabic proverb, the date palm prefers to grow "with its feet in water and its head in skyfire." A wetland with little rainfall, where temperatures frequently rise in excess of 100°F, provides an ideal habitat for wild date palms. Moreover, *P. datylifera* is surprisingly salt-tolerant, which suggests that it may have evolved near brackish water.[4] Thus, it could have originated near the site where the Tigris and Euphrates rivers empty into the Persian Gulf. Today, that area would be around the town of Basra, but it is likely that silting out has occurred over the centuries and that, in 4000 BCE, the Gulf extended farther inland, closer to the early cities of Uruk and Eridu. Indeed, the Gulf waters played, and continue to play, a large role in the hydrology of the area. Twice a day the tide rises, raising the levels of the rivers by up to 6 feet. In ancient times, the increase in the levels of the rivers helped to flood the numerous canals that were dug to irrigate crops in the dryer northern areas.

WHO DOMESTICATED THE DATE PALM?

If domestication had been achieved by around 4000 BCE, the cultivation of wild date palm trees, which preceded domestication, must have begun much earlier, perhaps during the fifth or sixth millennia BCE. This period corresponds to the Ubaid period.

During the Ubaid period, people in southern Mesopotamia subsisted on a diet rich in fish, turtles, crabs, and sea birds. Over time, they maintained livestock, including cattle, sheep, and goats. They also irrigated their fields and grew wheat, barley, and lentils. At the same time, they must have been exploiting wild date palms, which would have been abundant along the river banks[5] Wild date fruits contained less pulp and sugar, and were drier and pithier, than later domesticated varieties. During domestication, farmers probably selected from wild populations those trees producing sweeter, more succulent dates, and planted them in separate, irrigated orchards. Obviously, whole trees could not be uprooted, so a means of propagation had to be employed that would preserve the tree's desirable traits.

Date palms can reproduce either sexually by seed or asexually by offshoots sprouting from the base of the trunk. Ubaid date growers undoubtedly were selecting female trees from wild populations based on the quality of their fruits. However, orchards grown from seeds would have yielded equal numbers of males and females, defeating efforts to cultivate only the best fruit-bearing trees. This probably led to the practice of asexual propagation by means of offshoots. In fact, if the offshoots are not removed, they tend to grow horizontally around the main trunk, producing a spreading multitrunked tree. Because the trunks are effectively competing with each other, the multitrunked growth forms are less productive fruit bearers than single-trunked trees. Wild and feral date palms tend to produce fewer offshoots, and reproduction is mainly by seed. However, cultivated varieties produce abundant offshoots that must be removed. The practice of offshoot removal probably led to their use in propagating female trees producing the best fruits. Asexual propagation not only made it possible to plant orchards consisting only of female trees, it also shortened the time to maturity. Seed-grown trees typically take seven years to bear fruit, whereas trees grown from offshoots bear fruit in less than five years.

If the early date palm cultivators had excluded *all* male trees from their orchards, they would soon have discovered that the female trees failed to produce fruit, an observation that eventually would have led to the development of the technique of artificial pollination.

How long might it have taken the first wild date palm cultivators to produce the domesticated variety? Assuming that date palm domestication involved selecting for roughly a hundred gene mutations (a conservative estimate based on the many phenotypic differences between wild and cultivated varieties), full domestication probably took hundreds, if not thousands, of years. Thus, it is likely that the date palm growers of the Ubaid period began selecting for desirable traits quite early. Consistent with this theory, the archaeological evidence suggests that heavy exploitation of wild date palms in the Persian Gulf region may have begun as early as 5500 BCE, corresponding to the beginning of the Ubaid period in Southern Mesopotamia.[6]

URUK AND THE URBAN REVOLUTION

The Ubaid period ended around 4000 BCE, and a new archaeological phase called the Uruk period (4000–3000 BCE) began. Uruk is the ancient name of the city of Warka (the biblical city of Erech). Founded early in the Ubaid period, it was sited on the same branch of the Euphrates River about 75 miles north of Eridu. However, Uruk was larger than any Ubaid settlement. Estimates of its size range from 300 acres to nearly 500 acres, and it may have housed as many as 50,000 people. It functioned as an independent city-state, in competition with other neighboring city-states, such as Nippur.

The beginning of the Uruk period was marked by a vast increase in the number and size of settlements. Since there does not appear to be any marked cultural discontinuity between the Ubaid and Uruk cultures, much of the expansion appears to have been indigenous.[7] Many factors could have contributed to a population surge. The art of irrigation was mastered, allowing the cultivation of much previously nonarable land. The widespread use of the ox-drawn seeder plow also greatly increased agricultural efficiency. The consumption of milk and other dairy products provided an important nutritional boost. Commerce and trade were facilitated by the improvements in transportation brought about by the ox-drawn cart. This was important because Uruk's own agricultural output was insufficient to support its population and much of their food had to be obtained from neighboring villages, either through taxes, tribute, or trade. Wheeled transport facilitated this flow of materials to and from the city. Archeologist V. Gordon Childe called this phase of Mesopotamian history the "Urban Revolution," on a par with the agricultural revolution in terms of its impact on human society.[8] It was also during this period that the first script, called "proto-cuneiform," was invented.[9]

The Uruk period was succeeded by a 600-year period called the Early Dynastic period (3000–2350 BCE), in which the various Mesopotamian city-states were governed by local dynasties.[10] An important development during the Early Dynastic Period in Mesopotamia was the emergence, after 2400 BCE, of two separate languages: Akkadian in the north and Sumerian in the south. Akkadian is a Semitic language, related to the later Hebrew, Arabic, and Aramaic. Sumerian does not seem to be related to any other language. Although Akkadian was the main language of the North, there were many Sumerian speakers living in Akkad and vice versa. Apart from their different languages, both Akkadians and Sumerians had much in common thanks to the persistence of the earlier Uruk culture, which extended from Eridu to the Mediterranean.

THE AGRICULTURAL GODDESS INANNA

The Uruk period (4000–3000 BCE) saw the emergence of religious institutions that kept pace with the expansion of the settlement. Temples proliferated and became more elaborate in construction. In parallel with the rise in population and the number of cities, the number of deities seems to have increased exponentially. Sumerian cuneiform tablets from the Early Dynastic III Period (2600–2340 BCE) record the names of hundreds of gods and goddesses. In Uruk, two major religious centers stood out: Kullaba and Eanna. Among Kullaba's buildings was the ziggurat of An, the supreme sky god. According to the extant written records of neighboring Nippur, the sky god An, together with the earth goddess, Ki, comprised the divine elements that gave rise to the universe—*an-ki*.[11] However, other compositions, although written later, are thought to record an even earlier tradition. These state that both An and Ki were the children of Nammu, the primordial goddess of subterranean waters, who possessed parthenogenic procreative powers. The idea of Nammu is thought to have arisen from the experience of coming to water whenever deep holes were dug into the ground due to the high water table. Such an experience might have led to the belief that earth is supported by a vast subterranean ocean.

The other great religious center of Uruk, the Eanna temple complex, was where the Sumerian goddess Inanna (known in Akkadian as Ishtar) was worshipped. Inanna literally means "Lady of Heaven," and she is identified with the planet Venus—the morning and evening star. She was the tutelary (guardian) deity of Uruk and protector of the city's storehouses. In fact, the image of a storehouse gate, with its characteristic curled "ring-posts" made of reed-bundles, became her symbol in Sumerian relief art (Figure 5.1).

Inanna's role as protector of the city's storehouses connects her to agriculture and the harvest. In addition to grains, legumes, and other dried foods, dates were stored in the communal storehouses in Uruk. Thorkild Jacobsen proposed a possible alternative translation of Inanna's name as "Lady of the Date Clusters."[12] Inanna's association with agricultural abundance, and with dates in particular, is well-attested in the art and literature of the period. For example, in a third-millennium hymn, Inanna exclaims, "I am the one who makes the dates full of abundance."[13] We will return to Inanna's close association with dates later in the chapter.

Inanna's central importance at Uruk is best exemplified by the famous Warka vase (Figure 5.2A), which was found in the Eanna temple treasury at level II, dating to the Late Uruk period, around 3200 BCE. From the time of its discovery in the 1930s, the magnificent 3-foot tall alabaster vase was a centerpiece of the Iraqi National Museum.[14] Carved onto its surface is a rather sophisticated illustration of the different trophic (feeding) levels that

FIGURE 5.1 Relief of Sumerian storehouse with ringposts on surface of alabaster trough, third quarter of the fourth millennium BCE.
From Strommenger, E. (1964), *5000 Years of the Art of Mesopotamia*. H. N. Abrams, figure 23; British Museum, London.

make up the agricultural food chain. At the very bottom, a pair of wavy lines wrap around the base of the vase. These lines represent water, and their placement at the bottom of the vase expresses the scientifically sound observation that water is the basis of agriculture and, indeed, of all life on earth. The first register, or band, above the water sign, represents green plants—the ultimate source of food for all life on earth.[15] Wheat or barley alternate with some other type of stylized plant, which seems to bear small fruits on the ends of its three short branches.

The next register up shows a row of sheep, with horned males alternating with females. Since the sheep feed on the plants, we could label this register "Herbivores." The presence of both rams and ewes suggests the ideas of sexual reproduction and fecundity, foreshadowing the idea of the sacred marriage, which is to be the culmination of the procession. The plant register below contains no hint of sexuality, indicating that whatever notions of plant sex the farmers of Uruk may have gleaned from cultivating date palms, they did not apply it to other plants. The third register shows men bearing baskets, amphoras containing wine or beer, and other types of containers full of the produce of the field. The men are nude, indicating either that they are of low rank or that they are participating in a religious ceremony. This register represents the domain of ordinary people, agriculturalists who depend on both plants and animals for their sustenance.

In the upper register, the procession of tribute-bearers ends with the king (only partly visible due to damage) presenting, through his servant, the first basket of produce to the Goddess Inanna, who stands before a pair of ring-posts, her iconographic symbols (Figure 5.2B). Here, we leave the material realm of agriculture and enter the spiritual plane, where Inanna occupies the highest niche. Here, she represents two sacred realms, the astral and the vegetal. In her astral aspect, she embodies the all-embracing power of the heavens and celestial light. In her vegetal aspect, Inanna causes plants to grow, which, as the Warka vase illustrates, is the source of all crops and livestock.

Inanna is typically portrayed as a sexually charged young woman at the peak of her reproductive powers—a metaphor for the fertility of the earth. Her icon is the filled storehouse, the symbol of abundance. Before her awesome powers, kings must exhibit appropriate respect, humility, and submission, otherwise they risk bringing the goddess's wrath down upon themselves and their kingdom. Thus, the king leads a procession of tribute bearers in the Warka Vase. In turn, the goddess demonstrates her acceptance of the king's rule through the ritual of the "sacred marriage" (from the Greek, *hieros gamos*)[16] enacted by the king and a priestess who serves as the incarnation of Inanna.[17] In this way, the king takes the part of Inanna's lover, Dumuzi, and is thereby elevated from a mere mortal to a god-like status, capable of mediating between the earthly and heavenly planes.

A major source of information about the details of the sacred marriage ritual is the Old Babylonian royal hymn, usually referred to as Iddin-Dagan A, which was probably composed around 1974 BCE.[18] Iddin-Dagan A appears to describe a two-day festival and is divided into ten sections. After a joyous musical procession that includes cross-dressing and other types of carnival-like behavior, the storehouse is filled, the participants feast on the bounty of the harvest, engage in sexual activity, and fall asleep on the rooftops. Inanna appears to them in their dreams and pronounces judgment upon them. Just before daybreak, the participants rise, gather various offerings, and form another procession to a location beyond the city walls where the sacred marriage ritual takes place. The ninth section of the hymn describes the ritual as follows:

(a)

(b)

FIGURE 5.2 Details from the Warka vase, Early Sumerian Period, third quarter of the fourth millennium BCE. A. Lower registers. B. Upper register.
From Strommenger, E. (1964), *5000 Years of the Art of Mesopotamia*. H. N. Abrams, figures 5.21 and 5.22; British Museum, London.

At the New Year, on the day of the rites, in order for her to determine the fate of all the countries, so that during the day the faithful servants can be inspected, so that on the day of the disappearance of the moon the divine powers can be perfected, a bed is set up for my lady. Esparto grass[19] is purified with cedar perfume and arranged on that bed for my lady, and a coverlet is smoothed out on the top of it.

In order to find sweetness in the bed on the joyous coverlet, my lady bathes her holy thighs. She bathes them for the thighs of the king; she bathes them for the thighs of Iddin-Dagan. Holy Inanna rubs herself with soap; she sprinkles oil and cedar essence on the ground.

The king goes to her holy thighs with head held high, he goes to the thighs of Inanna with head held high. Ama-ucumgal-ana[20] lies down beside her and caresses her holy thighs. After the lady has made him rejoice with her holy thighs on the bed, after holy Inanna has made him rejoice with her holy thighs on the bed, she relaxes with him on her bed: "Iddin-Dagan, you are indeed my beloved!"[21]

EARLY FRUIT AND NUT TREE PLANTATIONS

Most of what we know about fruit and nut tree plantations in ancient Mesopotamia comes from documents written during the third millennium BCE. The sources include temple archives and the accounts of various officials and private merchants. J. N. Postgate has reviewed the types of fruit trees cultivated during this period.[22] He confirms that the date palm was the most abundant and valued fruit tree in Mesopotamia.[23] The Sumerians also grew grapes, figs, apples, pomegranates, and several other fruits whose identities have not yet been determined. Wild junipers may have been grown for their "berries," and some trees, such as pine, were apparently planted and managed as a source of lumber. Intercropping of fruit trees was commonly practiced, especially in date palm orchards, and specific mention is made of pomegranates. In addition, fig and apple trees were sometimes intercropped with vineyards. Annual herbaceous crops—such as wheat, barley, and sesame—were typically planted in date orchards as well.

Several kinds of nut trees were cultivated, including almond, walnut, hazelnut, and terebinth (*Pistacia atlantica*).[24] Given the limited number of fruit and nut trees that were cultivated in ancient Mesopotamia, it is significant that three of them—dates, figs, and terebinth—were dioecious.[25] For these fruit crops, plantation owners would have been strongly motivated to maximize their yield by planting only female trees. However, such a strategy would have required a method to ensure pollination of the female trees.

THE SEX LIVES OF DATE PALMS

In date palms, both the male and female inflorescences are surrounded by a flattened, woody bract or *prophyll* (leaf-like structure) commonly called a "spathe" (Figure 5.3A), and both also have a single, flattened, central *rachis* that bears numerous branches, or *rachillae,* resembling a whisk broom (Figure 5.3B and C).[26] They differ, however in their morphology. The male (staminate) flowers have three sepals, three petals, and six stamens (Figure 5.4, right). Three abortive carpels occur in the center of the male flowers. In contrast, the

(a)

(b)

FIGURE 5.3 Date palm reproductive structures. A. Male inflorescence still enclosed within its prophyll or spathe. The prophyll is starting to open. B. Female rachis and rachillae. C. Male rachis and rachillae.

(c)

FIGURE 5.3 Continued

female (pistillate) flowers are spherical in shape, with three small green sepals fused to form a cup-like structure containing three short, overlapping petals and three carpels with very short styles and small stigmas (Figure 5.4, left). The pistillate flowers also contain six abortive stamens. Note that visually and from an olfactory standpoint, the male flowers are much more "flower-like" because of their showy, white, fragrant petals and multiple filamentous stamens. In contrast, the female flowers, which are compact, greenish spheres with no apparent visual attractants for pollinators, resemble immature fruits.

EVIDENCE FOR ARTIFICIAL POLLINATION IN MESOPOTAMIA

By the end of the third millennium BCE, Mesopotamian city-states had made extensive contact with the Semitic Amorites to the west, and, by the early second millennium, many rulers of Mesopotamian city-states had Amorite names. One of these rulers of Amorite extraction was Hammurabi. Hammurabi was the king of Babylon from 1792 to 1750 BCE. Under Hammurabi, Babylon challenged and defeated many other city-states in Mesopotamia, unifying a vast region stretching from Eridu and Ur in the south to Mari in the northwest. Hence, historians use the term "Babylonia" instead of Sumer and Akkad beginning early in the second millennium BCE. In addition to his military and political victories, Hammurabi is best known for the legal code that bears his name. That code even included laws governing the artificial pollination of date palms.

FIGURE 5.4 Date palm flowers. (Left) Female (pistillate) flowers with green sepals and petals reduced to a cup-like structure. (Right) Male (staminate) flowers, with white petals.

Exactly when Mesopotamians first learned the technique of artificial pollination of date palms is irrecoverable, but it may have occurred as early as the Ubaid period during the fifth millennium BCE. Our first glimpse into the ancient practice during the historic period was provided by tablets written in Akkadian from the cities of Umma and Nippur. The tablets consisted of accounts of the number and types of trees in various date orchards. In the eight orchards that were listed, there was a total of 1,332 trees, of which 1,000 were described as "productive" and 332 were designated as "unproductive." Although the "unproductive" trees could refer to male trees, the literal translation of the term used to describe these trees was "not (yet) pregnant," which implies sexuality. The translator—the Dominican friar Jean Vincent Scheil, who in 1901 was part of the French team that discovered the stele bearing the code in the ancient city of Susa (the modern Iranian town of Shush)—interpreted the phrase "not (yet) pregnant" trees as meaning young female trees.[27] The tablet makes no specific mention of male date palms in any of the orchards. Because male date palms were not mentioned among the trees in the orchards, presumably they were being grown at different locations, suggesting that artificial pollination was being practiced.

Seven years after the publication of Scheil's paper, A. H. Pruesner reviewed the available evidence for artificial pollination of date palms in ancient Babylonia.[28] To the evidence Scheil presented, Pruesner added a startling new translation of two laws from the Code of Hammurabi, as well as additional translations of some agricultural contracts, all of which strongly supported Scheil's conclusion. According to the new translation, four paragraphs of the Code of Hammurabi specifically address the rights and responsibilities of date orchard owners and their gardeners. Two of these laws, numbers sixty-four and sixty-five, specifically addressed the question of artificial pollination. These laws had originally been translated by Robert F. Harper in 1904 as follows:

#64
If a man give his orchard to a gardener to manage, the gardener shall give to the owner of the orchard two-thirds of the produce of the orchard, as long as he is in possession of the orchard; he himself shall take one-third.

> #65
> If the gardener do not properly manage the orchard, and he diminish the produce, the gardener shall measure out the produce of the orchard on the basis of the adjacent orchards.[29]

As translated, Law #64 simply spells out the payment the owner of an orchard is entitled to receive for allowing a gardener to "manage" the orchard for him; he receives two-thirds of the fruit produced, while the gardener receives one-third. Law #65 states that if the gardener neglects his duty so that the harvest is reduced, the owner will still receive two-thirds of what the crop would have been based on the productivity of adjacent orchards.

The crucial word in these two laws, is "manage," which is how Harper translated the Akkadian word *rukkubu*. However, Pruessner pointed out that a secondary meaning for the related verb, *rakabu*, is "to fecundate."

An even more specific clue to the meaning of the word was obtained from a dictionary of the Talmud, where the cognate verb in Hebrew (Akkadian is a Semitic language) means "to graft, to place one branch upon another." Pruessner found a passage in the Mishna, a rabbinical treatise probably written in the third and fourth centuries AD, stating that it was lawful for the people of Jericho to "graft" date palms on the eve of Passover because otherwise they would spoil:

> The men of Jericho graft palms all day, even on the fourteenth of Nissan (Passover Eve). They would graft a branch of a [male] palm tree that bore poor fruits onto a barren [female] palm. In this manner, they grafted many branches, and subsequently, the whole female tree bore good fruit.[30]

However, "grafting" in the modern sense of the word cannot be what was meant because date palm trees cannot be grafted like most other fruit trees. Nor does it make sense to allow "grafting" on Passover eve to avoid spoilage. Pruessner found the answer to this conundrum in the commentary on this passage by Rabbi Rashi, an eleventh-century French rabbi. According to Rabbi Rashi, "grafting" in this context refers to the following practice:

> A soft branch [of the flower cluster] of the male date palm is placed in a split [of the flower cluster] of the female palm, because the female does not bear fruit, while the male does.[31]

Although Rabbi Rashi's explanation shows he was confused about which sex bears the fruit (a confusion perhaps traceable to the ancient Egyptians, as discussed later), his definition clearly describes the process of artificial pollination as it was practiced in the Middle Ages and as it is still practiced in many date orchards today. Based on Rashi's description of "grafting," Pruessner felt justified in substituting the word "pollinate" for "manage," which makes the law much more specific and also supports the idea that artificial pollination of date palms was routinely practiced in ancient Babylonia.

Although our understanding of ancient Mesopotamian languages has greatly improved since Pruessner's article appeared in 1920, his interpretation of the laws pertaining to pollination has withstood the test of time. According to Benno Landsberger,[32] the Akkadian word for the female flowers was *uhinnu*, and the name for the male flowers was *rikbu*. *Uhinnu* is literally translated as "fresh, green dates." Recall that the female flowers resemble tiny green fruits (see Figure 5.4B). Hence, the word *uhinnu* may refer to unfertilized female date flowers. Significantly, *rikbu* (male flowers) is the noun derived from the same verb *rakabu* that Pruessner connected to the word "graft." Modern dictionaries translate *rakabu* as "to ride, to mount, to be on top."[33] This translation is related to the Hebrew cognate word meaning "to graft." Moreover, a related Akkadian word, *rikibtu*, means "sexual intercourse." Thus, the male flowers of the date palm were referred to by a word that suggested both their use in artificial pollination and the sexual nature of the act.

Modern Akkadian dictionaries accept Pruessner's translation of *rukkubu* as "to pollinate." In the most recent version of the Code of Hammurabi, Martha Roth translates the two laws as follows:

> #64
> If a man gives his orchard to a gardener to pollinate (the date palms), as long as the gardener is in possession of the orchard, he shall give to the owner of the orchard two-thirds of the yield of the orchard, and he himself shall take one third.
>
> #65
> If the gardener does not pollinate the (date palms in the) orchard and thus diminishes the yield, the gardener [shall measure and deliver] a yield for the orchard to the owner of the orchard in accordance with his neighbor's yields.[34]

These translations make it clear that Babylonians understood the importance of pollination to fruit production and that they practiced artificial pollination. Pruessner provided additional evidence for artificial pollination in the form of a contract between the owner of an orchard and a gardener:

> A date grove of the god Amurru, in the field of the Arahtum-[canal],—(there are) dry leaves and offshoots,—the date grove of Hurazatum, from Hurazatum, Apil-ilišu, the son of Uraš-mubalit, has rented for caretaking. He shall pollinate the orchard; two-thirds (of the produce) the owner of the garden, one-third the renter shall take. Five talents of urê, ten male flower clusters he shall give (besides).[35]

Most of the contract simply restates law #64 of the Hammurabi Code. However, the contract also stipulates that the gardener shall, in addition to two-thirds of the harvest, give to the landowner "ten male flower clusters." This stipulation is important for several reasons. First, it indicates that the male flowers were used as a form of currency to pay for the use of the land. Second, we can infer that the male flowers provided in payment would have to have been collected prior to the opening of the flower buds (anthesis), since male date palm flowers begin shedding their pollen immediately upon opening. Third, we can assume that intact male rachillae were used in artificial pollination rather than pollen collected from the male rachillae, else the landowner would surely have specified pollen, *taltallu*, from the gardener, rather than male flowers. Fourth, the terms of the contract imply that male trees were grown on the same land as the female trees; otherwise, it would be difficult for the gardener to supply the male flowers at exactly the right stage. This last conclusion contradicts Scheil, who speculated that male trees were grown at some distance from the female orchards since they were not mentioned in the inventory tablets. Perhaps it was assumed that every orchard included a quota of male trees to provide a ready supply of male flowers for hand pollination.

THE DISCOVERY OF DATES IN THE QUEEN'S "DIADEM"

A fascinating example of date symbolism, which may shed light on the method of hand-pollination, has been identified among the jewelry found in the tomb of Queen Puabi of Ur. The Royal Cemetery at Ur, which belonged to the Early Dynastic III period (2650–2550 BCE), was excavated by British archaeologist Leonard Wooley in the 1920s and early 1930s. More than 1,850 intact burials were found. The bodies were either wrapped or enclosed in coffins at the bottoms of vertical shafts. A wealth of personal possessions was recovered, including jewelry, weapons, cylinder seals, and ceramic, stone, and metal pottery and vessels.

Sixteen of the tombs were much larger than the others. They consisted of multiroomed chambers with domed or vaulted roofs and ramps and passageways leading from one chamber to another.[36] One of the most impressive of these was the magnificent tomb of Queen Puabi. Among the finery she was wearing was an elaborate golden headdress of exquisite workmanship (Figure 5.5). Most of the gold ornaments were based on botanical motifs. At the top was a comb-like object bearing seven rosettes, which may represent flowers, stars, or both. Such rosettes are often associated with the goddess Inanna. Below this comb was a wreath of realistically rendered flowers, each with four sepals and four petals, alternating with narrow, lanceolate leaves, possibly willow, grouped in threes. Next were two wreaths of ovate gold leaves, which resemble poplar, and a final band of narrow gold rings. The sheer density of ornaments depicting plants in the Queen's headdresses is remarkable and attests to the strong symbolic association between women and plants as personified by the goddess Inanna/Ishtar. Botanically inspired jewelry for women (goddesses and priestesses) is also mentioned in various Mesopotamian literary texts and is thought to represent abundance.[37]

Also discovered near Queen Puabi's remains was an assemblage of thousands of tiny lapis lazuli beads and gold ornaments representing stags, gazelles, bulls, goats, and plants. Wooley assumed that the beads had once been sewn onto a cloth strip backing, and the gold ornaments had originally been fixed on top of the beads. He therefore assembled the ornaments onto a single "diadem," a detail of which is shown in Figure 5.6 A.[38] In the section shown, an upright fruit-bearing stalk is at the center flanked by a goat and a gazelle. Next are two twig-like structures and a pair of clusters of three spherical fruits. Wooley was unsure of the identity of the fruits at the center, but thought they resembled olives. He identified the two upright twig-like structures as wheat stalks.

Wooley's original reconstruction of Queen Puabi's diadem has since undergone a radical revision. Zettler and Horne have concluded that Wooley mistakenly combined several smaller pieces into a single diadem.[39] But, more importantly from our standpoint, Naomi Miller of the University of Pennsylvania has reinterpreted three of the botanical ornaments: the "fruit stalk" and the two so-called "wheat stalks."[40] Both of these had been fixed onto the cloth by Wooley in an upright orientation (Figure 5.6A), but Miller identified small loops at the ends of these structures, indicating that they are actually designed as pendants.

FIGURE 5.5 The larger of Queen Puabi's two headdresses, reconstructed by Wooley using hundreds of individual pieces.
From Zettler, R. L., and L. Horne (1998), *Treasures from the Royal Tombs of Ur*. University of Pennsylvania Museum, Philadelphia.

When hung as pendants, the supposed "wheat stalks" resemble the branches (rachillae) from a male date inflorescence (Figure 5.6B). Note that the male flowers of the rachillae are still in the bud stage, which is the proper stage for hanging in the female tree. The upright fruit stalk also revealed its true identity when hung as a pendant. Instead of olives, it now clearly resembles the fruiting branch (rachilla) of a date cluster (see Figure 5.6B).

The Sumerians had two words for date rachises: a_2*-an*, the date-palm rachis, and a_2*-an-sur,* the broom of the date-palm rachis.[41] They also had a word for an item of jewelry shaped like a date rachis, A_2*.AN su-sa-lal,*[42] although no physical example of such an ornament had ever been found prior to Miller's ingenious reinterpretation. Miller believes that both the male and female pendants of Queen Puabi's diadem represent the hitherto mysterious pieces of Sumerian jewelry known as A_2*.AN su-sa-lal.*

Jewelry shaped like date rachillae were icons of the goddess Inanna/Ishtar and in all likelihood would have represented agricultural abundance. Songs from the Inanna/Dumuzi cult often refer to head and neck decorations worn by the goddess that were based on flowers and leaves. In fact, they are regarded as "identifying symbols" of Inanna or some other vegetation goddess. The fact that the Queen or priestess buried at Ur bore these symbols on her headdress shows that she was identified with a goddess and indicates her semi-divine status. According to J. van Dijk, "these plant crowns are insignias of the goddesses and of

(a)

(b)

FIGURE 5.6 Orientation of the date rachis ornaments in Queen Puabi's "diadem." A. As originally reconstructed by Wooley. B. As corrected by Naomi Miller. The current consensus is that the ornaments actually belong to six or seven separate diadems.
From University of Pennsylvania Museum of Archaeology and Anthropology. See Naomi F. Miller (2000), Plant forms in jewelry from the royal cemetery at Ur. *Iraq* 62:149–155.

the priestesses, who personify these goddesses. Because they symbolize vegetation, they are regarded as decorative items specifically for women."[43]

It is possible that the *A2.AN su-sa-lal* pendants not only symbolized agricultural abundance, but that they may also depict the specific method of artificially pollinating date palms—by suspending male rachillae in the female tree. Although female rachillae eventually bend downward due to the weight of their fruit, male rachillae emerge from their woody prophylls and release their pollen while in an upright orientation. The pendant may depict the orientation of the male rachillae when it is hung in the female tree. This interpretation also explains why the male flowers are unopened. Male date flowers shed their pollen before anthesis (flower opening), and if the gardener waits until the corolla opens before hanging the rachillae in the female palm, most of the pollen will be immediately lost to wind or insects. The male flowers must be hung *prior to* anthesis, so that all the pollen will be released near the opened female inflorescence.

DATE PALMS AND GENDER IN MESOPOTAMIA

Queen Puabi's pendants are rare examples of the juxtaposition of male and female date flowers in Mesopotamian art, but they are not unique. Another striking example—an inscribed plaque—was found in the Temple of Inanna in the city of Nippur (Figure 5.7). The plaque dates to the latter part of the Early Dynasty period, around 2600 BCE. The upper register depicts a banquet scene in which a seated man (Figure 5.7, *left*) and a seated woman (*right*) are each being served by attendants. A musician facing the woman is playing an eight-stringed harp whose sounding box is shaped like a cow. Behind the harpist is an amphora held in a wooden holder, probably containing wine. The man and woman

FIGURE 5.7 Inscribed plaque from the Inanna Temple, Nippur. Left arrow points to the male date inflorescence; the right arrow indicates the female inflorescence.
From Hansen, D. P. (1963), New votive plaques from Nippur. *Near Eastern Studies* 22:145–166, plate VI.

(presumably husband and wife) are drinking from goblets, which they hold in their right hands. In her left hand, the woman appears hold a female date cluster (*right arrow*), while the man grasps in his left hand the whisk-like rachis of the male date palm (*left arrow*). If this interpretation is correct, it shows that the Sumerians correctly identified, at least metaphorically, the fruit-bearing tree as female and the pollen-bearing tree as male. In the same Temple of Inanna there is an inscription to Ninmu, goddess of plants (each temple was in fact a complex with shrines devoted to various deities). The association of the plaque with two vegetation deities, Inanna and Ninmu, turns this banquet scene into a celebration of agricultural abundance and prosperity.

The Goddess Ninmu may be the woman depicted on a relief vessel from Lagash, dating to around 2400 BCE (Figure 5.8A). Her horned crown indicates that she is a deity, and the lotus bud-like objects sprouting from her shoulders show she is a goddess of vegetation. In her right hand she holds a cluster of dates, the Mesopotamian symbol of fruitfulness. Similarly, in the Akkadian cylinder seal in Figure 5.8B, a winged Inanna/Ishtar is shown standing on a mountain next to a small tree, holding a bunch of dates. Images such as these clearly indicate that date fruits were closely associated with female deities. To her right, life-giving water gushes from the shoulders of the water god, Ea or Enki.

OTHER VEGETATION GODDESSES OF SUMER AND AKKAD

Inanna and Ninmu figure prominently in the myths of Sumer-Akkad, along with a host of other agricultural deities, most of which are female. We will return to the myths involving Inanna, but first we will introduce a few of the other vegetation goddesses of southern Mesopotamia.

Ninhursag and Uttu. Earlier, we noted that the goddess Nammu was the primordial deity representing subterranean waters. Enki, in one Sumerian text referred to as Nammu's son, is known as the "Lord of the Earth" as well as "king of the *abzu*" or "king of sweet water." Enki was thus also the god of the marshlands so widespread in southern Mesopotamia. As the god of sweet water, Enki was viewed as the ultimate source of all life, including plants, a function that the earlier goddess Nammu once held. Perhaps the transition from the goddess Nammu to the god Enki as the source of life reflects the transition from Neolithic "hoe-agriculture" to Bronze Age "plow-agriculture," when agriculture became increasingly male-dominated. Whatever the reason for the replacement of Nammu by Enki, some vestige of the primordial vegetation goddess remains in the Sumerian myth of Enki and Ninhursag.

The story begins in the land of Dilmun, the Sumerian paradise, where everything is clean and bright, but there is no life because there is only brackish water. In this pure, but lifeless place the god Enki lives with his wife, the goddess Ninsikil. One day, Ninsikil tells Enki that she wants sweet water for use in irrigation, which Enki promptly provides with the help of the sun god Utu. Sweet water bubbles up from the ground, the rivers are filled, and the earth produces crops of grain and dates. Enki's wife, Ninsikil, is functioning as a vegetation goddess in the first episode of the story.

Having provided the earth with water, which makes the grain and the dates grow, Enki next fathers a daughter and engages in a series of incestuous relationships. In this second episode, however, Enki has a different wife, the mother goddess, Ninhursag ("Lady of

(a)

(b)

FIGURE 5.8 Sumerian vegetation deities. A. Bas-relief of vegetation goddess, dating to 2500–2371 BCE; from Lagash, east of Uruk. B. Detail from an Akkadian greenstone cylinder seal impression, ca. 2,300 BCE, showing Inanna/Ishtar, winged, armed, and holding a cluster of dates (left), the Sun god Shamash cutting through the mountains of the east (below), and the water god Enki or Ea (right). A is from H. W. F. (2000), *Babylonians*. University of California Press, plate III, Berlin Museum. B is from the British Museum: WA 89115.

the Foothills"), also known as Nintu ("Lady of Birth"). Enki impregnates Ninhursag, and she gives birth to Ninnisaga, the goddess of mountain vegetation. (According to some versions, Ninhursag first gives birth to Ninmu, the goddess of plants.) No sooner does Ninnisaga reach puberty than Enki impregnates her as well, and she gives birth to Ninkurra, the goddess of tall, distant mountains. The libidinous Enki then mates with his own granddaughter, who gives birth to his great-granddaughter, Uttu, the spider goddess of weaving.

Ninhursag finally loses patience with her husband, but because there seems to be nothing she can do to stop him, she advises Uttu to resist Enki's advances *until* he brings her gifts of fruits and vegetables: cucumbers, apples, and grapes. Rather than walk by the riverbank where Enki usually makes his conquests, Uttu remains safe inside her house. When Enki comes to her door and finds it barred, he proposes marriage and asks what wedding gifts she wants. Uttu requests the gifts of cucumbers, apples, and grapes. Enki then finds a gardener and strikes a bargain with him. He will help the gardener by filling his irrigation ditches with sweet water, and, in return, the gardener will give Enki cucumbers, apples, and grapes. The gardener agrees, and Enki presents the fruits of the garden to Uttu. Uttu is delighted and allows Enki into her house. Once inside, however, Enki plies Uttu with beer and she becomes drunk. He rapes her:

> and she conceived the semen
> in the womb,
> very semen of Enki![44]

Ninhursag hears Uttu's cries and comes to her aid. The text is fragmentary at this point in the story, but it appears that Ninhursag removes the semen from Uttu's belly and plants it in the ground. As a result, eight different plants spring forth: the *tree* plant, the *honey* plant, the *vegetable* plant, and five other plants. Genealogically, these plants represent Enki's great-great-grandchildren. The incestuous cycle is now broken because Enki cannot copulate with plants. However, he violates them in another way—by eating them! This last act infuriates Ninhursag to such an extent that she curses him by giving him an ailment for each of the plants he has consumed. She then disappears, vowing never to return.

Soon Enki becomes mortally ill. Greatly distressed, the other gods intervene and eventually persuade Ninhursag to return. To cure Enki, she gives birth to eight healing deities, one for each of his afflictions. Presumably, these healing deities are connected with the Sumerian plant-based pharmacopoeia, traditionally a female area of expertise. In this myth, Ninhursag and Uttu serve as vegetation deities because of their roles in conceiving and planting the eight medicinal plants.

Goddesses of Grain and Grapes. Various goddesses presided over grain and activities related to grain. In the city of Isin, *Ashnan/Ezinu* (two different readings of the same sign) lived the Sumerian goddess of grain cultivation. In a poem lamenting the destruction of the temple,[45] she is described as "*Ezina, the laden, silver ear of grain.*"[46] In *The Debate Between Grain and Sheep*, the goddess Ezina-Kusu (Grain) is identified with the cereal crop:

> Grain standing in her furrow was a beautiful girl radiating charm; lifting her raised head up from the field she was suffused with the bounty of heaven.[47]

In the city of Nippur, *Ninshebargunu* was the goddess of barley, and in the city of Shuruppak, she was known as *Sud*, the goddess of cleaned, naked grains. The symbol of *Sala*, a Mesopotamian goddess of Anatolian origin, was a barley stalk, indicating her connection to agriculture.[48] *Bau* or *Baba*, a daughter of the sky-god An, was represented by a sign thought to be a winnowing fan, used for separating the grain from the chaff. Finally, the goddess *Ninkasi* presided over the important industry of beer-making, as described in the "Hymn to Ninkasi":

> Ninkasi, it is you who handle the . . . dough with a big shovel, mixing, in a pit, the beerbread with sweet aromatics.
>
> It is you who bake the beerbread in the big oven, and put in order the piles of hulled grain.
>
> It is you who water the earth-covered malt; the noble dogs guard it even from the potentates.
>
> It is you who soak the malt in a jar; the waves rise, the waves fall.
>
> It is you who spread the cooked mash on large reed mats; coolness overcomes.
>
> It is you who hold with both hands the great sweetwort, brewing it with honey and wine.
>
> You place the fermenting vat, which makes a pleasant sound, appropriately on top of a large collector vat.
>
> It is you who pour out the filtered beer of the collector vat; it is like the onrush of the Tigris and the Euphrates.[49]

The goddess of writing, *Nissaba*, may have originally been a grain goddess as well.[50] Thus, Sumerians continued the Neolithic tradition of associating the growing and processing of cereal crops with women.

Wine-making was also associated with a female deity, the goddess of grape vines, *Geshtinanna*. Geshtinanna was the sister of Dumuzi, the shepherd god who became the passionate lover of Inanna and who is Inanna's consort in sacred marriage rituals. Dumuzi dies and becomes an underworld deity, whose consummated marriage to Inanna is necessary for agricultural fertility. Geshtinanna's association with Dumuzi could suggest wine's reputation as an aphrodisiac.

Inanna as "Lady of Vegetation." According to tradition, it was the poet Enheduanna, daughter of the Akkadian King Sargon and priestess of the moon-god Nanna at Ur, who began her poem *Exaltation of Inanna* with the salutation, "Lady of all the divine powers, resplendent light, righteous woman clothed in radiance, beloved of An and Urash [Ki]!" The Inanna in Enheduanna's hymn is a warrior goddess, befitting Babylon under Sargon's imperial rule:

> At your battle cry, my lady, the foreign lands bow low. When humanity comes before you in awed silence at the terrifying radiance and tempest, you grasp the most terrible of all the divine powers.[51]

Yet, for the most part, the myths of Inanna dating to 3500 BCE or earlier emphasize her close connection to the natural world—to sexuality, fertility, and agricultural cycles. She is particularly associated with trees, fertility, and crops, as illustrated by three myths: *Gilgamesh, Enkidu, and the Netherworld; The Courtship of Inanna and Dumuzi*; and *Inanna and Shukaletuda*.

Gilgamesh, Enkidu, and the Netherworld. It is possible that the biblical Tree of Knowledge in the Garden of Eden had its origin in this story, which begins with an account of Inanna and the *huluppu* tree. According to the myth, soon after the creation of the world, there was a young tree, the huluppu tree, which had been washed from the banks of the Euphrates. "I shall plant this tree in my holy garden," said Inanna, plucking the sapling from the river. She hoped that the tree would eventually provide her with a throne to sit on and a bed to lie upon.

After ten years, the tree had grown thick, "but its bark did not split." Next, three demons took over the tree: a snake "who could not be charmed" made its nest in the roots, the *Anzu*-bird (a mythical mischief-causing bird) and its young roosted on the branches, and a "ghost maiden" built her home in the trunk. Barred access by the three demons, Inanna then approached the hero Gilgamesh, who agreed to help her take possession of the tree. Wielding his bronze axe, Gilgamesh dispersed the three demons and went on to carve a throne and a bed out of the trunk of the tree for Inanna. In return, she presented him with two unidentified gifts, a *pukku* and a *mikku*, made from the roots and crown of the tree, respectively. The remainder of the story concerns Gilgamesh and his gifts.

The identity of the huluppu tree is uncertain, although its origin on the banks of the Euphrates suggests a willow or poplar.[52] Alternatively, because every part of the huluppu tree is used by Gilgamesh and Inanna to make important objects, it could also be a date palm, which was celebrated for its many uses. The statement that the "bark did not split" is also consistent with the trunk of a date palm, which is made up of overlapping leaf bases. On the other hand, the huluppu tree supplies wood rather than fruit, which argues against a date palm. Whatever the huluppu tree's identity, its primary function is to provide the raw material for Inanna's throne and bed.

The exact nature of the Sumerian "ghost maiden" is unclear, but she may represent a precursor of the Akkadian *ardat-lili*, which "seems to have the character of a frustrated bride, incapable of normal sexual activity."[53] The fact that the ghost maiden blocks Inanna's access to the tree suggests that the huluppu tree represents Inanna's nascent sexuality and fertility. Gilgamesh exorcises the demons, and, from the wood of the huluppu tree, he builds a throne and bed for Inanna, the symbols of "her rule and womanhood."[54] She, in turn, presents Gilgamesh with gifts that symbolize his kingship and virility. Thus, through Inanna, the sacred huluppu tree becomes the instrument of sexual awakening, fertility, agricultural abundance, and kingship.

Adam and Eve also gain their sexual awakening through a tree, but in Genesis, the tree and garden belong to God, not to Eve. Adam's and Eve's sexual awakening is treated as a curse rather than a blessing, and the serpent facilitates Eve's transgression rather than blocking her way. Notwithstanding the differences between the two sacred garden myths, the original Sumerian association of trees, women, and sexuality is still evident in the later, derived story of Genesis.

The Courtship of Inanna and Dumuzi. If the myth of the huluppu tree describes Inanna's coming of age, the Courtship of Inanna and Dumuzi describes her sacred marriage. Although the shepherd Dumuzi becomes King of Uruk by marrying Inanna, he is still subservient to her. In fact, in the epic poem *The Descent of Inanna*, Inanna forces Dumuzi to take her place as a prisoner in the Underworld, thus freeing her to return above ground. Inanna still loves Dumuzi passionately, but she is the goddess of light and life, and Dumuzi's sacrifice, at least for half the year, is necessary to maintain the natural order.[55]

Inanna and Dumuzi's sacred marriage is a celebration of human sexuality and the harvest, in the same tradition as the biblical *Song of Songs* (discussed later in the chapter). It begins with Inanna's brother, the sun-god Utu, subtly hinting at her readiness for marriage by commenting on the ripened grain and flax in the fields, metaphors for her sexual maturity. Through a series of symbolic questions, he brings her to the choice of a bridegroom and announces that he has already chosen Dumuzi the shepherd to be her husband. Inanna objects strongly, expressing her preference for a farmer who will fill her storehouse with grain. Inanna's initial preference for a farmer shows that she is closely identified with agriculture. However, Utu argues that shepherds have their advantages as well; finally Dumuzi, himself, chimes in, boasting that he can provide her with anything that the farmer can provide. Eventually, Inanna yields and accepts Dumuzi as her bridegroom.

When Dumuzi calls on her at her house, Inanna "bathed and anointed herself with scented oil," and "covered her body with the royal white robe." She allows him into her house and they embrace passionately. The use of sexually explicit metaphors in the following exchange reveals the symbolic nature of their romance as an agricultural fertility ritual:

Inanna: My vulva, the horn,
The Boat of Heaven,
Is full of eagerness, like the young moon.
My untilled land lies fallow.

. . . Who will plow my vulva?
Who will plow my high field?
Who will plow my wet ground?

Dumuzi: Great Lady, the king will plow your vulva
I, Dumuzi the King, will plow your vulva.

Inanna: Then plow my vulva, man of my heart!
Plow my vulva!

Narrator: At the king's lap stood the rising cedar
Plants grew high by their side.
Grains grew high by their side.
Gardens flourished luxuriantly.

Inanna: He has sprouted; he has burgeoned;
He is lettuce planted by the water.
He is the one my womb loves best.
My well-stocked garden of the plain,
My barley growing high in the furrow,
My apple tree which bears fruit up to its crown,
He is lettuce planted by the water.

My honey-man, my honey-man sweetens me always.
My lord, the honey-man of the gods,
He is the one my womb loves best.
His hand is honey, his foot is honey,
He sweetens me always.

Dumuzi: O Lady, your breast is your field,
Inanna, your breast is your field.
Your broad field pours out plants.
Your broad field pours out grain.
Water flows from on high for your servant.
Bread flows from on high for your servant.
Pour it out for me, Inanna.
I will drink all you offer.[56]

Although Dumuzi begins as a shepherd, increasingly, he takes on the role of a gardener during their courtship. He is especially concerned with date orchards, indicated by his planting the "honey-covered seed":

Dumuzi: My sister, I would go with you to my garden.
Inanna, I would go with you to my garden.
I would go with you to my orchard.
I would go with you to my apple tree.
There I would plant the sweet, honey-covered seed.

Although Dumuzi is a gardener, it is clear that Inanna is the ultimate source of agricultural bounty. It is her womb that brings forth the plants of the field:

Inanna: He brought me into his garden.
My brother, Dumuzi, brought me into his garden.
I strolled with him among the standing trees,
I stood with him among the fallen trees,
By an apple tree I knelt as is proper,
Before my brother coming in song.
Who rose to me out of the poplar leaves,
Who came to me in the midday heat,
Before my lord Dumuzi,
I poured out plants from my womb,
I placed plants before him.
I placed grain before him,
I poured out grain before him.
I poured out grain from my womb.

Inanna's role as the primary agricultural deity—the "Lady of Vegetation"—is explicitly stated at the end of the poem, when Ninshubur, Inanna's faithful minister, leads Dumuzi to Inanna's "sweet thighs" and prays:

Ninshubur: In the orchards may there be honey and wine,
In the gardens may the lettuce and cress grow high,
In the palace may there be long life.
May there be floodwater in the Tigris and Euphrates,
May the plants grow high on their banks and fill the meadows,
May the Lady of Vegetation pile the grain in heaps and mounds.

Although Inanna also likens Dumuzi to a "blossoming garden of apple trees" and to a "fruitful garden of mes-trees," in general, Dumuzi is most strongly identified with his garden, while Inanna represents the vegetation goddess who makes agricultural crops grow.[57]

Inanna and Shukaletuda. Experts consider the text of *Inanna and Shukaletuda* to be very obscure and difficult to translate and therefore difficult to interpret. Nevertheless, the story clearly associates Inanna with date palms, and it is therefore of interest to us here.

The text includes three fragmentary topics, and the connection between them is not entirely obvious. In the first episode, we are told only that Inanna ascended the holy mountains "to detect falsehood and justice, to inspect the land closely, to identify the criminal against the just." The connection between this episode and the next is unclear.

The second episode tells how Enki instructed a raven to create the first date palm tree. He tells the raven to chop and chew up a magical substance used by priests, to mix it with sacred oil and water, and to plant it in a furrow in a vegetable garden. The raven does as he is told and soon a plant resembling a leek appears. The raven then irrigates the garden using a *shadouf*, a simple wooden device for lifting water with buckets. Voila! The leek becomes a date palm tree.

A plausible interpretation of this transformation is that the leek-like plant represents a basal offshoot used in the vegetative propagation of date palms. But since no date palm existed at the beginning, a magical substance had to be used to create the first offshoot, just as in the Bible the first humans had to be made from dust.

Next, this very unbird-like raven climbs up the date palm using a harness and smears the same magic substance from its beak onto something in the crown of the date palm, possibly the female flowers:

> Then the raven rose up from this oddity, and climbed up it—a date palm!— with a harness. It rubbed off the kohl [magic substance] . . . which it had stuffed into its beak onto the pistils.[58]

Although the precise meaning of this passage is ambiguous, it could very well refer to artificial pollination. The next few sentences enumerate the uses of the date palm, with special emphasis on the production of dates, which are fit for the "great gods":

> Its scaly leaves surround its palm heart. Its dried palm-fronds serve as weaving material. Its shoots are like surveyor's gleaming line; they are fit for the king's fields. Its branches [rachises? leaves?] are used in the king's palace for cleaning. Its dates, which are piled up near purified barley, are fit for the temples of the great gods.[59]

The implication of this passage is that the raven has fructified the tree so that it will produce dates.[60] The raven then returns to Enki, and the third episode begins.

Shukaletuda is a boy who is charged by his father with building a well to irrigate the family date orchard. But all the trees of the orchard have disappeared because Shukaletuda has "pulled them all up by their roots and destroyed them." Next, a storm wind blows dust from the mountain into Shukaletuda's eyes. Although his normal eyesight is impaired by the dust, his spiritual vision is somehow enhanced because he sees the exalted gods on the horizon:

> He raised his eyes to the lower land and saw the exalted gods of the land where the sun rises. He raised his eyes to the highlands and saw the exalted gods of the land where the sun sets. He saw a solitary ghost. He recognized a solitary god by her appearance. He saw someone who fully possesses the divine powers. He was looking at someone whose destiny was decided by the gods.[61]

Shukaletuda then spies a familiar poplar tree and lies down under its shade to rest. At that moment Inanna returns from her travels around the heavens. Exhausted, she lies down by the roots of a poplar tree near where Shukaletuda is resting and falls asleep. Shukaletuda notes that she is wearing a "loincloth of the seven divine powers" over her vulva. With the help of a passing shepherd, he unties her loincloth and has intercourse with her while she sleeps.

When Inanna awakens she discovers that she has been violated, and, in her rage, she fills all the wells of the land with blood and begins searching for the man who raped her. The boy Shukaletuda flees to the city and tries to escape by making himself as small as possible, but Inanna eventually finds him. He begs for mercy, blaming his crime on the dust that blew in his eyes, but she is unmoved:

> "So! You shall die! What is that to me? Your name, however, shall not be forgotten. Your name shall exist in songs and make the songs sweet. A young singer shall perform them most pleasingly in the king's palace. A shepherd shall sing them sweetly as he tumbles his butter-churn. A young shepherd shall carry your name to where he grazes the sheep. The palace of the desert shall be your home."[62]

One of the difficult aspects of this story is that it is made up of three fragmentary and apparently unrelated episodes. In Part I, Inanna ascends the mountains to judge humanity, to "identify the criminal against the just." Ironically, her abstract mission in Part I becomes personal in Part III when she herself becomes a victim of a crime.

Part II seems to refer to an earlier event, when the god Enki first created the date palm with the raven's help. There is also a hint of some sort of artificial pollination here. At least, the raven seems to carry out an important task in the crown of the tree, enhancing the tree's fertility.

In Part III, the theme of date palms begun in Part II is continued with the boy gardener Shukaletuda, who, instead of watering the trees as he was told, pulls them out of the ground and allows them to die. Then a storm wind blows and the boy gets "dust" in his eyes. The word for "dust" in Akkadian, *taltallu*, also means "date pollen." Date palms are thought to be wind-pollinated. While carrying out artificial pollination on a windy day, a date gardener might well get some "dust" in his eyes, which would be consistent with the interpretation of the myth as an allegory of the artificial pollination of date palms.

Shukaletuda, observing Inanna sleeping under a poplar tree, focuses on her vulva. References to Inanna's vulva in Sumerian poetry usually have agricultural significance. Inanna's vulva, in this context, may represent the female inflorescence of the date palm. Recall that the female inflorescence is enclosed within a prophyll, which must be unwrapped to insert the male inflorescence during hand-pollination. In the story, Inanna's "loincloth of the seven powers," which Shukaletuda must "untie" before he can consummate his lust, may symbolize the female prophyll. This would tie Parts II and III together, relating them both to the pollination of date palms.

According to our interpretation, the story of Inanna and Shukaletuda can be read as a meditation on the sexual, moral, and religious significance of artificial pollination. Sumerians clearly viewed the practice of opening the female prophyll to insert the male rachis in sexual terms, which takes on the character of a rape. Yet the metaphoric rape is, in reality, a routine agricultural practice. Therefore, something must be sacrificed to atone for Inanna's sexual violation, and Shukaletuda, like most of Inanna's lovers, must pay the ultimate price. She comforts him by assuring him that his name "shall not be forgotten," but will live on "in songs and make the songs sweet." The "sweetness" of the songs may refer directly to the date harvest. In the end, it was Inanna's destiny to be raped, just as it was Shukaletuda's destiny to be sacrificed so that the date harvest and the agricultural cycle can continue.

IMAGES OF TREE GODDESSES ON CYLINDER SEALS

Small stone cylinder seals for identifying property or notarizing clay documents first appeared in Uruk in the second half of the fourth millennium BCE.[63] The outer surface was carved with an illustration, usually of some mythological scene, which appeared as an impression when rolled onto wet clay. During the third millennium BCE, the new reverence for trees was manifested by the many cylinder seal illustrations of sacred trees, usually date palms, and their iconic goddesses. For example, a cylinder seal from the first half of the third millennium BCE shows a vegetation goddess seated on a throne of branches (Figure 5.9A). Two branches sprout from each shoulder, and she is holding a branch in her left hand. Three female servants attend her, and the tableau is bracketed by two small date palms. A similar scene from the same period shows the enthroned goddess in the center flanked on each side by two female attendants and date palm bearing fruit (Figure 5.9B). Sometimes the goddess is omitted and her physical incarnation, the female date palm, is shown providing its bounty, as in Figure 5.9C. Since the three women are tree-sized, we can infer that they are themselves tree spirits or priestesses. In other examples, the identity of the tree deity is indicated. For example, Ishtar can be identified in Figure 5.9D from her weapons and the fact that she is standing on a lion. Images of sacred trees were sometimes highly stylized, either in the shape of a pillar or a candelabra-like structure. Figure 5.9E is a cylinder seal from the middle of the second millennium BCE. A female adorant, perhaps a priestess, stands to the right of a highly stylized sacred tree, probably a date palm. A winged sun with a rosette,[64] the emblem of Inanna, is directly overhead, and a sword-bearing hero

FIGURE 5.9 Mesopotamian cylinder seals images of sacred trees and their goddesses. A. Early third millennium BCE. B. Early third millennium BCE. C. Late third millennium BCE. D. Mid-first millennium BCE. E. Mid-second millennium BCE. F. Assyrian, from mid-second millennium BCE. From Danthine, H. (1937), *Le palmier dattier et les arbres sacrés dans l'iconographie de l'Asie occidentale ancienne* Sacres. Librairie Orientaliste Paul Geuthner, Paris. 2 Vols., Text and Album: A. plate 15, no. 84; B. no. 87; C. no. 18; D no. 15; E. no. 90; F. no. 276.

FIGURE 5.9 Continued

seems to offer protection to the tree while naked male servants sacrifice a bull. An even more stylized date palm tree is shown in Figure 5.9F, in which the trunk is represented by a single line adorned with volutes and terminates in a tuft of strap-like leaves.

GENDERING TREES IN ANCIENT EGYPT

Judging from the two main figures on the upper register of the plaque from the Temple of Inanna (see Figure 5.7), Mesopotamians seem to have associated fruit-bearing date palm trees with women and pollen-bearing trees with men. Thus, it comes as a surprise to learn that in some city-states of ancient Egypt, the fruit-bearing trees were regarded as male and the pollen-producing trees as female. [65] Each Egyptian city also had its own distinct association of trees and goddesses. In the town of Kôm el Hisn, one of Hathor's names was "The Mistress of the Date Palms," which was written phonetically as *Nb.t jm3.w*.[66] She was given the name for the male date palms, *jm3.w*, rather than that of the fruit-bearing female date palms, *bnr.t*, which means "sweet." [67] Gendering pollen-producing trees as female seems counterintuitive because we feel that fruit-bearing, like child-bearing, is a female function. It suggests that in Egypt during the Bronze Age, designations of plants as either "male" or "female" were not always based on biological analogies but on cultural definitions of gender. If so, what qualities of the male palm connoted "femaleness" to the citizens of Kôm el Hisn?

Since male and female date palms are indistinguishable vegetatively, the only differences are in their reproductive structures. Here we encounter the possible cause of the confusion: male date flowers have prominent white petals that give off a sweet fragrance, whereas the female flowers lack visible petals, produce no scent, and resemble tiny green spheres. As will become increasingly evident in later chapters, throughout history there has been a tendency to associate attractive, sweet-smelling flowers with femininity. The ancient Egyptians had a thriving perfume industry, and the perfume was derived from flower petals. Mural paintings show that the collection and processing of flowers for making perfume was performed by women. This might account for the gendering of male date palm flowers as female.

Egyptians occasionally reversed the sex of animals as well. For example, the Theban god Amon was also known as the Great Cackler, an egg-laying goose, and Isis was born from one of Amon's eggs.[68] It has been proposed that the ancient Egyptians believed that the creation of new life was essentially a male attribute, and that women were responsible for nurturing that creation. Thus, the earth was represented by a male god, Geb, and the sky by the goddess, Nut.[69] Using the same logic, perhaps the citizens of Kôm el Hisn gendered the date palm tree that gave rise to the fruit as male, and the tree that "nourished" the flowers with pollen as female.

TREE GODDESSES OF ANCIENT EGYPT

According to Egyptologist Fekri Hassan, the iconography of the deities of Egyptians during Dynastic times is deeply embedded in their Neolithic past.[70] As far back as 7000 BCE, the people who were to become Egyptians lived as cattle-herding desert nomads. These nomadic societies tended to view the female as "the source of life and nurture" and the male as the hunter. Hassan argues that vestiges of this Neolithic gender dichotomy persisted into the Bronze Age, helping to explain why, in the Early Dynastic period (~3000 BCE), the Egyptian deity representing cattle, the source of life-giving milk and meat, is identified with the cow goddess Bat or Hathor, often depicted with horns and cow ears. Similarly, trees, which provide food, drink, shade, and shelter, were also associated with goddesses, a tradition probably dating to Neolithic times.

Many trees were considered sacred by the Egyptians. The date palm was revered throughout the region for its sweet fruit and its many other uses, and, as noted earlier, it was associated with the goddess Hathor in some Egyptian cities. The sycamore fig with its broad, spreading trunk, was worshipped for its shade, fruit, and wood. Architectural elements made of sycamore wood dating to the Old Kingdom more than 4000 years ago still survive. In fact, the sycamore tree was the only native Egyptian tree that was large enough and strong enough to be useful as lumber. Other sacred trees include acacia, persea, cedar, and willow. All of these sacred trees were associated with goddesses in the different cities of Egypt.

The sycamore fig has been identified with all three of the major Egyptian goddesses: Hathor, Isis, and Nut. Hathor, in particular, was known as "Lady of the Sycamore." In the Book of the Dead, the Sycamore deity is identified as Hathor or Nut. In chapter 59 of the Book of the Dead, called "The Chapter of Snuffing the Air, and of Having Dominion over the Waters in the Underworld," Nut is described as follows:

> Hail, thou sycamore tree of the goddess Nut! Grant thou to me of the water and of the air which dwell in thee. I embrace the throne which is in Unnu (Hermopolis), and I watch and guard the egg of Nekek-ur, the Great Cackler. It groweth, I grow; it liveth, I live; it snuffeth the air, I snuff the air, I the Osiris Ani, in triumph.[71]

Depictions of sacred trees in Egyptian funerary art leaves no doubt that Egyptians regarded the goddesses and trees as materially and spiritually equivalent. Tableaus frequently portrayed the deceased receiving nourishment and libations from a tree/goddess chimera. For example, the illustration accompanying chapter 59 in the Book of the Dead, referred to earlier, shows the deceased bending over a pool before a sycamore tree, from which the goddess emerges with her arms extended, offering food and drink (Figure 5.10A). Similar images showing tree goddesses pouring libations were often painted onto Shabti boxes,[72] as shown in Figures 5.10B–D. In these images, the goddess seems to be physically continuous with the fruit tree.

Taken to its logical extreme, the goddess is sometimes represented as arms extending from a tree, bearing trays of food and libation vessels (Figure 5.11A). Another image shows two fruit-bearing sacred trees, date palm and sycamore fig, superimposed. Two tree goddesses emerge from the crowns presenting food and drink to the deceased (Figure 5.11B).

The equivalency between the tree and the goddess is even more pronounced in the examples shown in Figure 5.12. In Figure 5.12A, the goddess is represented only by a pair of arms offering libation from what may be a poplar tree. One of the most striking images of a tree goddess was found in the burial chamber of Thuthmose III in Thebes. Here, Thuthmose is shown suckling from the extended breast of Isis, which emerges from a branch as if it were fruit. An arm extending from a branch directs the breast into Thuthmose's eager mouth.

Women of ancient Egypt were also associated with plants in their daily lives. Although men did the heavy work of plow agriculture, women helped with the sowing and harvesting of wheat, collected and processed flowers for perfume, and were primarily responsible for textiles. The strong association between women and flowers is expressed in the hieroglyphic symbol for "woman," in which a woman is shown smelling a flower (Figure 5.13).[73]

FIGURE 5.10 Representations of Egyptian tree goddesses. A. Funerary painting of the tree goddess Hathor emerging from a sycamore to pour a libation for the deceased. B. Shabti box showing two sons of Horus receiving a libation from the tree goddess. C. The deceased receives a libation from the tree goddess. D. The deceased and her *ba* (human-headed bird representing the spirit of the deceased) receiving a libation from the goddess Nut, ca. 1250 BCE.
A is from Antelme, R. S., and S. Rossini (2001). *Sacred Sexuality in Ancient Egypt,* J. Graham, Trans., Inner Traditions, p. 90. B–D are from the British Museum.

(c)

(d)

FIGURE 5.10 Continued

FIGURE 5.11 A. Bas-relief of date palm with one arm poised to pour a libation and the other offering a tray full of food offerings. Note the breast at top of the arm holding the food, indicating a tree goddess, Nineteenth Dynasty B. Bas-relief of date and sycamore tree goddesses, Nineteenth Dynasty. A is from Danthine, H. (1937), *Le palmier dattier et les arbres sacrés dans l'iconographie de l'Asie occidentale ancienne* Sacres. Librairie Orientaliste Paul Geuthner, Paris, figure 955 (originally from L. Borchardt, *Die aegiptische Pflanzensäule*, Berlin, 1897, figure p. 45); B is from Danthine, figure 956.

Identification of women with fruit trees can also be seen in the love poetry of the Turin Papyrus of the Twentieth Dynasty (1190–1077 BCE). Two secret lovers meet beside a pomegranate tree in a flower-filled garden. The pomegranate tree feels like the proverbial "wall flower" in their presence—utterly neglected. In an emotional soliloquy, she vows to reveal their secret unless she is ministered to as she deserves:

The Pomegranate: . . . My seeds are like her teeth,
My fruits like her breasts (. . .)
I remain constant in all seasons:
When the "sister" acts with the "brother"!
While they are intoxicated upon wines and liquors,
And liberally sprinkled with oil and balm (. . .)
Though I stand upright, shedding my flowers,
Those of next year are (already) in me.
I am the first of my companions,
(but) I have been treated like the second!
In future, if they again begin to act this way
I will not keep my silence on their behalf.[74]

(a)

(b)

FIGURE 5.12 Detail from an offering table showing the arms of a tree goddess extending from a tree. B. Drawing of Isis as tree goddess on a pillar in the burial chamber of Tuthmose III in Thebes, New Kingdom, 1479–1425, Eighteenth Dynasty.
A is from Baines, J. (2001), *Fecundity Figures*. Griffith Institute, Oxford. B is from Hassan, F. A. (1998), The earliest goddesses of Egypt, in L. Goodison and C. Morris, eds., *Ancient Goddesses: The Myths and the Evidence*. University of Wisconsin.

man as opposed to "woman"

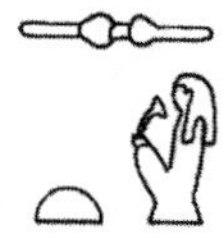

woman as opposed to "man"

FIGURE 5.13 Egyptian hieroglyphic symbols for "woman" versus "man."
From Schumann-Antelme, R., and S. Rossini (2001), Sacred Sexuality in Ancient Egypt the Erotic Secrets of the Forbidden Papyrus: A Look at the Unique Role of Hathor, the Goddess of Love, Inner Traditions.

She is, after all, a sacred tree, and sacred trees require offerings and libations!

Her statement that even as she sheds her flowers, "Those of next year are already in me," suggests the idea of parthenogenesis. The plant is represented as a self-fertile female, producing flowers and fruits within herself.

THE RISE OF NEO-ASSYRIA AND THE PALACE OF ASHURNASIRPAL II

The ancient city of Ashur, located north of Babylon on the Tigris river, was the capital of the small Bronze Age kingdom of Assyria, a region in northern Iraq whose people spoke an Akkadian dialect similar to that of the Babylonians. For much of its early history (from ca. 2000–1400 BCE), Assyria was overshadowed by its more powerful neighbors to the south, but, beginning around 1400 BCE, a series of ambitious rulers began converting the old city of Ashur—a trading post and cult center devoted to the god Ashur—into the urbanized political center of a powerful city-state encompassing all of northern Iraq.

From 900 to 600 BCE, Neo-Assyria, as historians refer to the Iron Age state, had developed a formidable war machine and had seized control of a vast empire stretching from Iran to Egypt. During this period, the Levant came under Neo-Assyrian control, and, as a consequence, rulers such as Sargon (721–705 BCE), Sennacherib (704–681 BCE), and Ashurbanipal (668–627 BCE) figure prominently in the Hebrew Bible.

Images of Sacred Trees were widespread among the palace relief sculptures of the Neo-Assyrian Period, especially during the reign of King Ashurnasirpal II, who ruled Assyria from 884 to 859 BCE. Ashurnasirpal II built his new capital (referred to as Calah in the Hebrew Bible) on the ruins of the old city of Nimrud. There, in the 1840s, British archaeologist A. H. Layard uncovered the Northwest Palace of Ashurnasirpal II and found colossal alabaster relief sculptures of the king engaged in rituals involving sacred trees and their winged protective deities, some with hawk-like heads, on the facades and inner walls of the palace compound (Figure 5.14). The sacred tree is depicted as an elongated trunk topped by a large palmette, surrounded by a sheath of interconnected smaller palmettes. Most scholar

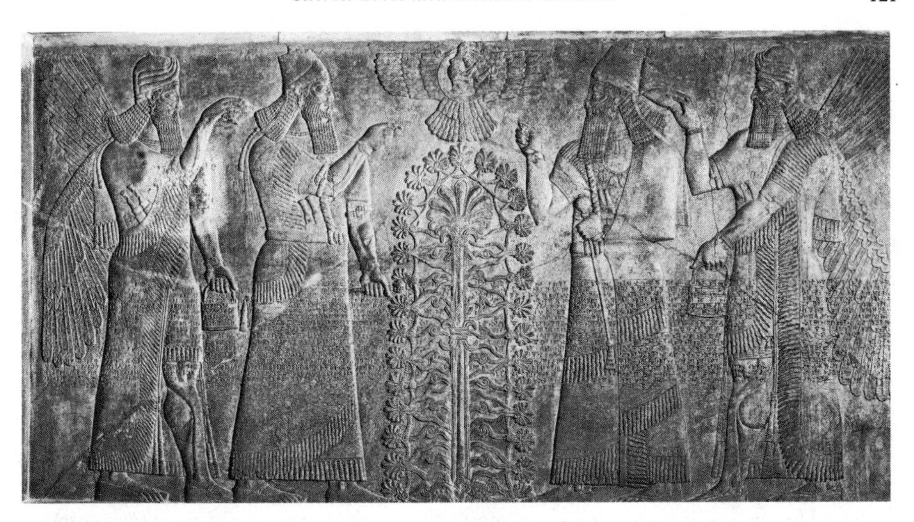

FIGURE 5.14 Alabaster relief of King Ashurnasirpal II (883–859 BCE) and the Sacred Tree, from the North-West Palace at Nimrud (Kalkhu). Two winged genii carrying cones and buckets stand behind and "purify" the king.
From the British Museum, London.

agree that the central trunk and apical palmette represents a date palm, and it has been suggested that the surrounding smaller palmettes could represent a date palm orchard derived from the vegetative offshoots of the mother trunk. The King is shown on either side of the tree, pointing his finger to the sacred tree as if directing the winged sun (emblematic of the sun-god Shamash) to shine on the tree.

The four-winged deities (only two wings are visible in profile view) behind the two images of the king are usually referred to as "genii."[75] They appear to be sanctifying the king by applying an object resembling a pine cone to his hair. According to Mallowan, two late Assyrian texts from Ashur provide clues regarding the functions of the bucket and cone borne by the "genii":

> In the Assur texts the bucket carried by certain figures is called *banduddû*, which appears in Akkadian texts as "ritual bucket." The other object described in the texts as carried by these figures is called *mullilu*. This word simply means "purifier," which tells us nothing about its appearance, but it may be the cone-shaped object carried by the "genii."[76]

The current consensus of opinion is that the objects known as *mullilu* were, as their appearance suggests, pine or cedar cones, which functioned as ancient "purifiers" in Neo-Assyria.[77] The bucket, according to this interpretation, was probably filled with "holy water" of some sort. However, as will be discussed in the next section, an alternative interpretation of the "cone" has been proposed, one that ties it directly to date palms.

Although early illustrations of sacred tree motifs from southern Mesopotamia were typically date palms, often in association with the goddess Inanna/Ishtar, the Neo-Assyrian sacred tree of northern Mesopotamia became more stylized as it gradually transcended its agricultural roots and took on cosmological and political significance. Whereas the King's authority in Sumer and Akkad was legitimized by Inanna/Ishtar or her representatives in the sacred marriage ritual, the Assyrian kings of the first millennium BCE owed their authority, as well as their responsibility to ensure the state's welfare, to their special relationship with the sacred tree and the sun god, without the direct intercession of the Goddess. Nevertheless, the presence of Inanna/Ishtar persisted implicitly in the guise of the sacred tree.

SACRED TREES AND ARTIFICIAL POLLINATION

In some of the Northwest Palace sacred tree reliefs, the genii are shown applying the *mullilu* to the surface of the sacred tree itself, as shown in Figure 5.15 As noted earlier, the currently accepted interpretation is that the genii are using a pine or cedar cone as an aspergillum (a device for sprinkling holy water) to anoint the king and the sacred tree. However, it is worth noting a competing theory because it bears directly on the question of when pollination was discovered. In 1890, Edward Burnett Tylor, a British anthropologist,[78] proposed a theory on the significance of the sacred tree ritual that was widely accepted for many decades. Tylor, Curator of the Royal Museum at Oxford and a Fellow of the Royal Society, presented photographic evidence that the inflorescence cluster of the male date rachis is very cone-like in appearance when it first emerges from its prophyll sheath.[79] Tylor photographed the male inflorescence cluster as it looks when held in a person's hand (his own) and compared it with the image of the "pine cone" held by an Assyrian winged deity (Figure 5.16). The similarity is intriguing. Even at this early stage of emergence from the prophyll, the flowers are beginning to open and pollen is being shed, so it is precisely this stage that would be used for artificial pollination.

Based on his findings, Tylor proposed that the cone-shaped object in the hands of the winged deities of Assyrian palace stone reliefs is not a pine cone, but the rachis of the male date palm. The entire sacred tree tableau, according to Tylor, represents the artificial pollination of the female date palm. Tylor believed that the Assyrians used the motif of artificial pollination of date palms as a metaphor for agricultural fertility and the general prosperity embodied by the king.

Tylor's theory has fallen into disfavor in recent years because of the lexical evidence cited in the previous section as well as the fact that at some locations in the palace the genius appears to be anointing the king, himself, instead of the sacred tree.

On the other hand, Tylor's pollination theory cannot be entirely ruled out. Date pollen in Neo-Assyria could have acquired a symbolic significance similar to the sacred maize pollen in Navajo religion.

THE PERSISTENCE OF VEGETATION GODDESSES IN MONOTHEISTIC RELIGIONS

Canaanite vegetation goddesses were common in the Levant during the Middle and Late Bronze Age (1,750–1,200 BCE). Images of naked Canaanite goddesses associated

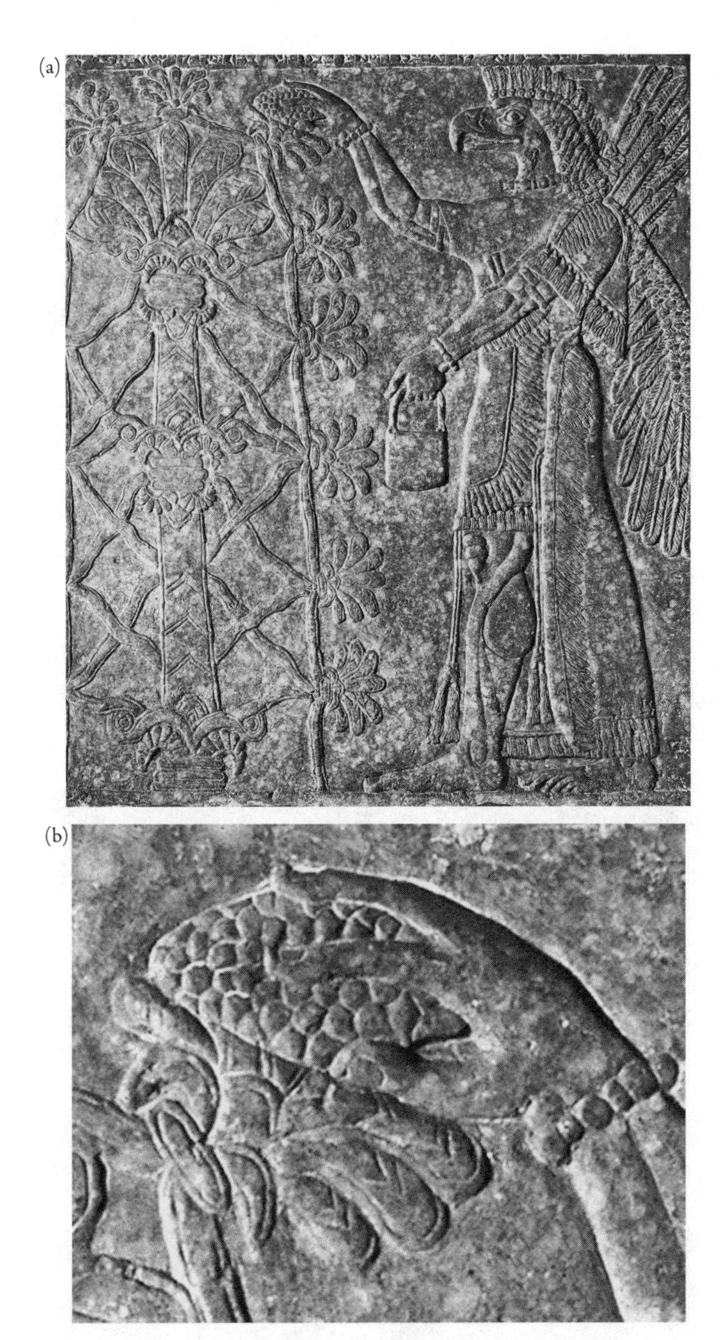

FIGURE 5.15 Possible Assyrian depiction of artificial pollination. A. Alabaster relief of winged, eagle-headed genius tending a stylized grove of palm trees the Sacred Tree, from the North-West Palace at Nimrud (Kalkhu). B. Detail showing hand of genius touching a rosette with a cone. From the British Museum.

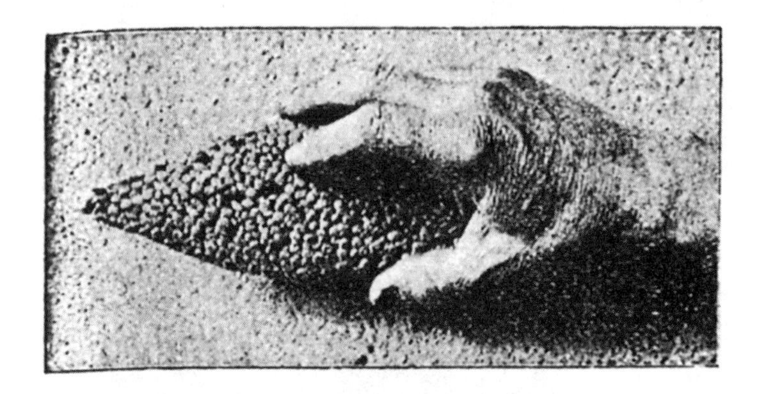

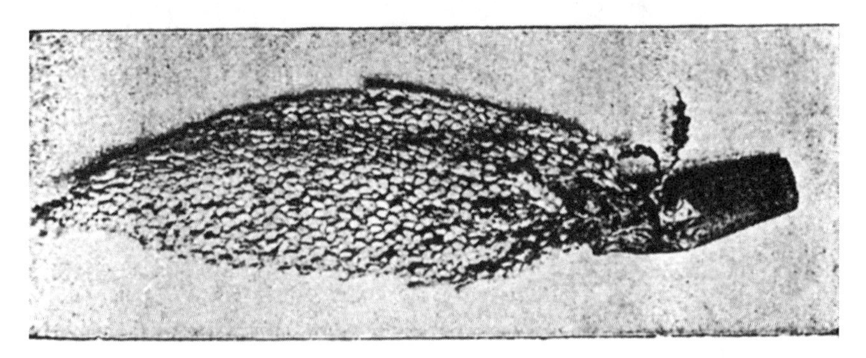

FIGURE 5.16 Photographs showing what an unopened male date palm rachis (bottom) looks like when held in the right hand (top) and its resemblance to a cone held by one of the winged genii of the Assyrian sacred tree reliefs (middle).
From Tylor, E. B. (1980, June), The winged figures of the Assyrian and other ancient monuments. *Proceedings of the Society of Biblical Archaeology*, 383–393, plus plates I–VI.

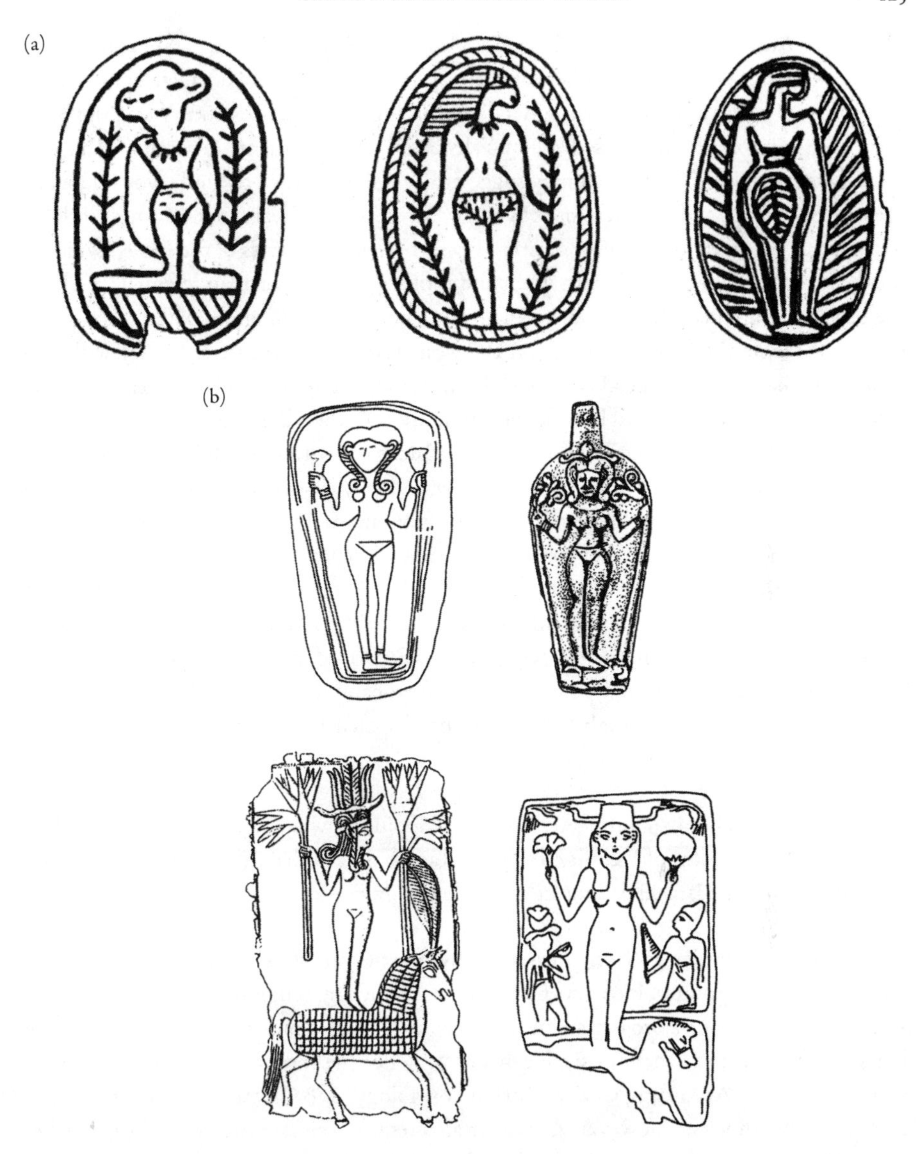

FIGURE 5.17 Canaanite vegetation goddess from the Middle and Late Bronze Age. A. Scarabs. B. Egyptian-inspired images of vegetation goddesses.
From Winter, U. (1983), Frau und Göttin: exegetische und ikonographische Studien zum weiblichen Gottesbild im Alten Israel und in dessen Umwelt, SLM Press, Jerusalem.

with vegetation were inscribed onto scarabs (Figure 5.17A), and Egyptian-inspired goddesses associated with the lotus, mandrake, and other sacred plants were also represented (Figure 5.17B).

The worship of female deities, especially among women, persisted in the Levant well into the Iron Age. Hundreds of nude female figurines have been found dating to the Late Monarchic period in Judah (seventh century BCE). The Hebrew Bible mentions the goddess "Asherah" forty times. In 1 Kings 15:13, King Asa removed the title of "Queen Mother" from his mother Maacah because she had made an "abominable image" of Asherah, and 1 Kings 18:19 makes mention of 400 prophets of Asherah who dine at Jezebel's table. Significantly, a fragment of a jar was found in the Sinai desert dating to the eighth century BCE site bearing the inscription, "I have blessed you by Yahweh of Samaria and his Asherah."[80] This and other archaeological findings strongly suggest that Asherah was originally worshipped as the wife of Yahweh and can therefore be characterized as a co-Creator and Mother Goddess whose attributes were deliberately omitted from the canonical texts of the Bible. Evidence for the association of Asherah with grain and probably wine appears in Jeremiah, wherein the prophet denounces the Hebrew women who "knead their dough to make cakes to the Queen of Heaven" (Jeremiah 7:18), and the women defiantly declare that they will continue to make these "sacrificial cakes in her image" and "pour out drink offerings to her" (Jeremiah 44:19) with the full knowledge and support of their husbands.[81]

Like Inanna/Ishtar, Asherah was called the "Queen of Heaven" by the ancient Israelites. She was also closely associated with trees. The word *Asherah* was used to denote the sacred tree or pole located next to her shrine. In Deuteronomy 16:21, God is said to have disapproved of Asherahs, whether they consisted of poles ("Do not set up any [wooden] Asherah [pole] beside the altar you build to the Lord your God") or of living trees ("You shall not plant any tree as an Asherah beside the altar of the Lord your God which you shall make"). Nevertheless, at least one monarch, King Manasseh, placed an Asherah pole in the Holy Temple (2 Kings 21:7). Sacred poles and pillars probably represent stylized date palms. Indeed, the prophetess Deborah held court and dispensed justice under her own sacred palm tree:

> At that time Deborah, a prophetess, wife of Lappidoth, was judging Israel. She used to sit under the palm of Deborah between Ramah and Bethel in the hill country of Ephraim; and the Israelites came up to her for judgment.[82]

Agricultural goddesses from Persian polytheistic religions from about the same period were incorporated into another great monotheistic religion from the Near East. Zoroaster, founder of Zoroastriansim in ancient Persia, is traditionally dated from the late seventh century to the mid-sixth century BCE, but the roots of the religion are probably much older. According to Zoroastrian cosmology, one Creator god, *Ahuramazda,* rules over the universe. However, Zoroaster also retained elements from earlier polytheistic religions in the form of a hierarchy of lesser divinities. Chief among these are the *Amesha Spentas,* or archangels. The female archangel *Ameretat* personifies immortality. She is also closely associated with plants and is regarded as their protector. According to a tradition written down in the twelfth century CE, when the earth was young, a demon caused the primordial plant to wither. *Ameretat* crushed the dried plant and mixed it with water. The rain god then spread the pulp—the life force of plants—over the earth, causing plants to sprout up. The story can be viewed on one level as an allegory of agriculture in general and vegetative propagation in particular. The association of *Ameretat* with both plants and immortality thus connects her to a widespread agricultural/religious complex. The Persian pantheon also included the goddess, *Azarvan,* a protectress of date palm trees. Zoroastrianism remained the dominant religion in Persia until 633 AD, when the Sassanid dynasty was overthrown by the Arabs under the Umayyad caliphate, and the Persian population was Islamicized. However, as we shall see in Chapter 10, echoes of ancient vegetation goddesses can even be heard in the literatures of Christianity and Islam.

THE *SONG OF SONGS* AND THE "GARDEN ENCLOSED"

The functions of the Sumerian love poems of Inanna and Dumuzi were both sacred and profane, combining the ritualized deification of the monarch with the practical function of assuring the abundance of the harvest and the resulting prosperity to the kingdom. Hence the love lyrics were suffused with agricultural metaphors related to a bountiful harvest. Egyptian love songs dating to the second millennium BCE, such as those in the Harris Papyrus 500, resemble the earlier Sumerian poems in some respects, but the tone is less triumphal, expressing tenderness and longing, as the following fragments illustrates:

The vegetation of the marsh is bewildering.
The mouth of my sister is a lotus,
her breasts are mandragoras,[83]
her arms are branches,
her —— are ——,
her head is a trap of "love-wood,"
and I—the goose!
The cord is my . . .,
her hair is the bait
in the trap to ensnare me.[84]

If only my sister were mine every day,
like the greenery of a wreath! . . .
The reeds are dried,
the safflower has blossomed,
the mrbb-flowers are (in) a cluster (?).
The lapis-lazuli plants and the mandragoras have come forth.
. . .
The blossoms from Hatti have ripened,
the bsbs-tree blossomed, . . .
the willow tree greened.
She would be with me every day,
like the greenery of a wreath.
All the blossoms are flourishing in the meadow,
. . . entirely[85]

In the Egyptian love poems, the aristocratic young lovers address one another as brother and sister (perhaps metaphorically, although consanguineous marriages were common among Egyptian royalty), and the woman is often associated with greenery in the form of wreaths and attractive, sweet-smelling flowers. Some of the poems suggest a garden scene, as when the girl sings:

I am headed to the "Love Garden,"
my bosom full of persea[86] branches
my hair laden with balm.
I am a noblewoman
I am the Mistress of the Two Lands,
when I am with you.[87]

The tradition of erotically charged marriage songs reached its apogee in the biblical *Song of Songs*, perhaps composed as late as the third century BCE. Consistent with such a late composition, the poem reflects the influence of Hellenism, and some scholars have pointed out similarities to the pastoral idylls of Theocritus of Alexandria.[88]

Like all great literature, and also because of its fragmentary and sometimes cryptic nature, the *Song of Songs* can be read on many levels. Several voices can be identified: the adolescent girl (the "Shulamite"), her male lover (sometimes identified as King Solomon), the Shulamite's supportive sisters (the "daughters of Zion"), and even, according to some scholars, her protective brothers. The poem takes the form of a bridal poem, or epithalamium—in the form of an antiphony involving a call and response between the Shulamite and her betrothed—while the sisters and brothers offer occasional commentary similar to a Greek chorus.

The *Song of Songs* takes place in a garden, and the Shulamite is identified with the garden itself, as seen in what is arguably its most famous line: "My sister, my spouse, is a garden enclosed, a garden enclosed, a fountain sealed" (4:12). In medieval literature, especially, the wall around the garden is usually interpreted as symbolizing a bride's virginity, but the descriptions in *Song of Songs* (1:12–14) are more suggestive of consummated love:

My king lay down beside me
and my fragrance
wakened the night.

All night between my breasts
my love is a cluster of myrhh,
a sheaf of henna blossoms
in the vineyards of Ein Gedi.[89]

The wall of the enclosed garden could signify the bridegroom's exclusive access to the bride's person. At another point in the poem, the Shulamite invites the lover to come into "his" garden, indicating his ownership: "Let my beloved come into his garden, and eat the fruit of his apple trees," and the bridegroom responded with, "I am come into my garden, my sister, my spouse," establishing "his proprietary rights over both woman and garden."[90]

In the famous line, "I am the rose of Sharon and the lily of the valleys" (2:1), the Shulamite compares herself to the two flowers, rose and lily, that would later be assimilated by medieval

Christian artists and theologians as symbols of the Virgin Mary. But in 4:13–15, she is also likened to sweet fruits and aromatic spices:

> Your branches are an orchard
> of pomegranate trees heavy with fruit,
> flowering henna and spikenard,
> spikenard[91] and saffron, cane and cinnamon,
> with every tree of frankincense,
> myrhh and aloes,
> all the rare spices.[92]

The Bridegroom also compares her to a stately date palm, tying her to earlier Near Eastern goddesses: "That day you seemed to me a tall palm tree, and your breasts the clusters of its fruit " (7:8–9). The bridegroom continues:

> I said in my heart,
> let me climb into that palm tree
> and take hold of its branches.
> And oh, may your breasts be like clusters
> of grapes on a vine, the scent
> of your breath like apricots,
> your mouth like good wine—[93]

Since date palms lack branches, the young lover presumably refers to the female rachis with its closed flower buds. Although botanical imagery is applied to both lovers in the poem, the majority of erotic, botanical metaphors, including the garden itself, are associated with the young woman, as summarized by Ariel and Chana Bloch:

> Throughout the Song the garden is the symbol of the Shulamite and her sexuality: she is the "locked garden" (4:12), inaccessible to anyone but her lover; he alone is invited to the garden (4:16), and he alone enters it (5:1). He describes her as a "garden spring" (4:15) and addresses her as "the one who dwells in the gardens" (8:13). Only she is associated with both vines and pomegranates in erotic contexts (1:6, 4:13, 7:9,13, 8:2); the only fruit associated with him is the apricot.[94]

According to a passage in the Mishnah, the first-century Rabbi Simeon ben Gamaliel stated that the *Song of Songs* was often sung by the young women of Jerusalem during harvest festivals as they danced and sang in the vineyards, trying to attract the notice of the young men (Ta'anit 4:8).[95] Perhaps the *Song* came to be so beloved it was included in the Bible by popular demand. It is widely believed that the Jewish biblical canon was probably decided around the end of the first century CE.[96] But before the *Song* could become scripture, something had to be done about its erotic content.

Philo (20 BCE–50 CE), the neo-Platonic Jewish philosopher from Alexandria, and other first-century Jewish scholars promulgated the idea that all the stories of the Bible had hidden meanings and that the *Song of Songs* was actually an allegory of God's love for his Chosen People. Thus, the eroticism of the *Song* was made acceptable to the rabbis. Rabbi Akiva declared that, far from being profane, the *Song of Songs* was the most sacred text in the Bible:

> God forbid! No man in Israel ever disputed the status of the *Song of Songs* . . . for the whole world is not worth the day in which the *Song of Songs* was given to Israel; for all the writings are holy, but the *Song of Songs* is the holiest of the holy.[97]

Nevertheless, the *Song of Songs* continued to be sung at Jewish wedding banquets for some time after. Rabbi Akiva acknowledged the practice and strongly disapproved of it: "He who recites a verse of the *Song of Songs* and treats it as a song, and one who recites a verse at a banquet, brings evil upon the world."[98]

As we shall see in Chapter 11, medieval Christian theologians also attempted to allegorize the *Song of Songs*, this time as the love of Christ (the bridegroom) for his Church (the bride) or the individual soul. By the ninth century CE, the term "enclosed garden" came to be identified in western literature with the Virgin Mary in her role as the "Second Eve."

NOTES

1. Barrow, S. (1998), A monograph of *Phoenix l.* (Palmae: Coryphoideae). *Kew Bulletin.* 53:513–575.

2. Tengberg, Margareta (2012), Fruit growing, in D. T. Potts, ed. *Blackwell Companions to the Ancient World: Companion to the Archaeology of the Ancient Near East.* Wiley-Blackwell, pp. 181–200.

3. Zohary, D., M. Hopf, and E. Weiss (2012), *Domestication of Plants in the Old World: The Origin and Spread of Domesticated Plants in Southwest Asia, Europe, and the Mediterranean Basin.* Oxford University Press.

4. Barrow, . A monograph of *Phoenix l.*

5. Beech, M. (2003), Archaeobotanical evidence for early date consumption in the Arabian Gulf, in ECSSR (ed.), *The Date Palm - From Traditional Resource to Green Wealth.* Emirates Center for Strategic Studies and Research, Abu Dhabi, pp. 11–31.

6. Ibid.

7. Postgate, J. N. (1992), *Early Mesopotamia: Society and Economy at the Dawn of History.* Routledge.

8. With the expansion of villages into cities came a radical restructuring of society that replaced the kinship-based social hierarchy of the Neolithic Ubaid culture with a new political hierarchy. Palaces were erected as well as monumental public buildings and sculpture. Taxes and tribute were collected, and laws were enforced by a professional military that formed part of the ruling elite. Religion adapted to the new realities and became strongly identified with the King, granting him divine legitimacy in exchange for royal sanction and financial support. Underpinning the brave new world of mass agriculture and public building projects were increasing numbers of specialized artisans and craftspeople, as well as an organized workforce.

9. Writing was invented, perhaps for the first time, during the Uruk period in Mesopotamia. The script evolved from a method for recording business contracts employing small clay tokens. These tokens, in the shapes of small cones, spheres, disks, and cylinders, were used to represent quantities of a specific commodity, such as bread or cloth. They were sealed into hollow clay spheres, called bullae, and broken open at the time of delivery to verify that the contract had been fulfilled. The bullae, which first appear around 3500 BCE, were often labeled on the outside by pressing the tokens into the soft clay of the sphere. This method of imprinting was soon applied to clay tablets, and, by around 3200 BCE, a reed was being used to form different configurations of wedge-shaped impressions into clay tablets—a type of writing called *cuneiform*, or "wedge-shaped." The earliest version of the new script is termed "proto-cuneiform." Proto-cuneiform employed about 700 different signs, including both numbers and words. Many of these signs are recognizable and were used strictly for accounting purposes. Although the proto-cuneiform script shows little relation to a spoken language, there is evidence that the people who developed it were Sumerian speakers. Van de Mieroop, M. (2004), *A History of the Ancient Near East: Ca. 3000–323 BC*. Wiley-Blackwell.

10. Saggs, H. W. F. (2000), *Babylonians*. University of California Press, p. 25.

11. Kramer, S. N. (1972), *Sumerian Mythology* (Revised Edition). University of Pennsylvania Press; Westenholz, Joan Goodnick (1998), Goddesses of the ancient Near East 3000–1000 B.C. in L. Goodison and C. Morris, eds., *Ancient Goddesses: The Myths and the Evidence*. British Museum Press, pp. 61–82.

12. Jacobsen, T. (1976), *The Treasures of Darkness: A History of Mesopotamian Religion*. Yale University Press. Many assyriologists are reluctant to accept this translation because it is based on a single lexical entry in which the Sumerian word *an* is equated with the Akkadian word *sissinnu*, which refers to the date inflorescence, or "date inflorescence *broom*." In contrast, there are numerous lexical entries in which *sissinnu* is equated with a longer Sumerian term for date inflorescence, *gish-an-na gish.nimbar*. Thus, the single lexical entry of *an* for *sissinnu* may represent an abbreviation for *gish-an-na*. If so, the correct expression for "Lady of the Date Clusters" would be In*gish*ana, not *Inanna* (Wolfgang Heimpel, personal communication, 2003).

13. Sjöberg, Åke W. (1988). A hymn to Inanna and her self-praise. *Journal of Cuneiform Studies* 40:165–185.

14. In the aftermath of the US invasion of Iraq in April 2003, the Warka Vase fell victim to the frenzy of looting during which much of the Iraqi National Museum's priceless collection was stolen or destroyed. Although the vase was subsequently recovered—shattered into fragments—and "restored" to some semblance of its pre-war state, it still exhibits considerable irreparable damage.

15. Broadly speaking, the first register encompasses all photosynthetic eukaryotes, from algae to higher plants. However, photosynthetic bacteria also make a huge contribution to the earth's total biomass, as do the chemosynthetic bacteria, which use chemical forms of energy in the environment instead of sunlight.

16. The term "sacred marriage" is a translation of the Greek term *hieròs gámos*, referring to the marriage of Zeus and Hera. However, J. G. Frazier in *The Golden Bough* also applied the term to certain rituals in earlier cultures in which marriages (often between a ruler and a deity) were symbolically consummated as a means of securing fertility and abundance for the group.

17. Sefati, Y. (1998). *Love Songs in Sumerian Literature: Critical Edition of the Dumuzi-Inanna Songs.* Bar-Ilan University Press.

18. Iddin-Dagan (1974–1954 BCE) was the third king of the first dynasty of Isin, a city in southern Mesopotamia near Uruk. Although the only surviving tablets containing the hymn date to the eighteenth century BCE, the hymn was probably composed during Iddin-Dagan's reign. See Jones, Philip (2003), Embracing Inana: Legitimation and mediation in the ancient Mesopotamian sacred marriage hymn Iddin-Dagan. *Journal of the American Oriental Society* 123:291–302.

19. Either of two species of gray-green needlegrasses (*Stipa tenacissima* and *Lygeum spartum*), or the fiber derived from them.

20. Identifies the king with the shepherd Dumuzi, Inanna's lover.

21. Translated by Black, J. A., Cunningham, G., Ebeling, J., Flückiger-Hawker, E., Robson, E., Taylor, J., and Zólyomi, G., *The Electronic Text Corpus of Sumerian Literature* (http://etcsl.orinst.ox.ac.uk/), Oxford 1998–2005.

22. Postgate, J. N. (1987), Notes on fruit in the cuneiform sources. *Bulletin on Sumerian Agriculture* 3:115–144.

23. Today, there are about 22 million date palm trees in Iraq producing about 600,000 tons of dates annually (Morton 1987). The date-growing region occurs in the lowland plains between 30° and 34° latitude. Starting at the Persian Gulf, dates are grown all along the Shatt-al-Arab River, including Basra. Beyond the Shatt-al-Arab, date-growing occurs throughout the region between the Tigris and Euphrates as far north as Baghdad and also occurs along the banks of the Euphrates river all the way to the Syrian border. Wheat and barley can also be grown in most of the date-growing areas, except the southernmost region near the Gulf. Wheat and barley are the main crops in the northern part of the country up to the Zagros Mountains.

24. The common names for the various types of pistachio trees are somewhat confusing. Sumeriologists usually refer to *Pistacia atlantica* as "terebinth," although this name more logically should be applied to the closely related *Pistacia terebinthus,* which goes by the common name of "turpentine tree." The common name for *Pistacia atlantica* is "Mt. Atlas mastic tree," to distinguish it from *Pistacia lentiscus*, the "mastic tree, " whose resin is used for chewing gum.

25. Although wild grapes were originally dioecious, the cultivated variety was hermaphroditic.

26. The rachis/rachillae complex of dates is often erroneously referred to as a "spadix" in the archeological literature. A spadix is an inflorescence borne on a single thick, fleshy spike, whereas the date rachis and rachillae are woody and branching.

27. V. Scheil (1913), De l'exploitation des dattiers dans l'ancienne Banylonie. *Revue d'Assyriologie et d'Archéologie Orientale* 10:1–9.

28. Pruessner, A. H. (1920), Date culture in ancient Babylonia. *American Journal of Semitic Languages and Literatures* 36:213–232.

29. Harper, R. E. (1904), *Code of Hammurabi King of Babylon About 2250 B.C.* University of Chicago Press.

30. Rav and Rashi say grafting was necessary because the "female" trees would otherwise be barren. Mishna (Pessachim, 4, 8).

31. Ibid.

32. Landsberger, Benno (1967), The date palm and its by-products according to the cuneiform sources. *Archiv für Orientforschung (Graz)* Beiheft 17.

33. Jason Klein, personal communication.

34. Roth, Martha T. (1997), *Law Collections from Mesopotamia and Asia Minor.* Society of Biblical Literature.

35. Pruessner, Date culture in ancient Babylonia.

36. Roaf, Michael (2000), *Cultural Atlas of Mesopotamia and the Ancient Near East.* An Andromeda Book, Oxford. Facts on File, Inc.

37. van Dijk, J. (1967), Ein Zweisprachiges Königsritual, in *Heidelberger Studien zum Alten Orient.* Adam Falkenstein zum 17, Septermber 1966, 233–268. Wiesbaden,. Harassowitz.

38. Woolley, Sir L. (1934), *Ur Excavations II, The Royal Cemetery.* Publications of the Joint Expedition of the British Museum and the Museum of the University of Pennsylvania to Mesopotamia.

39. Zettler, R. L., and L. Horne (1998), *Treasures from the Royal Tombs of Ur.* University of Pennsylvania Museum.

40. Pittman, H. and N. F. Miller (2015), Puabi's Diadem(s): The Deconstruction of a Mesopotamian Icon, in J. Y. Chi and P. Azara, eds., *From Ancient to Modern: Archaeology and Aesthetics.* Princeton University Press.

41. Miller, Naomi F. (2000), Plant forms in jewelry from the royal cemetery at Ur. *Iraq* 62:149–155; Gonzalo Rubio, personal communication.

42. According to Assyriologist Gonzalo Rubio, "A$_2$.AN šu-ša-lal (or A$_2$.AN šu-ša-la$_2$ la$_2$ — lal and la$_2$ are two readings of the same sign) refers to a part of a metallic vessel or container (written normally in various ways: šu-la$_2$; šu-ša-la$_2$; šu-še$_3$-la$_2$; šu-uš-la$_2$; šuš-)." This indicates that "part of this vessel must have looked like the spadix of a date-palm." [Note that the numbers or numerical indexes in Sumerian transliterations "serve to distinguish between homophonous signs; that is, signs that look different and have different meanings, but which "sounded" the same or, rather, have the same reading in the ancient Mesopotamian lexicographical tradition: a "water," a$_2$ "arm; strength" (G. Rubio, personal communication).]

43. van Dijk, Ein Zweisprachiges Königsritual.

44. Jacobsen, T. (1987), *The Harps That Once ... Sumerian Poetry in Translation.* Yale University Press.

45. The Temple of Isin was apparently destroyed by the Elamites (a pre-Persian group) at the end of the third millennium BCE when the Third Dynasty of Ur collapsed.

46. Jacobsen, *The Harps That Once.*

47. Black, J. A., Cunningham, G., Fluckiger-Hawker, E, Robson, E., and Zólyomi, G. (1998), *The Electronic Text Corpus of Sumerian Literature* (http://www-etcsl.orient.ox.ac.uk/). Oxford University Press.

48. Black, J., and A. Green (1992), *Gods, Demons and Symbols of Ancient Mesopotamia, An Illustrated Dictionary.* University of Texas Press.

49. *The Electronic Text Corpus of Sumerian Literature.* [http://etcsl.orinst.ox.ac.uk/section4/tr4231.htm]

50. Westenholz, *Ancient Goddesses*, pp. 61–82.

51. Black, et al., *The Electronic Text Corpus of Sumerian Literature.*

52. Kramer, S. N. (1972), *Sumerian Mythology*. University of Pennsylvania Press.

53. The *ardat-lili* bears some resemblance to the Jewish Lilith (Isaiah 34:14). Black, J., and A. Green, *Gods, Demons and Symbols of Ancient Mesopotamia*.

54. Wolkstein, D., and S. N. Kramer (1983), *Inanna, Queen of Heaven and Earth*. Harper & Row.

55. Dumuzi's sentence is commuted to half a year when his sister, Geshtinanna, volunteers to take his place during the other half. Echoes of this agricultural myth are evident in the classical period, as exemplified by myths surrounding Persephone and Hades.

56. Wolkstein and Kramer, *Inanna, Queen of Heaven and Earth*.

57. Sefati, Y. (1998), *Love Songs in Sumerian Literature. Bar-Ilan Studies in Near Eastern Languages and Cultures,* The Samuel N. Kramer Institute of Assyriology. Bar-Ilan University Press. The "mes-tree" is an unidentified sacred tree.

58. Quotes taken from: Black, et al., *The Electronic Text Corpus of Sumerian Literature*.

59. Ibid.

60. The raven's fructification of the date palm using a magical substance is reminiscent of a belief, attributed to the Amorites, that unpollinated female date palms are barren because they are *ill*. According to the Babylonian Talmud, barrenness was a device used by the palm tree to elicit pity and prayers from passersby. The Babylonian Talmud goes on to describe the Amorite cure for barrenness:

> —Said Rab Judah: They brought fresh myrtle, the juice of bay-fruit and barley flour which had been kept in a vessel less than forty days, and boiled them together and injected it into the heart [female inflorescence] of the palm tree; and every tree which stands within four cubits of this one, if that one is not treated, likewise immediately withers. R. Aha the son of Raba said: A male branch was grafted on to a female palm tree. (Pessahim 56a)

Note, however, that Rab Aha, the son of Rab Judah, added that, "A male branch was grafted on to the female palm tree," which means, as discussed earlier, that a male inflorescence was inserted into the female inflorescence. Presumably, artificial pollination was regarded by the later rabbi as a more effective means of curing the tree of barrenness. See Goor, A. (1967), History of the date through the ages. *Economic Botany* 21:332–334.

61. Quotes taken from Black, et al. *The Electronic Text Corpus of Sumerian Literature*.

62. Ibid.

63. Roaf, Michael (2000), *Cultural Atlas of Mesopotamia and the Ancient Near East*.

64. A pair of rays sometimes emanate from the ringed sun-disk terminating in fruit-like objects.

65. Abdel-Salam, M. (1937), From Palestine days to the Moslem Invasion, in *An Outline of the History of Agriculture in Egypt*. Grunberg, p. 35.

66. Because the vowels for ancient Egyptian words are not known, we can't tell how they were pronounced. However, the Russian letter *з* stands for a kind of glottal stop, whereas the dots indicate the boundaries between signs, which are essential for understanding the words. (Gonzalo Rubio, personal communication.)

67. Buhl, M. L. (1947),The goddesses of the Egyptian tree cult, *Journal of Near Eastern Studies* 6:80–97; Garner-Wallert, I. (1962), *Die Palmen im Alten Ägypten*. Verlag Bruno Hesserling, pp. 33–49.

68. Wasilewska, E. (2000), *Creation Stories of the Middle East*, Jessica Kingsley Publishers, p. 192.

69. Roth, A. M. (2000), Father earth, mother sky: Ancient Egyptian beliefs about conception and fertility, in A. Rautmanm, ed., *Reading the Body: Representations and Remains in the Archaeological Record*, University of Pennsylvania Press, pp. 227–232.

70. Hassan, F. A. (1998), The earliest goddesses of Egypt, in L. Goodison and C. Morris, eds., *Ancient Goddesses: The Myths and the Evidence*. University of Wisconsin, pp. 98–112.

71. The Project Gutenberg EBook of Egyptian Literature. *The Book of the Dead*. Dominion Over Elements. From Papyrus of Ani (British Museum No. 10,470, sheet 16).

72. Shabti are funerary figurines placed in tombs along with grave goods. They functioned as servants for the deceased, performing whatever tasks might be expected of the deceased in the afterlife. When multiple shabti were included, they were housed in special "shabti boxes."

73. This is not to say that Egyptian male deities were never associated with plants. Some male gods were identified with trees in Egypt, although they were never materially equated with the trees, as were the goddesses Hathor, Nut, and Isis. For example, Osiris, the god of death and regeneration, was identified with grain, and the agrarian God Min is associated with a ritual "kitchen garden."

74. Number 1966, Vernus, Chants d'Amour, pp. 83–36, cited in Antelme and Rossini, *Sacred Sexuality in Ancient Egypt*, p. 87.

75. "Genius" is a Latin term that, in pagan religions, referred to "the tutelary god or attendant spirit allotted to every person at his birth, to govern his fortunes and determine his character, and finally to conduct him out of the world; also, the tutelary and controlling spirit similarly connected with a place, an institution, etc." (Oxford English Dictionary).

76. Mallowan, B. P. (1983), Magic and ritual in the northwest palace reliefs, in P. O. Harper and H. Pittman, eds., *Essays on Near Eastern Art and Archaeology in Honor of Charles Kyrle Wilkinson*. Metropolitan Museum of Art, New York.

77. The cone of a conifer tree was called *terinnatu* by the Assyrians and is explained in the lexical text as "the seed of the ashuhu tree." During Ashurnasirpal's time, the wood of the ashuhu tree was used to make tall doors for temples or palaces. Assyriologists have concluded that the ashuhu tree is a pine, possibly the Aleppo pine, *Pinus halepensis* (Postgate 1992). For a comprehensive overview, see Giovino, M (2007), *The Assyrian Sacred Tree: A History of Interpretations*. Academic Press, Fribourg/Vandenhoeck & Ruprecht, Göttingen.

78. He was even credited with founding the field of anthropology itself (Lowie, R. H. [1917], *American Anthropologist* 19:262–268).

79. There is an interesting story behind the specimen shown in Tylor's photograph. Living in England, Tylor did not have direct access to male date palm inflorescences, so he sought the aid of his friend and fellow Quaker, Sir Thomas Hanbury. About twenty years earlier, the wealthy Sir Thomas had, on a whim, purchased the Orengo Palace during a vacation trip to the Italian Riviera. The building was located on a promontory on the Cape of Mortola overlooking the Mediterranean, a short distance from the town of Ventimiglia, just east of Monaco. Sir Thomas, who was an avid gardener and whose brother, Daniel, was a well-known pharmacist and botanist, had the idea of converting the property to a botanical garden that would grow exotic species from around the world. He began buying up the surrounding land and soon had acquired 45 acres of real estate, which was bordered on three sides by mountains covered with Allepo pines. By the late

1880s, Sir Thomas and his brother Daniel had amassed a goodly collection of exotic plant species, including a few male date palm trees. Thus, he was able to provide Tylor with several specimens of male date inflorescence clusters. (The Hanbury Botanical Garden is still in existence, although it is now owned by the Italian government and administered by the University of Genoa. It has a large collection of palm trees.)

80. Mazar, A. (1990), *Archaeology of the Land of the Bible, 10,000–566 BCE*, Doubleday York; Finkelstein, I., and N. Asher Silberman (2002), *The Bible Unearthed*. Simon and Schuster.

81. Jeremiah 44:15–19: "However, all the men who knew that their wives were burning incense to other gods, all the women standing by—a great assembly—and all the people who were living in the land of Egypt at Pathos answered Jeremiah, 'As for the word you spoke to us in the name of Yahweh, we are not going to listen to you. Instead, we will do everything we said we would: burn incense to the Queen of Heaven and offer drink offerings to her just as we, our fathers, our kings and officials did in Judah's cities and in Jerusalem's streets. Then we had enough food and good things and saw no disaster, but from the time we ceased to burn incense to the Queen of Heaven and to offer her drink offerings, we have lacked everything, and through sword and famine we have met our end.'

"And the women said, 'When we burned incense to the Queen of Heaven and poured out drink offerings to her, was it apart from our husband's knowledge that we made sacrificial cakes in her image and poured out drink offerings to her?'" (In some translation the word "drinks" appears as "wine.")

82. Judges 4:4–5. *The New Oxford Annotated Bible*, Augmented Third Edition (2007), M. D. Coogan, ed., Oxford University Press.

83. Mandrake (*Mandragora officinarum*) fruits, referred to as "love apples" in the *Song of Songs* 7:13, was used as an aphrodisiac in ancient Egypt. In an ancient Egyptian love song, the male lover sings: "If only I were her Nubian maid/her attendant in secret/[She would let me bring her love-apples/when it was in her hand, she would smell it/and she would show me] the hue of her whole body. Keel, O. (1994), *The Song of Songs: A Continental Commentary*. Augsburg Fortress, Minneapolis.

84. Translation from Fox, M. V. (1985), *The Song of Songs and the Ancient Egyptian Love Songs*. University of Wisconsin Press, p. 9.

85. Idem., p. 38

86. The ancient Egyptian fruit tree "persea" (*Mimusops schimperi*), belonging to the family Sapotaceae, is a smallish evergreen tree that produces small yellow fruit. According to Theophrastus, it was common in Upper Egypt.

87. Translation from Fox, *The Song of Songs and the Ancient Egyptian Love Songs*, p. 15.

88. Bloch, A., and C. Bloch (2006), *The Song of Songs: The World's First Great Love Poem*. Modern Library.

89. Ibid.

90. Augspach, E. A. (2004), *The Garden as Woman's Space in Twelfth and Thirteenth-Century Literature*. Edwin Mellon Press.

91. Aromatic essential oil used as a perfume and medicine since ancient times, from the flowering plant *Nardostachys jatamansi*, a member of the honeysuckle family.

92. Bloch and Bloch, *The Song of Songs*.

93. Ibid.

94. Ibid.
95. The *New Oxford Annotated Bible*, Augmented Third Edition, p. 959.
96. Bloch and Bloch, *The Song of Songs*.
97. The Mishnah (Yadaim 3:5).
98. Babylonian Talmud, Sanhedrin 12:10.

6

Mystic Plants and Aegean Nature Goddesses

THE HISTORY OF ancient Greece is inseparable from the histories of the other Aegean lands, including Crete and the roughly 100 smaller islands scattered throughout the Aegean Sea and the Ionian coast of Asia Minor (Anatolia or modern day Turkey). The Minoans of ancient Crete reached the height of their power and influence several hundred years ahead of the inhabitants of the Greek mainland, the Mycenaeans, perhaps because of their early trade relations with the older and more advanced civilization of Old Kingdom Egypt.[1]

From about 2000 to 1600 BCE, the Minoans were more unified, prosperous, and internationally connected than their Mycenaean neighbors on the mainland.[2] In addition, the Minoans had developed an exceptionally graceful style of painting, as well as a symbolically rich religion focused on a number of nature goddesses. Sacred trees and flowers, especially saffron lilies, were central to Minoan rituals and iconography, along with an emphasis on female spirituality.

Minoan wealth was derived from extensive commerce throughout the eastern Mediterranean, made possible by a mastery of the many-paddled Aegean longboat. They also founded settlements along the coast of Asia Minor. Among the artifacts found in Laconia on the southern edge of the Peloponnese was an abundance of Minoan-style pottery, suggesting the presence of Minoan artisans on the Greek mainland during this period.[3] Minoan artists may also have been recruited to fashion some of the art objects found in the Shaft Graves of Mycenae. Shaft graves, which are relatively rare on the Greek mainland, consist of a vertical shaft leading to an enlarged stone-lined *cist* grave with a roof made of timbers or reeds. At Mycenae, the presence of expensive grave goods and weaponry provides evidence for a strongly stratified warrior society by 1600 BCE.

Around 1525 BCE, a new, more extravagant form of burial for the ruling elite was initiated at mainland centers like Mycenae, which were possibly modeled on smaller versions on Crete.[4] These were the large, domed *tholos* or *beehive* tombs, filled with costly burial goods. Wealth of this magnitude could only have been accumulated through extensive international trade, which suggests that a relatively secure economic system was prevalent. Some regional fighting did occur, but the traditional image of constant warfare and violence on the mainland is at variance with many signs of stability among the numerous settlements not directly attached to the palaces.[5]

On Crete, ash and pumice from the volcanic eruption on the island of Thera—either ca. 1650–25 or 1550–25 BCE—may have disrupted life sufficiently to demoralize Minoan society, which was apparently becoming increasingly integrated with their Mycenaean neighbors. Whatever the cause, Crete's long history of independence came to an end around 1450 BCE. Mycenaeans took control of the island and rebuilt the Palace at Knossos.

Finally, around 1200 BCE, amidst widespread conflagration and population shifts throughout the eastern Mediterranean, which have yet to be satisfactorily explained, most of the Bronze Age kingdoms either collapsed or were severely weakened as city after city went up in flames. Perhaps the best examples of the violence that took place during this period are found on the Greek mainland, where many of the major Mycenaean sites were put to the torch. Mycenaean palace society, along with its Linear B script, disappeared abruptly, creating a social vacuum that would take several hundred years to fill.

THE MINOAN PALACE AT KNOSSOS IN CRETE (2000–1200 BCE)

Knossos is the home of the mythic King Minos of Crete and of the Minotaur—a fearful monster, half-human and half-bull,[6] whom King Minos imprisoned in an escape-proof Labyrinth constructed by Daedalus, the "cunning artificer" of wax-wing fame. The myth turns history on its head by portraying Crete as the cruel oppressor of Athens, annually sacrificing fourteen Athenian boys and girls to the Minotaur. According to the myth, Theseus—with the aid of King Minos's daughter, Ariadne, who provided him with both a sword and a ball of red wool thread, which he used to retrace his steps and escape the Labyrinth—put an end to the annual sacrifice by slaying the Minotaur.

The idea of the Labyrinth may have been suggested to the Greeks by the elaborate ruins of the Palace of Knossos, with its bewildering maze of hallways and compartments.[7] Whether there ever was a king named Minos is still unknown. When Sir Arthur Evans first unearthed the vast archaeological site at Knossos in 1900, he named the civilization that had produced it the "Minoan." The term was initially criticized on the grounds that it is inappropriate to name an entire civilization after a mythic king, but Evans countered that "Minos" was a royal title rather than a surname.[8] According to Evans, the term "Minoan" refers to the type of rule in Bronze Age Crete, just as "Pharaonic" does in Egypt.

Careful study of Minoan archaeological sites throughout Crete and the Aegean islands has revealed a civilization somewhat different from the one originally envisioned by Evans. Indeed, Crete may have been governed by a ruling elite rather than by a king. Instead of serving as a royal palace, archaeologists now believe the buildings at Knossos represented a complex with both administrative and religious functions. Administratively, Knossos was at the center of a "redistributive" economic system. In the absence of metal coinage, taxes in the form of agricultural products and other goods were collected by the ruling elite. The "palace" functioned as the central clearing house for the goods to be redistributed to the populace, and scribes kept detailed accounting records, first in Linear A, and later (after the Mycenaean takeover) in Linear B.[9] In this way, the ruling elite maintained tight control

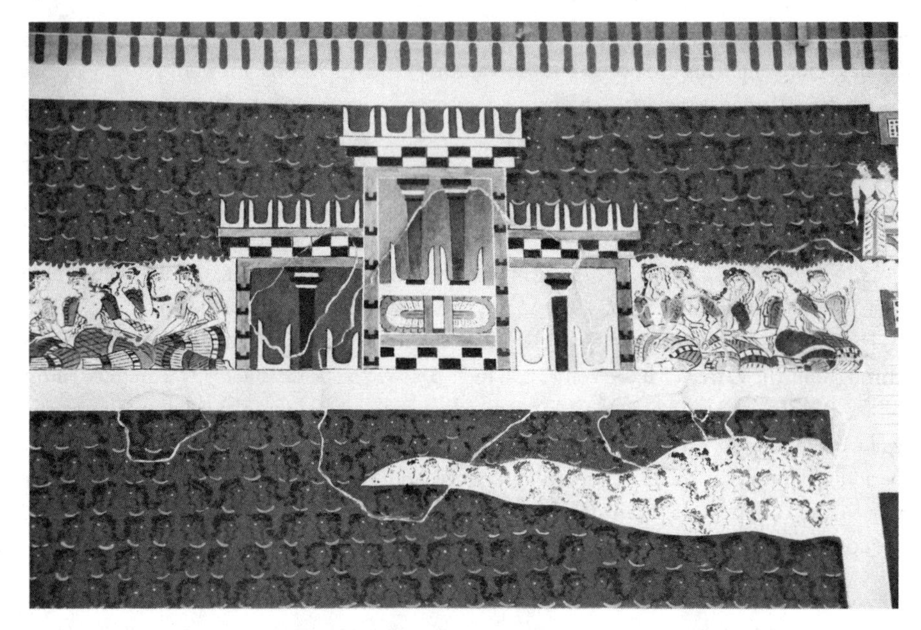

FIGURE 6.1 Reconstruction of the Grandstand fresco from the Second Palace Period in Knossos. Note Horns of Consecration on the tops of the tripartite shrine. Women and men are color-coded white and red, respectively. Priestesses preside over the ceremony on the main stage. From Herakleion Museum. Photo by Barry D. Kass@ImagesofAnthropology.com.

over economic resources while maintaining domestic tranquility by ensuring that the general population was adequately provided for.

THE HORNS OF CONSECRATION

That the palace at Knossos also served as a religious center is evidenced by the many beautifully painted murals portraying ritual activities within the palace–temple complex, as well as architectural elements and statuary consistent with sacred themes. The Horns of Consecration motif, reminiscent of the Neolithic bucrania of the Near East, is believed to be a sacred symbol representing the ritual practice of bull sacrifice. The reconstructed Grandstand fresco from the Second Palace period at Knossos shows a tripartite pillared shrine crowned with rows of the double-horn motif (Figure 6.1). However, the focus of the ceremony is unclear and may have nothing to do with the Horns of Consecration motif. The priestesses leading the ceremony are shown in white on the center stage. Their ceremonial garb, portrayal as complete figures, and position beside the shrine are consistent with their central importance as spiritual leaders. The audience consists of both men (the dark figures) and a smaller contingent of women (the white figures).[10]

GENDER AND SACRED TREES

Some rituals were performed near sacred trees, as shown in the Sacred Grove fresco from the West Court in Knossos (Figure 6.2). As in the Grandstand fresco, dancers perform in multicolored flounced dresses, with their right arms lifted in adoration toward the

FIGURE 6.2 Sacred grove frescoes from Knossos, Crete. A. Reconstruction of detail from upper portion of the Sacred Grove fresco. B. Reconstruction of the lower portion of the Sacred Grove fresco, showing women dancing.
Photographs by Barry D. Kass. See discussion by Preziosi, D., and L. A. Hitchcock (1999), *Aegean Art and Architecture*. Oxford University Press, p. 81.

"sacred grove," possibly an olive orchard, while larger priestesses observe from beneath the sacred tree. Again, the central role of women in Minoan religious life is evident, especially in rituals involving trees and, as we shall see, flowers. The West Court area also contains two large lined circular pits that may have been used to store grain.[11] In contrast to the Grandstand fresco, the women in the audience (in white) are more numerous and are seated closest to the trees, while the men (in red) are relegated to the back rows.

Another illustration of the association between women and trees can be seen on the North Wall frieze of the West House in Akrotiri on the island of ancient Thera,[12] an important island within the orbit of the Minoan thalassocracy (Figure 6.3). The entire frieze depicts a coastal settlement, but the detail shown here features men tending herds of animals on the right, and women with vessels on their heads walking past an enclosure with two fruit trees on the left. The fruit trees are almost certainly domesticated figs, *Ficus carica*, which were an important food staple for both Minoans and Mycenaeans. According to Lyvia Morgan, the fenced area is an animal pen, and the function of the trees is to provide shade for the animals.[13] However, there is a visual coupling of the two fruit-bearing fig trees with the two women, who appear directly below them. In addition, the outstretched arms of the women, which parallel the curve of the fence, are reminiscent of the dancers in the Sacred Grove fresco (Figure 6.2A), possibly indicating that the scene has ritual as well as aesthetic significance.

As in the Bronze Age societies of the Near East, orchards, more than cereal crops, came to symbolize the prosperity and spiritual well-being of the state in Minoan Crete. Minoans also followed the Near Eastern Bronze Age tradition of associating sacred trees with priestesses or goddesses. Among the cultivated fruit trees and berries were olives, figs, almonds, grapes, and possibly pistachios. Both figs and pistachios are dioecious,[14] so the Minoans would certainly have been aware of the existence of two sexes in these species, however they interpreted them. Whether or not they used caprification—attaching a branch of the caprifig to supply pollen to the female trees via the fig wasp (see Chapter 8)—is unknown. There is also some rather tenuous evidence that the Minoans cultivated date palms and practiced artificial pollination.[15] But despite the frequent illustrations of palm trees in Minoan and Mycenaean art, we are aware of none with hanging clusters of ripened fruit, as was typical in date palm depictions of the Near East. Thus it seems likely that Aegean palm tree illustrations represent the indigenous wild species of palm, *Phoenix theophrasti*, rather than the cultivated date palm, *P. dactylifera*.[16]

RING SEAL IMAGES OF MINOAN NATURE GODDESSES

Signet rings or seal rings bearing incised or carved images that can be pressed onto soft wax or clay were used throughout Mesopotamia, the Near East, and the Aegean for identification purposes during the Bronze Age. In Crete, these ring seal images often portrayed nature goddesses or their priestesses. The ring seal in Figure 6.4A, for example, which has been called "Mother of the Mountain," shows a goddess in a flounced skirt standing upon a symbolic mountain. It is clear from her commanding pose, the shrine behind her, the

FIGURE 6.3 Detail from North Wall frieze depicting a naval engagement, from the West House in Thera. Two women bearing jars pass by a fenced grove on the left, while men herd animals on the right.
From Marinatos, N. (1984), Art and Religion in Thera: Reconstructing a Bronze Age Society. Athens.

two apotropaic animals on either side of her, and the figure of the worshipper on the right, that the scene represents the epiphany, or manifestation, of a goddess. Although religious activities were directed by priests and priestesses from the palaces, the use of natural settings such as mountain tops and caves as sanctuaries was a distinctive feature of Minoan religion.

Figure 6.4B shows a female figure grasping the stem of a lily and tilting a blossom to her nose to smell it. The plant seems to be growing directly out of the horns of consecration on top of an altar. According to classicist Nanno Marinatos, the fact that she is smelling the flower and not just bringing it to the altar indicates that she is a goddess rather than a priestess.[17] Figure 6.4C shows a goddess or priestess in a flounced skirt flanked by a shield on the right and a dancing man worshipping a sacred tree growing from an undefined structure.

Perhaps the most famous of the Minoan seal-rings was found in a tomb at Isopata near Knossos (Figure 6.4D). It depicts four women wearing flounced skirts dancing (with visible feet) in a flowery meadow along with a snake-like form and a conical object. The presence of four lily plants corresponds to the four dancing women. The heads of the dancers are indistinguishable from their necks, which enhances the mystical, ecstatic quality of the scene. The strings of "beads" sprouting from the abbreviated heads probably represent the dancers' hair. The middle figure, although dressed like the others, may represent a goddess, based on her elevation relative to the others and her central position, while the three lower figures may be her priestesses. In addition, a much smaller fifth figure to the goddess's right has been interpreted as a second goddess descending to earth.[18] Archaeologist Christine Morris has argued that the odd heads and floating objects are manifestations of an ecstatic trance, indicative of a shamanic dimension in Minoan art.[19]

FIGURE 6.4 Minoan ring seals showing various nature and plant deities, c. 1500–1400 BCE. A. "Mother of the Mountains," reconstructed from fragments found at Knossos. B. Priestess or goddess grasping lilies; Peloponnese, probably of Minoan origin. C. Goddess and dancing man worshipping a sacred tree, from Vapheio near Sparta. D. Priestesses or goddesses dancing in a flowery field, Knossos, Isopata Tomb 1.

A and B are from Marinatos, N. (1993), *Minoan Religion: Ritual, Image, and Symbol*. University of South Carolina Press (figure 123); C and D are from Krzyszkowska, O. (2005), *Aegean Seals: An Introduction*. Institute of Classical Studies, University of London (figures 5a, 221b, and 215a.)

(d)

FIGURE 6.4 Continued

FRESCOES OF FLOWER RITUALS INVOLVING WOMEN

Crete is blessed with the richest flora in Europe, with more than 1,800 species, 10% of which are endemic. Crete's amazing floral diversity derives from two main factors: its former physical connection to both the European and Asian continents, and its extraordinary microclimatic diversity resulting from its many natural habitats, which include mountains, gorges, valleys, prairies, and coastal areas. Not only is Crete's flora unusually rich, but flowers can be found blooming at all times of the year: in the prairies in spring, in the mountains in summer and fall, and along the shore in winter. Thus, it is not surprising that flowers figure so prominently in Minoan art and religion.

In his book, *Minoan Religion as Ritual Action*, Peter Warren analyzed the ritual use of flowers in Crete based on the many painted frescoes of Crete, the island of Thera (located about 80 miles north of Crete), and Mycenae on mainland Greece.[20] Warren noted four distinct stages of the Aegean flower ritual: gathering, preparation, procession, and presentation. In each case, women were the primary participants, reinforcing the idea that women were strongly associated with flowers, as was evident in the ring seal images presented earlier. The stages of the flower ritual are discussed next.

Gathering

Two beautiful frescoes illustrate the first stage of gathering. The first example, from the villa of Ayia Triada in Crete, shows "a goddess seated upon a platform, supervising a scene of flower gathering conducted by women" (Figure 6.5). Whether or not the woman on the left is a goddess or a priestess is immaterial: what is important here is her close association with flowers. Lilies, crocuses, and ferns are abundant in the landscape, which also includes several types of animals: goats, cats, and birds. Interestingly, the animals and ferns are restricted to the right half of the panorama and are absent from the left half, where she is kneeling. Since lilies and crocuses do not bloom together in nature, the scene most likely depicts a symbolic or "magical" garden.[21]

The second example is from the Late Bronze Age Cycladic town near the modern village of Akrotiri on the volcanic island of Thera (modern Santorini), where an immense eruption

FIGURE 6.5 Fresco from Ayia Triada, showing kneeling goddess and her attendant gathering flowers. Reconstructed by M. Cameron.
From N. Marinatos (1993), *Minoan Religion: Ritual, Image, and Symbol*. University of South Carolina Press (figure 121).

took place, spewing enormous amounts of ash throughout the southern Aegean. Ancient Akrotiri was buried in ash, and, like Pompeii, it was thereby preserved nearly intact for future generations.

As early as 1939, archaeologist Spyridon Marinatos had speculated that something interesting was buried at the site, but he was prevented from excavating it until 1967 because of the outbreak of World War II and the Greek Civil War. Once the ash from the volcano was removed, Marinatos and his successors discovered part of a large town, complete with multistory buildings, streets, squares, and walls. The presence of a loom workshop showed that it was a center for the production and export of textiles. Mills and huge ceramic jars (pithoi), used primarily for storing grain and olive oil, were also found. Most spectacular of all are the many beautiful polychrome frescoes, some of which illustrate people and animals in natural settings, with others depicting ritual or mythological themes.

The Saffron Crocus Gatherers fresco (Figure 6.6A) can be identified as a ritual scene on the basis of two criteria. First, the stigmas of the flowers, which were used to make saffron,[22] are destined to be presented to a goddess or priestess. Second, the women's attire is similar to the ritual robes of the flower gatherers in the fresco from Ayia Triada, which also includes a goddess or priestess (see Figure 6.5). Such elaborate costumes take on the aura of religious vestments. The rugged terrain in the fresco suggests that the species the maidens are collecting is wild crocus, *Crocus cartwrightianus*, which typically grows on rocky hillsides up to 1,000 meters in altitude. A drawing of *C. cartwrightianus* is shown in Figure 6.6B.[23] Note that the single stigma is divided into three long branches.

The saffron crocus *Crocus cartwrightianus*, a member of the iris family, produces a bulb-like underground storage stem called a *corm* and blooms from October to December.[24] From this, we can identify the season depicted in the Crocus-Gathering fresco as autumn or early winter.

It is worth noting that the modern cultivated species of the saffron crocus, *Crocus sativum*, is a sterile triploid mutant (having three sets of chromosomes instead of two) that can only be propagated asexually. It is thought to be derived from the wild *Crocus cartwrightianus* of Crete. Presumably, growers initially selected for plants with elongated stigmas to increase the yield of saffron, which led to the selection of the triploid mutant plant. When this artificial selection took place has yet to be determined. However, based on the frequency of white (albino) crocus flowers in Minoan art, plant systematist Brian Mathew has speculated that the Minoan women of the Thera frescoes

(a)

(b)

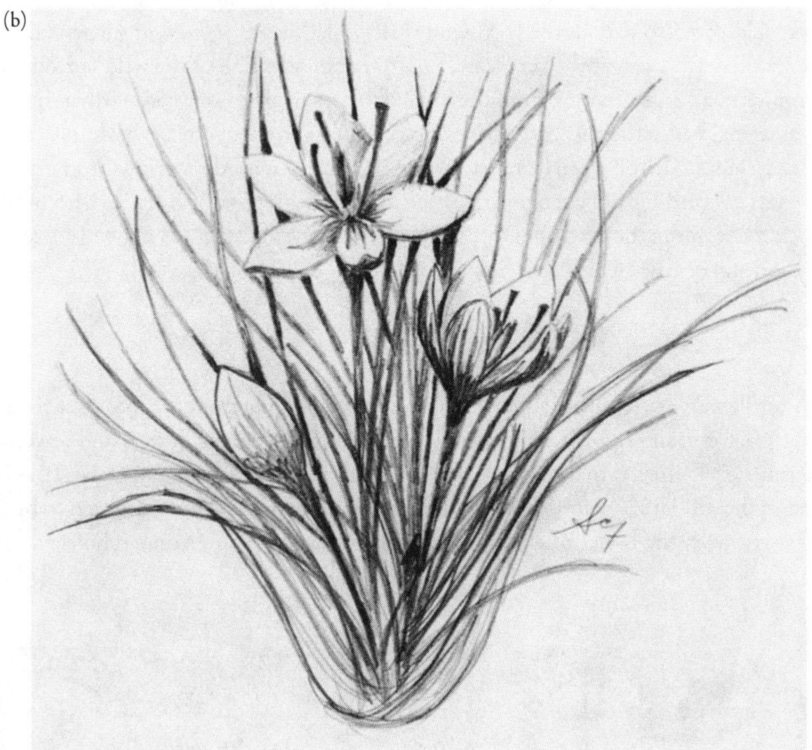

FIGURE 6.6 A. Detail of the crocus-gathering scene, from a painting above a lustral basin on the east wall of Xeste 3, Thera. B. *Crocus cartwrightianus*.
A is from Shaw, M. C. (1993), The Aegean garden. *American Journal of Archaeology* 97:661–685; B is from Ference, S. C., and G. Bendersky (2004), Therapy with saffron and the goddess at Thera. *Perspectives in Biology and Medicine* 47:199–226.

were most likely collecting a cultivated variety of *C. cartwrightianus* rather than the triploid mutant *C. sativus:*

> The origin of the *C. sativus* clone which exists today is unknown but it is highly probable that it is the same as that grown in England in the fourteenth century. There is even the possibility that it was known as long ago as 1600 B.C., for at Knossos in Crete there exist designs on Minoan frescoes and pottery in which a Crocus with simple long-exserted [protruding] red stigma branches is depicted. No Crocus exhibits this feature better than *C. sativus* although in some forms of *C. cartwrightianus* the branches are a little longer than the perianth segments. It is of course impossible to say whether the Minoans were cultivating the actual clone which still exists today, or a form of *C. cartwrightianus*, a species which occurs naturally on Crete. The argument is perhaps in favour of the latter case since some of the Crocus depicted are white-flowered and this species has a marked tendency to produce albinos. Whichever case is true, it is apparent that the Minoans possessed a plant which had exceptional stigma development, capable of a far better yield of saffron than the local wild *C. cartwrightianus.*[25]

Mathew's conclusion that the flowers "had exceptional stigma development" with a "far better yield of saffron" than the local wild variety of *C. cartwrightianus* suggests that the Minoans had achieved, at the very least, a partial domestication of the wild saffron crocus by around 1600 BCE. Judging from the Thera frescoes, women were primarily responsible for gathering, hence the *selection,* of the domesticated variety of crocus, both in Crete and in the Cyclades. Saffron was a lucrative crop for Crete, both for domestic consumption and for export. Accordingly, there must have been planted fields of saffron crocus with ordinary farmers performing the labor, but evidently mundane agriculture was not considered suitable for ritual or religious art.

Preparation

Plants or flowers are frequently shown on ring seals emerging from the tops of shrines (see Figure 6.4 for examples). This suggests that potted plants or cut flowers and leaves were sometimes used directly to decorate shrines. Flowers were also woven into garlands, as illustrated in the miniature painting shown in Figure 6.7. Their ritual use is attested to by their association with the bones of children and cult vessels in a Late Minoan house near the

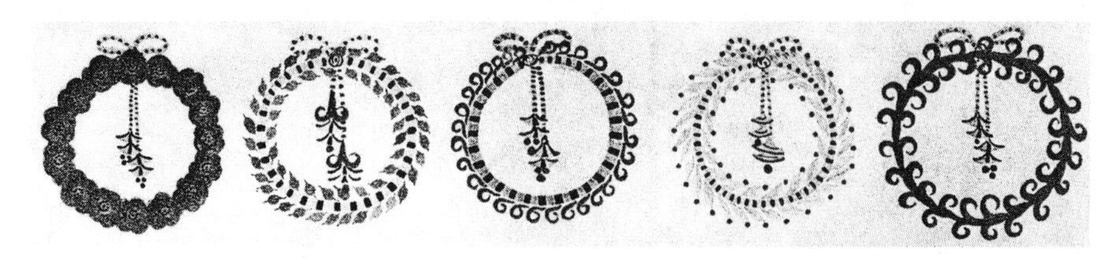

FIGURE 6.7 Fresco of garlands from Knossos (1450 BCE).
From Warren, P. (1988), *Minoan Religion As Ritual Action*. Studies in Mediterranean Archaeology and Literature. P. Astroms, Gothenburg.

Palace at Knossos. Warren identified candidates for the flowers that were woven into ritual garlands: anemone, poppy, ranunculus or rose, dittany, olive or myrtle, and red lily (Figure 6.7). Reconstruction of fragments of another garland wall painting from Knossos led to the identification of *Rosa canina* (Dog rose), *Cistus incanus creticus* (Cretan rockrose), and either the wild Crocus, *C. cartwrightianus*, or the sterile domesticated form, *C. sativus.*[26] As noted by Peter Warren, most of the flowers and plants in these garlands were either sacred to or associated with one or more deities in Crete.

Crocus blossom gathering was a ritual activity, but, as noted earlier, only the branched red stigmas were used to make saffron (Figure 6.8A,B). This raises the question whether the gatherers are collecting whole flowers or just the stigmas. Since about 60,000 stigmas are needed to produce a pound of saffron, the most efficient method of saffron production would be to gather the flowers first and remove the stigmas afterward (as is done today). However, as shown in Figure 6.6A, the girls appear to be pinching the stigmas out of the

(a)

(b)

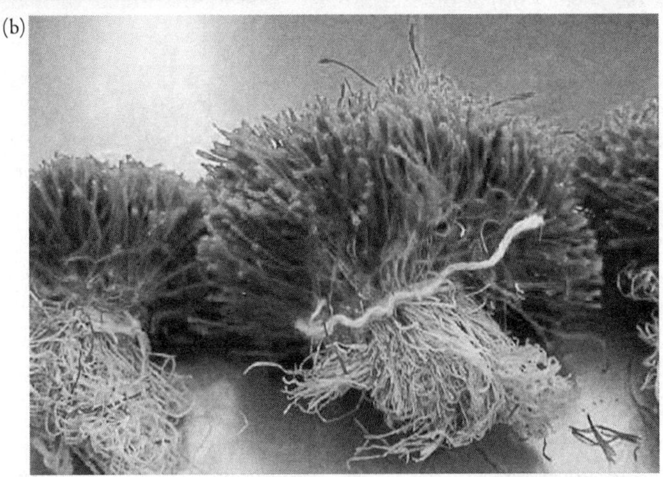

FIGURE 6.8 A. *Crocus sativus* flower. B. Bundles of stigmas from *C. sativus*. Note the intense maroon color of the stigmas.

corollas rather than collecting whole flowers. Either this reflects the actual way the stigmas were harvested in Minoan times, or the artist chose to emphasize the symbolic aspects of the scene rather than the agricultural details. The Theran frescoes were meant to be visual poems of sacred themes, with mythic figures set in dream-like landscapes, caught in moments of high drama and portent. Perhaps the labor-intensive preparative phase of stigma removal from harvested blossoms was omitted for aesthetic/dramatic reasons.

Processions

The gathering and preparation stages were followed by a procession of women, richly dressed and bearing the flowers. Such a procession involving many participants is symbolically represented by the lone woman carrying a basket of flowers on her shoulder, on the wall adjacent to the Crocus Gatherers fresco in Xeste 3 (Figure 6.9A). In the background one can see the hills and precipices from which the blossoms were collected. The ring seal from Mycenae (Figure 6.9B) shows three women approaching a shrine. According to archaeologist Bogdan Rutkowski, the women are part of a procession taking place inside a sacred enclosure:

> The cult ceremonies, especially the processions and dances, were held in the sacred enclosures, as we know from the iconographic finds. A gold ring from Mycenae, for instance, shows three women attired in skirts and short-sleeved bodices [see Figure 6.9B]. They are making their way, probably barefoot, in the direction of the sanctuary, and each of them has one hand held high, holding a flower, while in the other hand each is carrying a branch. The scene is meant to represent part of a ceremonial procession in which a large number of worshippers are taking part.[27]

Figure 6.9C shows two small faience robes and girdles from the Temple Repositories at Knossos. Their ornate floral patterns—again, featuring crocuses— suggest that robes such as these may have been worn by priestesses during flower processions. Their location in the Temple Repositories further suggests that they functioned as votive offerings and indicates the high, and possibly sacred, value placed on both the flowers and the textiles they adorn.

Yet another example of a ritual flower procession is a stately Mycenaean fresco from Thebes (Figure 6.10).[28] The reconstructed wall painting, which is more than 40 meters long, shows a procession of women wearing Minoan-style dresses, carrying offerings of flowers (lilies, papyrus, and either roses or rockroses), sealed ivory boxes (*pyxis,* pl. *pyxides*), and a serpentine jar. The background consists of wavy bands of color, which, in Minoan art, represents a mountainous terrain, but which may mean something different to Mycenaeans.

The depiction of a Mycenaean flower procession is suggestive of descriptions of the Eleusinian Mysteries during the Greek Classical Period, especially the processions honoring the grain goddess, Demeter. The day before the official opening of the Eleusinian Mysteries, priestesses of Demeter carried *kistai,* or sacred cists,[29] from Eleusis to Athens. Along the way they stopped at a Sacred Fig tree.[30] The *kistai* were returned to Eleusis as part of the grand procession, which was the high point of the festival for the general public. After arriving at Eleusis, the initiates spent several more days taking part in processions, sacrifices, fasts, and purification rites. At a crucial point in the initiation process, the priest

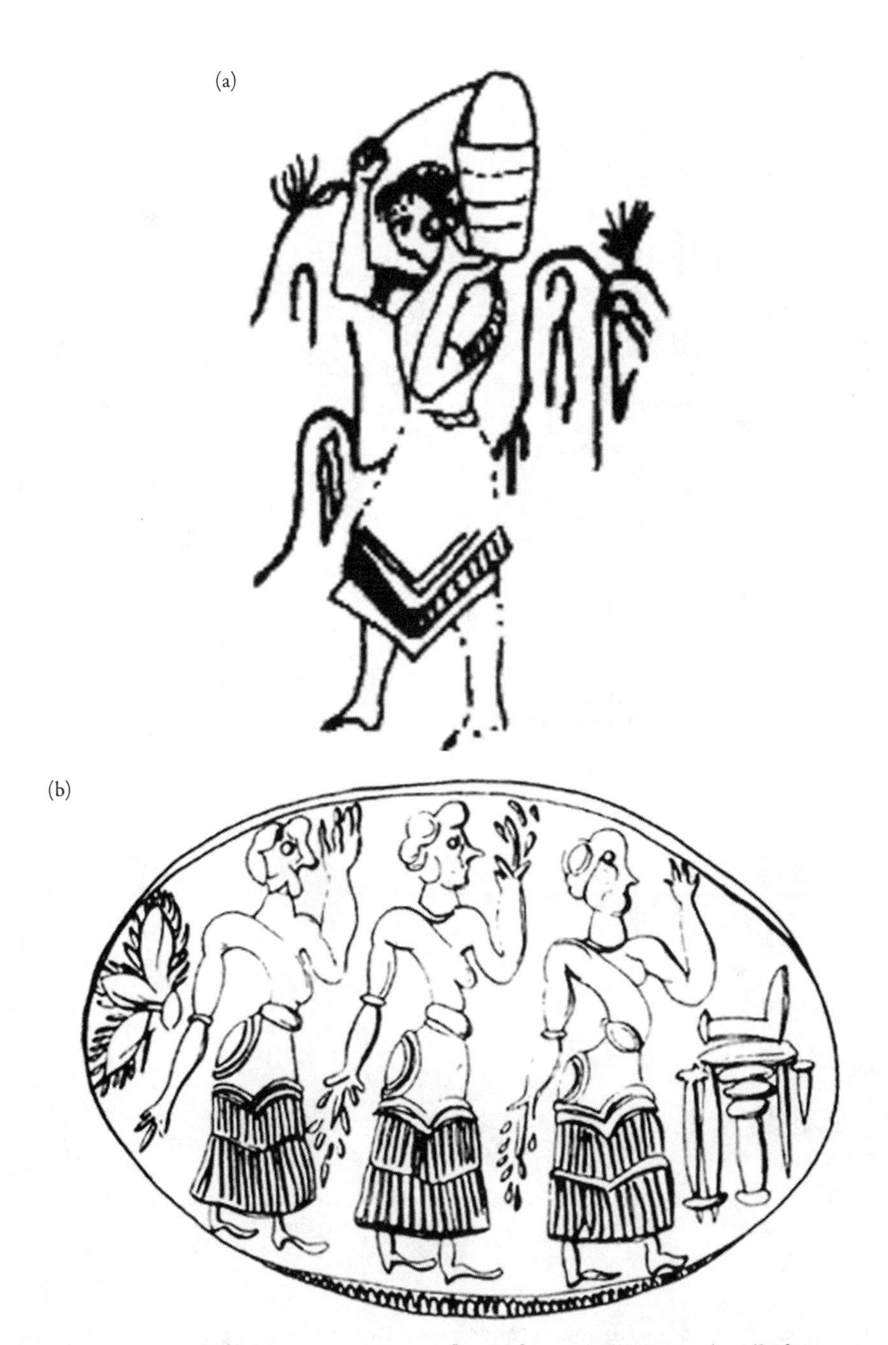

FIGURE 6.9 A. Detail of woman carrying crocus flowers from painting on north wall of Xeste 3, Akrotiri. B. Ring seal of women in flower procession, from Mycenae. C. Small faience robes and girdles decorated with crocuses from Temple Repositories Knossos.
A and B are from Rutkowski, B. (1986), *The Cult Places of the Aegean*. Yale University Place. C is from Marinatos, N. (1993), *Minoan Religion: Ritual, Image, and Symbol*. University of South Carolina Press (figure 123).

(c)

FIGURE 6.9 Continued

FIGURE 6.10 Mycenaean flower procession fresco from Thebes dating to around 1400 BCE. From Reusch, H. (1956), Die Zeichnerische Rekonstruktion des Frauenfrieses im Böotischen Theben. Akademie-Verlag, Berlin.

or priestesses removed sacred objects from the *kistai* and revealed them to the initiates. Very likely the *pyxides* of the Theban flower processions contained sacred objects, similar to the *kistai* of the Eleusinian Mysteries.

Presentation

The fourth stage of the Minoan flower ritual is the presentation or offering of the flowers to a deity or shrine, which occupies the center of the north wall of the house at Akrotri

FIGURE 6.11 Detail of painting from north wall of Xeste 3 in Akrotiri illustrating the presentation phase of the flower ritual.
From Porter, R. (2000), The flora of the Theran wall paintings: Living plants and motifs—sea lily, crocus, iris, and ivy, in S. Sherrat, ed., *The Wall Paintings of Thera: Proceedings of the First International Symposium*. Thera Foundation, pp. 585–618.

named Xeste 3 (Figure 6.11). This stage is clearly the culmination of the entire sequence. We see a goddess seated in the foreground on a pile of cushions on the upper level of a series of platforms representing a sanctuary or shrine. The background consists of a field of crocuses, and the goddess herself has a crocus blossom painted on her cheek. Her divinity is indicated by two attendants: a mythological winged griffin rearing up behind her and a monkey with a basket of stigmas at his foot before her. Behind the monkey, a female crocus gatherer empties her pail of flowers into a large basket. The monkey's basket contains the precious thread-like stigmas, the basis of saffron. The omission of the preparation stage (stigma removal) from the tableau invites us to imagine that the tedious and time-consuming operation of removing the stigmas is accomplished by the magical monkey. The goddess graciously accepts the tangled mass of stigmas from the monkey's cupped hands, but not all at once. With the fingers of her right hand she delicately selects a few at a time and arranges them neatly into a bunch in her left hand, similar to the bundle of stigmas shown in Figure 7.10B. There is a touch of humor here, contrasting the monkey's naive messiness with the goddess's refined neatness, which is also the contrast between the wildness of nature and the decorum of the religious sanctuary.

The frescoes of Xeste 3 clearly illustrate a flower ritual of the type described by Warren. According to one interpretation, the Xeste 3 frescoes represent an initiation ceremony for young girls, coupled with a celebration of the renewal of spring. [31] However, the season shown in the fresco is unclear. Although some of the flowers shown are spring flowers (lilies, iris, and wild roses), crocus blossoms can only be gathered in the autumn. As in the garden of the Ayia Triadha fresco (see Figure 6.5), the season is indeterminate, existing outside of time in the realm of the divine.

Although less conspicuous than the plants, animals—for example, birds, insects, and cats—are also attributes of the goddess. However, the goddess of Xeste 3 may not represent a generalized goddess of nature or of vegetation as is often assumed. According to Ferrence and Bendersky, the seated deity may represent a goddess of healing. This conclusion is based on the following three criteria:

> (1) the unusual degree of visual attention given to the crocus, including the variety of methods for display of the stigmas; (2) the painted depiction of the line of saffron production from plucking blooms to the collection of stigmas; and (3) the sheer number (ninety) of medical indications for which saffron has been used from the Bronze Age to the present. [32]

Of the ninety medical uses of saffron that have been recorded, 14% of them are gynecological, and one of these uses (as an analgesic for menstrual pain) has been confirmed in clinical trials and is still in use on the islands. [33] Although most of saffron's medicinal uses were recorded *after* the Bronze Age, it seems likely, given the prominence of crocus flowers in their art, that the Minoans were aware of some of them. Moreover, the minor role of men in the Xeste 3 frescoes is consistent with the rituals' being focused on women's health and healing, or women as healers.[34]

The so-called Adyton fresco at Akrotiri provides further support for the general theme of women's health (Figure 6.12). It was located on the ground floor immediately below the Crocus Gatherers room and directly above the "adyton" or "holy of holies"—the Greek term for a restricted area within a temple, often housing a statue of a deity, which was accessible only to priests and priestesses.[35] The Adyton fresco probably illustrates a well- known story about a young woman who steps on a thorn while gathering crocus flowers. She sits on a rocky outcrop nearly identical to the one in the Crocus Gatherers fresco in the level

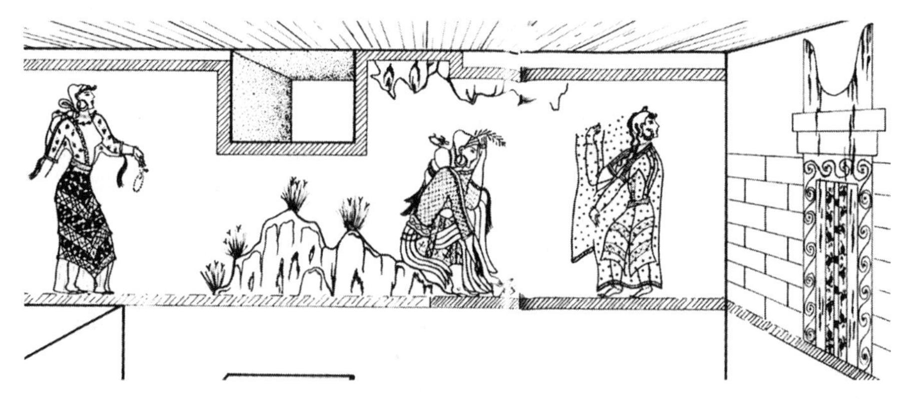

FIGURE 6.12 Sketch of the Adyton fresco on the south wall of room 3 on the ground floor of Xeste 3, Akrotiri. The delineation of different ages of the girls and women in the fresco based on body form, hair, and dress is consistent with an interpretation based on a female rite of passage. It is probably significant that the wounded figure is located directly below the seated goddess shown in Figure 6.11.
From Marinatos, N. (1984), Art and Religion in Thera: Reconstructing a Bronze Age Society. Athens.

above (see Figure 6.6). To show her pain she presses her left palm against her forehead, while rubbing her injured foot, which appears to be bleeding. A crocus blossom next to her foot shows that she has been gathering crocuses. A small, leafy branch projects from a diadem in her hair, and the fingers of her left hand touch it at its base. The shape and arrangement of the leaves are consistent with olive, myrtle, or willow, but the association with pain points to willow. Willow bark and leaves both contain salicylic acid, the basic ingredient of aspirin, which had been used to cure headaches in the Near East since the early Bronze Ages. However, olive oil and myrtle leaves also have medicinal uses, and it's possible that the leafy branch was considered symbolic of several medicinal plants.

As noted earlier, one among the many interpretations of the Xeste 3 frescoes is that they represent a puberty ritual for adolescent girls. The injury to the young woman's foot, which causes bleeding, could be construed as symbolic menstruation. The rocky outcropping on which the woman sits is nearly identical to the terrain in the Crocus Gatherers fresco, perhaps connecting the two frescoes in a single narrative. Symbolic menstruation would be consistent with the hypothesis of Ferrence and Bendersky that the goddess in Xeste 3 is a goddess of healing especially concerned with women's health. However, because saffron was a well-known and valued pharmaceutical in the Mediterranean region, this scene could also reference the function of women as healers with particular knowledge of the uses of botanical medicines.

THE TABLE DANCERS OF PHAISTOS

Two of the most intriguing examples of the association of women and plants in ancient Crete were painted onto two ceramic objects found at the Palace at Phaistos dating to the Old Palace period (c. 1900–1700 BCE). The first is a reconstructed pedestal table with a basin used for ritual offerings (Figure 6.13). All of the painted anthropomorphic figures on the offering table are bird-women. Three dancers are illustrated on the upper surface—a larger central figure with a flower in each hand, and two smaller figures on each side, each with one arm curved downward and the other raised over her head. Three blossoms with prominent stamens attached to the corollas enclose the dancing bird-maidens on three sides. Additional female figures are painted along the rim of the table and also around the base. The figures along the rim are all bending over to the ground, possibly picking flowers, while those on the base are shown with arms akimbo. According to Goodison and Morris, the ritual significance of the table is suggested not only by the meticulous craftsmanship required to make it, but by the figures themselves:

> Several features of the design itself also suggest more than an everyday scene of people dancing. The all-female cast suggests a special dance performed by a particular group. The repeated bending gesture around the rim shows no practical purpose; since repetition can be a defining feature of ritual action, this repeated gesture may be symbolic. Moreover, the heads of the figures seem to have beaks like birds—a feature noticeable in other Cretan designs, perhaps indicating a bird-mask or some ritual headgear. Lastly, the gesture of the central figure with two arms raised is one which in later images usually indicates a goddess. . . . Whether she is a priestess or a goddess, the scene seems clearly focused both on her and on what she is holding up in her hands: two pieces of vegetation in flower.[36]

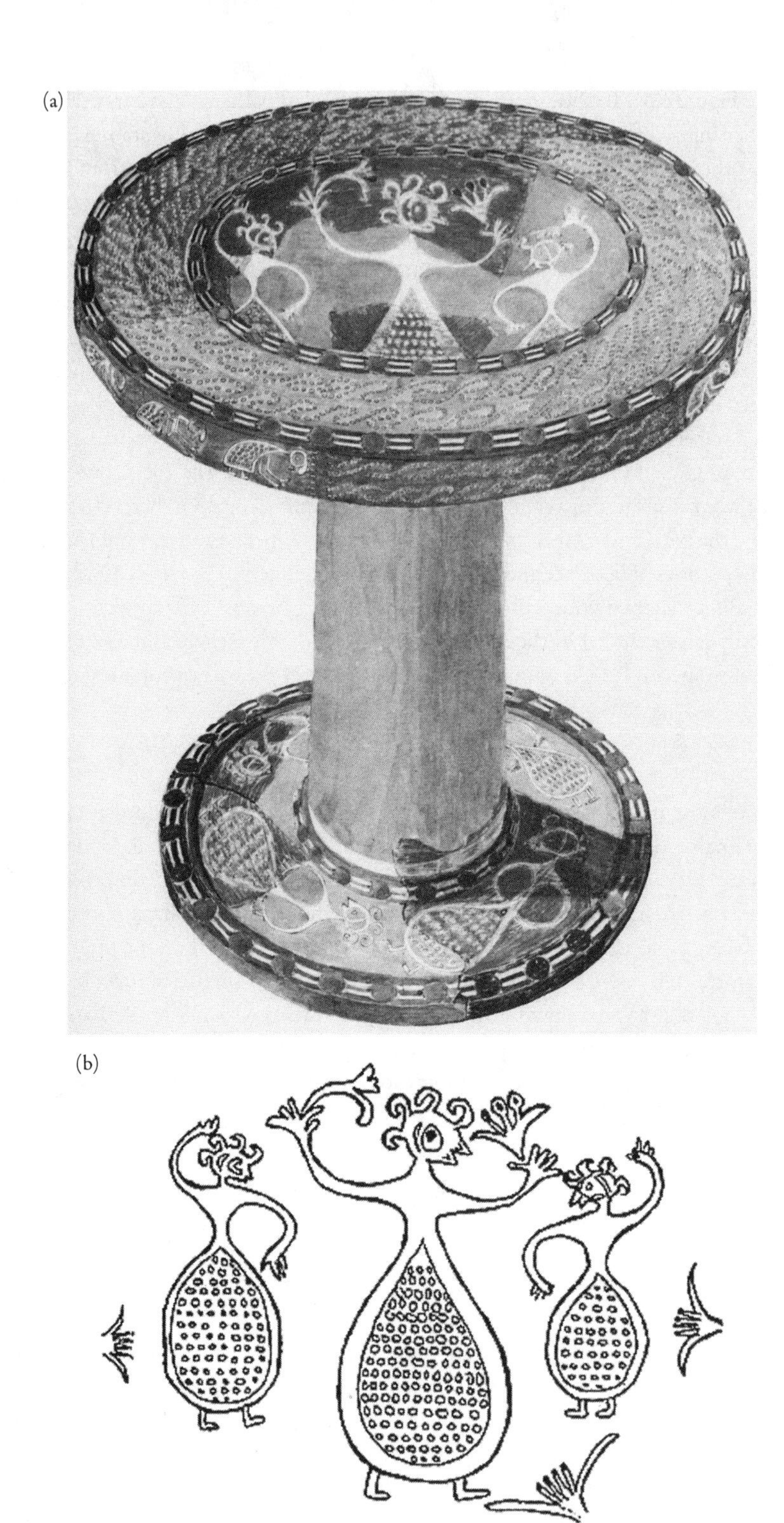

FIGURE 6.13 Ceramic offering table from Phaistos. A. Painting of reconstructed table. B. Sketch of the design on the table top.
From Levi, D. (1976), *Festòs e la civiltà minoica*. Vol. I, Edizioni dell'Ateneo, Roma.

The presence of numerous small seed-like dots within the lower, swollen halves of the figures suggests the idea of fruits, perhaps pomegranates. The inclusion of multiple lilies in the design may indicate that this is a spring ritual.[37] The "bird-masks" have comb-like structures reminiscent of chickens. According to Rodney Castleden, chickens may have been introduced into Crete as early as 1500 BCE, around the time they were introduced into Egypt.[38] Alternatively, the curly headdress could represent feathers or even hair. The blending of bird and fruit features in these dancing figures may have been intended as a dual metaphor for fertility. Birds lay eggs, while plants produce abundant seeds. The rows of small dots within the figures may represent *both* seeds and eggs. The pre-Socratic philosopher Empedocles, writing in the fifth century BCE, employed the same metaphor: "so first tall trees lay olive eggs," a famous quotation cited by both Aristotle and Theophrastus.[39]

THE BOWL OF PHAISTOS: A MINOAN PERSEPHONE

A scene related to that of the three dancing bird-maidens was painted onto the inside of a ceramic bowl, also found in Phaistos (Figure 6.14). Once again, there are three female figures with bird-like heads, but only the two smaller figures resemble the dancing flower maidens from the offering table (see Figure 6.15A). The large central figure lacks arms and legs and is lower than the two smaller dancers.[40] Some scholars believe the scene depicted in the Phaistos bowl may represent an early version of the Greek myth of Demeter and Persephone.[41] Walter Burkert offers the following description:

> [T]wo female forms dance on either side of a similar, but armless and legless, figure who seems to grow out of the ground. . . . Her head is turned towards a large stylized flower. . . . The association with the flower-picking Persephone and her companions is compelling.[42]

The myth of Demeter and Persephone opens with a calamity. While picking flowers in a field with her friends, the maiden Kore, daughter of Demeter, goddess of agriculture and the harvest, is suddenly seized by Hades, the god of the underworld, and dragged down into the earth. This traumatic experience—represented by the "sinking" figure in the Phaistos bowl—transforms Kore, the maiden, into Persephone, the wife of Hades. Eventually, with the help of the other gods, Demeter tracks down Persephone and frees her, but just as Persephone is leaving the underworld, Hades tricks her into tasting a pomegranate seed. Tasting the fruit of the underworld forces her to spend part of each year (4–6 months) underground with her husband. The identification of the dancing figures in the bowl with pomegranates could reflect the crucial role that pomegranate seeds play in the myth, which ends on a joyful note with Persephone's rescue and return to the surface.

In classical Greece, Demeter and Persephone were strongly identified with grain. In Homer's Iliad, "blond Demeter" is the one who "separates fruit and chaff in the rushing of the winds." The myth of Demeter and Persephone was the basis for several important religious festivals concerning grain, the most renowned of which was performed at Eleusis. According to Burkert, the myth of Demeter and Persephone was understood by the Greeks "as a piece of transparent nature allegory." Kore's descent into the underworld has been

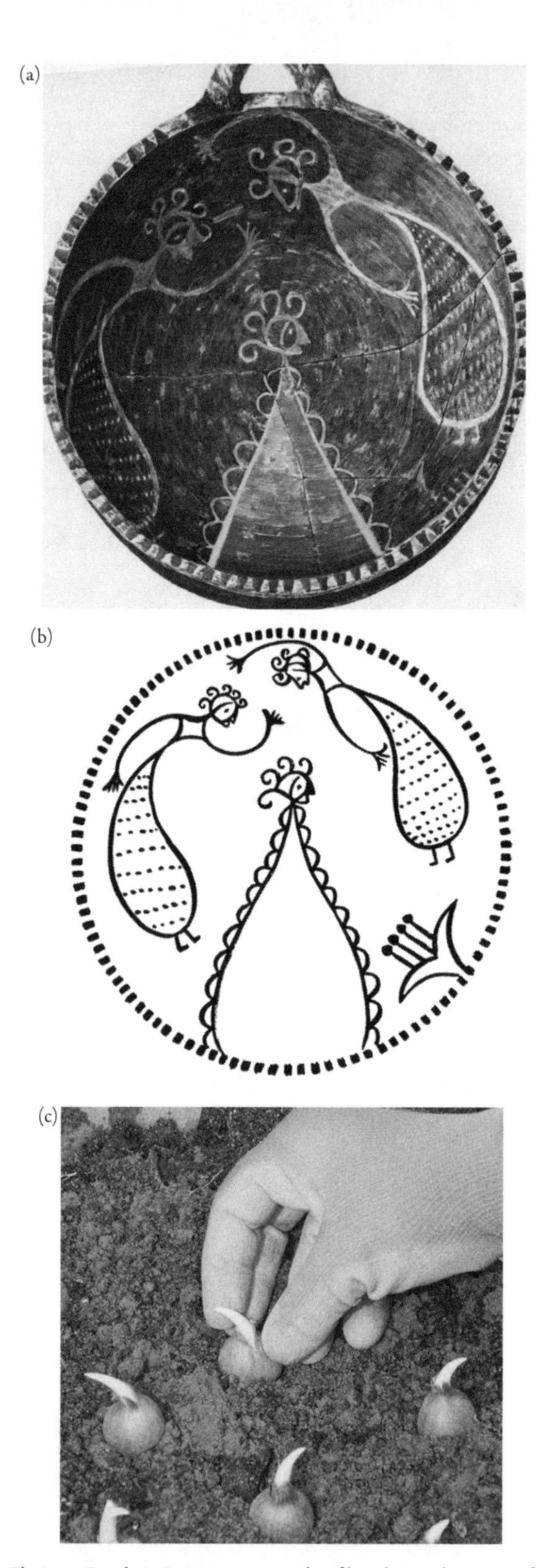

FIGURE 6.14 The Phaistos Bowl. A. Painting on inside of bowl. B. Schematic of painting inside Phaistos bowl. C. Saffron crocus corms planted flat side down.
A is from Levi, D. (1976), *Festòs e la civiltà minoica*. Vol. I, plate LXVIIa.

interpreted as a metaphor for planting grain. If correct, Kore's annual return to the earth's surface represents the seasonal growth of the cereal crop (mainly wheat and barley), which occurs, according to the Homeric *Hymn to Demeter*, "when the earth blooms with spring flowers."

Both events, Persephone's descent into the underworld and her re-emergence on the surface, occur in the spring. But as Burkert pointed out, the temporal cycle implied by the myth is inconsistent with the growing season in Mediterranean climates, where wheat and barley germinate shortly after being sown in the autumn and continue to grow throughout the winter, producing grain in the spring with the onset of the dry season. In Mediterranean climates, the planting and emergence of grain occur in the autumn rather than in the spring.

Because of the apparent seasonal discrepancy between the narrative and the cereal planting cycle, Nilsson proposed an alternative interpretation of the myth based on actual Greek agricultural practices. According to Nilsson, Kore's descent represents the storing of grain in underground silos during the dry season, a method practiced by the Greeks to protect the grain from desiccation. Kore's return to the surface might symbolize the digging up of the stored grain for planting in the autumn, when the rains begin. This interpretation is a better fit with Greek cereal cultivation, but there's no evidence that the Greeks themselves understood the myth in this way.

The discrepancies apparent in both interpretations of the Demeter/Persephone myth probably derive from the fact that the Greeks of the Iron Age adapted a much older myth and reinterpreted it in a way that reflected the high value they placed on wheat, which they regarded as the basis of civilization. Such is the conclusion of Walter Burkert, who speculated that the original meaning of the myth probably lies buried in "pre-Greek, perhaps Neolithic times." Because Bronze Age Minoan society exerted such a powerful influence on the Mycenaeans, it is possible that a precursor of the Demeter/Persephone myth originated in Crete. If so, the Phaistos bowl painting could represent an earlier version of the myth.

REIMAGINING PERSEPHONE AS A MINOAN CROCUS GODDESS

The most important known vegetation deities of the Minoans were not associated with cereal crops, but with the saffron crocus and other flowers. Indeed, there is little evidence for a grain goddess in Minoan religion. Of the many goddesses noted in the Linear B tablets, there are none from Crete identified with grain, although a single reference to a "Grain Mistress" (*sitopotnija*) has been found in Mycenae on the mainland. [43] With this in mind, we propose an alternative explanation for the large central figure often identified with Persephone in the Phaistos bowl. The large central figure is clearly differentiated from the two smaller dancers above her—in size, shape, and location. Her internal anatomy is also distinct. Instead of being fruit-like, containing numerous small seeds, she is painted a solid color inside. It's therefore possible that she represents an entirely different botanical structure from the dancing figures. In our opinion, the triangular central figure could be interpreted as symbolizing a crocus corm, as illustrated in Figure 6.14C. The orientation is correct, with the pointed end, from which the shoot emerges, oriented upward. Although superficially it may also resemble an onion bulb, corms lack bulb scales and are solid internally, consistent with the painting on the Phaistos bowl. In addition to its triangular shape, Peter Warren has called attention to the fact that the central figure is surrounded by "a group of little flowers with maroon tops," which "could be taken as schematic crocuses."[44] Recall that crocus stigmas are an intense maroon color (Figure 6.8B), which is the same color as the figure.

Another feature of the putative crocus corm provides a potential clue to her identity. Unlike the surrounding figures, the partially sunken central figure has a series of small bumps on its outer surface. Interpreted as snakes by Burkert, these enigmatic bulges have long baffled observers. We propose that their significance lies in their functional equivalence to the dots ("seeds") inside the two attending dancers. Just as seeds function as the reproductive units of fruits, the external bumps on the corm-like figure may represent *cormlets*, which are involved in the asexual propagation of crocus corms. Dividing crocus corms into cormlets, a practice that may have been invented in Crete, would have been familiar to Minoans but probably not to the archaeologists and historians who have studied them. To begin to understand the mindset of the Minoans, we need to look at the different stages of saffron crocus cultivation, which typically follows a three- to five-year planting cycle.

THE SAFFRON CROCUS PLANTING CYCLE

As discussed earlier, the Minoans were likely growing a domesticated variety of the wild crocus, *C. cartwrightianus*, which produces fertile seed. However, propagation by corms has several advantages over seed propagation. Crocuses grown from corms grow faster and flower sooner than seed-grown plants: seed-grown plants flower in three years, whereas corm-grown plants flower in the first year. Propagation by seed would also require setting aside a portion of the flower crop for seed production. Finally, vegetative propagation by corms minimizes genetic variability and ensures a uniform crop.

Flowering occurs in the fall (October–November) during the rainy season. The cooler temperatures associated with autumn also promote flower emergence.[45] The flowering period is brief, lasting only two to three weeks, after which the plant continues to produce leaves and roots. Throughout the fall and winter growing season, the sugars produced by the leaves are translocated to the "mother corm" and stored there in the form of starch. Some of the sugars are also used for the production of branch shoots, which sprout from numerous buds on the surface of the corm (Figure 6.15A). Note that during the first year of growth, the branch shoots only grow vegetatively, producing only leaves.

The branch shoots develop swollen bases, which turn into "cormlets" (Figure 6.15B,C). Each "mother corm" produces as many as ten branch shoots with basal cormlets.

The first growing season ends with the onset of the dry season in the early spring (April–May). The leaves wither and die, and the mother corm and cormlets become "dormant." This is not a true dormancy, however, because the cells in the apical meristems (growing tips) of the corm and cormlets continue to divide, although the structures they produce fail to enlarge. This is the most critical stage in the process, for during the hot summer months when the "dormant" cormlets remain in the ground, they undergo the transition from the vegetative to the reproductive stage of

FIGURE 6.15 Three stages in crocus cormlet formation. A. Branches emerge from the mother corm in the fall during the rainy season. B. The bases of the side branches swell to form cormlets. C. Mature cormlets ready to be divided from the mother corm.
From Molina, R. V., et al. (2004), The effect of time of corm lifting and duration of incubation at inductive temperature on flowering in the saffron plant (*Crocus sativus* L.). *Scientia Hort.* 103:79–91.

development. This developmental change occurs within the apical meristems, or growing tips, of the cormlets, which begin producing tiny flower buds instead of leaves. In the autumn, these flower buds grow into flowers, which produce the valuable stigmas. This is why a period of dormancy underground is absolutely essential for saffron production, because without it the cormlets would never make the transition from the vegetative to the reproductive stage.

It is inconceivable that the Minoans could have been ignorant of the necessity for a "dormant" underground phase of the crocus planting cycle. If they ever planted a corm that had been removed from the ground prematurely, which had not gone through a normal dormancy process, the resulting plants would have produced leaves but no flowers—a disaster for the saffron crop. The "mother corm" only flowers for three to five seasons, after which "she" dies. But her death is not in vain, for her cormlets are now fully matured and ready to become "mothers" themselves. To start the cycle up again, the gardener removes the depleted mother corm from the soil at the end of the growing season, divides the individual cormlets, and replants them in the ground. The flowering shoot emerges when the rains begin in the autumn, and another three- to five-year cycle begins.

WAS HADES A CONTRACTILE ROOT?

The Greek version of the Demeter/Persephone myth involves a violent abduction of Persephone by Hades, which appears to have no parallel in the religious art and iconography of Minoan Crete. Some have suggested that Hades's barbarous act was an embellishment tacked onto an earlier, more benign, agricultural myth by the violence-prone Greeks.[46] But perhaps there was an agricultural basis for the abduction scene in the original myth that has escaped notice.

Crocus corms, and many other types of corms and bulbs, possess contractile roots—thickened specialized roots that have the unusual ability to contract irreversibly in length. A diagram of the root system of a saffron crocus corm is shown in Figure 6.16. The mother corm produces a set of thin fibrous roots from a basal root plate, while the cormlets produce only thickened contractile roots.[47] These contractile roots grow downward during the autumn, winter, and spring seasons, anchoring themselves in the soil. During the dry summer months, they contract vertically and pull the corm down through the soil to its optimum

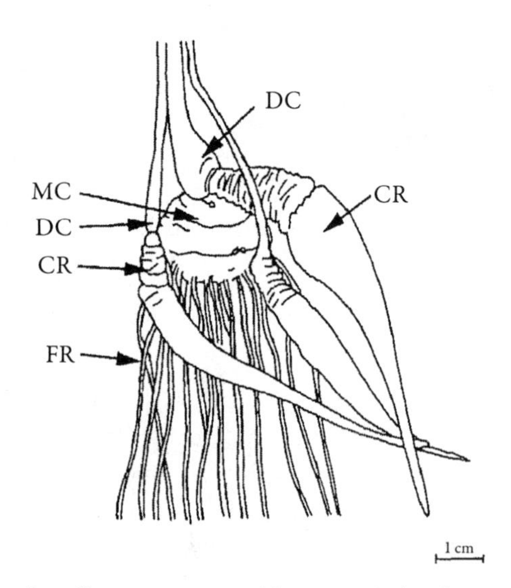

FIGURE 6.16 Diagram of a saffron crocus corm with roots. DC, daughter corm; CR, contractile root; MC, mother corm; FR, fibrous roots.

From Negbi, M., et al. (1989), Growth, flowering, vegetative reproduction, and dormancy in the saffron crocus (*Crocus sativus L.*). *Israel Journal of Botany* 38:95–113.

planting depth. Although Minoan farmers may have planted their crocus corms at the appropriate soil depth, it would not have escaped their notice that corms planted on the surface in the spring were slowly dragged down into the ground over the summer. The gradual disappearance of crocus corms below the soil surface could thus have inspired a myth based on a young "cormlet maiden" who is dragged underground by the Lord of the Underworld. It is also possible that since the corm actually pulls itself underground, coercion may not have been involved, as suggested by Eleanor Gadon.[48] In any event, after spending several years in the Underworld, the cormlet maiden, now transformed into a "mother corm," returns to the surface with her cormlet daughters, allowing the cycle to continue. The return of the corm mother in the spring would have provided an obvious raison d'etre for a Minoan agricultural festival celebrating the fecundity of the corm mother and the fertility of the soil.

Although purely speculative, it would be surprising if the Minoans, who pioneered the domestication of the saffron crocus and to whom the plant was sacred, had no myths inspired by its particular means of propagation. Assuming such a myth existed, perhaps it represents the elusive "pre-Greek" template for Demeter and Persephone postulated by Burkert and others.

MULTIPLE PARALLELS BETWEEN PERSEPHONE AND CROCUSES

The Greek myth of the mother and daughter grain goddesses, Demeter and Persephone, has no obvious antecedant in the Mediterranean world. Although comparable myths involving agricultural deities are known from the Near East (e.g., the Sumerian Inanna's descent to the Underworld and the Hittite Telepinu's angry retreat to the steppe),[49] none of these myths involves a mother and daughter.

Several parallels between the Greek myth of Demeter/Persephone and the asexual life cycle of cultivated saffron crocuses are noteworthy:

1. *It is a story about a mother and daughter.* The "mother crocus corm" with her asexually generated "daughter cormlets" correspond to Demeter and Persephone, who are not really separate individuals but constitute a mother–daughter *continuum*. In the earliest versions of the myth, Kore is the *parthenogenic* daughter of Demeter. In the *Homeric Hymn to Demeter*, composed around 650–550 BCE, Demeter refers to Persephone as an "offshoot":

   The daughter I bore, a sweet offshoot noble in form—[50]

The use of the term "offshoot" for progeny suggests a noncereal agricultural origin for the myth. If so, the practice of *dividing* the crocus cormlets and placing them in the ground in the spring may have provided a metaphor for Demeter's separation from Persephone.

During her search for Persephone, Demeter disguises herself as a barren old woman:

> ... a very old woman cut off from childbearing
> and the gifts of garland-loving Aphrodite.[51]

This, too, has its parallel in the crocus planting cycle, which terminates with the depletion and death of the mother corm.

2. *It is a young girl's "coming of age" story.* In the Greek myth, Persephone goes through the transition to womanhood in the Underworld as the wife of Hades. Persephone's coming of age myth has its parallel in the development of crocus cormlets, which undergo a transition from juvenile to reproductive stages while lying dormant beneath the ground. Because flowers were closely identified with women in Crete, Minoans may have perceived the underground dormant phase of the crocus planting cycle in sexual terms analogous to puberty because, without it, the flowers with their valuable stigmas would not develop.
3. *Both the Persephone myth and the crocus corm propagation cycle are tied to the spring.* According to the Homeric Hymn to Demeter, Persephone descends into the earth in the spring and re-emerges in the spring after an unspecified period underground. Crocus corms, like Persephone, are planted in the ground in the late spring and are removed from the soil in the spring three to five years later. The planting cycle of crocus corms is thus a better temporal fit with the Greek myth than the wheat planting cycle.
4. *Like Persephone, corms are pulled into the ground.* In this interpretation, the contractile root becomes the original inspiration for Hades, who carries Kore to his underground kingdom.
5. *Pomegranate seeds play a pivotal role in the myth.* Pomegranates ripen in the fall, just when crocus flowers bloom. Likewise, Hades must have tempted Persephone with pomegranate seeds in the autumn, when the fruit is ripe. Fruit may have been offered as "bloodless sacrifices" to Minoan flower deities. The geographer Pausanias (second century CE) described some curious features of the cult of "Black Demeter" that are reminiscent of the "pomegranate" figures of the Phaistos table and bowl. Festivals to Black Demeter were held annually in a grove before a sacred cave in the southwest corner of Arcadia. Upon an altar were placed "fruits from various trees, grapes, honey-combs, and uncarded wool."[52] Although pomegranates are not specifically mentioned in the list of offerings, they belong to the general categories of "tree fruits." The absence of any reference to grain among the offerings to Demeter is significant and could indicate that the cult of Black Demeter had preserved features that predate the goddess's identification with grain. Indeed, Walter Burkert commented that the cult is "strangely reminiscent of Bronze Age religious practices, be it Minoan, Mycenaean or Anatolian."[53]
6. *One of Persephone's attributes is a bird.* Persephone is usually associated with the pomegranate, but a lesser known attribute of the goddess is the chicken, as illustrated in a famous marble relief of Persephone and Hades in the Underworld, in which Persephone holds a chicken in her right hand and a clump of ripened wheat in her left hand (Figure 6.17). Note the comb on the chicken's head, which is reminiscent of the headdresses of the dancers on the Phaistos table and bowl.

Finally, there is a parallel between the yellow color associated with the Minoan saffron goddess and the yellow color of ripened wheat, emblematic of Demeter and Kore. Yellow was gendered female by both the Minoans and the Greeks, a tradition that clearly arose from the yellow colors of two plants sacred to the goddesses.[54]

FIGURE 6.17 Persephone and Hades in the Underworld (~470 BCE), by Greek colonists in southern Italy.
From Reggio Calabria in southern Italy, Museo Nazionale, Italy.

THE STIGMA-SKIRTED PRIESTESS OF AYIA TRIADA

The sarcophagus of Ayia Triada, which dates to about 1450–1340 BCE, is a magnificent painted coffin found in a minor tomb in a settlement not far from Phaistos. The plaster-covered limestone structure was painted on all sides with funerary frescoes in the Egyptian style. From the left of one side of the sarcophagus a procession of women approaches a sacrificed bull trussed on a table. Next to the bull on the right, a priestess wearing a patterned skirt places a bowl of fruit on the altar. A libation vessel seems to hover over the bowl, but may be resting on a support that is no longer visible (Figure 6.18A). According to J. A. Sakellarakis, former Director of the Herakleion Museum, the altar is located "near the fence around the sacred tree, which is crowned with sacred horns, and in front of a tall column supporting a double axe."[55] The priestess is clearly associated with the sacred tree.

To our knowledge, the significance of the curious psi-shaped forms on the priestess's skirt has not yet been addressed. The Greek letter psi (ψ) had not been invented in the fifteenth century BCE, so we must look to earlier sources for the meaning of this particular symbol. Poseidon's trident is sometimes cited as the origin for the letter psi, but this particular symbol of Poseidon, like the Greek alphabet itself, was not introduced until the Archaic Period. We propose that the psi-like figures were meant to indicate saffron stigmas, an example of which is shown in Figure 6.18B. If this interpretation is correct, it

(a)

(b)

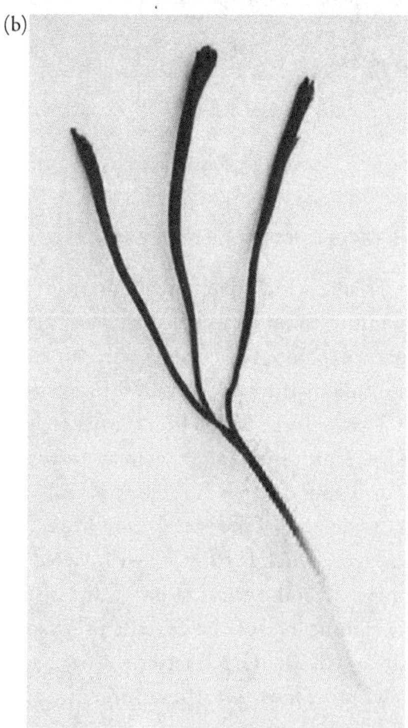

FIGURE 6.18 A. Detail from the sarcophagus of Ayia Triada showing priestess with a psi-like pattern on her skirt standing before a shrine. The sarcophagus dates to the Third Palace Period, during the Mycenaean occupation of Crete. B. Close-up of saffron crocus stigma with same psi-like pattern.
A is from the Heraklion Museum; B is from the Internet.

would suggest that by the fifteenth century BCE, saffron stigmas had evolved into a symbol for a saffron goddess. If so, the presence of a saffron priestess standing at the altar at the head of the procession may indicate that a saffron goddess enjoyed primacy among the vegetation goddesses of Crete.

THE SHAFT GRAVES OF MYCENAE

There is little artistic evidence for warfare in Minoan society during the Bronze Age, although scholars are divided on the question of just how peaceful Minoan society really was.[56] It was once thought that Crete lacked any fortifications suggestive of warfare. However, later excavations have shown that some early settlements were either built on hilltops or included some degree of fortifications. Weapons, such as swords, have been found in male burials, and their excellent quality suggests that they became the prototypes for two important sword types of the Mycenaeans.[57] Whether or not the Minoan swords were actually used in warfare or were ceremonial is still being debated.[58] Compared to their Near Eastern neighbors, however, Minoans were far less warlike, and perhaps because of this they were able to evolve a relatively peaceful society that persisted for nearly 2,000 years. In addition, their strong navy may have served to discourage potential invaders.

The Minoans engaged in widespread international trade facilitated by the development of the longboat in the third millennium BC. In the course of this trade, they encountered another Bronze Age civilization in the northeastern Peloponnese, on the Greek mainland. This civilization was called the "Mycenaean" by its discoverer, Heinrich Schliemann, after its most famous archaeological site, the city of Mycenae. Schliemann, a German businessman who had retired to pursue his passion for archaeology, had already excavated the site of Troy from 1871 to 1873. In 1876, he began excavating the ancient site of Mycenae, where he hoped to discover the Palace of Agamemnon, the Greek king who led the assault on Troy in Homer's *The Iliad*.

Schliemann literally struck gold when he discovered a series of vertical, or "shaft," graves filled with gold jewelry, silver vessels, bronze weapons, and finely painted pottery. Among these was a stunning golden funeral mask found on the face of a skeleton in one of the burial shafts, which Schliemann identified as the death mask of Agamemnon.[59] However, even if one assumes that Agamemnon was a historical figure rather than a fictional hero, the shaft graves of Mycenae have been shown to date to the sixteenth century BCE, whereas the Trojan War, if it occurred, was fought in the thirteenth century BCE. The Mycenaean shaft graves thus predate the events that inspired the *Iliad*—which weren't written down until the eighth or seventh century BCE—by 400 to 500 years.

WHO WERE THE MYCENAEANS?

The sudden appearance of so much gold in the shaft graves, without apparent antecedents in the region, has long puzzled archaeologists and has led to various theories to account for such a rich hoard. First, there is the question of where the shaft grave rulers obtained the gold, which is relatively scarce on the mainland. International trade, possibly with Transylvania, has been suggested as the source of the gold.[60]

The second question concerns the identities of the shaft grave rulers themselves. Were they indigenous chieftains or foreign invaders? According to scholars, the weight of the evidence indicates that the rulers of Mycenae in the early sixteenth century BCE were indigenous chieftains who spoke an early form of Greek. Oliver Dickinson has suggested that the high quality of the artifacts found in the shaft graves may "represent a burst of exuberant innovation," which was "the work of only one or two workshops," rather than part of a generalized cultural tradition spread throughout the mainland.[61] Beginning in the fifteenth century BCE, however, a more widely prosperous period seems to have emerged, one characterized by a different kind of royal burial chamber, the *tholos* tombs (large, underground domed chambers). Tholos tombs, which may have been modeled after smaller vaulted tombs found on Crete, suggest increasing contacts with other Mediterranean societies in the course of vigorous trade and population expansion to far-flung settlements.

Although the term "Mycenaean" has been adopted for the entire culture of the mainland during this period, it is important to remember that Mycenae itself was but one of many competing palace-based societies scattered throughout the mainland during the Bronze Age. Like Mycenae, some of these large settlements were well fortified and rich in art and artifacts. According to Linear B inscriptions on clay tablets from Pylos, the palaces took in flax, linen, woolen textiles, livestock, ox hides, perfumes, raw materials, and certain spices, and distributed quantities to workers, craft workers, shrines, and the like. [62] However, most agricultural products, including wheat, legumes, and orchard fruits, were not processed through the palace, but were obtained from independent suppliers (farmers and private estates), although the nature of the arrangement is unclear. This is consistent with the presence of many smaller, unfortified farming settlements scattered throughout the mainland, which apparently co-existed with their local palaces according to some type of social contract. As noted earlier, co-existence of the outlying communities and the palaces suggests that the warlike reputations of Mycenaean kingdoms have been exaggerated. A relatively stable economic system seems to have prevailed on the mainland during much of the Late Bronze Age. Only at the cataclysmic end of the Bronze Age do we encounter evidence for the sort of widespread violence often associated with Mycenaean society in general, a misconception ultimately derived from the Homeric epics.

MINOAN VERSUS MYCENAEAN ART

The art and religious iconography of the Mycenaeans was strongly influenced by the Minoans during the early part of their history. Just as Rome ceded the realms of scholarship, religion, and art to the Greeks, the Mycenaeans seemed content to defer to Minoan tastes and styles in art, pottery, clothing, and religious iconography. This includes the Minoan tendency to associate sacred plants with women. For example, a gold pin found in the shaft graves of Mycenae shows a sacred tree emerging from the head of a woman, presumably a tree goddess (Figure 6.19A). This may, in fact, be a Minoan import, but its presence in the shaft graves suggests that Mycenaeans had assimilated the iconography.

Mycenaean ring seal images are also strongly reminiscent of Minoan ring seals. In Figure 6.19B, a central goddess is flanked by a woman bent over a table (right) and a male

(a)

(b)

FIGURE 6.19 Mycenaean pin and ring seals. A. Gold pin from shaft graves of Mycenae depicting a sacred tree emerging from the head of a woman. B. Gold signet ring, Mycenae CT 91. C. Gold signet ring, Mycenae Acropolis treasure.

A is From Nilsson, figure 158; The Acropolis in Mycenae, Berlin, ca. fifteenth century BCE. B is from *Heinrich Schliemann: The 100th Anniversary of His Death* (1990). Athens National Archaeological Museum; Troy, Mycenae, Tiryns, Orchomenos.

(c)

FIGURE 6.19 Continued

acolyte grasping a sacred plant (left). In the gold ring seal from Mycenae in Figure 6.19C, a goddess sits beneath a sacred tree whose fruits resemble clusters of grapes or perhaps mulberries. Mulberries are dioecious, so if the tree represented is a mulberry, it would have to be a female tree, which may have symbolic significance.[63] In her right hand she holds up three poppy capsules by their stems, while her female attendant extends her left hand as if to receive them.

One notable difference between Minoan and Mycenaean religions is represented by a single reference in the Linear B tablets of Mycenae to a "Grain Mistress" (*sitopotnija*) among the various deities receiving offerings. No grain goddesses are known from Crete, so this seems to have been either a Mycenaean invention or a Near Eastern import. A fresco from the cult center of Mycenae, dating to the thirteenth century BCE, displays a woman holding sheaves of wheat in each hand (Figure 6.20A). On her head she wears a crown or headdress, indicative of her exalted status. A similar motif occurs on an ivory pyxis lid from Syria dating to the same period (Figure 6.21B). This, too, depicts what appears to be a goddess or priestess of grain. Her bared breasts and Aegean-style flounced skirt are indicative of a Mycenaean origin, but the rampant goats are a Mesopotamian motif. Perhaps these figures represent forerunners of Demeter, the Greek goddess of grain.

Judging from the wealth of their burials and imposing citadels fortified by cyclopean walls, the Mycenaean elite seems to have been more inclined toward lavish displays of wealth and power than their Minoan counterparts. Given the harsh realities of power politics, it was probably inevitable, therefore, that the balance of influence between the two neighboring peoples eventually tilted in favor of the Mycenaeans.

The Mycenaeans may have begun their gradual dominance of Crete in the wake of a devastating natural disaster. Around 1530 BCE, the island of Thera exploded in a titanic volcanic eruption—larger than the Krakatoa explosion of 1883. All that remained after the blast was a circular archipelago of smaller islands (the largest of which is modern day Santorini) surrounding the huge sunken caldera of the former volcano. Miraculously, the city of Akrotiri not only remained intact, but, as noted earlier, it was preserved for posterity by being buried under a layer of pumice and ash. The impact of the eruption was felt far beyond Thera. Tidal waves, blackened skies, cooling temperatures, and falling ash may have caused widespread panic and insecurity throughout Crete. Agriculture must surely have

(a)

(b)

FIGURE 6.20 Mycenaean images of women with plants. A. Fresco of goddess or priestess holding clusters of wheat, from Mycenae. B. Ivory pyxis lid from Syria showing a goddess wearing a Mycenaean-style skirt holding clusters of grain.
From the Louvre Museum, Paris: http://history2701.wikia.com/wiki/File:7017_xl.jpg.

been adversely affected, although the extent of the impact is unclear. Mainland Greece, located farther away from the eruption, apparently emerged unscathed. In contrast, the ensuing economic and social stress in Crete is blamed by some historians for initiating the decline of the Minoan Second Palace Period, which paved the way for the eventual domination of the island by the Mycenaeans.

Just as Rome's embrace of Greek culture reached its ultimate fulfillment through the physical assimilation of its more venerable mentor, so, too, did the Mycenaeans' embrace of Minoan culture eventually culminate in the loss of Crete's independence in the late fifteenth century BCE, ushering in the Last Palace Period of Minoan society. Under Mycenaean domination, Greek replaced the Minoan language, and the palace at Knossos was rebuilt with a strong Mycenaean flavor.

For the duration of the Last Palace period, Minoan religion seems to have become more decentralized and more variable in its expression. By the thirteenth century BCE, benched shrines (small, rectangular, one-room buildings with a bench along one wall for offerings) were located within individual villages. Such shrines often contained large clay female figurines, which were called "goddesses-with-upraised-arms" by Spyridon Marinatos, the archaeologist who first discovered them in 1937.[64] In contrast to the more generalized "nature goddesses" of the ring seals, the terracotta goddesses, or priestesses, with upraised arms of the Last Palace Period of Crete typically wear tiaras or crowns adorned with sacred objects—birds, disks, horns, double axes, snakes, and plants—which are thought to represent either aspects of a single goddess or attributes of individual goddesses.[65]

The difference in ethos between the terracotta goddesses-with-upraised-arms and the Minoan religious art of earlier periods is profound. Gone are the self-confident exuberance and ecstatic states of the earlier Minoan goddesses of the seals and frescoes. Gone, too, are the stunning artistry and skilled craftsmanship that characterized the works of the First and Second Palatial Periods in their glory days. Stiff and formal-looking, mute and expressionless, they nevertheless speak eloquently of the enduring power of religious devotion. The largest of the goddesses-with-upraised-arms found at Gazi is about 2.5 feet tall and is known as "The Poppy Goddess, Patroness of Healing," the name originally given to her by Spyridon Marinatos (Figure 6.21A). The slits of the poppy fruit capsules on her tiara (Figure 6.21B) are painted a darker color, resembling the color of dried poppy juice.[66] The Minoan Poppy Goddess may, in fact, represent the earliest evidence for the use of psychoactive drugs as part of a religious ritual and/or healing. Her closed eyes and parted lips suggest a trance-like state, while the deep creases around her mouth could be interpreted as an incipient frown or smile, or perhaps a self-canceling hybrid of the two.

The poppy plant (*Papaver somniferum*) was well-known to the Greeks of Homer's day for its pain-relieving and sleep-inducing qualities. In Book IV of *The Odyssey*, Homer refers to a certain pain-negating drug (*nepenthes pharmakon*). The reference occurs in the scene in which Telemachus, distraught over his father's absence, visits Helen and Menelaus at Sparta to obtain information about Odysseus's fate. To alleviate his pain, Helen prepares an herbal anodyne:

> Presently she cast a drug into the wine whereof they drank, a drug to lull all pain and anger, and bring forgetfulness of every sorrow. Whoso should drink a draught thereof, when it is mingled in the bowl, on that day he would let no tear fall down his

(a)

(b)

FIGURE 6.21 Minoan poppy goddess-with-upraised-arms from Gazi, Crete. A. General view. B. Close-up of head showing slits on the poppy fruits (*arrows*).

cheeks, not though his mother and his father died, not though men slew his brother or dear son with the sword before his face, and his own eyes beheld it.[67]

Helen's expertise in the use of botanicals in medical treatment is consistent with representations of the role of women as healers in the Bronze Age, as exemplified by religious and healing rituals showing female cultic figures and centered on crocus, poppies, and other flowers and plants in Cretan and Mycenean art.

Although Homer never cites poppies as the source of *nepenthe pharmakon* (literally, "no-pain drug"), its effects are suspiciously like those of opium, the active ingredients of which are morphine and codeine. On the other hand, Homer's poetic and wildly extravagant claims of nepenthe's power to nullify all grief, no matter how horrific the provocation, may explain why, several hundred years later, the literal-minded botanist, Theophrastus, insisted that there was no known plant with the properties of Homer's nepenthe. Nevertheless, Theophrastus, a student and colleague of Aristotle's who is generally recognized as the "Father of Botany," did report that a mixture of the juices of hemlock and poppy could induce an easy and painless death.[68] Based on the latter description, we can infer that Theophrastus, like both Hippocrates and Aristotle before him, was well aware of opium's potent narcotic properties, a tradition that extended at least as far back as the Bronze Age. Indeed, there is some evidence that, during the Bronze Age, poppy juice was being traded in parts of the Near East. Poppy-shaped vessels dating to 1500 BCE have been found in Cyprus and Egypt, and they may have been used as containers for the transport of "pharmaceutical preparations" based on poppy juice.[69]

THE CATACLYSMIC END OF THE BRONZE AGE

The hegemony of the Mycenaean Palace societies over the Greek mainland, Crete, and the Cyclades lasted for about 200 years. When the end came around 1200 BCE, it came swiftly and conclusively—not just weakening the Palace social order, but effectively ending it. However, in those parts of the mainland outside the control of the palaces, the general decline of Mycenaean society was more protracted and patchy.[70]

The cause of the Mycenaean Palace implosion has been the subject of much speculation and debate. Greek historians interpreted the ruins visible at numerous locations as the direct result of invasions by Dorians from the north. However, there is no evidence for any significant invasion of the mainland at the end of the Bronze Age. Based on the characterization of Mycenaean society as extremely war-like, later historians postulated that constant internecine warfare between feuding princes brought about the destruction of the palaces, causing large-scale dislocations of peasants suddenly deprived of their livelihoods. But the archaeological evidence suggests that a much more stable situation existed on the mainland until the very end of the Bronze Age.

In recent years, a more complex explanation has emerged, one that attributes the end of the Bronze Age to "systems collapse," a breakdown of the Mycenaean economic and social systems triggered by multiple adverse factors, such as prolonged drought, overpopulation, soil exhaustion, reliance on too few crops, earthquakes, and other adversities, which the slow-acting bureaucracies were unable to mitigate. The effects were not limited to Greece. Currently, the collapse of the Mycenaean civilization is believed to have been part of a larger, regional collapse caused by multiple interconnected failures, including civil wars and foreign invasions, the cutting of trade routes, and physical factors such as earthquakes and drought—spelling the end of most Bronze Age palatial societies.[71]

Whatever the exact causes of the Mycenaean collapse, all are agreed on the outcome: widespread violence, the burning of palaces, and a certain amount of migration as well. Troy VIIA, the *Iliad*-era city, was besieged and burned in the late thirteenth century BCE, and Mycenaean Greeks may have taken part in its sacking, as described by Homer in the *Iliad*, although the story of the abduction of Helen is fictional. The violence associated with the end of the Bronze Age triggered a decline in the population on the mainland by as much as 75–90%, as displaced Greeks sought more stable and secure conditions elsewhere.[72] Some dispersed to remote or secluded sites on the mainland, but most sought safe haven overseas. The latter sailed eastward and founded new settlements on Cyprus, Crete, and other Aegean islands, eventually reaching the western coast of Anatolia, a region that later came to be known as "Ionia." Here diaspora Greeks encountered older, more advanced Near Eastern civilizations—Assyrians, Phoenicians, and others—and the resulting cultural cross-fertilization was to prove crucial to the development of a new Greek identity during the so-called "Dark Ages" of Greek history, which lasted from c.1200 to c. 800 BCE.

One of the earliest consequences of the turmoil that preceded the Greek Dark Ages was the complete disappearance of Linear B writing. Linear B had always been restricted to the Mycenaean ruling elite and their scribes. Thus it is not surprising that Linear B use was discontinued soon after the fall of the palaces. However, the oral transmission of the Greek cultural heritage continued through story-telling, music, singing, and the recitation of poetry. In this way, the most basic ideas the Greeks had about themselves were preserved and passed on from generation to generation.

NOTES

1. A simplified chronology of some of the major events of the Aegean Bronze and Iron Ages is presented in Table 6.1.

2. Strictly speaking, there were no "Mycenaeans" until 1600 BCE.

3. Dickinson, O. (1996), Minoans in mainland Greece, Mycenaeans in Crete? *Cretan Studies* 5:63–71.

4. Hood, M. S. F. (1960), Tholos Tombs of the Aegean. *Antiquity* 34:166–176.

5. Dickinson, O. (1994), *The Aegean Bronze Age*. Cambridge University Press.

6. The Minotaur was the monstrous offspring of the King's wife, Queen Pasiphaë, and a white bull, originally presented to the King by the god, Poseidon. When the king failed to sacrifice the bull to Poseidon, the god caused the queen to lust after the bull, the fulfillment of which was enabled by yet another technical innovation by Daedalus: a hide-covered wooden framework in the shape of a cow.

7. The term "Labyrinth," from the Greek word *labyrinthos*, meaning maze, may be related to the Lydian word *labrys*, meaning "double-headed axe," an important religious and royal symbol that is ubiquitous in the art of Minoan Crete.

8. Hood, S. (1971), *The Minoans*. Praeger.

9. Warren, P. (1975), *The Aegean Civilizations*. Elsevier-Phaidon.

10. The apparent predominance of men in the audience in the reconstructed sections of the fresco is purely hypothetical.

11. See Marinatos, N. (1987), Public festivals in the west courts of the palaces, in R. Hägg and N. Marinatos, eds., *The Function of the Minoan Palaces*, Stockholm, pp. 135–142.

12. Thera is the ancient name of this Cycladic Island, but it is commonly known today as Santorini (or officially as *Thíra* in Greek). The ancient name Thera is still used for the volcano that destroyed most of the island in the Bronze Age and buried the town of Akrotiri under pumice and ash. The archeological site of the preserved town is called Akrotiri after a modern village not far away, but the Bronze Age name of the encapsulated town is not known.

TABLE 6.1

Chronology of the Aegean Bronze and Iron Ages

Date (BCE)	Period	Crete	Mainland Greece
7000–2600	Neolithic	Permanent farming villages; domesticated plants and animals in use; PIE dialects spoken by earliest settlers from Anatolia.	
3000–2000	Bronze Age: Prepalatial Period	Destruction of some Minoan settlements ca. 2200 BCE	Social ranking and specialization continues to develop; Greek language evolves from PIE dialects.
2000–1650	Aegean First Palace Period	Palaces at Knossos, Phaistos, and elsewhere; rise of artistic, urban Minoan civilization	
1900		Linear A script; widespread commerce with Egypt, Cyclades, and Anatolia	
1700	Aegean Second Palace Period	Earthquake destroys old palaces at Knossos, Phaistos, etc. Larger palaces built and new settlements established.	
1650–1450		Height of Minoan power and prosperity; Minoan settlement in Miletos (Asia Minor)	Mycenaean Palace society emerges as a political and economic force.
~1600		Eruption of Thera volcano ca. 1530 BCE; possible disruption of agriculture throughout Crete.	Shaft Grave Circle B (earlier); Shaft Grave Circle A (later) in Mycenae; evidence of wealth (gold) and social stratification.
1525			Mycenaean palaces and tholos tombs with lavish grave goods; height of Mycenaean power and prosperity.
1450–1200	Aegean Third Palace Period	Increased Mycenaean influence over Crete; Cyclopean walls on mainland; first appearance of Linear B script in late fifteenth to early fourteenth century BCE; Linear B replaces Linear A in Crete; Greek replaces Minoan language; decline of Crete.	

TABLE 6.1 Continued

Date (BCE)	Period	Crete	Mainland Greece
1200–1100	End of Bronze Age; Aegean Postpalatial Period.	Destruction of many important Mycenaean sites on mainland (but apparently not Athens) and other Bronze Age settlements throughout eastern Mediterranean. Crete less affected. Linear B disappears.	
1200–800	Greek "Dark Ages"		Some population movement to Cyprus and Ionia (coast of Asia Minor).
1050–700	Iron Age		
900–700			Geometric Period; Greek alphabet developed; expansion of trade and colonization in Asia Minor, southern Italy, and Sicily.
750–480			Archaic Period; Homeric epics recorded; Greek city-states evolve. Continued expansion via colonization throughout Mediterranean as far west as Spain.
480–323			Classical Period

13. Morgan, L. (1988), *The Miniature Wall Paintings of Thera: A Study in Aegean Culture and Iconography*. Cambridge University Press, p. 18.

14. About half of fig species are gynodioecious, comprised of some individuals that have enclosed inflorescences (syconia) with only long-styled pistillate flowers (females) and other plants with staminate flowers in addition to short-styled pistillate flowers (hermaphrodites).

15. The Minoans would have had some exposure to date palm cultivation from their extensive trade contacts with the Egyptians. Hood speculated that the Minoans imported date palms from Egypt and considered the tree sacred. R. M. Dawkins proposed that the illustrations on a painted vase from Knossos represent three female date palm trees with their rachises beginning to emerge from the axils of their upper leaves. According to Dawkins, the small "branches" alongside the female trees are actually detached male rachises used for artificial pollination. However, the smaller detached branches do not look much like male date palm rachises, which are shaped like whisk-brooms. Perhaps they represent either basal shoots of the palm, as Morgan has suggested, or fern fronds. See Dawkins, R. M. (1945), The cultivation of date-palm in Minoan Crete. *Man* 45: 47; Morgan, L. (1988), *The Miniature Wall Paintings of Thera: A Study in Aegean Culture and Iconography*. Cambridge University Press.

16. According to the *Red Data Book of Rare and Threatened Plants of Greece* (Hellenic Botanical Society, 2009), the distribution of *Phoenix theophrasti* is restricted to Crete.

17. Marinatos, N. (1984), *Art and Religion in Thera: Reconstructing a Bronze Age Society*. Athens.

18. Nilsson, M. P. (1949), *The Minoan-Mycenaean Religion and Its Survival in Greek Religion*, second edition. C. W. L. Gleerup, Lund.

19. Morris, C. (2004), "Art makes visible": An archaeology of the senses in Minoan elite art, in Neil Brodie and Catherine Hills, eds., *Material Engagements: Studies in honour of Colin Renfrew*. McDonald Institute Monographs, pp. 31–43; and personal communication.

20. Warren, P. (1988), *Minoan Religion As Ritual Action. Studies in Mediterranean Archaeology and Literature*. P. Astroms, Gothenburg.

21. Chapin, A. P. (2004), Power, privilege, and landscape in Minoan art. *ΧΑΡΙΣ*, 47–64.

22. Numerous authors erroneously refer to the source of saffron as the stamens, rather than the stigmas, of the crocus flower.

23. Ference, S. C., and G. Bendersky (2004), Therapy with saffron and the goddess at Thera. *Perspectives in Biology and Medicine* 47: 199–226.

24. Sfikas, J. (1992), *Wild Flowers of Crete*, second edition. Efstathiadis Group.

25. Mathew, B. (1977), *Crocus sativus* and its allies (*Iridaceae*). *Plant Systems Evolution* 128: 89–103.

26. Warren, *Minoan Religion As Ritual Action*.

27. Rutkowski, B. (1986), *The Cult Places of the Aegean*. Yale University Press.

28. Reusch, H. (1956), *Die Zeichnerische Rekonstruktion des Frauenfrieses im Böotischen Theben*. Akademie-Verlag, Berlin; Immerwahr, S. (1990), *Aegean Painting in the Bronze Age*, Pennsylvania State University Press.

29. Kistai were, according to Mylonas, "cylindrical, pyxis-like receptacles with close-fitting covers that sealed the contents firmly." See Mylonas G. E. (1961), *Eleusis and the Eleusinian Mysteries*. Princeton University Press.

30. Mylonas, *Eleusis and the Eleusinian Mysteries*.

31. Marinatos, *Art and Religion in Thera*.

32. Ference and Bendersky, Therapy with saffron and the goddess at Thera.

33. Ibid.

34. In the introductory chapter, we noted the changes in the gender associations of pink and blue in the United States since World War II. In Classical Greece, the color yellow was traditionally regarded as a feminine color. For example, the peplos, or robe, that was woven for the statue of Athena each year was dyed yellow with saffron. In his comedy, *The Frogs*, Aristophanes satirized Dionysus, the god of drama, as a feminized man by having him wear a yellow chiton (sleeveless tunic)—the equivalent, in modern terms, to having him wear pink. The gendering of the color yellow as female by the Greeks of the Iron Age probably originated with the medicinal and religious uses of saffron in Minoan society.

35. Although "adyton" is the term used by archaeologist Nanno Marinatos (*Art and Religion in Thera*), it should not be assumed that this room had the same function as the Greek adyton.

36. Goodison, L., and C. Morris (1999), Beyond the "Great Mother": The sacred world of the Minoans, in L. Goodison and C. Morris, eds., *Ancient Goddesses*. University of Wisconsin Press.

37. Marinatos, N. (1993), *Minoan Religion: Ritual, Image, and Symbol*. University of South Carolina Press.

38. Castleden, R. (1990), *Minoans: Life in Bronze Age Crete*. Routledge.

39. The verb used in the original Greek text is ωοτοκεί ("giving birth to eggs"), which ultimately became the term "biology." (Sophia Rhizopoulou, personal communication.)

40. The shape given in the line drawing from the original publication (figure 7.18B) is somewhat misleading because the sides of the figure are straight, as indicated in the painted reconstruction (figure 7.18A), not curved at the base as depicted. The curved base presumably reflects the concave dimensions of the bowl's inner surface, which becomes curved in flat projection, as well as the curved shape of the lip of the bowl.

41. Burkert, W. (1985), *Greek Religion: Archaic and Classical*, trans. John Raffan, Harvard University Press; Kerényi, C. (1967), *Eleusis: Archetypal Image of Mother and Daughter*. Princeton University Press.

42. Burkert, Greek Religion.

43. Ibid.

44. Warren, P., personal communication.

45. Molina, R. V., et al. (2004), The effect of time of corm lifting and duration of incubation at inductive temperature on flowering in the saffron plant (*Crocus sativus L.*). *Scientia Horticulturae* 103:79–91.

46. For example, Gadon, E. W. (1989), *The Once and Future Goddess: A Symbol for Our Time*. HarperCollins.

47. Negbi, M., et al. (1989), Growth, flowering, vegetative reproduction, and dormancy in the saffron crocus (*Crocus sativus L.*). *Israel Journal of Botany* 38:95–113.

48. Gadon, *The Once and Future Goddess*.

49. According to the Hittite myth, the god Telepinu's disappearance causes the crops to fail. He reappears only after the goddess of magic transfers his anger to the Underworld.

50. Translated by Helene P. Foley (1994), in *The Homeric Hymn to Demeter: Translation, Commentary, and Interpretive Essays*. Princeton University Press, p. 4.

51. Idem., p. 6.

52. Burkert, W. (1979), *Structure and History in Greek Mythology and Ritual*. University of California Press.

53. Ibid.

54. For example, in Aristophanes' *Lysistrata*, a woman cites the wearing of "saffron robes" as a part of an initiation ceremony for young girls:

I bore the holy vessels
At seven, then
I pounded barley
At the age of ten,
And clad in saffron robes,
Soon after this,
I was Little Bear to
Brauronian Artemis;
Then neckletted with figs,
Grown tall and pretty,
I was a Basket-bearer . . .
(Lysistrata 641–647)

Note the association of the feminine with the carrying of "vessels" and "baskets," as well as the botanical associations of femininity with grain, saffron (yellow), and fruit. Such associations could have been originally derived from Minoan and Theran sources.

55. Sakellarakis, J. A. (1995), *Herakleion Museum, Illustrated Guide*. Ekdotike Athenon S. A., Athens.

56. For an important forum on this subject, see Laffineur, Robert, ed. (1999), Polemos: *Le Contexte Guerrier en Egee a L'Age du Bronze*. Actes de la 7e Rencontre égéenne internationale de l'Université de Liège, 1998. Université de Liege, Histoire de l'art d'archeologie de la Grece antique.

57. Peatfield, A. (1999), The paradox of violence: Weaponry and martial art in Minoan Crete. In *POLEMOS, Le contexte guerrier en égée à l'âge du bronze*. Liege, Aegaeum (vol. 19) http://www2.ulg.ac.be/archgrec/aegaeum19.html

58. http://en.wikipedia.org/wiki/Minoan_civilization

59. Although questions about the mask's authenticity have been raised, archaeologist Oliver Dickinson has argued persuasively that the mask was neither faked nor planted (Dickinson [2006], *The Aegean from Bronze Age to Iron Age: continuity and change between the twelfth and eighth centuries B.C.* Routledge). Dickinson pointed out that several gold burial masks were found at the site. In a telegram Schleimann sent to a colleague at the time of the discovery, he stated that the mask of the "dead man with the round face" reminded Schleimann of Agamemnon more than the more elongated face with the mustache usually cited. Indeed, Schliemann chose the round-faced mask as the frontispiece for the chapter in his book, *Mycenae*, in which he identifies the burials as those of Agamemnon and his followers. See Dickinson, O. (2005), The "Face of Agamemnon." *Hesperia* 74:299–308.

60. Davis, E. N. (1983), The gold of the shaft graves: The Transylvanian connection. *Temple University Aegean Symposium* 8:32–38; Dickinson, O. T. C., personal communication.

61. Dickinson, O. (1994), *The Aegean Bronze Age*. Cambridge University Press.

62. Ibid.

63. Mulberries were well known in the ancient world. For example, Homer mentions them in *The Iliad* when he describes Hera as "fashioning now/ In her pierced ears the pendants, triple pearled,/ Hued like the mulberry, With grace agleam." *The Iliad of Homer* (1911), trans. Arthur Gardner Lewis. The Baker and Taylor Company.

64. Although the terracotta statues in the shrines are all female, the "Goddesses-with-upraised-arms" probably represent only a fraction of the total pantheon of Postpalatial Minoan deities. It is also possible that deities of the shrines reflect the folk religion of the native Minoan population.

65. See Gesell, G. C. (2004), The popularizing of the Minoan household goddess, in A. Chapin, ed., *Charis: Essays in Honor of Sara A. Immerwahr*. American School of Classical Studies at Athens, p. 144, for a discussion of the number of goddesses represented.

66. Kritikos, P. G., and S. P. Papadaki (1967), The history of the poppy and opium and their expansion in antiquity in the eastern Mediterranean area. *Bulletin on Narcotics* 19:3.

67. *The Odyssey*, Book IV; Harvard Classics 1909–1914, trans. S. H. Butcher and A. Lang. Collier & Son.

68. *Historia Plantarum*, ix 9.16.8

69. Kritikos and Papadaki, The history of the poppy and opium.

70. Dickinson, *The Aegean from Bronze Age to Iron Age.*

71. Cline, E. H. (2015), *1177 B.C.: The Year Civilization Collapsed*, revised edition. Princeton University Press.

72. Desborough, V. R. d'A. (1972), *The Greek Dark Ages*. Ernest Benn Limited.

7

The "Plantheon" of Greek Mythology

THUS FAR, WE have seen that it was an agricultural innovation in Mesopotamia—the artificial pollination of female date palms—that provided the first demonstration that at least one kind of tree required pollen for fruit production, although there is no evidence that this finding was ever generalized to other plants. Rather, in the Near East, the productivity of date palms became a metaphor for human and agricultural fertility, as portrayed in religious iconography as well as in myths and poetry invoking the goddess Inanna. In Egypt, the goddesses Hathor, Nut, and Isis were strongly identified with fruit trees and were often depicted providing nourishment to the deceased in the afterlife. In the Aegean, sacred trees, poppies, saffron crocuses, and other plants often associated with goddesses acquired symbolic significance in rituals preceded over by priestesses. As mentioned earlier, there are intriguing parallels between the planting cycle of the saffron crocus and the Demeter/Kore myth of Greece. Such examples indicate that, prior to the emergence of a competing scientific paradigm, agricultural knowledge was transformed into myth and integrated into a religious worldview that associated agricultural abundance with women and goddesses.

We now turn to the Greeks, the first people of antiquity to develop cosmologies based on principles of causation to explain the substance and behavior of the material universe. However, the transition from myth-based to logic-based belief systems was never complete, nor did it occur linearly. Contingency played an important role. The sudden collapse of the Mycenaean palace societies at the end of the Bronze Age allowed previously marginalized, smaller, and more egalitarian societies to proliferate, and the subsequent "Dark Age" that engulfed Greece after the twelfth century BCE can be thought of as a latency period during which the Greeks were reinventing themselves.

The primary form of government during this period was rule by chieftains called *basileis* (singular, *basileus*)—the real life models for the Homeric "kings" described in the *Iliad* and the *Odyssey*.[1] The mainland population was divided into independent mini-states, each of which was called a *demos* (plural, *demi*). A demos typically included several settlements, each of which consisted of a town or village plus adjoining farm and pasture lands. The largest town within a demos was referred to as the *polis* (plural, *poleis*), from which the words "policy" and "politics" are derived. The basileus resided in the polis and ruled with the aid of a council of lesser chiefs. Policy decisions were usually discussed with a citizen assembly to ensure their cooperation. These assemblies were restricted to men of fighting age or older and can be viewed as Dark Age precursors to the more complex democratic institutions of Classical Greece.

An important technical advance around this time, one that facilitated communication both within poleis and between distant poleis, was the adoption of the Greek alphabet around 750 BCE. Greeks improved the consonant-only Phoenician script by adding five vowels, making it the first true alphabet capable of recording everyday speech. The Greek alphabet soon stimulated a creative outpouring in the arts and sciences, providing posterity with unprecedented access to the thoughts of some of the greatest minds of the ancient world.

The lasting cultural achievements of the Mycenaeans were primarily material in nature—cyclopean walls, palaces, and tombs laden with gold. In contrast, the enduring legacy of the Dark Age in Greece was political. For the first time, ruling oligarchs were forced to consult small, but assertive, citizen assemblies before implementing public policies or engaging in battles with neighboring groups. Over the centuries, such citizen assemblies grew into formidable democratic institutions. The prickly tendency of Greek citizens to question, challenge, and debate public policy (dramatized by Homer in the war councils of the *Iliad*) helped to define a new, self-confident Greek character, which eventually manifested itself in every facet of Greek life, leading some Greeks—those who called themselves "philosophers" or "lovers of learning"—to risk heresy by questioning the very existence of the gods.

Yet, despite the best efforts of the philosophers, the vast majority of Greeks continued to adhere to traditional beliefs, praying to the same deities and celebrating the same religious festivals as their ancestors had done. Even philosophers participated in important religious festivals, and although some rejected the concept of personified deities, they retained the ancient belief in the divinity of heavenly bodies. Their proto-scientific views on an array of topics, including plant sex, must therefore be considered against the backdrop of Greek mythology and religion.

HESIOD'S *THEOGONY* AND THE GENEALOGY OF THE GODS

We know from translations of the Linear B tablets that the Mycenaean pantheon included such familiar names as Zeus, Hera, Poseidon, Hermes, Athena, Artemis, and possibly Apollo, Ares, and Dionysus.[2] Additional deities were probably absorbed from Near Eastern sources between 750 BCE and 650 BCE during the artistic period known as the "orientalizing" phase.[3] By the beginning of the eighth century BCE, the Greek pantheon was essentially complete.

Hesiod's *Theogony*, written near the end of the eighth century BCE, has been called "a basic textbook of Greek religion." [4] It presents a genealogy of the Greek gods and goddesses and chronicles the ascension of Zeus to the position of ruler of the universe. This collection of myths also provides important insights into Greek ideas about gender and a wide array of other topics. For example, the idea that female sexuality is inherently dangerous and threatening to male authority is a recurring theme, providing Athenian men with religious justification for the many restrictions placed on women's freedom.[5] In Homer's *Iliad*, Helen of Troy is the classic example of the terrible power of female beauty. The satirist Semonides, who lived in the seventh century BCE, cited the example of the bee, believed to be asexual, as a symbol of the ideal woman. According to Aristotle, bees are disgusted by sex and obtain their young chastely from flowers, reeds, or olive trees, which are themselves symbols of chastity. The ideal Greek wife, Semonides asserted, spends all her time toiling away at her household chores like a bee, thus keeping her appetites and passions in check.[6]

According to Hesiod, in the beginning, the sky-god Ouranos (Uranus), who was married to Gaia, the earth goddess, imprisoned his three sons, the one-eyed Cyclopes, in his wife's womb (the Underworld). His motive for doing so was to prevent them from reaching adulthood and overthrowing him. Ouranos and Gaia then had additional children: the Titans. Fearing that Ouranos would imprison the Titans as well, Gaia persuaded them to attack their father. She armed one of them, Cronos, with an adamantine sickle. The choice of a sickle, the implement used to harvest grain, alerts us to the myth's connection to agriculture. The Titans attacked Ouranos, and Cronos used the sickle to cut off his father's genitals, throwing them into the sea. But the severed genitals performed one last generative act. From the white foam that formed around them, Aphrodite, the Goddess of Love, was born. This is Hesiod's first example of a male deity's parthenogenic powers.

Cronos married his sister, Rhea, the Mother of the Gods, who bore him several children, among them Demeter, Hera, Hades, and Poseidon. Hoping to avoid his father's fate, Cronos swallowed them all. However, Rhea managed to save her infant son, Zeus, by hiding him deep inside a cave on the island of Crete. When he had grown to a man, Zeus fought and overthrew Cronos and forced him to disgorge the gods and goddesses he had swallowed. Upon ascending the throne, Zeus ensured he would never be overthrown by his sons by swallowing his pregnant wife, Metis. Zeus thus acquired the ability to give birth himself, as he did in the case of Athena, who emerged fully grown from his head (Figure 7.1).

A monument dating to the middle of the fourth century BCE depicts Zeus's maternal aspect, with multiple breasts as a symbol of fertility.[7] Figure 7.2A shows a relief of *Zeus Labrandeus*, which was dedicated by King Idrieus and his wife/sister Ada of the Kingdom of Caria, located in the southwestern corner of Anatolia. Zeus is shown bearded and holds a spear and a labrys, or double-headed axe. Six breasts are arranged in an inverted triangle on his chest, reminiscent of the Ephesian Artemis. According to Herodotus, Zeus Labrandeus was worshipped "in the sacred grove of plane trees," the traditional domain of Bronze Age vegetation goddesses.[8] Multiple breasts indicating fruitfulness and abundance are a standard feature of statues of Artemis of Ephesus (Figure 7.2B).

Although the *names* of many important Greek deities were carried over from the Bronze Age, we have no way of knowing whether the myths were carried over as well. Linear B was not used for literary purposes, and no known Mycenaean artwork depicts the later Greek deities listed in the Linear B tablets. Thus, while it is tempting to assume that the myths in Hesiod's *Theogony* are directly related to those of the Mycenaean deities, it is also possible that the identities and myths were changed substantially during the Dark Age. Only in the

FIGURE 7.1 Athena's birth from Zeus's head depicted on a vase.
Two-handled jar (amphora) depicting the birth of Athena, Accession Number 00.330; from Digital Image Resources, Boston Museum of Fine Arts: 617-369-4338/mfaimages@mfa.org.

case of Zeus, whose role as the sky-god can be traced to Indo-European origins, has continuity with the Bronze Age been demonstrated.

DEMETER/KORE AND THE ELEUSINIAN MYSTERIES

In the highly stratified monarchies of Bronze Age Mesopotamia, sacred marriage rituals were enacted in which a priestess, as proxy for Inanna, symbolically granted the Goddess's imprimatur to the King's legitimacy. In Aegean religions, groups of priestesses appear to have presided over rituals involving powerful female deities, indicative, perhaps, of increased agency and autonomy compared to their Mesopotamian counterparts.[9]

From girlhood to old age, Greek women played important roles in religious festivals and rituals. The most famous and enduring of all the Greek religious festivals, which lasted until the end of the Roman Empire, took place at the temple of Demeter at Eleusis, a settlement near Athens on the west coast of Attica overlooking the island of Salamis. Held annually from late August to early September, the Eleusinian Mysteries coincided with the end of the harvest season, prior to the fall planting, when the grain was stored in underground containers and the fields looked desolate and barren. Already well-established by the seventh or sixth centuries BCE, the Eleusinian Mysteries was a major religious festival, attracting seekers and dignitaries from all over the ancient world. Participation in the mysteries was

(a)

(b)

FIGURE 7.2 Multiple breasts as symbol of fertility and abundance. A. Relief of Zeus Labrandeus, c. 350 BCE, with six breasts arranged in an inverted triangle on his chest. British Museum, London. B. Statue of Artemis of Ephesus.
A is from Delcourt, M. (1961) Hermaphrodite: Myths and Rites of the Bisexual Figure in Classical Antiquity, trans. Jennifer Nicholson. Studio Books.

thought to place the individual in direct contact with the divine, and initiates swore never to reveal the details of the proceedings.

So scrupulously did the initiates adhere to their vows of silence that the actual events of the Eleusinian Mysteries remain obscure. However, rituals similar to the Eleusinian Mysteries were practiced throughout the Mediterranean region, and possible scenarios have been pieced together from a few fragmentary accounts and from descriptions of these other festivals. Greeks, as polytheists, were generally receptive to foreign deities, and they readily incorporated many aspects of foreign religions into their own.

Festivals in honor of the Phrygian goddess Cybele were held in Greek city-states in Ionia on the west coast of Anatolia as early as the seventh century BCE. Cybele was known as the "Great Mother," or simply as "Meter," by the Ionian Greeks. A cult of Meter was established on the mainland in southern Thessaly in 464 BCE, and this foreign goddess soon came to be identified with the Greek Demeter.[10] Both were mother goddesses, and their devotees achieved transcendence through ecstatic rituals and dance.

There are also similarities between the myths of Demeter and Isis that suggest parallels, if not a direct relationship, with Egyptian rituals. Isis was a mother goddess, often depicted with her infant son Horus, and she was also closely associated with agriculture, especially the annual flooding of the Nile River. According to tradition, it was Isis who taught the Egyptians how to make linen from flax, and she was also known as "the Lady of bread, of beer, of green fields."[11] Under Ptolemy III, the rising of Isis's star, "Sothis" (probably Sirius), was celebrated with a festival marking "the new year, the summer solstice, and the beginning of the inundation." [12] Herodotus equated her with Demeter, and, over the centuries, Isis was assimilated in the Graeco-Roman pantheon and merged with both Demeter and Persephone.

The festival that was the most closely related to the Eleusinian Mysteries was the *Thesmophoria*, which literally means "law-bearer." The name "law-bearer" may refer to the belief that Demeter brought the knowledge of planting crops to the wandering Greeks, which caused them to settle down and institute laws. Alternatively, "law-bearer" might refer to the rules of the ritual that Demeter was supposed to have taught to Greek women. The Thesmophoria were celebrated by women throughout Greece in late October, at the time of the olive harvest and the fall grain planting. The central ritual of the Thesmophoria was the sacrifice of a pig. It was believed that mingling the flesh of the pig—an animal symbolizing fertility—with grain, would guarantee an abundant harvest. Only married women participated in the Thesmophoria, emphasizing the connection of the ritual with reproductive powers, and sexual abstinence was required to ensure that their sexual energy was focused on the fertility of the crops.

Both the Eleusinian Mysteries and the Thesmophoria took as their main text the "Hymn to Demeter," probably written sometime after Homer in the seventh century BCE. As outlined in Chapter 6, the hymn recounts the story of Demeter and her daughter, Kore ("The Maiden" or "Virgin") who became separated from each other when Hades abducted Kore while she was picking a flower "wondrous and bright" and carried her down to the underworld to become his bride. In the Homeric "Hymn to Demeter," the "flower-faced maiden" describes her abduction to her mother as follows:

> We were all in a beautiful meadow . . .
> playing and picking lovely flowers with our hands,
> soft crocus mixed with irises and hyacinth,
> rosebuds and lilies, a marvel to see, and the
> narcissus that wide earth bore like a crocus.
> As I joyously plucked it, the ground gaped from beneath,
> and the mighty lord, Host-to-Many, rose from it
> and carried me off beneath the earth in his golden chariot
> much against my will.[13]

Disconsolate over the loss of her daughter, Demeter searches all over the world for her in vain. She finally learns of her daughter's fate from the goddess Hekate and the sun-god Helios, who confess to her the role of the Olympian deities in Kore's abduction.

There follows a lengthy interlude in which Demeter travels to Eleusis and establishes her temple there. At Eleusis, Demeter continues to mourn for her daughter. To exact revenge, she refuses to make the seeds sprout from the soil, and, as a result, famine spreads throughout the world. Struggling to survive, the people stop making gifts and sacrifices to the gods. To restore the normal order, Zeus sends Hermes to the underworld to convince Hades to release Kore and allow her to return to the surface. Hades reluctantly agrees, but, just as she is leaving, Hades ensures her return with one final trick. In the "Hymn to Demeter," Kore tells her mother what happened after she learned of her release:

> Then I leapt for joy, but he stealthily put in my mouth a food honey-sweet, a pomegranate seed, and compelled me against my will and by force to taste it. [14]

Having tasted the seed, Kore is now obliged to spend one third of every year (the winter) underground with Hades, returning to the surface each spring. Upon returning to the surface, Kore is referred to by her adult name, Persephone, indicating that her maidenhood is over. In the fall, she again becomes Kore in anticipation of her return to the underworld.

Hades tricks Kore twice, the first time with a beautiful flower, the symbol of virginity; the second time, after she has become his bride, with a seed, the symbol of fertility. The blood-red juice of the pomegranate suggests menstruation. The reproductive cycle of flowering plants thus serves as a metaphor for the female reproductive cycle, from her flower-like life as a virgin to her fruit-like life as a wife and mother.[15]

Demeter's ultimate power over crops (and, indirectly, the well-being of the Olympian gods) establishes her as the primary vegetation deity of Greece. It was Demeter who first taught the boy, Triptolemos, how to plant and harvest grain, considered by the Greeks to be the foundation of civilization. With the help of Demeter and Persephone, Triptolemos flew all over Greece in a winged chariot, sowing grain and educating his fellow Greeks in the art of cereal cultivation (Figure 7.3).

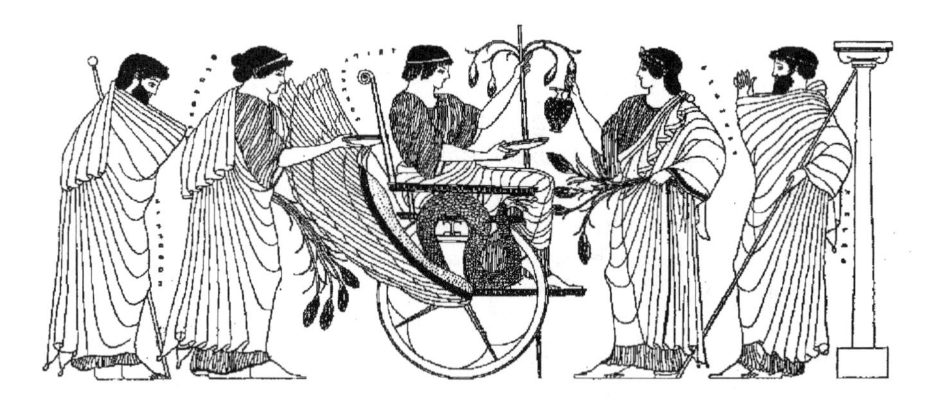

FIGURE 7.3 Triptolemos seated in his winged chariot, flanked by Demeter (left) and Kore (right). Krater from Agrigentro, Sicily; Museo Archeologico, Palermo.

The myth of Demeter and Persephone resonated with the Greeks on multiple levels. The most widely cited benefit to initiates of the Eleusinian Mysteries was the hope for a better fate in the afterlife. In Greek mythology, the dead initially went to the Underworld, a dark, misty place ruled by Hades. An elect few, the heroic and virtuous, could be rewarded by being sent to a paradise called Elysium, located at the "margins" of the world. Those who were judged to have committed crimes against the gods were sent to Tartarus, the Greek version of hell, a place of fire and brimstone. The afterlife for the vast majority of ordinary people, however, consisted of milling about disconsolately in the crowded, gloomy Underworld of Hades, yearning for their former lives on the surface. The Eleusinian Mysteries promised mitigation of this fate—an upgrade, if you will, from coach to first class.

According to later myths associated with the Eleusinian Mysteries, it was Demeter's and Persephone's protégé Triptolemos who judged the dead and decided their fate in the afterlife. Successful completion of the training and initiation rites of the Eleusinian Mysteries all but guaranteed a positive verdict from Triptolemos. As noted by Sophocles in a fragment from his play, *Triptolemos*: "Thrice blessed are those mortals who have seen these rites and thus enter into Hades: for them alone there is life, for the others, all is misery."[16] In a sense, the salvation conferred by Triptolemos in the Underworld was a mirror image of his role in spreading grain cultivation, which, according to Greek belief, rescued them from a bestial state.

APHRODITE'S GARDEN, ARTEMIS'S GARLAND

Aphrodite, the goddess of love, is also strongly associated with flowers. In a fragment from the *Cypriot Epics*, the seventh-century BCE epic poet Stasinos described the floral decorations on Aphrodite's garments:

> She clothed herself with garments which the Graces and Hours had made for her and dyed in flowers of spring—such flowers as the Seasons wear—in crocus and hyacinth and flourishing violet and the rose's lovely bloom, so sweet and delicious, and heavenly buds, the flowers of the narcissus and lily. In such perfumed garments is Aphrodite clothed at all seasons. Then laughter-loving Aphrodite and her hand-maidens wove sweet-smelling crowns of flowers of the earth and put them upon their heads—the bright-coiffed goddesses, the Nymphs and Graces, and golden Aphrodite too, while they sang sweetly on the mount of many-fountained Ida.[17]

The flowers associated with Aphrodite's garments and crown—so "lovely," "sweet," "delicious," and "heavenly"—are clearly metaphors for female beauty and attractiveness and are thus the appropriate sartorial expression for human love in "all seasons."

Aphrodite had her own temple in Athens, Aphrodite in the Garden, which underscores the goddess's association with flowers and trees. In midsummer, the temple was the site of the festival of Arrephoria, which, according to Bruce Thornton, was "a ritual connected both with the olive tree, one of the most important crops for the ancient Greeks, and with the initiation of girl-citizens, called 'dew-carriers,' into puberty."[18] The ritual involved the bearing of secret objects between the temple of Athena on the Acropolis to the nearby temple of Aphrodite in the Garden via an underground passageway, evoking fertility and the Underworld. After depositing their burdens at the temple of Aphrodite, the girls returned to the Temple of Athena, carrying some equally mysterious items. According to Thornton, the function of the ritual was to "harness the fecundating power of Aphrodite so that the olive and the Athenian people alike, especially its girls, were fruitful and multiplied."[19]

The goddess Artemis was a Near Eastern import and was referred to as the "Mistress of Animals" in the *Iliad*. In Archaic Art, she was often depicted as a huntress who "triumphantly slays her prey with bow and arrow." [20] Beginning with Homer, however, Artemis's Near Eastern fierceness was attenuated, and she came to be represented as an adolescent girl associated with female virginity.[21] In Euripedes's *Hippolytus*, the hunter Hippolytus gains partial access to the elusive goddess by presenting her with a garland of flowers plucked from "an inviolate meadow."

> For you, lady, I bring this plaited garland I have made, gathered from an inviolate meadow, a place where the shepherd does not dare to pasture his flocks, where the iron scythe has never come: no, it is inviolate, and the bee makes its way through it in the spring-time. Shamefaced Awe tends this garden with streams of river-water, for those to pluck who have acquired nothing by teaching but rather in whose very nature chastity in all things has ever won its place: the base may not pluck.[22]

By associating her with the purity and chastity of flowers, Euripedes symbolically associates Artemis with these ideal feminine qualities.

MINOR FEMALE VEGETATION DEITIES OF ANCIENT GREECE

Flowers were important attributes of Aphrodite, but Demeter and Persephone were by far the most important vegetation/agricultural goddesses of ancient Greece. Locally, they were sometimes worshipped by emphasizing different aspects. For example, *Demeter Chloe* (Demeter of Green Shoots), *Demeter Chthonia* (Underground Demeter), *Demeter Anesidora* (Demeter Who Provides Gifts from the Earth), and *Demeter Malophoros* (Demeter the Apple Bearer) emphasize four different attributes of the goddess. Greek mythology also included many lesser vegetation goddesses, consistent with a strong identification of plants with women.

Chloris, from the Greek word for pale green or greenish yellow, was a nymph associated with spring flowers, whom the Romans renamed Flora. Like Persephone, Chloris was abducted and then later married to her abductor. But instead of being abducted by Hades, Chloris was captured by Zephyrus, the West Wind. It was Zephyrus who gave Chloris dominion over spring, and together they had a son, Carpus ("fruit"). It is tempting to read into this marriage between flowers and wind a metaphor for wind as a "fertilizing" (pollinating) agent, which causes Chloris to produce fruit. Cereal crops, so central to Greek agriculture, are all wind-pollinated, but if wind pollination is being referred to, a more likely source for the idea is the date palm, as the role of wind in date palm pollination was well-known in the Near East. However, there is no direct evidence that the Greeks imagined that Zephyrus's role in fertilizing Chloris had anything to do with pollination. Instead, Zephyrus was likely associated with rain, which encourages the fruit to grow.

Fast-forwarding to Roman times, Ovid, in his epic poem *Metamorphoses*, compiled Greek myths in which "forms are changed into new bodies." Such metamorphoses symbolically associate the attributes of a person or deity with those of some object in nature. Not surprisingly, women were often changed into plants, usually plants whose stories concern sexuality.

For example, the nymph Mintha was changed into a mint plant. Mintha's story is intricately connected to Persphone's journey to the Underworld. Prior to Persephone's arrival, Mintha had been the mistress of Hades. When Hades announced his intention to wed Persephone, Mintha flew into a rage and provoked Persephone's anger by insulting her. According to one version of the myth, Persephone tore Mintha limb from limb, prompting Hades, as an act of mercy, to turn her into a mint plant so she would not feel the pain. Mint is sweet-smelling and freshens the breath, and Greeks considered it an aphrodisiac. At the same time, many garden mints are sterile hybrids. According to one myth, Demeter cursed the mint so it would not bear fruit, and eating mint before intercourse was therefore thought to prevent conception.[23]

Leucothoe was a princess, the daughter of King Orchamus, King of the Persians. When the sun-god, Apollo, fell in love with the king's daughter and slept with her, the king had Leucothoe buried alive in a trench and covered over with sand. By the time Apollo arrived to rescue her it was too late. Unable to revive her, he covered her with a fragrant nectar and vowed that she would rise again. Apollo's nectar melted Leucothoe's body away, and in its place grew the shrub, *Boswellia sacra*, from which the aromatic resin frankincense is obtained. [24]

The girl Myrrha (or Smyrna), had strong Oedipal feelings toward her father, Cinyras. One night, while her mother was away at a Thesmophoria festival, Myrrha disguised herself as Cinyras's mistress, slipped into his bed, and had sexual intercourse with him. She then repeated the illegal act twelve consecutive nights. When her father discovered her true identity, he chased her with a sword, and when he caught her threatened to kill her. Myrrha prayed to the gods to make her invisible, and they responded by changing her into a fragrant myrrh tree. Ten months later, the god *Adonis* was born from the Myrrh tree, tying him closely to the vegetable realm. The physical identification of the tree with Myrrha's body is graphically illustrated in the sixteenth century illustration shown in Figure 7.4.

In the satirical play *Lysistrata* by Aristophanes, the Greek women attempt to bring an end to the Peloponnesian war by withholding sex from their husbands. Lysistrata urges the women first to inflame their husband's sexual appetites before refusing to consummate their desires. The character Myrrhina ("Little Myrtle") provides the perfect demonstration of this technique:

> Pretending to yield to her husband's desire, Little-Myrtle allows herself to be coaxed into the grotto of Pan ... but she then finds a thousand excuses for procrastination. ... Finally, just as she uncovers her breasts, Myrrhina remembers they have both forgotten to rub themselves with perfumes. Despite the protestations of her husband, she runs off to fetch a flask of balm and carefully rubs herself with it, inviting her husband to do likewise. Then just as the wretched man, by now consumed with desire, thinks he will clasp his wife in his arms, Myrrhina slips away, this time for good.[25]

FIGURE 7.4 Birth of Adonis from myrrh tree. The infant Adonis is shown exiting a large vulva at the base of the tree. Sixteenth-century Italian plate.
Francesco Durantino, Italian (Urbino, 1543–75). Philadelphia Museum of Art, Accession Number: 2001-148-3.

According to Marcel Detienne, Aristophanes's use of the name "Little Myrtle" is highly appropriate for a seductress because of its erotic overtones:

> The branches of this aromatic shrub are used, in Attica, to weave the crowns worn by marrying couples. The name of this plant, which is consecrated to Aphrodite, is used to refer to either the clitoris or the pudenda of the woman. Thus the perfume in which Myrrhina smothers herself is simply the ultimate expression of the seductive attraction emanating from a woman totally committed to Aphrodite. . . . The resemblance between Myrrhina and the Myrrha who seduces her father is all the greater in that in one of the versions of the myth of Adonis, his mother is transformed into not a myrrh tree but a sprig of myrtle. [26]

Although not a deity, Aristophanes's Myrrhina is emblematic of the strong association of myrtle with female sexuality.

Several other minor female vegetation deities underwent metamorphoses into plants to escape sexual advances. Lotis was the name of a Naiad Nymph, a water nymph associated with the springs of the river Sperkheios in northern Greece. Lotis metamorphosed into a lotus flower to escape the god Priapus, of the enormous phallus, who was pursuing her.

The myth of the goddess Dryope reprises the same theme of escape from a sexual predator by metamorphosis into a plant. In this case, Apollo at first succeeds in raping Dryope by disguising himself as a turtle and then changing into a snake when Dryope places him on her lap. Later, Apollo, disguised as a snake, appears to Dryope again as she drinks from a spring. However, Dryope frustrates his lascivious intent by turning herself into a poplar tree.

The goddess Daphne ("laurel") provides yet another example of escape from unwanted sexual advances by transformation into a plant. Once again, Apollo is the pursuer, chasing the frightened goddess to the river of her father, the god Peneus. Peneus saves her by turning her into a laurel tree, which prompts the saddened Apollo to make a laurel wreath for himself (Figure 7.5). As a result, the laurel tree became sacred to Apollo. Metaphorically, the transformed plants are still gendered female, but they no longer have "sexuality."

Three other female agricultural/nature deities, the Horae ("the hours"), are worthy of note. The Horae represented the orderly progress of time throughout the agricultural year, which the Greeks divided up into the three (not four) seasons: autumn, spring, and summer. The three earliest Horae were associated with different stages of plant development. Auxo ("one who increases") was worshipped as the goddess of plant growth, Thallo (literally "one who flowers") was the goddess of spring blossoms, and Carpo ("bringer of food") was associated with fruit ripening.

Finally, in Book Six of the *Odyssey*, Odysseus encounters the fair *Nausicaa* after being shipwrecked on the coast of Phaeacia. Dazzled by her appearance, he sings her praises. "I never yet saw any one so beautiful," he tells her, "neither man nor woman, and am lost in admiration as I behold you." He goes on to compare her, somewhat incongruously, to a splendid palm tree he once saw on the isle of Delos:

> I can only compare you to a young palm tree which I saw when I was at Delos growing near the altar of Apollo—for I was there, too, with much people after me, when I was on that journey which has been the source of all my troubles. Never yet did such a young plant shoot out of the ground as that was, and I admired and wondered at it exactly as I now admire and wonder at yourself.[27]

Since there is no obvious resemblance between a beautiful woman and a palm tree, Odysseus may be referring to an epiphany of a tree goddess. Presumably, the reference to the palm tree of Delos refers to one of the many sacred groves of date palms, typically dedicated to a goddess and tended by a priestess, which were quite common in the ancient Mediterranean world. Although not a deity herself, Nausicaa's association with date palms identifies her with the ancient tradition of tree goddesses or their priestesses.

FIGURE 7.5 Statue of Daphne and Apollo (1622–24) by Gianlorenzo Bernini. Daphne is metamorphosing into a laurel tree to escape Apollo's amorous pursuit.
Galleria Borghese, Rome; photo by Alinari, Florence. From Delcourt, M. (1961) *Hermaphrodite*, trans. Jennifer Nicholson. Studio Books.

YOUTHFUL MALE VEGETATION DEITIES

Although most Greek vegetation deities were female, some were male. In Bronze Age societies throughout the Mediterranean, the withering of flowers, garden vegetables, and cereal crops under the blazing heat of the sun during the dry season was commemorated in various myths involving young male gods who were the consorts of important goddesses, cut down in the prime of their youth. Some examples include the Babylonian Tammuz (Sumerian Dumuzi), the consort of Ishtar/Inanna; the Phrygian Attis, consort of Cybele; and the Egyptian Osiris, the brother-consort of Isis. A Babylonian hymn employs plant drought-stress metaphors to describe Tammuz:

> A tamarisk that in the garden has drunk no water,
> Whose crown in the field has brought forth no blossom.
> A willow that rejoiced not by the watercourse,
> A willow whose roots were torn up.
> An herb that in the garden has drunk no water.[28]

The Greeks, too, had several versions of what James Frazer referred to as "dying and reviving gods" of vegetation, including Adonis, Hyakinthos, and Dionysus. According to Frazer, the deaths of the young male gods were recognized by the Greeks as tragic but necessary aspects of the agricultural cycle, comparable to Persephone's abduction to the Underworld. Indeed, flowering plants are said to have sprouted from the blood of the dying youths: anemones from the blood of Adonis, hyacinths from the blood of Hyakinthos, and pomegranates from the blood of Dionysus.

As we have already seen, Adonis was the "fruit" of the myrrh tree, identifying him as a vegetation deity from the spice-growing lands of the East. So great was Adonis's beauty, even as an infant, that the love goddess Aphrodite jealously hid him from the other gods in a chest, which she entrusted to Kore. But when Kore beheld Adonis, she, too, coveted him, and an argument between Aphrodite and Kore ensued. The two goddesses compromised by allowing Adonis to spend eight months of the year above ground with Aphrodite and four months of the year below ground with Kore. The fact that Adonis spends most of the year with Aphrodite indicates that he is predominantly associated with sexual love, although his relationship with Kore suggests that he is also identified with the earth. Frazer, in his interpretation, emphasized the similarities between the Adonis and Persephone myths, both of which involve annual cycling between the surface and the Underworld.

The cult of Adonis was a privately sponsored mystery religion primarily attracting women. The festival of Adonia was held in mid-July on the rooftops of private homes. According to Bruce Thornton, the participants in the Adonia festival included both men and women:

> especially prostitutes, courtesans, and their lovers: women, that is, whose sexuality was its own pleasurable end rather than directed to the procreative needs of the city. . . . The celebrants ate and drank liberally, exchanged bawdy jokes, and the women, bedecked and perfumed, made love.[29]

The festival focused on the carnal aspects of Aphrodite's obsession with Adonis. Plants played a key role in the festivities in the form of "Adonis Gardens." The women planted seeds of wheat, barley, lettuce, and fennel in pots, baskets, and pottery shards, providing them with just enough water to allow germination. After growing for about eight days on the root-tops of the houses, the seedlings were allowed to wither under the hot summer sun. The women then carried them to the shore and cast them into the sea, wailing over the death of Adonis. The withered seedlings symbolized the youthful god, who dies without "flowering"—a virgin, but an erotically charged virgin. In Frazer's view, Adonis's death is necessary for the natural cycle because his virility and eroticism help to make the earth more fertile and the cereal crop more productive, just as Kore's female sexuality is brought underground each year to replenish the earth. According to Frazer, the fact that Adonis becomes Kore's lover for part of the year reinforces the connection between the two myths.

Marcel Detienne disputes Frazer's assertion of a link between the Adonis gardens and the grain harvest.[30] Having been born from a myrrh tree, Adonis is strongly associated with aromatic spices. Spices in Greek mythology are associated with seduction, and Adonis was the ultimate seducer. Aphrodite and Persephone became infatuated with him when he was only an infant, hence their love can hardly be sexual. According to Detienne, Adonis represents the antithesis of the virile lover. Greeks regarded him as an effeminate god, and although men participated in the Adonia festival, the cult of Adonis was practiced almost exclusively by women. Even his manner of death—being gored by a boar—was by Greek standards an unmanly and unheroic death. As he lay on the ground bleeding to death, his blood gave rise to anenomes, delicate flowers whose petals fall off at a single touch. In a warrior culture in which men were trained to be ruthless in battle and domineering at home, it would not be surprising if women found the idea of the beautiful and completely nonthreatening lover, Adonis, irresistibly attractive and erotic, like the aromatic perfume from which he was born. Contrary to Frazier, Detienne argues it was not the infant Adonis's virility that made him so attractive to Aphrodite and Kore, but his very *lack* of virility. Seen in this light, the "Adonis gardens" withering in the sun before they reach sexual maturity are not symbols of agricultural fertility, as Kore's marriage to Hades seems to have been, but of the lack of sexual fulfillment resulting from the tragic death of the youthful lover. The gender qualities that made Adonis irresistible to Greek women—innocence and delicacy—were the opposite of those manly qualities held in highest esteem by a warrior society. In Detienne's interpretation, Adonis's gender identification tilts toward the feminine.

Two other male vegetation deities, Narcissus and Hyacinthus, exhibit a similar type of gender ambiguity. Like Adonis, both Narcissus and Hyacinthus were exceptionally beautiful youths. Narcissus was the son of a river god and a forest nymph, connecting him to vegetation. Although the forest nymph Echo was attracted to him, Narcissus rejected her and, instead, fell in love with himself, spending his days admiring his own reflection in pools formed by the river. As punishment for his vanity, the gods caused him to fall into the river and drown. Narcissus's rejection of Echo underscores his lack of sexuality, and his self-love carries with it no sexual connotations. From the perspective of a warrior society, his obsessive vanity points to a feminine rather than a masculine character. His transformation into a sweet-smelling flower, usually associated with women, is consistent with his gender ambiguity.

The myth of Hyacinthus, best known in Sparta, also involves the premature death of a handsome youth. In Ovid's version, Apollo was Hyacinthus's lover, in keeping with the Greek—and particularly the Spartan—practice of pederasty, whereby an older male served as the lover and mentor of an adolescent boy. One day Apollo and Hyacinthus were taking turns throwing the discus. Hyacinthus tried to chase and catch the discus thrown by Apollo, but the object glanced off a rock and struck the boy in the face, killing him. The Spartan festival Hyacinthia was held every summer to commemorate his death and celebrate his rebirth as the flower Hyacinth. Frazer observed that because Hyacinth's lover was a male, he didn't fit into the mold of an agricultural deity symbolizing fertility. Therefore, he hypothesized that the original pre-Greek myth involved a heterosexual relationship between Hyacinth and a goddess and that the Greeks had altered it to a pederastic relationship. However, the Greeks themselves knew the myth as Ovid described it. As Apollo's young lover, Hyacinthus's gender is ambiguous and effeminate. The fact that he was killed by his own ineptness in catching a discus also makes his death, like the death of Adonis, unmanly and unheroic. Thus, his transformation into a flower is consistent with his lack of masculinity. Ovid compares his death to the picking of a flower:

> As in a garden, if one breaks a flower,
> crisp violet, or poppy, or straight lily
> erect with yellow stamens pointing high,
> the flower wilts, head toppled into earth.[31]

By comparing Hyacinthus's death to the death of a plucked garden flower, Ovid emphasizes his delicate, fragile beauty. While it is true that Ovid does mention "yellow stamens pointing high," which, if he were writing in the late eighteenth century, one might interpret as a sexual reference, there is no evidence that the sexual function of stamens was known in Ovid's time. The "pointing" stamens in Ovid's metaphor function to illustrate the turgor pressure of a fresh, living flower as opposed to a plucked, wilted one.

DIONYSUS/BACCHUS: WINE AND ECSTASY

The one male vegetation deity who clearly represents sexuality in its frankly phallic form is Dionysus, whom the Romans referred to as Bacchus. Dionysus is the god of wine and drama, and he was also associated with a transcendent state. Unlike Adonis and the other youthful vegetation gods who died young, Dionysus attained adulthood after a life of travel and adventure in which he survived numerous close calls with death. According to the Theban myth, his mother was a mortal, Princess Semele of Thebes, and his father was Zeus. When Semele was seven months pregnant with Dionysus, Hera, jealous of Zeus's mistress, told her that Zeus was *not* the true father. Semele then demanded that Zeus show himself to prove he was the father. Knowing that the sight of him would be lethal to a mortal, he resisted at first, but finally yielded to her persistent demands. As expected, when he materialized before her arrayed in lightning bolts, Semele was instantly incinerated. Zeus somehow managed to save the fetus, however, sewing him into a slit in his thigh. Two months later, Dionysus was born and was thenceforth known as the "Twice-Born."

FIGURE 7.6 Birth of Dionysus from Zeus's Thigh (Proto-Apulian, c. 390 BCE).
Proto-Apulian red-figure volute krater from Ceglie del Campo, late fifth to early fourth centuries BCE. Museo Nazionale, Taranto.

The story of Dionysus's birth reprises the theme of Zeus's procreative power, first demonstrated by Athena's birth from his head. But instead of emerging from his father's head, Dionysus emerges from a slit in his thigh—suggesting a vulva-like opening (Figure 7.6). Unlike Athena, who represents an extension of Zeus's head (i.e., masculine intellect), Dionysus draws his godlike power from his father's faux-vulva (i.e., feminine passion).

Dionysus is therefore a male deity with a feminine orientation (he was often depicted wearing women's robes), and his psychological qualities are those that Greek writers, from Hesiod onward, attributed to women. Whereas male attributes included reason, sobriety, and the rule of law, women were associated with emotionalism, unbridled sexuality, and a tendency toward indulgence of the appetites.[32] In general, women were considered to be closely allied with the dangerous forces of nature, which it was necessary to rein in and subdue if civilization was to flourish. Dionysus thus came to be identified with wine, whose power to subvert normal social inhibitions and induce wild, emotional, and disorderly conduct was closely associated with the god.

Two types of Dionysian rites and festivals were held, one appealing to men and the other to women.[33] The male rituals were essentially drinking parties where politics, literature, and philosophy were discussed, as in the Symposia, or where men attempted to drink each other under the table, as on the second day of the Anthesteria, the spring festival of Dionysus.

FIGURE 7.7 Offerings made before the image of Dionysus: Attic vase from Campania, fifth century BCE.
From The State Hermitage Museum, St. Petersburg, Russia.

The women's rites of Dionysus were ecstatic spiritual experiences that went well beyond the confines of state religion and were viewed with suspicion by the male religious establishment. The participants were known as Maenads (*mainades* or "mad women"), suggesting strong masculine disapproval (Figure 7.7). Little is known about their activities, but the wearing of animal skins, the use of wands or *thyrsoi* (staffs tipped with pine cones), ritual chants, and "frenzied dancing to the music of drums and flutes" were probably typical elements.[34] In contrast to the masculine rites of Dionysus, little drinking seems to have occurred during the women's Dionysian rites. Symbolic sexuality in the form of a sacred marriage was a part of the Athenian Anthesteria, and, in some Dionysian festivals attended by both men and women, phallus-shaped objects were carried in street processions. For men and for women, the rites of Dionysus afforded a socially acceptable release from the oppressive restrictions of everyday life.

Just as Demeter was said to have brought grain cultivation to the Greeks, Dionysus was credited with teaching the Greeks the art of wine-making. According to the poet Nonnus, Dionysus fell in love with a beautiful satyr youth named Ampelos. Ampelos was killed when the bull he was riding was stung by a gadfly, but the Fates transformed him into a grape vine, from which Dionysus is said to have squeezed the first wine. In Ovid's version of the myth, Ampelos was killed when he fell from an elm tree while attempting to pick grapes that had grown on its branches.[35] Ampelos thus belongs to that class of minor dying young male vegetation deities who avoid the Underworld by being transformed into flowers or fruits. Ampelos's identity as an immature, feminized boy is suggested both by his pederastic relationship with Dionysus and by his unheroic death caused by falling either from a bull or a tree.

AESCHYLUS'S FURIES: ANCIENT AGRICULTURAL GODDESSES

Beginning in the sixth century in Ionia, ancient myth and religion co-existed with the new rational/secular philosophy in Greek culture. As is true in any age, the latter way of thinking impacted ordinary citizens hardly at all. Not only did it fail to displace religion, the new natural philosophy was sometimes erroneously deployed to shore up cherished beliefs.

The presumed physical and mental superiority of men over women, fortified by myth, was a central tenet of the Greek patriarchal belief system and was used to justify an entire corpus of law depriving women of equal rights. A self-serving example of the use of reason and logic to validate legal arguments in favor of patriarchal rule can be found in Aeschylus's *The Furies*, the Roman word for *Erinyes*, or "angry ones," the last play in his *Oresteian* trilogy. In this famous sequence, Aeschylus invokes both myth and pseudo-science to assert that fathers are the sole biological parents of their children.

To propitiate the wind gods so that the Greek fleet can sail to Troy and achieve victory, Agamemnon has sacrificed his daughter, Iphigenia. Ten years later, Agamemnon returns home from Troy with his concubine, the captured Trojan princess Cassandra, and is murdered by his wife, Clytemnestra, in revenge for sacrificing their daughter. Clytemnestra is then murdered by her son, Orestes, in revenge for killing his father. Orestes believes that his act of matricide was sanctioned by Apollo and that he will be forgiven by the gods. Hounded by the Furies—ancient female spirits—who demand his punishment for the crime of matricide, Orestes seeks protection from Apollo. In true Athenian fashion, Apollo calls upon Athena and a panel of judges to decide Orestes's fate.

A panel of judges, including Athena as chief judge, is soon assembled, with the Furies acting as prosecutors and Apollo speaking on Orestes's behalf. In the testimony that follows, Orestes claims that his action was justified and that the Furies should have hounded his mother for killing her husband. The Furies counter that a child's bond with its mother takes precedence over a wife's bond with her husband because "[t]he man she killed was not of her own blood." Orestes then asks, "But am I of my mother's?" The Furies express outrage at Orestes's sacrilege and denigration of the mother's role: "Vile wretch, she nourished you in her own womb. Do you disown your mother's blood?"

Orestes then appeals to Apollo to testify on his behalf, and Apollo obliges with the following "scientific" argument, employing logic in the manner of the rationalistic philosophers:

> The mother is no parent of that which is called
> her child, but only nurse of the new-planted seed
> that grows. The parent is he who mounts. A stranger she
> preserves a stranger's seed, if no god interfere.
> I will show you proof of what I have explained . . .
> There she stands, the living witness, daughter of
> Olympian Zeus.[36]

Here, Apollo expresses the conventional wisdom, traceable to the Bronze Age, that the female merely incubates the seed from the male, as the soil incubates and nourishes the growth of grain in the plowed field. Apollo then cites Athena's own birth from Zeus's head as further proof that men can father children on their own. At the end of the trial, the judges are divided, leaving it to Athena to cast the deciding vote. Her own birth having been cited in favor of Orestes's acquittal, Athena's verdict comes as no surprise:

It is my task to render final judgment here.
This is a ballot for Orestes I shall cast.
There is no mother anywhere who gave me birth,
and, but for marriage, I am always for the male
with all my heart, and strongly on my father's side.[37]

And so it was that Athena, the male-identified Goddess of Justice, was entrusted with the vindication of Orestes. But the Furies, who represent an even more ancient tradition, do not accept defeat gracefully. In their rage at the "younger gods," they hurl deadly threats of pestilence and agricultural disasters:

Gods, of the younger generation, you have ridden down
the laws of the elder time, torn them out of my hands.
I, disinherited, suffering, heavy with anger
shall let loose on the land
the vindictive poison
dripping deadly out of my heart upon the ground;
this from itself shall breed
cancer, the leafless, the barren
to strike, for the right, their low lands
and drag its smear of moral infection on the ground.[38]

Mindful of their power to inflict disaster, Athena offers the Furies a sacred dwelling where they will be worshipped as guardians of the law, and, with their acquiescence, the Furies become the Eumenides ("gracious ones"), and the reign of the old female-centered religion based on nature and agriculture is symbolically ended, thanks to the happy congruence of religion and pseudoscience.

HERMAPHRODISM AND THE GODDESS HERMAPHRODITE

Most flowering plants are hermaphroditic, a concept never explicitly applied to plants by the Greeks. This raises the question of whether or not Greeks were aware of hermaphrodism in nature, and, if so, what they thought of it.

One of the earliest references to hermaphroditic beings occurs in the poem *On Nature* written by the pre-Socratic philosopher Empedocles. According to Empedocles, humans were first generated in a sexually undifferentiated state, having neither male nor female qualities:

First there came up from the earth whole-natured outlines
Having a share of both water and heat;
Fire sent them up, wanting to reach its like,
And they did not yet show any lovely frame of limbs,
Nor voice nor again organ specific to men.[39]

The few surviving fragments of the poem leave out most of the details of the story. However, it appears that one of these presexual humanoids, upon seeing another of its own kind, was filled with sexual desire: "upon him comes also, through sight, desire for intercourse." But intercourse was impossible between sexless humanoids. To remedy the situation, Aphrodite, who presided over all sexual matters, tore up their bodies and deposited the pieces on the ground. Those falling onto cold, damp earth became women and those falling onto warm, dry earth became men. The Greeks believed that men had more heat than women, and, as a result of this, according to Empedocles, "men are dark and sturdier of limb and more shaggy."

In Empedocles's poem, hermaphrodism is regarded as a rather inconvenient, awkward, and unsatisfying intermediate stage in human development. The story presages Plato's tongue-in-cheek treatment of the origin of the sexes described in "The Symposium." However, in Plato's version of the story, human hermaphrodites begin as exalted god-like creatures whose pride and arrogance lead to their downfall. The story occurs in a famous passage in which Aristophanes, the comic dramatist, recounts the origin of heterosexual and homosexual love. In the beginning, says Aristophanes, there were three types of humans, male, female, and

> a third which partook of the nature of both, and for which we still have a name, though the creature itself is forgotten. For although "hermaphrodite" is only used nowadays as a term of contempt, there really was a man-woman in those days, a being which was half male and half female.[40]

Aristophanes describes these three types of humans as spherical, each with two sets of the appropriate body parts:

> [E]ach of these beings was globular in shape, with rounded back and sides, four arms and four legs, with one face one side and one the other, and four ears, and two lots of private parts, and all the other parts to match.

They could move backward or forward as they pleased, but when they broke into a run,

> they simply stuck their legs straight out and went whirling round and round like a clown turning cartwheels. And since they had eight legs, if you count their arms as well, you can imagine that they went bowling along at a pretty good speed.

Aristophanes is clearly pulling our leg—or legs, as the case may be. The males, he tells us, were descended from the sun, the females from the moon. Because of their strength and energy they soon became arrogant and tried "to scale the heights of heaven and set upon the gods." Annoyed by their aggressive behavior, Zeus called a meeting of the gods:

> At this Zeus took counsel with the other gods as to what was to be done. They found themselves in rather an awkward position; they didn't want to blast them out of existence with thunderbolts as they did the giants, because that would be saying good-bye

to all their offerings and devotions, but at the same time they couldn't let them get altogether out of hand.

Finally, Zeus devised a way to weaken them without destroying them by simply cutting them in half. Bisecting the original humans would also have the salutary effect of doubling their number, resulting in twice as many offerings to the gods. Pleased with his idea, Zeus lost no time in implementing it.

Aristophanes concludes his parable by explaining how Zeus's surgical solution gave rise to heterosexuals as well as to lesbian and gay lovers:

> And so, gentlemen, we are all like pieces of the coins that children break in half for keepsakes. . . . The man who is a slice of the hermaphrodite sex, as it was called, will naturally be attracted by women. . . . But the woman who is a slice of the original female is attracted by women rather than by men—in fact she is a Lesbian—while men who are slices of the male are followers of the male, and show their masculinity throughout boyhood by the way they make friends with men, and the delight they take in lying beside them and being taken into their arms.

Around 400 BCE, a new cult arose in which the androgynous Hermaphrodite was worshipped. According to the first-century BCE Greek historian Diodorus Siculus,[41] the deity was the child of Hermes and Aphrodite, whereas in Ovid's *Metamorphosis* Hermaphrodite is strongly associated with vegetation. According to Ovid, at the age of fifteen, Hermaphrodite, who was born a boy, wandered from his home in Mount Ida's caves to the land of the Carians, where he came upon a tempting pool of water surrounded by delicate grasses and ferns. This pool was the home of Salmacis, a water nymph who spent her time combing her hair and picking garlands of the sweet-smelling flowers that grew nearby. When she beheld Hermaphrodite by her pool, she was instantly smitten. Upon recovering her composure, she declared her love to him and proposed marriage. Having spoken, she approached him to give him a kiss, but the boy blushed and threatened to run away. Retreating at once, Salmacis begged him not to leave. The boy relented and Salmacis departed; after she was gone, he took off his clothes and dived into the pool. But Salmacis was only hiding. From behind a hedge she saw his nude body and was overcome with passion. "I've won, for he is mine!" she cried. Stripping naked, she dived into the pool, and embraced the struggling youth, "as though she were quick ivy tossing/her vines around the thick body of a tree":

> "Dear, naughty boy," she said, "to torture me;
> But you won't get away. O gods in heaven,
> Give me this blessing; clip him within my arms
> Like this forever." At which the gods agreed;
> They grew one body, one face, one pair of arms
> And legs, as one might graft branches upon
> a tree, so two became nor boy nor girl,
> Neither yet both within a single body.[42]

Significantly, Salmacis's embrace of Hermaphrodite is compared to the grafting of trees. Salmacis is, in fact, a vegetation deity. Once grafted to Salmacis, Hermaphrodite loses his virility and becomes "tamed" and "impotent." This loss of sexuality probably reflects Salmacis's dominance over Hermaphrodite in their relationship.

In Greek sculpture, Hermaphrodite is always represented with a female body and male genitals. As a cult figure, she is effectively a goddess who has acquired male procreative powers. The earliest known mention of the cult of Hermaphrodite in Greek literature is in a book called *Characters* by Theophrastus, a student of Arisotle's, widely regarded as the founder of the field of botany. This is particularly significant for our purposes since it establishes unequivocally that Theophrastus was familiar with the concept of hermaphrodism, even though he never applied it to plants.

Theophrastus wrote on a wide range of subjects, little of which has survived. In his book on *Ethics*, Theophrastus distinguished various human virtues and their related vices. In *Characters*, he personifies these vices in brief, satirical sketches of various Athenian types. Among the thirty characters Theophrastus lampoons are "The Flatterer," "The Surly Man," "The Officious Man," "The Offensive Man," and "The Loquacious Man." The reference to Hermaphrodites is found in the sketch of "The Superstitious Man":

> Also, on the fourth and the seventh days of each month he will order his servants to mull wine and will go out and buy myrtle-wreaths, frankincense, and smilax; and, on coming in, will spend the day in crowning the Hermaphrodites.[43]

The reference to the fourth day, considered the luckiest day for a wedding, suggests a possible association with the institution of marriage, in which men and women figuratively become one. Some examples of sculptures representing Hermaphrodite are shown in Figure 7.8. Often, the female figure is raising a long skirt revealing male genitals underneath.

Why did explicit sculptures of hermaphroditic deities suddenly appear in Athens in the fourth century BCE? Just prior to this time, in the latter half of the fifth century BCE, Athens was wracked by political and social disruptions brought about by the Peloponnesian War and two outbreaks of plague.[44] In addition, a major earthquake may have caused considerable damage around 430 BC. Many commentators have pointed out that an interest in foreign divinities coincided with this series of disasters. Cults related to healing arose throughout Attica during the plague and its aftermath. Other foreign divinities, such Isis, Cybele, Attis, and Adonis, all entered the Athenian pantheon late in the fifth century BCE as well. By 415 BCE, the population of Athens had been seriously depleted by the combination of war and plague. Thus one can imagine that a potent new fertility deity like Hermpahrodite would have been warmly embraced by Greek households during this bleak period.

But while androgyny and hermaphrodism were perfectly acceptable in gods and goddesses, biological hermaphrodites—humans who possessed both male and female genitals—were considered to be monstrosities. When a child was born either with real or imagined signs of hermaphrodism, the entire community felt threatened by the gods. The child was usually left exposed outdoors and died. In Roman times, even worse fates fell to those unfortunate few who manifested hermaphroditic characters after reaching adulthood. They were reportedly burned alive to appease the gods.[45]

(a)

FIGURE 7.8 Statues of Hermaphrodite. A. Hellenistic or Roman replica in marble of bronze original, possibly dating to the early fourth century BCE. The folded towel-like cloth on his head may refer to Salmacis's pool. B. Hermaphrodite offering fruit (grapes) to a bird. C. Hermaphrodite in an attitude of "unveiling," exposing the genitals, thus emphasizing their magical power. On her head is a basket of fruit. D. Hermaphrodite with one hand resting on a phallic statue of Hermes called a "herm" (left) and the other resting on the head of a young Pan.
From Ajootian, A. (1995), Monstrum or daimonin B. Berggreen and N. Marinatos, eds., *Greece and Gender.* The Norwegian Institute of Athens. A. Staatliche Museen, Berlin; B. British Museum, London; C. National Museum, Stockholm; D. Louvre Museum, Paris.

FIGURE 7.8 Continued

(d)

FIGURE 7.8 Continued

In general, then, we can say that the Greeks held two views about hermaphrodism. Hermaphrodism in gods was not only acceptable, it also symbolized fertility because it combined male and female sexual powers into one being. Such attributes in a human, however, were regarded with horror and as an affront to the gods. This dual attitude toward hermaphrodism comes into play when we discuss Aristotle's and Theophrastus's views on sex in plants in the next chapter.

NOTES

1. Dickinson, O. T. P. K. (2006), *The Aegean from Bronze Age to Iron Age: Continuity and Change Between the Twelfth and Eighth Centuries* B.C. Routledge.
2. Burkert, W. (1985), *Greek Religion*, trans. John Raffan. Harvard University Press.

3. Burkert, W. (1992), *The Orientalizing Revolution: Near Eastern Influence on Greek Culture in the Early Archaic Age*, trans. M. E. Pinder and W. Burkert. Harvard University Press.

4. Burkert, *Greek Religion*.

5. Thornton, B. S. (1997), *Eros: The Myth of Ancient Greek Sexuality*. Westview Press.

6. Idem., p. 78.

7. Delcourt, M. (1961), *Hermaphrodite: Myths and Rites of the Bisexual Figure in Classical Antiquity*, trans. Jennifer Nicholson. Studio Books.

8. Book V, paragraph 119.

9. Dickinson, *The Aegean from Bronze Age to Iron Age*.

10. Roller, L. E. (1999), *In Search of God the Mother: The Cult of Anatolian Cybele*. University of California Press.

11. Witt, R. E. (1971), *Isis in the Graeco-Roman World*. Cornell University Press.

12. Ibid.

13. Foley, H. P. (1994), *The Homeric Hymn to Demeter: Translation, Commentary, and Interpretative Essays*. Princeton University Press.

14. Ibid.

15. To historian Sue Blundell, The Demeter/Persephone myth may also reflect a more immediate, less benign experience of women in ancient Greece: "Marriage to a stranger, arranged by her father and against her mother's wishes, and envisioned as a kind of rape, would have been a reality and not a fanciful tale for many Greek women. That the event was also seen as bringing with it a kind of death—a loss of individual identity—can be easily imagined. Indeed, the fear would also be present that marriage might be fatal in a very real sense, for many women died in childbirth. The link commonly made in myth between death and marriage can thus be seen to have its roots in a shared feminine experience. Persephone, of course, provides the ultimate example of this response, for she marries Death himself". Blundell, S. (1995), *Women in Ancient Greece*. British Museum Press.

16. Cited by Foley, *The Homeric Hymn to Demeter,* p. 70.

17. Evelyn-White, H. G. (2008), *Hesiod, The Homeric Hymns, and Homerica*. Digireads.com Publishing.

18. Thornton, *Eros*.

19. Ibid.

20. Burkert, *Greek Religion*; Kerényi, C. (1967), *Eleusis: Archetypal Image of Mother and Daughter*. Princeton University Press.

21. A notable exception to this trend was the violent Spartan cult practiced at the Sanctuary of Artemis Orthia. The cult is thought to have a pre-Olympian origin.

22. Euripides (1995), *Hippolytus*, trans. D. Kovacs. Harvard University Press.

23. Detienne, M. (1994), *The Gardens of Adonis: Spices in Greek Mythology*, trans. J. Lloyd. Princeton University Press.

24. Ibid.

25. Ibid.

26. Ibid.

27. Homer's *Odyssey*, Book VI, trans. Samuel Butler. http://classics.mit.edu/Homer/odyssey.html

28. Frazer, J. (1959), *The New Golden Bough*, ed. T. H. Gaster. Mentor.

29. Thornton, *Eros*, p. 152.

30. Detienne, *The Gardens of Adonis.*

31. Ovid, *Metamorphosis,* Book X, trans. Horace Gregory. https://archive.org/stream/OvidHersiodVirgil22/Ovid__Horace_Gregory_The_Metamorphoses_djvu.txt

32. Thornton, *Eros.*

33. Blundell, S. (1995), *Women in Ancient Greece.* British Museum Press.

34. Ibid.

35. Nonnus, *Dionysiaca,* Books 10 (lines 230–425) and 11 (lines 1–325) (trans. Rouse; Greek epic, fifth century AD) http://www.theoi.com/Text/NonnusDionysiaca1.html; Ovid, *Fasti,* Book 3. (trans. Frazer). http://www.theoi.com/Text/OvidFasti1.html. Lines 403–414. (Roman poetry, first century BC to first century AD).

36. Aeschylus (1953), *Oresteia,* Vol. 1, trans. Richmond Lattimore, D. Grene and R. Lattimore, eds; University of Chicago Press.

37. Ibid.

38. Ibid.

39. Inwood, B. (2001), *The Poem of Empedocles: A Text and Translation with an Introduction by Brad Inwood.* University of Toronto.

40. Hamilton, E., and U. Cairns, eds. (1961), *Plato: The Collected Dialogues; Timaeus,* trans. Benjamin Jowett. Princeton University Press.

41. *Bibliotheca historica,* Book IV, 4.6.5.

42. Ovid, *Metamorphosis,* trans. Horace Gregory. Signet Classics.

43. Theophrastus (2006), Characters in *Arcana Mundi: Magic and the Occult in the Greek and Roman Worlds,* trans. George Luck. Johns Hopkins University Press.

44. Ajootian, A. (1995), Monstrum or daimonin B. Berggreen and N. Marinatos, eds., *Greece and Gender.* The Norwegian Institute of Athens, pp. 93–108.

45. Ibid.

8

Plant Sex from Empedocles to Theophrastus

THE SIXTH CENTURY BCE marked a turning point in intellectual history throughout Eurasia. Charismatic religious philosophers such as Confucius and Lao-Tsu in China, Buddha in India, Zoroaster in Persia, and Pythagoras and the pre-Socratic Ionian philosophers in Greece all rose to prominence during this period. In the same century, the first five books of the Hebrew Bible, the Torah, are thought to have been compiled.[1] These free-thinking scholars all had one thing in common: for the most part they abandoned the older, polytheistic religions, with their ever-expanding and increasingly unwieldy and implausible pantheons of personified deities, in favor of a universe operating according to a single, all-embracing, divine order. Eastern philosophers (Confucius, Lao-Tsu, Buddha, and Zoroaster) taught that the divine order was ineffable and unknowable. They emphasized personal, spiritual, and moral perfection as the only paths to enlightenment. In contrast, the Pythagoreans and pre-Socratic Greek philosophers taught that the universe was governed by a divine order that could only be understood through mathematical or physical laws. "Natural laws" were discoverable through direct observation and the application of logic. This is the tradition that eventually gave rise to modern science.

Greek faith in the ability of logic to solve the deepest questions of existence can perhaps be traced to the early city-states, or *poleis*, where civic laws and policies were hammered out in democratic citizen assemblies by arguments and persuasion rather than by fiat from monarchs or oligarchs. During public debates, statements backed by evidence were considered more compelling than unsupported assertions. Similarly, in the legal system, court cases were decided by juries composed of hundreds of citizens after listening to the arguments and weighing the evidence presented by both sides of a dispute. In his play *The Birds*, Aristophanes makes fun of the Greek passion for legal arguments:

For grasshoppers sit only for a month
Chirping upon twigs; but our Athenians
Sit chirping and discussing all the year
Perched upon points of evidence and law.[2]

Over time, Greeks developed faith in the collective wisdom of their citizen assemblies and juries to arrive at the truth of any question. As such, the decisions of the democratic assemblies and juries came to be regarded as expressions of the will of the gods. Conditions were thus ripe for the earliest Greek philosophers to apply the rules of logic and persuasion to answer questions about both natural phenomena and the divine order.

THE PRE-SOCRATIC PHILOSOPHERS

The Ionian city of Miletus provided the backdrop for the first speculations of the pre-Socratic philosophers. Miletus was by then a large and powerful city bordering Lydia, a remnant of the former Hittite empire. The principal founder of the Milesian school of philosophy was Thales. Thales and the other Ionian pre-Socratic philosophers were strongly influenced by their more technologically advanced neighbors in the Near East, especially Babylonia and Egypt. Thales is best known for his belief that water is the fundamental building block of all matter, giving rise to air. Thales water theory was probably based on Babylonian cosmology, in which water (Apsu, the goddess of salt water, and Tiamat, the god of fresh water) preceded the creation of earth, sky, and the heavens.

In the latter half of the sixth century BCE, the Milesian philosopher Anaximenes, a student of Anaximander, asserted that air, not water, was the fundamental element that gave rise to all things. Then, in the early fifth century BCE, Heraclitus of Ephesus claimed that fire was the fundamental element. Finally Empedocles, a citizen of Agrigentum in Sicily, added the fourth basic element, earth. According to Aristotle, Empedocles was the first to propose that all material things on earth are composed of the four basic elements—water, air, fire, and earth.[3]

The pre-Socratic philosophers also formulated theories that became the basis of classical Greek cosmology. In the early sixth century BCE, Anaximander presented the first mechanical model of the universe as a series of moving, concentric rings. The outermost ring containing the stars was crystalline and divine.

Anaximander understood the creation of material things to be the result of the opposition of "justice" and "injustice." Heraclitus proposed that matter was driven by the forces of "crime" and "revenge." During winter, for example, cold commits a crime against heat, and heat exacts revenge in the summer. Such theories were clearly inspired by the Greek system of justice. In the words of historian Stephen Mason:

> The notion that there was a principle of retribution in natural processes was derived by analogy from the customs of human society in which the practice of vengeance preceded that of the due process of law. . . . Such a notion was replaced ultimately by the conception that nature, like human society, was governed by laws.[4]

Pythagoras was born on the island of Samos and studied with Thales and Anaximander. According to tradition, Pythagoras, a talented young mathematician with a mystical bent, traveled to Egypt to learn geometry and to be initiated into the secret religious societies of the Egyptian priestly caste. Eventually, he moved to Croton, Italy, and set up a community based on the study of mathematics, the observance of a simple lifestyle, and the practice of secret religious rites involving "divine" numbers.[5]

Among other things, the Pythagoreans studied the physics of musical instruments, such as the lyre, and discovered that musical harmonies are determined by numerical ratios. Pythagoras applied the notion of harmony to cosmology, proposing that the concentric spheres of celestial bodies rotated harmonically to produce the "music of the spheres." In the Pythagorean system, earth was not at the center, but occupied the innermost sphere, circling around a mysterious sacred fire. Aristarchus modified this scheme by substituting the sun for the Pythagorean "sacred fire" at the center of the universe, thus arriving at the first heliocentric model of the solar system.

The pre-Socratic philosophers concerned themselves with the microcosm as well as the macrocosm. Leucippus of Miletus, who taught during the first half of the fifth century BCE, is credited with the theory that all matter is composed of tiny indivisible particles called "atoms." His student, Democritus, elaborated a theory in which the properties of solids and liquids are determined by the shapes and densities of their component atoms, anticipating modern chemical ideas about the shapes of molecules and their chemical interactions.

EMPEDOCLES'S EPIC POEM, *ON NATURE*

Empedocles was one of the most influential of the pre-Socratic philosophers. As noted earlier, all that remains of his writings are fragments of his epic poem, *On Nature*, which we know only from quotations in the writings of later Greek authors. Aristotle states in *Metaphysics* that Empedocles was the first to formalize the principle of the four material "roots," or elements: water, air, fire, and earth. Like Heraclitus, he believed that matter was in constant motion, driven by attractive and repulsive forces. But instead of *crime* and *punishment* Empedocles called the forces *love* (emanating from the goddess Aphrodite) and *strife* (from the god Ares). At first, love dominated and drew all matter unto itself, forming a perfect divine sphere, "rejoicing in its joyous solitude." However, strife gradually gained in strength, shattering the sphere and forming the universe.

Empedocles proposed that life evolved on earth by a process akin to natural selection. However, rather than new organisms evolving from more ancient ones, as in Darwinian evolution, Empedocles proposed that new species, including humans and other animals, arose from novel combinations of free-living body parts. These isolated body parts (arms, legs, feet, noses, ears, foreheads, eyeballs, etc.) had originally sprouted from the ground by spontaneous generation:

> as many heads without necks sprouted up
> and arms wandered naked, bereft of shoulders,
> and eyes roamed alone, impoverished of foreheads.[6]

Whenever one body part saw another it fell in love with it, and the parts would mate and become fused. After repeated couplings, various random combinations arose, but only the most successful survived, as when arms and legs came together with a body. Humans came into existence when all the necessary body parts came together, although some combinations were less successful than others. Among the earliest humans were ox-headed "androids," half man and half woman, which later had to be separated because of strife.

Aristotle and many other later Greek philosophers viewed Empedocles's evolutionary theory with considerable skepticism. Aristotle was firmly committed to a static, unchanging universe, in which all of earth's species were created intact at the beginning, each with its particular degree of perfection that determined its proper place in Nature's hierarchy. Aristotle dismissed the attempts of the pre-Socratic philosophers to develop an evolutionary scheme for the origin of species. For example, he mocked Empedocles for failing to apply his evolutionary scheme for animals based on the random assembly of body parts to plants:

> So, did there come into being in plants " 'vine-like [plants] with olive faces," ' like the " 'oxlike [animals] with men's faces," ' or not? For it is absurd. But they should have, if it also happened in animals.[7]

A clue to Empedocles's ideas about reproduction in plants can be found in an oft-cited fragment in which he likens the production of fruit by olive trees to the laying of eggs by birds: "so first tall trees lay olive eggs."[8]

According to an interpretation attributed to the Roman author Aëtius, the quotation refers to Empedocles's belief that trees were the first living beings to appear on earth and that the olive was the first tree. Trees (and presumably other plants) sprouted from the earth intact before the evolution of animals. The cosmos was still in an undifferentiated state, and the two sexes had not yet been separated. Thus, trees contained both male and female principles:

> Trees first of living beings sprang from the earth, before the sun was unfolded in the heavens and before day and night were separated; and by reason of the symmetry of their mixture they contain the principle of male and female.[9]

As we shall see, although Aristotle rejected Empedocles's theory of animal evolution, he adopted Empedocles's idea that plants, unlike animals, are not differentiated into sexes.

HIPPOCRATES AND THE "FOUR HUMORS"

Hippocrates was among the first to apply the new materialistic approach of the pre-Socratic philosophers to the practice of medicine. He was born on the Greek island of Kos off the southern coast of Asia Minor, and it was there, under a large oriental plane tree, that he established his famous medical school.[10] Hippocrates believed that all remedies should be firmly grounded on direct experience rather than folklore, and he made no reference to supernatural causes of illness—a clear break with traditional practices. He is also the source of the idea, popular during the Middle Ages, that disease is the result of imbalances of four fluids, termed *humors*: blood, phlegm, yellow bile, and black bile.[11] Hippocrates integrated the four humors

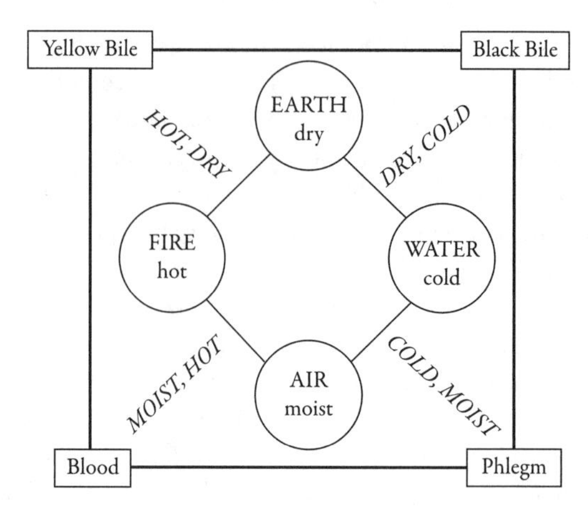

FIGURE 8.1 Diagram of the four humors in relation to the four elements and properties of matter.

into a scheme including the four elements and the four qualities of matter (Figure 8.1). This scheme was used by Aristotle to arrange all organisms into a hierarchy, from primitive (cold/moist) to advanced (hot/dry), and to differentiate between the sexes.[12] In general, things that were hot and dry, like the sun, were considered superior to things that were cold and wet, like the earth. His hierarchy of Nature placed cold/wet organisms at the bottom (plants and invertebrate animals) and hot/dry organisms at the top (quadruped mammals and humans).

ARISTOTLE ON SEX, GENDER, AND PLANTS

What set Aristotle apart from his intellectual forebears was the vast scope of his investigations and his preference for direct observation as the surest way to discover nature's secrets. Nowhere were these qualities more on display than in his biological treatises. Aristotle was the first to study living organisms in a systematic way, from their behaviors in the field to their internal anatomy.

Because Aristotle was primarily interested in animals, it was left to his student and colleague, Theophrastus, to apply his mentor's meticulous methods to plants. Together, they collected and systematized a prodigious body of biological knowledge, supplementing their own observations with reports from a wide range of informants, including farmers, *rhizotomai* (herbalists), woodsmen, fishermen, and craftspeople. They also incorporated reports from far-flung regions of the Hellenistic world sent back to them by scientific observers, many of whom were undoubtedly their own students accompanying Alexander's army.

But despite their considerable resources, neither Aristotle nor Theophrastus succeeded in advancing the problem of sex in plants much beyond what ancient Mesopotamians had learned through their mastery of artificial pollination in dioecious date palms and the use of the caprifig and fig wasp to pollinate edible figs. The failure of these two luminaries to recognize the more general role of pollination in hermaphroditic flowers has long puzzled historians of botany.

Whereas Aristotle's celebrated teacher, Plato, had viewed the physical world as mere "shadows" of a perfect celestial plain, knowable only through mathematics and abstract philosophy, Aristotle and Theophrastus regarded everything in the natural world that could be perceived by the senses as both real and knowable. Both believed that direct observation combined with reason and logic were the most reliable guides to truth. In his treatises on animals, Aristotle referred to other works he had written about plants, but these, unfortunately, have been lost. However, we can reconstruct Aristotle's general thoughts about plants from the frequent references to them in his writings about animals. Most of his ideas about sex in plants can be found in *De Generatione Animalium.*

Aristotle defined four essential life processes: nutrition, growth, movement, and feeling. By "movement," Aristotle meant not only kinetic movement through space, but also the changes in shape brought about by development. "Feeling," according to Aristotle, included both "thinking" and "desiring" in addition to "sense perception." The possession of any one of these four functions was indicative of life, but the highest category of living beings possessed all four. Plants were included among living beings because they had the properties of nutrition, growth, and movement (development), but were lower than animals because they lacked feeling. The idea that plants lacked feeling contradicted Empedocles, who attributed "strife" and "love" to everything in the universe, including plants and minerals.

According to Aristotle, all living things possessed a soul. The soul was thought of as the indivisible essence of the organism and the source of "the powers of self-nutrition, sensation, thinking, and motility." Plant souls, however, were only capable of self-nutrition.

Reasoning that the soul is required by every part of the body, Aristotle inferred that it must be present throughout the entire organism. He then posed the question: "Is each of these a soul or a part of a soul?" To Aristotle, the fact that the higher animals die when cut in half suggested that they possess a single, large soul that is extinguished by division. Plants and "certain insects,"[13] on the other hand, are able to survive being cut into segments, suggesting that each of the segments contains its own soul:

> It is a fact of observation that plants and certain insects go on living when divided into segments; this means that each of the segments has a soul in it identical in species, though not numerically identical in the different segments, for both of the segments for a time possess the power of sensation and local movement.[14]

Because plants can regenerate whole plants after being divided into segments, the single soul of an individual plant must also be capable of forming multiple souls when cut into pieces:

> [P]lants which when divided are observed to continue to live though removed to a distance from one another . . . show . . . that in their case the soul of each individual plant before division was actually one, potentially many.[15]

As noted earlier, Aristotle also believed that living things were ordered from their creation into a hierarchy, which in the Middle Ages came to be called the *Scala Natura* (Ladder of Nature) or "Great Chain of Being." Celestial beings—divine and therefore perfect—occupied the uppermost rung, with people just below them. Large animals were regarded as degenerate forms of humans, and smaller, simpler animals were degenerate forms of the larger more complex animals. Plants were degenerate forms of the simpler animals, and minerals were degenerate forms of plants.

Because Aristotle believed that humans represented the most perfect form of life on earth, he assumed that it was possible to understand the biology of lesser animals and plants by drawing analogies to humans. For example, Aristotle accepted the conventional wisdom that plants obtained all their nourishment from the soil. He knew nothing about photosynthesis—the fixation of carbon dioxide from the air in the presence of light. Because plants seemed to absorb their food from the soil via their roots, Aristotle concluded that the root performed the rudimentary functions of a mouth and digestive system. In contrast, the plant's reproductive structures, fruits and seeds, were located at the top of the plant rather than the bottom. He therefore concluded that plants were analogous to upside-down animals.

ARISTOTLE ON "GENERATION" IN PLANTS

Aristotle recognized two types of generation in biological organisms. In the usual case, plants and animals arose from pre-existing plants and animals. However, some organisms were thought to arise by "spontaneous generation" from decaying organic matter, such as dung, forest litter, and rotting garbage, as in the cases of certain insects and seeds.[16] The theory of spontaneous generation persisted well into the nineteenth century, until Louis Pasteur's sterilization experiments in 1859 finally disposed of it.

Among the animals that arose from pre-existing animals, Aristotle contrasted those that move with those that do not move, such as barnacles, sponges, and other marine invertebrates, "which live by clinging to something else." Animals that move have two sexes and reproduce by copulating. The complementary roles of the sexes during copulation defines which animal is male and which is female:

> For by a male animal we mean that which generates in another, and by a female that which generates in itself.[17]

Nonmoving animals, on the other hand, reproduce without copulation. According to Aristotle, such animals resemble plants both in their lack of motility and in the absence of sexual reproduction:

> But all those creatures which do not move . . ., inasmuch as their nature resembles that of plants, have no sex any more than plants have, but as applied to them [plants] the word is only used in virtue of a similarity and analogy. For there is a slight distinction of this sort, since even in plants we find in the same kind some trees which bear fruit and others which, while bearing none themselves, yet contribute to the ripening of the fruits of those which do, as in the case of the fig tree and the caprifig.[18]

Here, Aristotle admits that in the case of the fig and the caprifig there is "a slight distinction of this sort" resembling the two sexes of animals, but in actuality the caprifig is merely hastening the ripening of the edible fig by supplying it with heat,[19] not by copulating with it. In this passage, Aristotle seems to imply that plants are asexual.

> However, Aristotle was clearly dissatisfied with his characterization of plant reproduction as asexual. Greek philosophers generally preferred principles that were both simple and universal. Aristotle was therefore reluctant to make an exception to the general rule of sexual reproduction in the case of plants. Thus he adopted Empedocles view that plants do, in fact, have two sexes—but they are "mingled": In all animals

which can move about, the sexes are separated, one individual being male and one female, though both are the same species, as with human and horse. But in plants these powers are mingled, female not being separated from male. Wherefore they generate out of themselves, and do not emit semen but produce an embryo, what is called the seed. Empedocles puts this well in the line: " 'at first tall trees lay olive eggs."' For as the egg is an embryo, a certain part of it giving rise to the animal and the rest being nutriment, so also from part of the seed springs the growing plant, and the rest is nutriment for the shoot and the first root.[20]

In this passage, Aristotle seems to come close to the idea of hermaphrodism in plants. Echoing Empedocles's doctrine that plants were formed *prior to* the differentiation of male and female in nature, he makes the startling argument that a plant can be compared to a *pair* of animals in the act of copulation:

In a certain sense the same thing happens also in those animals which have the sexes separate. For when there is need for them to generate the sexes are no longer separated any more than in plants, and their nature desiring that they copulate and are united, that one animal is made out of both.[21]

The analogy of plants to copulating animals derives from Empedocles's conception of animal evolution by the random assemblage or copulation of free-living body parts. The image also evokes the passage in Plato's Symposium about the divine hermaphrodites:

In all this Nature acts like an intelligent workman. For to the essence of plants belongs no other function or business than the production of seed; since, then, this is brought about by the union of male and female, Nature has mixed these and set them together in plants, so that the sexes are not divided in them.[22]

By suggesting that plants are like copulating animals, Aristotle solved the problem of the universality of sex, but he also introduced a logical problem because it implied that the original plants were divided into two sexes, which subsequently fused by copulation. Such animal-like behavior would be inconsistent with the immobility of plants and their lowly place on the *Scala Natura*.

Later, in *De Generatione Animalium*, Aristotle backs away from the idea that plants contain both sexes. He begins by contrasting the female's contribution to the growing embryo versus that of the male. The female provides the "material" out of which the embryo is constructed, and the male supplies the organism's essence or "soul":

The female always provides the material, the male, that which fashions it, for this is the power that we say they each possess, and this is what is meant by calling them male and female. Thus while it is necessary for the female to provide a body and a material mass, it is not necessary for the male, because it is not within the work of art or the embryo that the tools or the maker must exist. While the body is from the female, it is the soul that is from the male.[23]

The female "semen," which forms the embryo, corresponds to the menses. The male semen, according to Aristotle, is composed of hot air and water; after transferring the immaterial soul to the female it evaporates, leaving no material trace behind.

In a related passage, Aristotle states that, in animals, it is the male that provides the one quality that differentiates animals from plants: sense-perception:

> And yet the question may be raised why it is that, if indeed, the female possesses the same soul and if it is the secretion of the female which is the material of the embryo, she needs the male besides instead of generating entirely from herself. The reason is that the animal differs from the plant by having sense-perception. If then, when the sexes are separated, it is the male that has the power of making the sensitive soul, it is impossible for the female to generate an animal from itself alone, for the process in question was seen to involve the male quality."[24]

Although Aristotle stated that plants have a soul, it is a smaller and simpler *vegetative* soul. Presumably this is the soul that the female aspect of the plant provides in full. The *sensitive* soul, which the male provides in animals, is absent in plants. But if the sexes are indeed "mingled" in plants, as Aristotle stated earlier, it would imply that plants have both vegetative *and* sensitive souls. This would place plants on a par with animals on the *Scala Natura*. To be consistent with their lowly station, plants must lack sensitive souls, which, in animals at least, are derived from the male sex. That being the case, plants should have no need for the male sexual contribution during reproduction as implied by Empedocles's metaphor, "tall trees lay olive eggs," which clearly associates the tree with a female (egg-laying) function. In this context, Aristotle seems to suggest that a plant is like a female who is able to "generate entirely from herself"—that is, parthenogenically. This is equivalent to a "one-sex model" for plants in which plants are regarded as female.

ARISTOTLE ON HYBRIDIZATION

The word "hybrid" is derived from the Latin word *hybrida*, meaning "the offspring of a tame sow and a wild boar," hence "mongrel" or "half-breed." It is also presumed to be related to the Greek word *hubris*, referring to a criminal action, often sexual, that shames the victim.[25] As we shall see in later chapters, the shameful aspect of plant hybridization persisted until well into the eighteenth century.

Early Greek philosophers, despite their reservations about hybridization, were convinced that it occurred frequently in nature. The most familiar example was the mule.

Classical writers described many other types of interspecies crosses—both real and imaginary. According to Aristotle, hybridization usually occurred in Africa near lakes:

> It would appear that in [Libya] animals of diverse species meet, on account of the rainless climate, at the watering-places, and there pair together; and that such pairs will often breed if they be nearly the same size and have periods of gestation of the same length. For it is said that they are tamed down in their behavior towards each other by the extremity of thirst.[26]

Aristotle never mentioned hybridization of plants for the simple reason that hybridization required two sexes, and, in Aristotle's view, plants were effectively unisexual and female.

INTIMATIONS OF PLANT SEX IN THE BIBLE

Was the Greek view of the absence of plant sex shared by their Mediterranean neighbors in the Levant? There is some tantalizing evidence from the Hebrew Bible that the early Jews may have had some vague notion of plant hybridization. For example, in Leviticus 19:19, we find the admonition against various types of mixing:

> Thou shalt not let thy cattle gender with diverse kind; thou shalt not sow thy field with mingled seed; neither shall a garment mingled of linen and wool come upon thee.

Because of the inclusion of both living and nonliving objects in the list, it is impossible to say whether the injunction against "mingled seed" is equivalent to the "gendering" of cattle or to the mingling of linen and wool fibers in a garment. However, in Deuteronomy 22:9, a similar injunction against intercropping, the mixing of seed in vineyards, is given that implicates some type of hybridization process:

> Thou shalt not sow thy vineyard with diverse seeds; lest the fruit of thy seed which thou hast sown, and the fruit of thy vineyard be defiled.

In this case, the author seems to suggest that some type of "defilement" can occur as the result of mixing grape vines and the seeds of other crops. According to Flavius Josephus, a historian of the Jews in Roman times, the preceding two biblical passages should be interpreted as follows:

> The seeds are also to be pure and without mixture, and not to be compounded of two or three sorts, since nature does not rejoice in the union of things that are not in their own nature alike.[27]

This interpretation of the mingling of seed is very similar to descriptions of animal hybridization, consistent with the idea of sex in plants. But since no physical basis for the mixing of plant species is ever given in the Hebrew Bible, it is probably a metaphysical concept rather than a biological one. In Genesis 1:11–12, God says,

> Let the earth bring forth vegetation: plants yielding seed of every kind, and fruit trees of every kind on earth that bear fruit with the seed in it. . . .
>
> And it was so. The earth brought forth vegetation: plants yielding seed of every kind and trees of every kind bearing fruit with the seed in it.[28]

Note that the passage specifies that the newly created trees bear fruit "with the seed in it," without any mention of sexual reproduction.

God next commands "every living thing that moveth" (animals) to "be fruitful and multiply." In contrast, plants are not encouraged to "multiply," apparently because there is no need—they already contain the seeds of the next generation within themselves:

> See, I have given you every plant bearing seed that is upon the face of all the earth, and every tree with seed in its fruit; you shall have them for food.[29]

Plants are exempted from the divine command to "multiply" because plants were thought to produce seeds asexually.

THEOPHRASTUS ON PLANT SEX

We now turn to the writings of Aristotle's most famous pupil, Theophrastus, whose botanical works represent the first comprehensive treatises on plants. Fortunately, Theophrastus's writings about plants have come down to us relatively intact, allowing us to reconstruct his thinking on the subject in some detail. Two of his botanical works have survived: *Historia Plantarum* and *De Causis Plantarum*.

Theophrastus's conception of plants differed from those of Aristotle in many respects. Most importantly, he did not consider plants to be merely degenerate forms of animals, as implied by Aristotle's *Scala Natura*. He argued that the anatomy, growth, and reproduction of plants were qualitatively different from those of animals. He rejected the idea that the root could be compared to the mouth and digestive system of animals. Because of the many differences between plants and animals, Theophrastus warned against trying to draw analogies between two qualitatively different types of organisms.

Although Theophrastus's views on sex in plants are basically similar to those of Aristotle, he studied plants in far greater depth and detail than did his mentor. For example, he wrote extensively about plant anatomy and developed a general system of classification based on three main groupings: herbs, shrubs, and trees. Trees, perhaps because of their size, were considered to be the most advanced. He also distinguished between plants based on the structures of their flowers, anticipating the Linnean sexual system of classification. However, Theophrastus's definition of the flower was restricted to the sepals, petals, and probably the stamens:

> Thus of flowers some are downy, as that of the vine mulberry and ivy, some are leafy, as in almond, apple, pear, and plum. Again some of these flowers are conspicuous, while that of the olive, though it is leafy, is inconspicuous. Again it is in annual and herbaceous plants alike that we find some leafy, some downy.[30]

He regarded the carpel, or pistil, as a structure separate from the flower that developed into the fruit, and he associated the persistence of the pistil after the corollas had been shed with higher crop yields:

> [T]here are . . . differences in the way of growth and the position of the flower [corolla]; some plants have it close above the fruit [pistil], as vine and olive; in the latter, when the flowers [petals] drop off, they are seen to have a hole through them, and

> this men take for a sign whether the tree has blossomed well; for if the flower is burnt up or sodden, it sheds the fruit along with itself, and so there is no hole through it.[31]

Theophrastus correctly observed that cucumbers have "sterile" (male) flowers at the tips of the shoots, which he believed inhibited the growth of the fruit:

> Again some flowers are sterile, as in cucumbers those that grow at the ends of the shoot, and that is why men pluck them off, for they hinder the growth of the cucumber.[32]

Of course, the male flowers of cucumber are required for fruit production in cucumber. However, Theophrastus is correct insofar as too many fruits on a vine can compete with each other, thus exerting an inhibitory effect on fruit growth. If the male flowers at the tip are removed *after* pollination of the existing female flowers has occurred, fewer additional fruits will be produced, and the growth rates of the existing fruits will be proportionately increased.

In the case of citron, Theophrastus correlated the presence of the pistil with fruit production:

> And they say that in the citron those flowers which have a kind of distaff [i.e., the pistil] growing in the middle are fruitful, but those that have it not are sterile.[33]

The use of the term "distaff"—a staff used to hold bunches of flax or wool fibers during the spinning process—to describe the shape of a pistil is significant. The term has strong gender connotations, being strongly associated with textiles and with women. In fact, the archaic usage of "distaff" was synonymous with women and women's work. The phrase "distaff side" refers to the female branch of a family. By his choice of "distaff" as a metaphor to describe the shape of the citron pistil, Theophrastus consciously or unconsciously gendered the pistil as female, no doubt because of its role in producing seed. No other gender-specific term is used to describe any other floral structure.

Theophrastus was aware that in date palms the trees came in two types: sterile (male) and fruitful (female). He noted that the males produced flowers but no fruits, while the females seemed to produce fruits without flowers:

> Some say that even of plants of the same kind some specimens flower while others do not; for instance that the " 'male" ' date-palm flowers but the " 'female" ' does not, but exhibits its fruit without any antecedent flower.[34]

Recall from Chapter 5 that male date palm flowers have showy white petals, while the females do not, which accounts for Theophrastus's erroneous conclusion that the female tree produces fruit without any "flowers." The "fruit" in this case is actually the female flower, which comprises the pistil and its scale-like petals.

In several passages, Theophrastus reported that in Greece it was customary to refer to fruit-bearing trees as "females" and trees without fruit as "males."[35] But he also cites several exceptions to this rule, resulting in a confusing picture:

> A difference common to all [trees] is the way in which people distinguish female and male, of which the former are fruit-bearing and the latter are fruitless in some kinds [of trees]. In other kinds both bear fruit but the female bears nicer fruit and more prolifically, except that some people call these trees male. The distinction is analogous in character to the way in which domestic and wild are differentiated.[36]

> Generally, the trees people call male are the unfruitful ones of a particular variety, and of these they say some flower a lot, some a little, and some not at all. In contrast, in some kinds [of trees] the males alone bear fruit.[37]

> And generally speaking, all those of any given kind which are called "male" trees are without fruit. . . . On the other hand they say that in some cases it is only the "males" that bear fruit, but that, in spite of this, the trees grow from the flowers [of the female tree], [just as in the case of fruit-bearing trees that grow from fruit].[38]

Theophrastus's attempts to understand sex in plants were made more challenging by the long-standing practice among Greek farmers of gendering trees on the basis of their growth habit and wood properties rather than on their floral structures.

According to historian Lin Foxhall, such ideas about gender in trees reflected prevailing Greek gender biases:

> Generally, male-gendered plants/trees were believed to be wild, rough, dense, dry, compact, knotty and . . . often less fruitful or unfruitful (Theophr. *Hist. Pl.* 5.4.1.). Wood from male trees was inferior, being shorter, more twisted, easily warping and difficult to work. Male plants were thought to need less good soils and growing conditions, and were generally perceived as more vigorous, tougher, and hardier. (Theophr. *Caus. Pl.* 1.15.3–4; 1.16.6). In contrast, female plants (of both fruiting and non-fruiting varieties) were generally seen as more amenable in all senses (in many cases, particularly, in terms of fruit production and wood working) (Theophr. *Caus. Pl.* 1.15.3–4). Essentially, they were felt to be more easily mastered and controlled by humans. The qualities which defined plants as male made them 'by nature' difficult to control.[39]

Theophrastus himself never assigned sexes in plants based on wood properties or vegetative growth habits. Indeed, Theophrastus ultimately rejected the existence of sex in plants. Yet in his discussion of the most favorable time for the grafting of trees Theophrastus adopted the traditional custom of referring to spring budding in feminine terms as "pregnancy":

> The arguments in favor of each season are much like the arguments in favor of each as a time for planting. Some persons recommend spring, the trees being still pregnant at the time of the vernal equinox, since the graft in that case will sprout at the time of pregnancy, and meanwhile the bark grows over the graft and encloses it.[40]

Clearly, Theophrastus did not hesitate to use metaphors that reinforced the plants-as-female gender bias. The closest he came to recognizing actual sex in plants occurred during his discussions of fruit shedding and the phenomenon of "degeneration" in fruit trees.

THEOPHRASTUS ON THE CAUSES OF FRUIT SHEDDING IN FIGS

Theophrastus had noted that the premature shedding of fruit in fruit tree orchards was a chronic problem for farmers. He assumed that the causes of premature shedding were the same in all fruit trees. In fact, there are several potential causes of premature fruit drop. One type may occur when the fruits are still small. This type of fruit drop is thought to be caused by competition among the fruits, resulting in self-thinning. A second type of fruit drop is triggered by adverse external factors, such as disease, pests, temperature, and water stress. A third cause is lack of pollination.

Fruit drop in cultivated figs is typically caused by the absence of wasp-mediated pollination. Caprification—the practice of tying branches of caprifig syconia to the intact branches of the edible fig female trees— prevents this type of fruit drop by supplying pollen-bearing fig wasps to the immature syconia of the female tree. However, Theophrastus adhered closely to Aristotelian doctrine by concluding that the beneficial effect of caprification was an asexual process. Having ruled out sexuality, he felt obliged to provide an alternative physiological explanation, the so-called "open fig theory."

In Theophrastus's view, the critical step in the caprification process occurs when the "gall-insects" (fig wasps), which "are engendered from the seeds" of the caprifig, emerge and "eat the tops of the cultivated figs." By so doing, the wasp stimulates the growth of the syconium:

> This is the reason for the process called 'caprification'; gall insects come out of the wild figs which are hanging there, eat the tops of the cultivated figs and so make them swell.[41]

According to the "open fig theory," fig wasps stimulate the "swelling" of the cultivated fig syconia in two ways: (1) by creating a hole in the fig that allows accumulated vapors to escape and (2) by consuming the "excess fluid" in the fruit. Based on the humors theory, cold, damp things grow slowly, while hot, dry things grow rapidly. The cultivated fruit's internal cool moisture prevents it from actively growing, so it is shed from the tree. The gall insect prevents this by gnawing holes in the syconium and consuming the liquid, enabling it to dry.

THEOPHRASTUS ON ARTIFICIAL POLLINATION IN DATE PALMS

Theophrastus thought he had succeeded in explaining the role of the wasp in caprification by the open fig theory. The historian Herodotus, whose work Theophrastus knew well, mistakenly believed that the method of fruit production in cultivated date palm trees was the same as that for the fig in every respect, including the participation of a "gall-fly":

> Palm-trees grow [in Babylonia] in great numbers over the whole of the flat country, mostly of the kind which bears fruit, and this fruit supplies them with bread, wine, and honey. They are cultivated like the fig-tree in all respects, among others in this. The natives tie the fruit of the male-palms, as they are called by the Greeks, to the

> branches of the date-bearing palm, to let the gall-fly enter the dates and ripen them, and to prevent the fruit from falling off. The male-palms, like the wild fig-trees, have usually the gall-fly in their fruit.[42]

Although fruit production in date palms resembled caprification in figs in some respects, such as the existence of "male" and "female" trees, Theophrastus recognized that the two processes were not identical. For example, in dates, unlike figs, "gall insects" did not seem to play a role, and there indeed seemed to be a true "union of the two sexes," which, he maintained, was not the case in figs:

> With dates it is helpful to bring the male to the female; for it is the male which causes the fruit to persist and ripen, and this process some call, by analogy, " 'the use of the wild fruit."' The process is thus performed: when the male palm is in flower, they at once cut off the rachis on which the flower is, just as it is, and shake the bloom with the flower and the dust over the fruit of the female, and, if this is done to it, it retains the fruit and does not shed it. In the cases both of the fig and the date it appears that the " 'male"' renders aid to the " 'female,"' for the fruit-bearing tree is called " 'female"'—but while in the latter case [dates] there is a union of the two sexes, in the former [figs] the result is brought about somewhat differently.[43]

In this passage, Theophrastus seems to teeter on the edge of recognizing sex in date palms. He notes that the phenomenon of artificial pollination in date palms seems to be somewhat different from the caprification of figs:

> That the fruit does not remain on the female date-palm unless you shake the flower of the male over it together with the dust . . . occurs only in the date-palm, but is similar to the caprification of fig trees. From these instances one would be most inclined to infer that even a female tree cannot by itself bear completely formed fruit; except that this should hold not of just one or two female trees but of all or most of them, since this is how we decide the nature of the class of females. And in the cases before us that of the date-palm is very strange indeed, since caprification is considered to have a clear explanation.[44]

Theophrastus wrestled with the dilemma that the date palm and the fig tree seemed to be sexual, demonstrating that "even a female tree cannot by itself bear completely formed fruit." Against such a proposition, "all or most" other female trees do not seem to require a male tree for fruit production. Curiously, although Theophrastus was well aware that pistachio nut trees (*Pistachio atlantica*) are also dioecious, he never compared them to either dates or figs.[45] In the very next paragraph, he describes mulberry trees, but fails to mention their dioecism. Had he included these two other examples in his treatise, he might have been forced to conclude that all female trees required the dust from the male tree, which might have led him to generalize pollination to hermaphroditic flowers.

Aristotle had indeed noticed that hermaphroditic flowers also produce "dust" that was collected by bees for food, but such "bee bread" did not appear to have any function for the flower.

Rather than apply a sexual theory based on date palms to figs, Theophrastus chose the reverse strategy. He applied his open fig theory to date palms. According to the four humors theory, the "dust" from the male tree must somehow cause drying of the moist female date, stimulating it to grow. Thus, according to Theophrastus, the dust of date palms performs the same physiological function as the fig wasp does during caprification, causing the drying of the fruit.

In view of his ultimate rejection of the sexual theory, it is surprising that Theophrastus concluded his discussion of date palms by making a very apt biological analogy between pollination and the sprinkling of milt by male fish onto the newly laid eggs of the female:

> "What occurs in the date-palm, while not the same as caprification, nevertheless bears a certain resemblance to it, which is why the procedure is called *olyntházein* (from *ólynthos*, the edible wild fig). For the flower and dust and down from the male date-palm, when sprinkled on the fruit, effect by their heat and the rest of their power a certain dryness and ventilation, and by this means the fruit remains on the tree. Something similar in a way to this seems to happen with fish, when the male sprinkles his milt on the eggs as they are laid. But resemblances can be found in things widely separate.[46]

The reference to milt shows just how close Theophrastus came to correctly interpreting date palm pollination as a sexual process. Backing away from brink of apostasy to safer ground, he hastily added: "But resemblances can be found in things widely separate."

THEOPHRASTUS ON "DEGENERATION" IN FRUIT TREES

Cultivated plants were thought to have been given to the Greeks by the gods. However, the belief arose that under certain conditions, one species could transform (or "degenerate") into another. Thus, when a few wheat plants grew up in a field of barley, it was thought that barley had "degenerated" into wheat. The possibility that a few wheat grains might have become mixed in with the barley grains during sowing was apparently not seriously considered.

Although the supposed "degeneration" of barley into wheat or some other cereal was entirely specious, there was one example of crop "degeneration" in Theophrastus's day that actually had a basis in fact: the "degeneration" of cultivated fruit trees grown from seed. Here is how Theophrastus describes the phenomenon:

> All trees grown from seed are as a rule inferior, at least among the cultivated fruit trees (as in pomegranate, fig, vine, and almond); some indeed often undergo a mutation of their entire kind and become wild.[47]

It is still true today, as it was in Theophrastus's time, that most cultivated fruit and nut trees do not breed true from seed and must therefore be propagated from cuttings. Seeds of cultivated trees, instead of producing progeny with the same phenotype as the parent tree, give rise to progeny with a diverse array of phenotypes. Because most of the desirable characteristics of the cultivated tree (from a human perspective) were usually lost in seed-grown trees, the Greeks interpreted such "degeneration" as a return to the wild state:

> [T]hose that . . . are planted from slips are all held to breed true. But those that propagate from [seeds] are practically all inferior, and some depart completely from their kind, as vine, apple, fig, pomegranate and pear; for from the fig seed no cultivated tree at all is produced, but either a wild fig or fig gone wild . . ., and from the noble vine comes an ignoble one, and often one of a different kind, and sometimes no cultivated tree at all but a wild one, and occasionally of such a sort that it cannot bring its fruit to concoction, and some cannot even form fruit but only get as far as flowering. . . .
>
> From the stones of the olive grows an olive run wild, and from the berries of the sweet pomegranate ignoble pomegranates, and from those of the stoneless kind hard ones, and often sour ones. . . . The almond too becomes inferior both in flavor and in turning from soft to hard.[48]

Theophrastus believed that the cause of fruit tree degeneration was the inability of the seed to assimilate the nutrient-rich soils provided in cultivated orchards.

Of course, the correct explanation of fruit tree "degeneration" is that all the cultivated fruit and nut trees used by the Greeks were *hybrids,* the end results of multiple crosses between different natural varieties of wild trees. Such crosses had occurred spontaneously in the wild, and early farmers had selected these wild hybrids for cultivation because of their desirable traits. The two sets of chromosomes in such hybrid individuals may have come from different varieties, or even from different species, and such hybrids are sometimes sterile. Cultivated fruit and nut trees are fertile hybrids because they can produce seed, but they cannot breed true because during sexual reproduction the genes from the two different parental chromosomes randomly reassort to form new combinations of alleles (alternate versions of genes) on the chromosomes of the progeny. These new gene combinations give rise to new types of individuals. Theophrastus could never have guessed the true biological cause of fruit tree "degeneration" because, having rejected the possibility of sex in plants, the idea that different varieties of plants could cross with one another was outside his frame of reference. Intentional plant breeding, which depends on a knowledge of sex in plants, was still centuries away.

Theophrastus died in 288 BCE. In accordance with his will, he was buried in a modest coffin in a corner of the botanical garden he loved so well. His funeral was attended by a large numbers of his former colleagues and pupils, as well as by dignitaries and ordinary citizens. After his death, the main center of Greek science moved from Athens to Alexandria. The Alexandrian period of Greek science lasted from about 250 BC to 200 AD, during which time the emphasis shifted to engineering, mathematics, and astronomy. Galen, from Pergamum on the Anatolian coast, was the last great Greek biologist of antiquity.

NOTES

1. The earliest evidence for Hebrew inscriptions of proto-biblical texts dates to the tenth century BCE. See Garfinkel, Y., et al. (2015), The ʾIšbaʿal Inscription from Khirbet Qeiyafa. *Bulletin of the American Schools of Oriental Research* 373: 217–233.

http://www.livescience.com/51223-king-david-era-inscription-discovered.html.)

2. *The Plays of Aristophanes* (1949), Vol. 1, trans. J. H. Frere, Everyman's Library, Dent & Sons, p. 141.

3. Aristotle's *Metaphysics*, Book 1, Parts 3–4.

4. Mason, S. F. (1962), *A History of the Sciences*. Collier Books.

5. Wertheim, M. (1997), *Pythagoras' Trousers: God, Physics, and the Gender Wars.* Fourth Estate.

6. Empedocles, and Inwood, B. (1992), *The Poem of Empedocles: A Text and Translation with an Introduction*. University of Toronto Press. Fragment 64/57, p. 235.

7. Aristotle, *Physics*, 199b7–13.

8. Empedocles and Inwood, *The Poem of Empedocles*. Fragment 79/79 p. 241.

9. Aëtius, *Vetusta Placita,* v. 26; 440.

10. *Platanus orientalis,* as opposed to the western or American plane tree (*Platanus occidentalis*).

11. Fahråeus, a Swedish physician, suggested that the four humors may have been based on the separation of blood into four discrete color zones when allowed to settle in a transparent container: a blackish clot at the bottom ("black bile"), a layer of red cells above it ("blood"), a whitish layer ("phlegm"), and clear yellow serum in the upper zone ("yellow bile"). Fahråeus, R. (1921), The suspension stability of blood. *Acta Medica Scandanavica* 55:1–228.

12. Singer, C. (1922), *Greek Biology and Greek Medicine, Chapters in the History of Science.* Clarendon Press.

13. Aristotle did not cite any specific examples in the case of "insects." Among the animals that can regenerate whole organisms, including both heads and tails, after being cut into pieces are the flatworms (planarians) and some segmented worms used in composting, neither of which is an insect (arthropod).

14. Aristotle, *On the Soul,* Book I, Part 5.

15. Aristotle, *De Anima*, Book II Part 2.

16. Aristotle, *De Generatione Animalium,* Book I, Part I, 715b.

17. Idem., Book I, Part 2, 716a. Aristotle notes in the same sentence that this manner of distinguishing the sexes accounts for gender assignments in myths about cosmology: "wherefore men apply these terms to the macrocosm also, naming Earth mother as being female, but addressing Heaven and the Sun and other like entities as fathers, as causing generation."

18. Idem., Book I, Part 1, 715b.

19. See Aristotle, *Meteorology*, Part 3.

20. Aristotle, *De Generatione Animalium,* Book I, Part 22, 731.

21. Ibid.

22. Ibid.

23. Idem., Book II, Part 4.

24. Ibid.

25. Cohen, D. (2012), Law, society and homosexuality or hermaphrodity in Classical Athens, in R. Osborne, ed., *Studies in Ancient Greek and Roman Society.* Cambridge University Press, p. 64.

26. Aristotle, *History of Animals*, Book 8, ch. 28.

27. Flavius Jospehus (1737), *Antiquities of the Jews* (4.228), trans. William Whiston. Willoughby & Co., London.

28. The New Oxford Annotated Bible, third edition. Oxford University Press.

29. Ibid.

30. Theophrastus (1968), *Enquiry into Plants*, trans. A. Hort. Harvard University Press, Vol. I, Book XIII, p. 91.

31. Idem., Vol. I, Book XIII, pp. 91–93.

32. Idem., Vol I, Book XIII, pp. 93–95.

33. Idem., Vol. I, Book XIII, p. 95.

34. Ibid.

35. Negbi, M. (1995), Male and female in Theophrastus's botanical works. *Journal of the History of Biology* 28:317–332.

36. Theophrastus, *Enquiry into Plants*, Vol. III, Book VIII, p. 203.

37. Ibid.

38. Idem., Vol. III, Book III, p. 177.

39. Foxhall, L. (1998), Natural sex: The attribution of sex and gender to plants in ancient Greece, in L. Foxhall and J. Salmon, eds., *Thinking Men: Masculinity and Its Self Representation in the Classical Tradition*. Routledge, pp. 57–70.

40. Theophrastus, *De Causis Plantarum*, trans. B. Einarson and G. K. K. Link. William Heinemann Ltd., Vol. I. Book 6, Section 3.

41. Theophrastus, *Enquiry into Plants*, Vol. III, Book III, pp. 151–155.

42. Herodotus (1942), *Persian Wars*, trans. Rawlinson. Modern Library, Book 1, ch. 193, p. 105.

43. Theophrastus, *Enquiry into Plants*, Vol. II, Book VIII, Section III, p. 155.

44. Theophrastus, *De Causis Plantarum*, Vol. III, Part 18, Section 1, pp. 135–137.

45. Theophrastus, *Enquiry into Plants*, Vol. III, Part 15, Section 3–4 and Vol. IV, Part 4, Section 7.

46. Theophrastus, De Causis Planarum, Vol. III, pp. 135–137.

47. Idem., Vol. I, Degeneration from Seed, pp. 67–71.

48. Theophrastus, *Enquiry into Plants*, Vol. II, Part 2. On Degeneration, 4–6.

9

Roman Assimilation of Greek Myths and Botany

JUST AS GREECE had earlier absorbed many of the cultural traditions of its more advanced neighbors, so, too, did Rome assimilate much of Greek culture. However, the influence of Greece is more apparent in the art, architecture, and religion of Rome than in its philosophy or science. Rome, like Greece, was an agricultural society, and, not surprisingly, some of its best minds—including Cato the Elder, Varro, Virgil, and Columella—wrote about agriculture. An ample supply of agricultural deities was regarded as essential to the prosperity of the empire, and they were regularly celebrated in art, poetry, and festivals. Not surprisingly, given their antecedents, Roman myths reinforce the prevailing gender bias in which plants, particularly flowers, were associated with women. The study of theoretical botany, including the role of sex in plants, was largely abandoned, and, in the absence of any countervailing scientific evidence, the one-sex model of plants, implicit in Aristotelian botany, became even more firmly entrenched during the Roman period.

ROMAN AGRICULTURAL DEITIES

By 900 BCE, Rome, the future seat of one of the largest and longest lasting empires the world has ever known, consisted of a sprinkling of small, agricultural, ethnically diverse villages. Greek colonies began to crop up in Sicily and southern Italy by the middle of the eighth century BCE, and Roman assimilation of Greek and Phoenician customs and traditions, a process already underway as the result of earlier trade contacts, intensified during this period. The indigenous religion of early Rome, like most ancient religions, was polytheistic. However, the Romans never developed a complex set of myths surrounding their own deities as did the Greeks and Near Eastern societies. Exposed early in their history to the fully realized gods and goddesses of the Greek pantheon, with their exciting narratives and awe-inspiring cosmic lineages, there was little incentive to construct an elaborate competing body of myths for Roman deities.

Accordingly, Rome's founder was said to have been the Trojan warrior Aeneas, the son of Venus, who was conceived on Mount Ida in Anatolia. Although the Trojans and Achaeans[1] were portrayed as bitter enemies in the *Iliad*, Greek deities presided over the fates of the combatants on both sides of the conflict, so despite their enmity toward the Greeks in that great Homeric epic, Romans understood themselves to be co-religionists who once shared a common language.[2] Following the Trojan War, Aeneas, like Odysseus, set out on a voyage and, after many wanderings patterned on the *Odyssey*, arrived in Italy where he defeated the Latins and founded the city of Rome. Virgil's *Aeneid*, composed in the latter part of the first century BCE, gave the myth its final, enduring form.

It has often been said that the Romans were the first Hellenized society in the Mediterranean world. Among the earliest of the Roman deities were Jupiter, his wife, Juno, and their daughter, Minerva. The Jupiter-Juno-Minerva triad was originally assimilated from the Etruscans during the period of their hegemony over Rome. Jupiter, whose name indicates that he was derived from the same Indo-European weather god whom the Greeks called Zeus, was a storm god whose worship promoted favorable weather conditions for crops. Juno represented youthful vigor as well as sexual maturity. She was strongly associated with the summer harvest, and the month of June was named after her. Minerva was later identified with Athena, the protector of Athens, born from the head of Zeus.

Mars, later identified with the Greek god of war, Ares, was originally a Roman tutelary deity who safeguarded farmlands and the wilderness. Hence he was invoked both to protect the crops and to assist in waging wars outside of the borders of the city. Mars's initial function as an agricultural deity accounts for the fact the month of March, associated with the return of spring and the beginning of the growing season, bears his name.

Ceres (Figure 9.1), the oldest and most important agricultural deity in the Roman pantheon, was the goddess of cereals and agricultural abundance in general, who, by the beginning of the Republic, was already identified with Demeter and was often associated with Tellus, Liber, and Libera (discussed later). Her antiquity is attested to by the fact that she, like Jupiter, Juno, and other native Roman deities, was assigned a *flamen*, or sacred priest, dedicated to her worship. Her name, probably derived from the verb "to grow" (*crescere*) or "to create" (*creare*), first appears as an adjective coupled with the names of earlier Italic deities. There are other examples in Roman religion in which an adjectival attribute (in this case "grower" or "creator") belonging to a more generalized older deity splits off and becomes a separate deity in its own right.[3] The Cerialia, held on April 19, were games and rituals staged in Ceres's honor. As Ovid explains in Book IV of *Fasti*, a Latin poem describing the practices on different religious holidays, Ceres, like Demeter, represents the grain crop. During the Cerialia, the farmers made offerings to the goddess of spelt wheat and salt (perhaps in the form of cakes) as well as milk, honey, and wine, in order to ensure "a bounteous harvest." According to Ovid, white was the appropriate color for Cerialia attire. White garments were also worn by initiates of the Eleusinian Mysteries in Greece, which is probably the source for the dress code at the Roman Cerialia.

(a)

FIGURE 9.1 Images of Ceres. A. Marble statue of Livia Drusilla, wife of Augustus, depicted as the goddess Ceres, wearing a flower garland and holding wheat sheaves in one hand and a cornucopia in the other. First century CE. (Louvre, Paris.) B. From wall painting at Pompeii.
A is from Wikimedia Commons. B is from Yaggi, L. W., and T. L. Haines (1881), *Museum of Antiquity (A Description of Ancient Life)*, Western Publishing House.

(b)

FIGURE 9.1 Continued

Liber and Libera were originally a Roman divine couple representing the fertility of men and women, respectively, but the two deities also had power over the natural cycle and the growth of crops. Under Greek influence, Liber became more narrowly identified with Dionysus, the Greek god of wine, or Bacchus, the Roman equivalent. In 493 BCE, in the aftermath of a plague, a temple was built on the Aventine to honor the triad of Ceres, Libera, and Liber, the Roman counterparts to the Greek Demeter, Persephone, and Iachhos—an epithet of Dionysus. In other contexts, Libera was equated with either Ariadne or Venus, as the consort of Liber. Libera was gradually subsumed by another Roman goddess, Proserpina, the Roman equivalent of Persephone in the myth of the rape and abduction by Pluto (Hades), the god of the underworld.

Closely allied with Ceres was Tellus (later Terra Mater) the goddess of the earth. Tellus was equated with the Greek goddess Gaia or Ge. Tellus may have arisen from a Roman spirit that inhabited farmland and helped to bring forth crops.[4] Subsequently, she was generalized to an earth mother and identified with Gaia.

The goddess Venus does not appear to have been among the original Roman deities. She seems to have arisen later, outside of Rome, as the Italian equivalent of Aphrodite, the mother of the Trojan hero Aeneas. Like Aphrodite, Venus was initially worshipped as the protector of gardens before acquiring her more famous attribute as the goddess of love. She was worshipped during two wine festivals: the *Vinalia rustica* (country wine festival), held in the fall, and the *Vinalia urbana* (city wine festival), celebrated in the spring.[5]

The fact that Venus, rather than the wine god Liber-Dionysus, was venerated during the two Vinalias may reflect the origin of these holidays as military/agricultural festivals rather than as ecstatic mystery rites. According to tradition, the *Vinalia rustica* began as a feast day in which libations of wine were poured in Jupiter's honor in gratitude for Aeneas's victory over the Etruscan king, Mezentius. Because the festival coincided with the date of the founding of Rome's first temple to Venus in 295 BCE, the two feast days were combined into one. Thereafter, in addition to the pouring of libations to Jupiter, offerings were also made to Venus for the protection of both vegetable gardens and vineyards. The *Vinalia urbana*, held in late April, was a feast celebrating the arrival in the city of the previous year's wine, stored in jars and wineskins. It was appropriate for Venus to be associated with the *Vinalia urbana* because of her strong ties to spring:

And no season was more fitting for Venus than spring.
In spring the landscape glistens; soft is the soil in spring;
Now the grasses push their blades through the cleft ground;
Now the vine-shoot protrudes its buds in the swelling bark.[6]

Venus's specific identification with sexual love is attested by the fact that during the Vinalia she was worshipped by Rome's prostitutes:

I will now tell of the festival of the Vinalia;
But there is one day interposed between the two.
You street women celebrate the divinity of Venus:
Venus favors the earnings of ladies of the liberal profession.
Offer incense and pray for beauty and popular favor;
Pray to be charming and witty;
Give to the Queen her own myrtle and the mint she loves,
And wicker baskets filled with clustered roses.[7]

In this passage, Venus is identified with flowers, blurring the line between Venus and Flora, the goddess of flowers. As noted in Chapter 7, Chloris, the Greek goddess of flowers, was assimilated as the Roman Goddess, Flora. Flora (Figure 9.2) was also among the early Roman deities, predating Venus, as evidenced by the special priest (*Flamen Florialis*) assigned to her worship. A festival with games, the Floralia, was held on April 28, the day of the dedication of her first temple in 240 BCE. Women wore multicolored garments at the Floralia, and grain and legumes were scattered as fertility symbols. A general mood of merriment and sexual license prevailed, encouraged, no doubt, by the presence of prostitutes who worshipped Flora at the Floralia as they did Venus at the Vinalia.

FIGURE 9.2 Flora, Roman goddess of flowers, filling her basket with blossoms. Detail from Spring, first century AD Roman fresco from Stabiae, a seaside resort near Pompeii that was largely destroyed by the eruption of Mount Vesuvius (Museo Archeologico Nazionale di Napoli, Naples). From Wikimedia Commons: [https://commons.wikimedia.org/wiki/File:Flora_mit_dem_Füllhorn.jpeg].

Ceres and Venus were often conflated with Flora because, as she explains in *Fasti* V, her dominion over flowers also gives her power over the crops:

Perhaps you think my rule is confined to dainty wreaths.
My divinity touches the fields, too.
If crops flower well, the threshing-floor churns wealth;
If the vines flower well, Bacchus flows;
If the olives flower well, the year shimmers
And the season fills with bursting fruit.
Once their bloom is damaged, vetches and beans die,
Your lentils die, too[8]

Here, Ovid speaking through Flora recognizes the developmental relationship between flowers and fruits, including the less obvious case of grain. This is far less equivocal than Theophrastus's tentative musings on the necessity of the "distaff" for fruit production in citron, which we noted in Chapter 8. Paradoxically, Theophrastus was handicapped by the Greek penchant for thoroughness and precision. He knew that there were examples of plants, such as the fig, which seemed to produce fruits without any apparent flower, and the existence of such exceptions prevented him from formulating a universal principle about the development of fruits from flowers. Roman poets and writers on agriculture, who preferred pragmatism to universal principles, emphasized the rule of thumb and ignored exceptions. The fact that figs and a few other plants appeared to lack flowers in no way invalidated the general principle that fruits arise from flowers.

Regarding sex in plants, Ovid in *Fasti* V relates the story of an encounter between Flora and Juno that symbolically connects flowers to parthenogenesis. The story begins with Juno's grieving because her husband Jove has borne a daughter, Minerva, from his head, without her participation. She would love to beget a child on her own to get even, but despairs because she thinks it's impossible. She sets out on a journey "to complain to Ocean," the goddess of the sea (always a good listener), but stops at Flora's house along the way to seek the flower goddess's advice:

> If Jove became a father without using a spouse
> And possesses both titles by himself,
> Why should I not expect a spouseless motherhood,
> Chaste parturition, untouched by man?
> I'll try every drug on the broad earth and empty
> Ocean and the hollows of Tartarus.[9]

Having finished her plea she observes "a look of doubt" on Flora's face and pleads for her help. At first, Flora expresses her fear of Jove's wrath, but when Juno promises to keep her identity a secret Flora yields and tells Juno about a magic flower:

> "A flower," I said, "from the fields of Olenus[10]
> Will grant your wish. It's unique to my gardens.
> I was told: 'Touch a barren cow; she'll be a mother.'
> I touched. No delay; she was a mother."[11]

Flora then plucked the magic flower from her garden and touched Juno with it, whereupon the goddess immediately conceived. Having succeeded in her quest, Juno departed and made her way to "Thrace and West Propontis," where she gave birth to Mars. As a token of her gratitude, Juno decreed that henceforth, Flora would always have a temple in Rome.

The story of Flora's role in the conception of Mars, which is based on a Greek story of Hera's conception of Ares,[12] invites speculation about the symbolic significance of the magic flower from "the fields of Olenus." Based on Olenus's reputation for natural beauty and abundance, the flower is probably a symbol of fecundity, but of what type—sexual or parthenogenic—is not clear. The Romans were familiar with the role of the male date palm in fecundating the female tree, so it is possible the magic flower is patterned on the male rachis of the date palm. According to this interpretation, Flora "pollinated" Hera with the magic flower, just as Babylonian farmers hand-pollinated their female date trees. Alternatively, the magic flower might symbolize the female parthenogenic powers of plants. When Hera is touched by the flower, she acquires the ability to conceive parthenogenically, like plants. The latter interpretation would be more in keeping with the classical understanding of reproduction in the plant kingdom as a whole.

Pomona (from the Latin *pomum*, meaning fruit) was the Roman goddess of fruit trees, especially those with showy flowers. Like Flora and Ceres, she was one of the early Roman deities and thus had her own *flamen* dedicated to her worship. Unlike these two goddesses, however, Pomona was never merged with a comparable Greek deity, and she retained her original attribute, the pruning knife.

Two other important nature deities that the Romans imported from Anatolia and Egypt by way of Greece were Magna Mater and Isis. Magna Mater, or Cybele, as she was called in Rome, was originally the Phrygian goddess *Matar kubileya* ("Mother of the Mountain"). The Greeks worshipped her as Kybele or Meter Thea ("Mother Goddess")—the mother of the gods, and the Etruscans also depicted her on pottery as early as the mid-sixth century BCE (Figure 9.3).[13] However, Roman worship of Magna Mater greatly intensified around 204 BCE, when a stone artifact representing the goddess was removed from her temple in Pergamon and borne triumphantly back to Rome, marking the beginning of her status as a mainstream Roman deity. Several factors contributed to the importation of the Phrygian mother goddess at this particular time. During the preceding year, a series of meteorite showers had exacerbated a feeling of insecurity and religious anxiety brought on by the lingering presence of Hannibal's army in Italy. As was the practice whenever a crisis or a perplexing situation arose, Roman leaders sought guidance from a collection of oracular sayings known as the "Syballine books." Based on the oracle's interpretation of certain passages, most likely selected at random, officials were directed to seek and bring back to Rome a new deity from Anatolia. Magna Mater was already revered by Romans based on her association with Mount Ida near Troy, the birthplace of Aeneas. Indeed, upon her arrival in Rome in 204 BCE in the form of a sacred stone, she was welcomed as the primordial Roman mother goddess from Troy, rather than as a newly adopted foreign deity.

Like the Greeks, Romans revered Magna Mater as the "womb of the gods." Once firmly installed in Rome, however, she took on new attributes that differed significantly from those of either her Greek or Phrygian counterparts. Both the Phrygian and Greek versions of the deity had been concerned mainly with "wild and unstructured mountain landscapes," and, despite their roles as mothers of the gods, they had little or nothing to do with human fertility or agricultural abundance.[14] However, both the Magna Mater of Rome and her male consort Attis were strongly associated with human sexuality, fertility, and bountiful harvests. According to the Roman statesman and encyclopedist, Pliny the Elder, Magna Mater's arrival in Rome led to bumper crops the following year. Her festival, the Megalensia Ludi Comitialis, was celebrated on April 4.

The Egyptian goddess Isis did not enjoy the same warm welcome into Rome as Magna Mater because she was viewed as a potentially subversive Egyptian influence. During the reign of Antony and Cleopatra in the Hellenistic period, her Roman devotees received rough treatment from the Roman Senate, as well as from Rome's first two emperors, Augustus and Tiberius. From the reign of Caligula onward, however, Isis was welcomed

(a)

(b)

FIGURE 9.3 Greek and Roman images of Cybele. A. Greek statue (~350 BCE) of Kybele with a patera (dish), tympanon (hand drum), and a lion on her lap. B. Roman Magna Mater/Cybele with her attributes: cornucopia with pine cone, grapes, and fruit; lion; *polos* (in the shape of city walls); poppies and wheat sheaves reminiscent of Demeter (~50 CE).
A is from Wikimedia Commons, Louvre Museum, Photographed by Marie-Lan Nguyen. B is from the Getty Museum: [http://www.getty.edu/art/collection/objects/6511/unknown-maker-statue-of-a-seated-cybele-with-the-portrait-head-of-her-priestess-roman-about-50/?dz=0.5000,0.7217,0.39].

into Rome's pantheon, and she assumed many of Demeter's attributes. In his satirical novel *The Golden Ass*, written around 155 AD, Lucius Apuleius describes his personal encounter with the great goddess. After being accidentally turned into an ass by magic, he was miraculously restored to his human form by Isis, the Queen of Heaven. On her head she wore "many garlands interlaced with flowers" and she held "sheaves of wheat" in one arm. She informs him that she is known by many names throughout the world, among which she lists "Demeter" and "Persephone." Lucius attends her sacred processional in Rome in which she is led by blossom-strewing women dressed in white, reminiscent of both the Elusinian Mysteries and the Cerialia:

> Some women, sparkling in white dresses, delighting in their diverse adornments and garlanded with spring flowers, were strewing the ground with blossoms stored in their dresses along the route on which the sacred company was to pass.[15]

TELLUS AND THE ARA PACIS AUGUSTAE

From the foregoing it should be clear that Romans, along with most of their neighbors and forebears throughout the Mediterranean, believed that plants, agriculture, and, indeed, all of the natural world, were primarily the domain of goddesses. From the earliest stages of urbanization in ancient Mesopotamia to the height of the Roman Empire, absolute rulers associated themselves with powerful agricultural goddesses who represented the very foundation of civilization. This tradition arguably reached its most elaborate expression in Rome, in the relief panel located at the southeast corner of the Ara Pacis Augustae, an altar dedicated in 9 BCE in honor of the Pax Romana, or era of peace, established under the rule of Augustus.

Like the Mesopotamian palace relief sculptures of sacred trees during the Neo-Assyrian Period, the monumental Ara Pacis Augustae, which was originally located north of Rome on the former floodplain of the Tiber River, embodied the Roman civil religion and symbolized Augustus's power and prestige. The upper panels of the exterior walls depict various figures ranging from processionals of Roman citizens to deities representing the state. The elaborate floral friezes of the lower panels are filled with complex nature symbolism.[16] These floral symbols are symbolically connected to the stunning relief sculpture on the southern end of the eastern wall, shown in Figure 9.4, the so-called "Tellus panel."

The identity of the goddess depicted in the "Tellus panel" is still unresolved. In addition to the earth goddess, Tellus, several other related goddesses have also been suggested, including Ceres, Italia, Venus, Rhea Silvia, and Pax, to which we might add Flora and Magna Mater.[17] The presence of several kinds of fruit (grapes, pomegranates, and nuts) on the goddess's lap, as well as the prominent display of wheat, poppies, and other flowers on her left are all primarily associated with Ceres, but, as we have seen, the same botanical attributes have also been attached to several other Roman agricultural goddesses, thus raising the possibility that the figure is an amalgamation of two or more of them. The two figures holding billowing cloth could be Aurae (breezes), nymphs, Horae, or other spirits of nature. The two infants clearly identify the central figure as a mother goddess, drawing a parallel between human and agricultural fertility.

FIGURE 9.4 The so-called "Tellus" panel of the Ara Pacis Augustae. The upper panel depicts the goddess with two infants, flanked by a spair of nymphs. The lower panel is part of the floral frieze that surrounds the monument.
Photographed by the authors.

The floral frieze below the Tellus panel extends around the entire structure, making it the dominant decorative motif of the altar. Ultimately, its symbolic origin can be traced to the sacred tree motifs of Mesopotamia, which were strongly associated with Inanna/Ishtar. From their early incarnations as palm trees, sacred trees evolved into highly stylized, mythical plants, often bearing a variety of fruits at the tips of their branches. It could be said that the floral frieze of the Ara Pacis Augustae takes the sacred tree motif to its most complex apotheosis. From a basal cluster of acanthus leaves sprouts a complex pattern of curling stems and leaves of various species, including ivy, grape, laurel, oak, and olive, terminating in a variety of flowers and fruits rendered in strikingly naturalistic detail.[18] Nestled among the foliage of the mythical plant are animals: snakes, birds, frogs, lizards, and grasshoppers.

Each of the plants and animals is rich with symbolic significance. However, the general impression one has of the entire floral frieze is that it represents the fertility and profusion of wild nature, which underpins the agricultural abundance and prosperity necessary to the peace and security of the Roman empire. The goddess of the Tellus panel, who seems to combine the traits of many different nature goddesses, provides the focal point for all the nature symbolism. In a sense, the entire monument can be viewed as an updated symbolic version of the "sacred marriage" ceremony between the ruler, Emperor Augustus, and the symbolically unified goddess of Nature.

THE ROMAN WRITERS ON AGRICULTURE

Four principal Roman writers on agriculture—Cato the Elder (234–49 BCE), Varro (117–27 BCE), Virgil (70–19 BCE), and Columella (4–70 CE)—represented a continuous tradition spanning the three centuries after the death of Theophrastus in 288 BCE. In addition to their thorough grasp of agricultural practices, a feature common to them all was their avoidance of any discussion of the basic mechanisms of plant growth and reproduction, so central in the writings of Theophrastus. In their view, Theophrastus spent too much time pursuing useless theoretical questions instead of focusing on the practical concerns of maximizing crop productivity. According to Varro, the works of Theophrastus, although containing valuable information, were more suitable for philosophers than for practical agriculturalists.[19]

Notwithstanding their jaundiced view of Greek philosophy, Varro, Virgil, and Columella were well-acquainted with Theophrastus's writings on agriculture. They assimilated his discussions of arboriculture and soil classification as well as the principles of manure application and plant pathology. Most importantly, they absorbed the important insight that each plant species or variety is adapted to a particular set of environmental conditions. The practical significance of this insight was that farmers took pains to learn the requirements of specific crops in relation to soil, climate, water, and habitat and used the information to maximize their yields. This ecological insight, together with a systematic analysis of different methods of plant propagation, was the lasting legacy of Theophrastean botany to agriculture.[20] Virgil discussed it in his poem *Georgics*. The principle was further elaborated in the work of Columella, which retained its authority throughout the Middle Ages. Columella's treatise was the longest and most detailed account of Roman agriculture ever written. Among the new cultivation techniques he described was a "hotbed" heated from below by a layer of fermenting manure and shielded by panes of glass, which provided both sunlight for plant growth and supplementary heating during the winter.

Columella frequently enlivened his treatise with poetic images and quotations, which often reveal his gender biases. In particular, Columella regularly employed female metaphors for the earth and its fruits. In the following passage, he compares a bountiful grape harvest to Nature's breasts:

> [I]f he should enter a field at the proper time, [the vintner] would marvel most pleasurably at the benevolence of nature, . . . whereby the fostering earth each year, as if delighting in never-ending parturition, extends to mortals her breasts distended with new wine.[21]

Although Bacchus is the Roman god of wine, grapes themselves are identified with the breasts of the nature goddess. Columella also compared flowers to maidens, as when he describes the rose as "full of maiden blush."[22]

The vexing question of sex in plants was never broached by any of the major Roman agricultural writers, who were content to accept the pronouncements of Aristotle and Theophrastus on the subject. However, as the following dialogue illustrates, Varro's comparison of fruit production to female pregnancy clearly genders plants female:

> "Tell us now," said Agrius, "of the third step, the nurture and feeding of the plant."
>
> "All plants, " resumed Stolo, "grow in the soil, and when mature conceive, and when the time of gestation is complete bear fruit or ear, or the like; and the seed returns whence it came. Thus, if you pluck the blossom on an unripe pear, or the like, no second one will grow on the same spot in the same year, as the same plants cannot have two periods of gestation. For trees and plants, just as women, have a definite period from conception to birth."[23]

The phrase "and when mature conceive" implies that Varro regarded conception as occurring at a specific point in development. In the absence of any mention of pollen as a fertilizing agent, we infer that Varro, like his Greek predecessors, thought of seed production as female parthenogenesis, which occurs naturally when the plant reaches maturity.

PLINY'S *HISTORIA NATURA*

Caius Plinius Secundus, or Pliny the Elder, was born in Gaul (the region of France and Belgium), but his father took him to Rome at an early age to be educated. As a young man, he practiced law before beginning his military career as a junior officer in Germany during the reign of Claudius. Two years after the volatile and unpredictable Nero came to power (~56 CE), Pliny prudently left military service to live a quiet, unassuming life in Rome. More than a decade later, under the Emperor Vespasian, Pliny served as Procurator in Spain and elsewhere. In the course of his duties, he traveled throughout much of the ancient world, including Gaul, Africa, and probably Judaea and Syria. Pliny the Younger, his nephew, described his uncle as an incessantly voracious reader. Whether dining at home or, more comically, while being carried through Rome in a sedan chair, Pliny always arranged to have one servant read aloud to him while another took dictation of passages he wished to excerpt.

Precisely when Pliny began organizing and synthesizing his readings into his great compendium, *Natural History,* is not known. This encyclopedic compilation, comprising 37 volumes (of which Books XII–XXVI deal with plants) has been characterized as a "storehouse of ancient errors," for Pliny was a compulsive and uncritical compiler of information—both valid and apocryphal. He was, to use a modern metaphor, a one-man Internet. Until the rediscovery in the Vatican library of Theophrastus's lost works, *Historia Plantarum* and *De Causis Plantarum*, in 1453, Pliny's error-prone accounts of plant biology remained the only available records of the great synthesis of Greek botany produced by Theophrastus.

A wonderful example of Pliny's use of the fabulous to spice up his factual reporting is his oft-quoted description of love among the date palms:

> [I]t is stated that in a palm-grove of natural growth the female trees do not produce if there are no males, and that each male tree is surrounded by several females with more attractive foliage that bend and bow towards him; while the male bristling with leaves erected impregnates the rest of them by his exhalation and by the mere sight of him, and also by his dust; and that when the male tree is felled the females afterwards in their widowhood become barren. And so fully is their sexual union understood that mankind has actually devised a method for impregnating them by means of the flower and down collected from the males, and indeed sometimes by merely sprinkling their dust on the females.[24]

Pliny, who cannot resist a good anecdote however implausible, seems torn between the fantasy of courtship among the palms and a dry, scientific explanation. On the other hand, Pliny, unlike Theophrastus, didn't hesitate to equate pollination with sexual reproduction. In fact, in another passage, Pliny appears to generalize the sexuality of palm trees to other trees and plants, a conceptual leap neither Aristotle or Theophrastus dared to make:

> The more diligent enquirers into the operations of Nature state that all trees, or rather all plants, and other productions of the earth, belong to either one sex or the other; a fact which it may be sufficient to notice on the present occasion, and one which manifests itself in no tree more than in the palm.[25]

A quick gloss of this sentence might lead one to conclude that Pliny, however fallible, was the first to discover the universality of sex in plants. However, upon closer reading it is clear that the generalization is based on the erroneous idea that all plants are dioecious, belonging to "one sex or the other." Elsewhere in *Natural History* Pliny describes the properties of male and female plants strictly in terms of their wood, growth habits, the sizes of their fruits, and other vegetative properties—spurious notions that Theophrastus had dismissed as folklore. Pliny's distinction between male and female plants was not based on the sexual structures of the flower, but on the same culturally defined gender associations we encountered earlier in Greek botany. Yet Pliny deserves an honored place in the history of botany for giving us in his description of the Madonna Lily the term stamen (*staminum*, Latin for the warp of an upright loom).[26] Pliny also used the term "stamens" in his description of rose blossoms:

> Gradually acquiring a ruddy tint, this bud opens little by little, until at last it comes into full blow, developing the calyx, and embracing the *yellow-pointed stamens* which stand erect in the center of it.[27]

Pliny thus takes a small, but significant step beyond Theophrastus by naming the pollen-containing structures of flowers. However, he makes no mention of the pollen contained in the anthers and makes no connection to the "dust" of palm trees, so he missed the functional significance of the structure he had identified. Pliny's generalization of sex to all plants was therefore not based on the identification of pollen as the male reproductive structure, and thus we cannot credit him with discovering sex in plants.

In another chapter on "chaplet flowers"—flowers used for garlands and wreaths—Pliny explicitly states that such flowers are merely decorative, created by Nature to delight people or perhaps to warn them of the transience of life:

> Cato bade us include among our garden plants chaplet flowers, especially because of the indescribable delicacy of their blossoms, for nobody can find it easier to tell of them than Nature does to give them colors, as here she is in her most sportive mood, playful in her great joy at her varied fertility. To all other things in fact she gave birth because of their usefulness, and to serve as food, and so has assigned them their ages and years; but blossoms and their perfumes she brings forth only for a day—an obvious warning to men that the bloom that pleases the eye most is the soonest to fade.[28]

Pliny's anthropocentric view of the "purpose" of garden flowers was assimilated into the literary traditions in the Middle Ages and Renaissance, and it proved to be very durable. In the mid-nineteenth century, Darwin himself complained in *Origin of Species* about the persistence of the idea that attractive flowers were "created for the sake of beauty, to delight man or the Creator." Thus Pliny's interpretation of the "purpose" of ornamental flowers prevailed throughout much of the Christian era until it was finally deposed by the new paradigm of Darwinian natural selection.

NICOLAUS OF DAMASCUS AND *DE PLANTIS*

In addition to Pliny's *Natural History*, the only other work on plants derived from Greek philosophical writings that survived into the Middle Ages was *De Plantis*—a brief treatise on plants that was originally attributed to Aristotle. Thought to be Aristotle's lost treatise on plants, this work was greatly revered as the definitive text on plants—a disaster for botany since it was actually either a truncated, bowdlerized version of Aristotle's writings or a crude summary of fragments taken from Aristotle and earlier writers.

Not only was *De Plantis* greatly inferior to the authentic works of Aristotle and Theophrastus, it had also become garbled and incomprehensible in many places in the course of repeated copying and sequential translations (Greek → Syriac → Arabic → (Hebrew, Latin, Greek) → English).[29] Thus, the authority of Aristotle came to be affixed to a totally inadequate and muddled version of Greek botany, which retarded the progress of botanical science for hundreds of years. The actual author of *De Plantis* is now known to be Nicolaus of Damascus, who probably intended it as a brief summary of Aristotle's ideas about plants.[30] He appears to have had little or no familiarity with Theophrastus's botanical writings.

According to his biographer, B. Z. Wacholder, the little we know about Nicolaus of Damascus is derived from "remnants of his autobiography, the account of Herod in Josephus, and scattered references in secondary sources."[31] Nicolaus states that he was born into a prominent family in Damascus in 63 BCE.[32] Nicolaus's parents were pagans, but whether they were Greeks or Syrians in origin is unknown. Trained in rhetoric by his father, Nicolaus belonged to a new school of Peripatetic scholars whose intellectual horizon encompassed both East and West. Anticipating the attitudes of medieval scholars, these self-described "Aristotelians" believed that Aristotle had acquired a god-like, universal understanding of nature and that only the limitations of time had prevented him from writing it all down.

In 36 BCE, Nicolaus, who had earned a reputation as an excellent teacher, had the good fortune to meet and impress Cleopatra, who was on her way to meet Antony during the prolonged Parthian War.[33] She hired him to tutor her twins by Antony—which afforded him access to the elite of Alexandrian society. After Cleopatra's death in 30 BCE, Nicolaus entered the service of Herod I (Herod the Great), the Roman client king of Judea. A skilled rhetorician, Nicolaus accompanied Herod abroad on numerous occasions and was a favorite with Emperor Augustus of Rome. He frequently sent Augustus gifts of dates, prompting Augustus to name a variety of date after him (nicolai). He also undertook many delicate missions on Herod's behalf and was successful at mediating conflicts.[34]

Nicolaus's memory was cherished as late as the seventh century, and his philosophical writings were among the first to be translated into Semitic tongues when Greek learning gained the ascendancy in the East.

Much of *De Plantis* is taken up with such metaphysical questions as whether plants are living or inanimate, whether they have souls, and whether they experience desires and feelings. But he also expresses the wish to "reach some conclusion" about the question of plant sex:

> Now Anaxagoras and Empedocles say that [plants] are influenced by desire; they also assert that they have sensation and feel sadness, deducing this from the fall of their leaves; while Empedocles held the opinion that sex has a place in their composition. Plato indeed declares that they feel desire only on account of their compelling need of nutriment. If this be granted, it will follow that they also feel joy and sadness and have sensation. I should also like to reach some conclusion as to whether they are refreshed by sleep and wake up again, and also whether they breathe, and whether they have sex and the mingling of the sexes or not.[35]

Nicolaus identifies the question of plant sex as "the most important and appropriate subject of inquiry" in all of plant biology:

> The most important and appropriate subject of inquiry which arises in this science is that proposed by Empedocles, namely whether female and male sex is found in plants, or whether there is a combination of the two sexes.

Nicolaus begins his analysis of plant sex by paraphrasing Aristotle's dictum that males generate "in another," while females generate "from another." However, this is "not found to be the case in plants," he states, because:

> in a particular species the produce of the male plant will be rougher, harder and stiffer, while the female will be weaker but more productive.

This explanation (from the English translation of the Greek retranslation) seems to be a *non sequitur*, indicative of an omission or mistranslation of the original text. He seems to say that both the male and female produce progeny, but they differ in their characteristics according to gender.

Next, he considers the possibility that the two sexes are "mingled" in plants, a clear reference to Aristotle. He points out that, logically speaking, things can only be said to "mingle" if they were once separated, and if, in plants the sexes are mingled, it ought to be possible to find individuals at various intermediate stages of the mingling process. This, he states, one never observes, so the two sexes cannot be "combined" in plants:

> We ought also to enquire whether the two kinds are found in combination in plants, for things which mingle together ought first to be simple and separate, and so the male will be separate and the female separate; they afterwards mingle, and the mingling will only take place when it is produced by generation. A plant, therefore, would have been discovered before the mingling had taken place, and it ought therefore to be at the same time an active and a passive agent. The two sexes cannot be found combined in any plant.

Indeed, Nicolas states, if individual plants were hermaphroditic, this would make them "more perfect" than animals:

> if this were so, a plant would be more perfect than an animal, because it would not require anything outside itself in order to generate.

If plants do not reproduce by the copulating ("mingling") of the two sexes—how do they reproduce? Here, Nicolaus, like Aristotle and Theophrastus before him, invokes the old metaphor of Empedocles comparing the fruit of an olive tree to a bird's egg. The seed, like the egg, contains the embryo within it, and supplies it with nourishment:

> But we must suppose that the mingling of the male and the female in plants takes place in some other way, because the seed of a plant resembles the embryo in animals, being a mixture of the male and female elements. And just as in a single egg there exists the force to generate the chicken and the material of its nutriment up to the time when it reaches perfection and emerges from the egg, and the female lays the egg in a short space of time; so too with the plant. And Empedocles is right when he said the tall trees do not bear their young; for that which is born can only be born from a portion of the seed, and the rest of the seed becomes at first the nutriment of the root; and the plant begins to move as soon as it is born.

To Nicolaus, the egg-like nature of seeds is proof enough (by analogy to birds) that two sexes must be involved in the making of seeds by plants. However, since the only activity we observe when plants produce seeds is the "generation of fruits," sex may only play a role in plant reproduction under certain "circumstances":

> This then is the opinion which we ought to hold about the mingling of the male and female in plants, similar to that which we hold about animals. This process

is the cause of plants under a certain disposition of circumstances; for in the case of an animal when the sexes mingle the powers of the sexes mingle after they have been separated, and a single offspring is produced from them both. But this is not the case with plants . . . in plants the only operation which we find is the generation of fruits.

Nowhere in the discussion of the two sexes are flowers specifically referred to. According to the Syriac translation of *De Plantis*, which is earlier than the Latin and Greek translations, "flowers" are comparable to the "hair and feathers" of animals, thus reinforcing the impression that Nicolaus did not consider the flower (i.e., the corolla) to be a reproductive structure.[36]

How could the two sexes of plants "mingle" without copulating? To explain the apparent lack of copulation in plants, the Syriac version of *De Plantis* includes the bold assertion that the penis of plants is the flower stalk, or pedicel, and the immature fruit, or carpel, serves as the plant uterus. In other words, the male and female organs of plants are joined one on top of the other in a state of permanent coitus:

In plants there exists something similar to a uterus and a penis *joined* together, viz. the forms called *moschoi*. A *moschos* is a stalk by which leaves or fruit are suspended. In this the conjunction of the two capacities resides, and there is the generative capacity acting upon that which is generated. Seeds resemble eggs, because in them the active and the passive capacities begin to move together.[37]

This was a logical inference, which neatly solved the apparent paradox that, like birds, plants need two sexes to reproduce but, unlike birds, are unable to move in order to copulate. On this basis, there would be no reason to suspect stamens as the male sexual structures because they are physically separated from the carpel. The *pedicel-as-penis hypothesis*, which has been overlooked by historians of botany, deserves to be recognized as the earliest known anatomical hypothesis for the male sexual structure of hermaphroditic flowers. It appears to be present only in the Syriac translation of *De Plantis*.[38] This raises the possibility that the hypothesis was introduced by the Syriac translator but dropped by subsequent translators, perhaps on aesthetic or moral grounds. As we shall see later in the book, the *stamens-as-penis hypothesis*, when proposed by the British physician Nehemiah Grew in 1684, was rejected by many contemporary naturalists on moral grounds. Perhaps the pedicel-as-penis analogy in the Syriac version of *De Plantis* was similarly received.

Apart from the anomalous Syriac translation that was not generally available in the Middle Ages, Nicolaus's *De Plantis* provides no new insights into the question of plant sex, and its truncated and corrupted presentations of Aristotle's writings on the subject only succeed in muddying the waters. We are left with a paraphrase of Empedocles's oft-cited bird's egg metaphor: "the hen lays the egg . . . so too with the plant." This was the sum total of knowledge about plant reproduction that was available to scholars of the Middle Ages from the Greco-Roman philosophers.

THE DEGENERATION OF OLIVE TREES AS PROOF OF ORIGINAL SIN

As discussed in Chapter 8, because of the hybrid nature of most cultivated fruit trees (that is, their *heterozygosity*), Greek farmers knew that such crops did not breed true when propagated by seed, but underwent a deterioration in quality. Roman farmers were similarly aware of this phenomenon, as illustrated by the following passage from Virgil's *Georgics*:

> Some seeds I've seen, though chosen with time and care,
> Degenerate still, unless with human hand
> The largest were selected every year.
> But so it is; it is the will of fate
> that all things backward turn, all things deteriorate.[39]

Fortunately, desirable varieties of wild olive and other fruit trees could be propagated vegetatively from cuttings, allowing for "instant domestication."

The degeneration of seed-grown olive trees was such a well-known phenomenon that St. Augustine, the Bishop of Hippo, cited it as "palpable evidence" for the truth of original sin:

> It is, no doubt, very wonderful that what has been forgiven in the parent [original sin] should still be held against the offspring; but nevertheless such is the case. That this mysterious verity, which unbelievers neither see nor believe, might get some palpable evidence in its support, God in his providence has secured the example of certain trees. For why should we not suppose that for this very purpose the wild olive springs from the [domesticated] olive? Is it a wonderful thing, then, how those who themselves have been delivered by grace from the bondage of sin, should still beget those who are tied and bound by the self-same chain, and who require the same process of loosening? Yes, and we admit the wonderful fact. But that the embryo of wild olive trees should latently exist in the germs of true olives, who would deem credible, if it were not proved true by experiment and observation? In the same manner, therefore, as a wild olive grows out of the seed of the wild olive, and from the seed of the true olive springs nothing but a wild olive, notwithstanding the very great difference there is between the wild olive and the [true] olive; so what is born in the flesh, either of a sinner or of a just man, is in both instances a sinner, notwithstanding the vast distinction which exists between the sinner and the righteous man.[40]

At a later point, Augustine traces the source of original sin to Adam, who "changed from a pure olive . . . into a wild olive" and thereby converted the entire race into a "wild olive stock." Only God's grace can turn a bad olive into a "good olive," which constitutes proof of the necessity of baptism:

> That, however, which in the case of a regenerate parent, as in the seed of the pure olive, is covered without any guilt, which has been pardoned, is still no doubt retained in the case of his offspring, which is yet unregenerate, as in the wild olive, with all its guilt, until then also it be remitted by the self-same grace. When Adam sinned, he was changed from that pure olive, which had no such corrupt seed whence should

> spring the bitter issue of the wild olive, into a wild olive tree; and inasmuch as his sin was so great, that by it his nature became commensurately changed for the worse, he converted the entire race of man into a wild olive stock. The effect of this change we see illustrated, as has been said above, in the instance of these very trees. Whenever God's grace converts a sapling into a good olive, so that the fault of the first birth (that original sin which had been derived and contracted from the concupiscence of the flesh) is remitted, covered, and not imputed, there is still inherent in it that nature from which is born a wild olive, unless it, too, by the same grace, is by the second birth changed into a good olive.[41]

The idea that Adam, rather than Eve, was responsible for condemning all future generations of Christians to original sin can be traced to Aristotelian notions about conception, according to which the male seed provides the embryo's "psyche," or soul, while the female seed contributes only the physical matter. Since original sin is an affliction of the soul, it was Adam's immaterial essence, not Eve's egg, that passed on the taint of original sin to future generations.

NOTES

1. Greeks from the north-central part of the Peloponnese.

2. The current consensus is that the language spoken by the historical Trojans was probably Luwian, an Anatolian branch of the Indo-European language family.

3. Forsythe, G. (2006), *A Critical History of Early Rome: From Prehistory to the First Punic War*. University of California Press.

4. Scullard, H. H. (1981), *Festivals and Ceremonies of the Roman Republic*. Cornell University Press. First U. K. Edition.

5. Ibid.

6. Ovid, *Fasti* IV, trans. James G. Frazer.

7. Ibid.

8. Ovid (2000), *Fasti* V, trans. A. J. Boyd and R. D. Woodard.

9. Ibid.

10. Olenus was an ancient Greek city in Achaea, located on the northwest corner of the Peloponnese peninsula. Olenus is mentioned in a tragedy by Sophocles known only in fragments. The story concerns Periboea, the daughter of the King of Olenus, who is captured by Oeneus, the King of Calydon. In the fragment, Periboea says, "For I am being brought from the rich land of Olenus." Sophocles (1996). *Fragments*, trans. H. Lloyd-Jones, Vol. III, Loeb Classical Library, Harvard University Press, p. 137.

11. Ovid, *Fasti* V.

12. Recall from Chapter 7 that Dionysus had been borne from a slit in Zeus's thigh. Jealous of Zeus's giving birth to a mortal woman's child, Hera is touched by a magic herb and becomes pregnant with Ares.

13. Roller, L. E. (1999), *In Search of God the Mother: The Cult of Anatolian Cybele*. University of California Press.

14. Ibid.

15. Apuleius (1994), *The Golden Ass* 11.9–10, trans. P. G. Walsh. Oxford University Press.

16. Castriota, D. (1995), *The Ara Pacis Augustae and the Imagery of Abundance in Later Greek and Early Roman Imperial Art*. Princeton University Press.

17. Spaeth, B. S. (1996), *The Roman Goddess Ceres*. University of Texas Press.

18. Castriota, *The Ara Pacis Augustae and the Imagery of Abundance.*

19. Varro, M. T. (1913), *Res Rustica*, Book I, in *Roman Farm Management: The Treatises of Cato and Varro*, trans. "A Virginia Farmer" [pseudonym of F. Hamilton]. Macmillan.

20. Hughes, J. T. (1985), Theophrastus as ecologist. *Environmental Review* 9:296–306.

21. Lucius Junius Moderatus Columella (1941), *De Re Rustica*, Book III, trans. H. B. Ash. Loeb Classical Library edition, p. 343.

22. Lucius Junius Moderatus Columella (1941), *On Agriculture*, Book X. Loeb Classical Library doi: 10.4159/DLCL.columella-agriculture.

23. Marcus Terentius Varro (1960), *On Agriculture*, trans. William Davis Hooper, Revised by Harrison Boyd Ash. Harvard University Press, p. 277.

24. Pliny (1968), *Natural History*, Vol. III, Book XIII, ch. VII, in *Pliny, Natural History in Ten Volumes*, trans. H. Rackham. Loeb Classical Library, William Heinemann Ltd., Harvard University Press.

25. Pliny (1855), *Natural History*, Vol. III, Book XIII, ch. VII, in *The Natural History of Pliny*, trans. J. Bostock and H. T. Riley. Taylor and Francis.

26. Idem., Book XXI, 56, 96.

27. Idem., Book XXI, x.

28. Pliny, *Natural History*, Book XXI. i.

29. Wacholder, B. Z. (1962), *Nicolaus of Damascus*. University of California Press.

30. According to B. Z. Wacholder, the medieval scholar Albertus Magnus was the first to attribute *De Plantis* to Nicolaus of Damascus rather than Aristotle. In 1841, the German scholar E. H. F. Meyer published *De Plantis* and followed Albertus's example by ascribing it to Nicolaus as well. Then, in 1923, an Arabic manuscript of *De Plantis* dating to the ninth century was found in a library in Istanbul. The title page of this Arabic version was written: "The book of Plants by Aristotle, the commentary of Nicolaus, translated by Ishak ibn Hunayn, with the corrections of Thabit ibn Kurra." Nicolaus's authorship of *De Plantis* is further supported by Syriac fragments of Nicolaus's writings, which include a page of *De Plantis*. In the words of Wacholder, "It is now certain that the Hebrew, Latin and Greek translations are based on the Arabic version, itself a translation from the Syriac. There is no longer any reason to doubt that De plantis, in its present form, must be credited to Nicolaus."

31. Wacholder, *Nicolaus of Damascus*.

32. The Seleucid Empire, ruled by the Selucid dynasty from 312 BCE to 63 BCE, was a major Hellenistic state centered in Selucia in modern-day Iraq. At its peak, it included Anatolia, the Levant, and Mesopotamia. The Empire was finally defeated in by the Roman army under Pompey in 63 BCE.

33. Periodic clashes that occurred from 66 BCE to 217 CE between the Roman Republic and the Parthian Empire centered in Iran.

34. According to Josephus (*l.c.* xii. 3, § 2), in 14 BCE Nicolaus successfully negotiated with M. Agrippa to restore the privileges of Jews in Ionia. He also mediated a conflict between Herod and Augustus in 7 BCE and restored harmony between them.

35. Nicolaus of Damascus (1878/1984), *On Plants*, trans. J. Bussemaker. In J. Barnes, ed., *The Complete Works of Aristotle*, Vol. 2, Princeton/Bollingen Series LXXI, p. 1251.

36. Drossaart Lulofs, H. J., and E. L. J. Poortman (1989), *Nicolaus Damascenus De Plantis: Five Translations*, North Holland Publishing Co., p. 56.

37. Idem., p. 76.

38. Ibid.

39. Virgil, *Georgics* I.197–200.

40. Augustine of Hippo (2012), *Marriage and Concupiscence*, trans. Philipp Schaff. Jazzybee Verlag, Book I, ch. 21.

41. Idem., Book I, ch. 37.

10

From Herbals to Walled Gardens

PLANT GENDER AND ICONOGRAPHY

IT IS DIFFICULT to say precisely when classical antiquity ended and the Middle Ages began. In the West, the Roman Empire ended decisively in 476 AD with the fall of Ravenna to the Ostrogoths, an East Germanic people. This seismic event marked the founding of Western Europe, whose inhabitants spoke Latin and Germanic languages and whose rulers over time adopted Catholicism and accepted papal authority. In contrast, the Eastern, or Byzantine, Empire centered in Constantinople (the "second Rome") was insulated by distance from the Germanic tribes and avoided a catastrophic collapse. Instead, it was gradually absorbed into Western Asia until little remained of its former Roman identity. Citizens of the Byzantine Empire spoke Greek instead of Latin, belonged to the Eastern Orthodox Church instead of the Catholic Church, and adopted eastern styles of art and architecture.

In the seventh century, a new Abrahamic religion—Islam—arose and spread throughout the Arabian Peninsula. By the middle of the eighth century, the new Islamic Empire had displaced the old Sassanid Persian Empire and had expanded westward to include all of northern Africa, Spain, and Portugal, and eastward to encompass modern-day Iraq, Persia, Armenia, parts of Afghanistan, and northwestern India.

Despite their numerous cultural differences, the educated classes of the three civilizations—Western Europe, the Byzantine Empire, and the Islamic Empire—shared a common scientific vocabulary based on Greek physics and cosmology, including theories about the four elements and the Hippocratic humors theory. In the field of botany, medieval scholars deferred to the authority of *De Plantis*, believing it to be the work of Aristotle. Likewise, they continued the tradition of the illustrated herbal based on Greek models. This shared Hellenistic heritage of the medieval world, combined with the rapid diffusion of ideas that followed in the wake of conquest and commerce, allows us to discuss medieval botany as a relatively integrated whole.

All told, the Middle Ages lasted approximately a thousand years, ending definitively with the defeat of a tottering Constantinople by the Ottoman Turks in 1453. Even before its demise, early Renaissance scholars like Petrarch already were referring to it as the "Dark Ages" because of its domination by religion. This view is now regarded as incomplete, if not obsolete. Far from being a mere place-holder between the Classical and Renaissance periods, the Middle Ages transformed human society in fundamental ways that set the stage for the Renaissance and the Scientific Revolution. The period is usually divided into three main phases: the Early Middle Ages (500 AD–1000 AD), the High Middle Ages (1000 AD–1299 AD), and the Late Middle Ages (1300–1453), with the latter, in some regions, overlappping the Early Renaissance.

In this chapter, our focus will be on the Early Middle Ages, a formative period during which the three dominant monotheistic faiths—Catholicism, the Orthodox Church, and Islam—vied for hegemony and borrowed freely from one another and from paganism to enhance their mass appeal. Sometimes called the "Age of Faith," the Early Middle Ages was not a favorable period for rationalistic philosophy. Neoplatonism, with its emphasis on otherworldliness, submerged Aristotelian empiricism, while in art, iconography and symbol replaced naturalism. Nowhere are these trends more apparent than in the decline of the medieval herbal from practical field guide to decorative status symbol.

THE EARLY GREEK HERBALS

The Greek pharmacopoeia had been compiled largely from more ancient traditions, including Minoan, Egyptian, Babylonian, and other sources. By the time Theophrastus began teaching at the Lyceum around 335 BC, Greek pharmacological knowledge had already been gathered and systematized by his contemporary, the Athenian physician Diocles. Diocles wrote two medically oriented works on plants, one dealing with their nutritional value and the other describing their medicinal properties. The latter was titled *Rhizotomikon*, from the term *rhizotomist*, or "root-cutter"—referring to mostly illiterate men and women who made their living by collecting medicinal plants. Theophrastus, who regarded Diocles as the pre-eminent authority on pharmacology, quoted him extensively in Book IX of his *Historia Plantarum*. Book IX is sometimes referred to as the oldest extant Greek herbal.[1] Many of Diocles's plant descriptions were incorporated into other herbals during the Roman period.

With the death of Theophrastus in 287 BCE, theoretical botany, the study of plants as organisms, came to an end in the classical world. However, Greek physicians continued to distinguish themselves in pharmacology, an applied branch of botany. One of the most important of these was Krateuas, the court physician to King Mithradates VI of Asia Minor, who was himself a skilled herbalist legendary for his knowledge of poisons.[2] Krateuas authored two works that had long-lasting influence on the history of herbals. The first was a treatise on medicinal plants (*Rhizotomikon*), which was designed for practicing physicians. Many features of Krateuas's *Rhizotomikon* were retained in the herbals of the Middle Ages, including the alphabetical arrangement of the plants. a written description of each plant's appearance, a list of synonyms, and an account of each plant's medical uses. Although the text of Krateuas's *Rhizotomikon* has been lost, it was likely derived from Diocles's book by the same name. Later, it served as the primary source of the herbal of Sextius Niger, a

Roman physician during the time of Augustus, on which Pliny and Dioscorides (discussed later) based much of their information.[3]

Krateuas's second work on plants was a simplified version of *Rhizotomikon* intended for the lay public. Instead of a written treatise, Krateuas took the innovative step of producing a series of painted illustrations of plants, with the name and medical properties listed below. Although artists of the region had been representing plants in paintings, frescos, and reliefs throughout antiquity, Krateuas's herbal is the first known scientific use of botanical illustrations. These illustrations are termed "portraits" because they were painted from live models. The obvious value of plant portraits, as opposed to drawing from memory, lies in their greater accuracy, making them more useful for physicians. Although Krateuas's original illustrated herbal has been lost, some of the early copies of some of his portraits, possibly dating back to the second century AD, were preserved in an early Byzantine herbal.

Prior to the advent of printing, the only means of replicating books was by copying, often by artists with little or no understanding of plants and with minimal artistic skill. Inevitably, errors were introduced, and the more cycles of copying a manuscript underwent the less accurate the copies became. Even Pliny the Elder, whose writings on medicinal plants lacked illustrations, complained about the inaccuracy of the plant portraits in his copies of the illustrated herbals of Krateuas.[4]

The physician Pedanius Dioscorides was the last of the great Greek herbalists during the Roman period. Dioscorides's herbal, written in Greek around 65 CE and subsequently translated into Latin as *De Materia Medica*, displaced Krateuas's *Rhizotomikon* and remained the standard pharmacology of the West for more than 1,500 years. Born in the Roman province of Cilicia in Anatolia, he is thought to have been a surgeon in the Roman army who traveled widely and studied a broad sampling of Mediterranean flora.[5] His pharmacopoeia consisted mainly of plants from Greece, Anatolia, and the Near East, but also included species from Italy, Sardinia, Spain, Gaul, Britain, and even India. Although he was careful to cite the work of other herbalists, he expressed pride in "knowing most herbs with my own eyes."

As a physician writing for other physicians, Dioscorides restricted himself to descriptive botany, supplying only that information helpful for field identification of medicinal plants. Theoretical questions, such as the mechanism of plant reproduction, were of no interest to him. On the other hand, he frequently distinguished between "male" and "female" plants in his herbal. Dioscorides's gender assignments were based entirely on anthropomorphic, culturally based criteria (hardness, toughness, height, etc.), which Theophrastus had previously rejected as fanciful.[6] Yet, such was Dioscorides's reputation that his spurious gender assignments continued to be employed by herbalists through the sixteenth century.

THE JULIANA ANICIA CODEX

Sometime during the second century AD, the plant portraits of Krateuas and other illustrated herbals were combined with the text of Dioscorides to produce the ancestor of all subsequent "Dioscoridean" herbals spanning many centuries and many regions. The magnificently illustrated *Juliana Anicia Codex*, completed around 512 CE, is the earliest extant herbal of Dioscorides. It resides at the Austrian National Library in Vienna and is sometimes referred to as the *Vienna Dioscorides* or *Codex Vindobonensis*.[7] It was a gift to the Roman imperial princess, Patricia Juliana Anicia from the people of Honoratae, a suburb of Constantinople, in gratitude for the construction of a vast, elaborately decorated church dedicated to St. Polyeuktos, a third-century martyr. Princess Juliana was descended from two of the most illustrious and aristocratic families in the Roman Empire. Her father had been one of the last Emperors of the West, whereas her mother was descended from both Theodosius I, who briefly united the Western and Eastern Empires, and Valentinian III, another late emperor of the Western Empire. Disappointed that her son had not succeeded to the throne, Juliana built the Church of St. Polyeuktos both as an expression of her extreme piety and as an assertion of her family's prestige and importance. So impressive was the Church of St. Polyeuktos that Emperor Justinian, who reigned from 527 to 565 CE, is believed to have ordered the construction of the Hagia Sophia expressly to outshine it.

The walls and foundation of St. Polyeuktos were excavated in 1960, and an inscription was found on a marble slab lavishly eulogizing Princess Juliana:

> What choir is sufficient to sing the work of Juliana, who after Constantine—embellisher of his Rome, after the holy golden light of Theodosius, and after the royal descent of so many forebears, accomplished in a few years a work worthy of her family, and more than worthy? She alone has conquered time and surpassed the wisdom of renowned Solomon, raising a temple to receive God, the richly wrought and graceful splendor of which the ages cannot celebrate.[8]

The inscription continues with rapturous praise of the architectural and artistic wonders of the church. Archaeologist Martin Harrison has speculated that the decorations for the Church of St. Polyeuktos (Figure 10.1) were inspired by Solomon's Temple, which, according to biblical descriptions, was decorated with botanical imagery: palm trees, open flowers, pomegranates, capitals in the form of lilies, and, in the Herodian Temple, grapevines. Most of the decorations of the columns, walls, and arches of St. Polyeuktos were also botanically inspired, a pastiche of Near Eastern plant iconography and more naturalistic botanical motifs reminiscent of the *Ara Pacis*, such as grape vines (see Figure 10.1D).

It is no accident that Juliana, a powerful and wealthy aristocrat, chose botanical themes for the decorations of her magnificent church. As we saw in earlier chapters, the association of women and plants in ancient religions predates Christianity. Indeed, the Church of St. Polyeuktos might easily be mistaken for a temple devoted to Flora, Ceres, or some other agricultural deity. What better gift of appreciation to the Princess than a stunningly illustrated herbal, a graphic reminder of the church's botanical themes? Note that the church's plant motifs range from the highly stylized (Figure 10.1C) to the naturalistic (Figure 10.1D). The illustrations in the *Juliana Anicia Codex* display the same broad range in style, from the iconographic to the naturalistic, reflecting the openness of sixth-century Constantinople to a wide range of artistic styles from diverse cultural traditions. It is likely that the floral decorations of St. Polyeuktos inspired those adorning the columns and arches of Hagia Sophia.

Significantly, Juliana's enthroned, goddess-like portrait in the frontispiece of the codex is painted in the more naturalistic Greco-Roman style rather than in the prevailing Byzantine style (Figure 10.2A). The Byzantine style had arisen when Greco-Roman art, notable for its

(a)

(b)

(c)

FIGURE 10.1 Decorations from the Church of St. Polyeuktos. A. Pier-capital. B. Pier-capital with central date palm. C. Detail of cornice showing two kinds of palmette, flowers, and leaves. D. Detail of an arch showing grapevine.
From Harrison, M. (1989), A Temple for Byzantium: The Discovery and Excavation of Anicia Juliana's Palace-Church in Istanbul. Harvey Miller.

(d)

FIGURE 10.1 Continued

naturalistic portraits, fused with Eastern artistic traditions, becoming more stylized and decorative—as exemplified by the mosaic of Empress Theodora (wife of Justinian) and her retinue at the church of San Vitale in Ravenna (Figure 10.2B).

Princess Juliana is clearly a transitional figure, representing the old guard of the Roman Empire, while the younger Justinian and Theodora embody the new Byzantine spirit of the Early Middle Ages. According to Minta Collins, the *Juliana Anicia Codex* was intended to serve more as a "volume of antiquarian, literary, and even sentimental interest" than as a practical medical text.[9] Collins's hypothesis should be kept in mind when evaluating the extreme range of styles of the plant "portraits" contained in the herbal.

As previously noted, the 435 plant illustrations of the *Juliana Anicia Codex* are highly variable in style and quality, ranging from the naturalistic and recognizable (Figure 10.3A–D) to the stylized and unrecognizable (Figure 10.3E). It has long been an article of faith that the plant illustrations in the *Codex* must have been copied from earlier Greek herbals because Byzantine artists, steeped in the mystical, otherworldly philosophy of the Eastern Orthodox Church, regarded the natural world as inherently sinful and unworthy of study by devout Christians, whose minds ought to be focused on eternity, scripture, and the purity of their souls.

Byzantine artists employed an iconographic style patterned after the stylized religious icons of saints that were, and still are, prevalent in the Eastern Church. They made no attempt to portray likenesses, and, in any event, the facial features of most saints were unknown. Religious symbols were therefore added to differentiate one saint from another. The goals of the Byzantine iconographic artists were thus antithetical to the basic function of the Greek illustrated herbal, which was to serve as a practical guide for the identification of medicinal plants in the field. Charles Singer has argued that the Byzantine artists who illustrated the *Juliana Anicia Codex* had lost the ability to paint directly from nature.[10] Based on the assumption that the plant illustrations of the *Juliana Anicia Codex* must have been copied from previous herbals, Singer postulated that the most naturalistic paintings

(a)

(b)

FIGURE 10.2 Comparison of Greco-Roman style illustrations in *Juliana Anicia Codex* with Byzantine style. A. Frontispiece of *Juliana Anicia Codex*. Her enthroned figure is shown flanked by personifications of "Magnanimity" and "Wisdom." Her open hand rests on the open volume, held aloft by a *putto*, or winged baby, while a kneeling figure symbolizing the "Gratitude of the Arts" kisses her feet. B. Mosaic of Empress Theodora (c. 500–547 AD), wife of Justinian I, Byzantine emperor of Constantinople, with her entourage of attendants. (Church of San Vitale in Ravenna, Italy, from the south wall of the apse. ~547 AD).
Source: Wikimedia Commons.

(a)

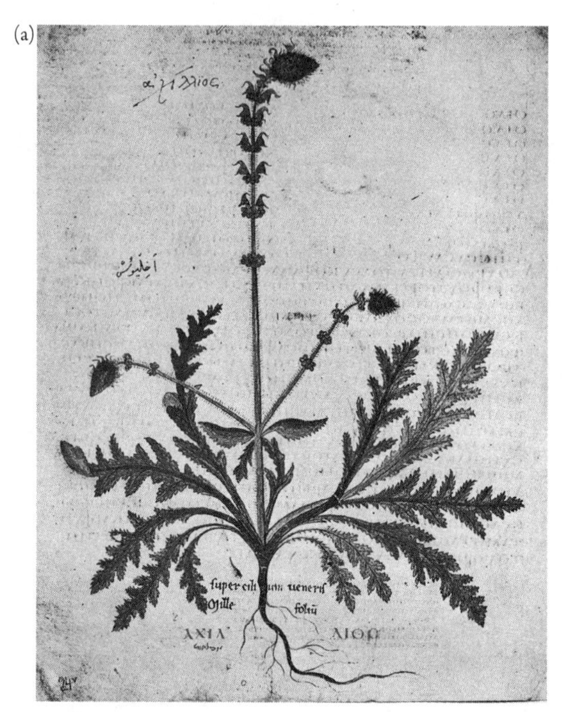

(b)

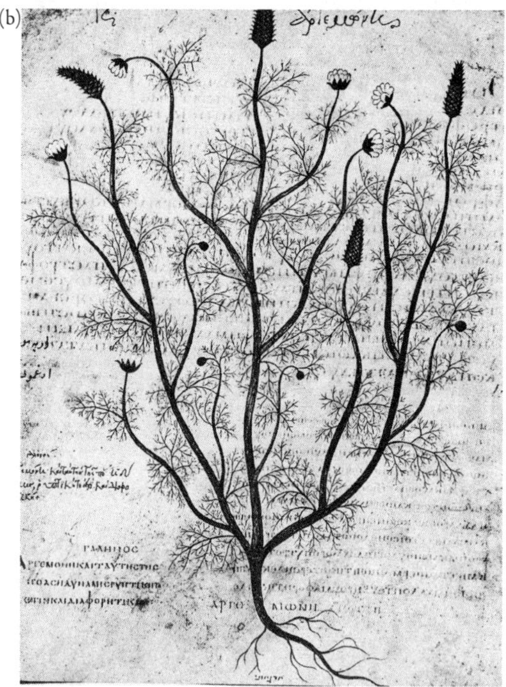

FIGURE 10.3 Two examples of unusually naturalistic illustrations from the *Juliana Anicia Codex*, thought to be based on early (second-century AD) copies of Krateuas. A. "Achilleios" (*Salvia multifida*—mint); B. "Argemone" (*Adonis aestivalis*— a member of the buttercup family); Oregano, showing broken stem (*Origanum creticum*—a member of the mint family); D. Yarrow? (*Achillea magna*, Asteraceae) showing broken stems and cut stems. E. "Lochitis" Example of a highly stylized and unrecognizable plant illustration (probably a fern-bearing sporangia).
From *Dioscurides De Materia Medica*, Codex Neapolitanus Graecus I of the National Library of Naples. (facsimile). (2000? undated) ΜΙΛΗΤΟΣ, Alimos.

(c)

(d)

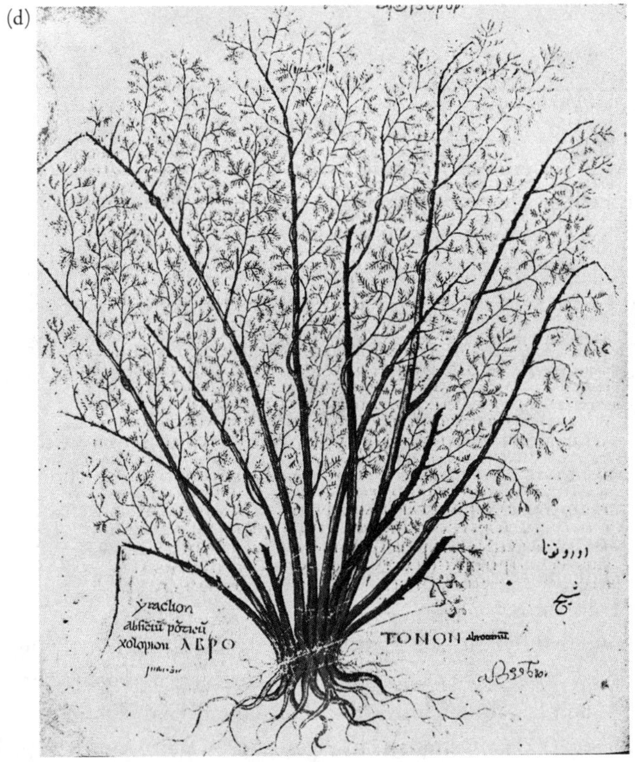

FIGURE 10.3 Continued

(e)

FIGURE 10.3 Continued

(Figures 10.3A, B, for example) are those that underwent the fewest rounds of copying from previous herbals. Conversely, the least naturalistic and most schematic illustrations, such as Figure 10.3E, are those that were copied repeatedly, accumulating numerous errors. Accordingly, the least copied, most naturalistic illustrations should most closely resemble their original Greek models. Singer believed that eleven of the most accurate portraits, whose written descriptions cite Krateaus as the source for the text, were copied directly from a second-century copy of Krateuas's illustrated herbal. If so, these paintings provide us with a close approximation of Krateaus's original portraits.

However, as discussed by Minta Collins, Singer's hypothesis may be too simplistic. For example, the variations in style and scientific accuracy of the plant images could simply reflect variations in the artistic abilities or tastes of the copyists rather than the antiquity of the models from which the images were copied.[11] As we have seen, a comparable diversity of artistic skill and aesthetic orientation is apparent in the botanical decorations of the Church of St. Polyeuktos.

Singer's dictum that Byzantine botanical artists never painted from living models is also open to question. Consider the plant illustrations identified as oregano and yarrow shown in Figures 10.3C and D. Although the overall quality of these two images is not as colorful and naturalistic as the illustrations in Figures 10.3A and B, note the presence of broken or cut branches—obvious defects in the specimens that suggest that the artists were

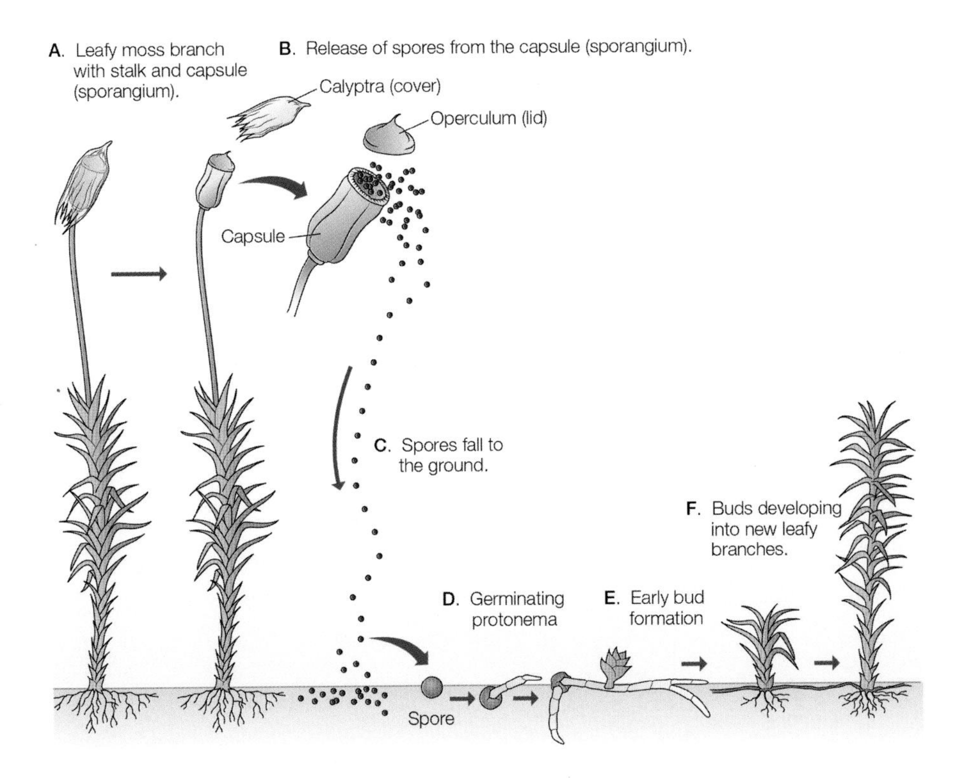

FIGURE 17.1.

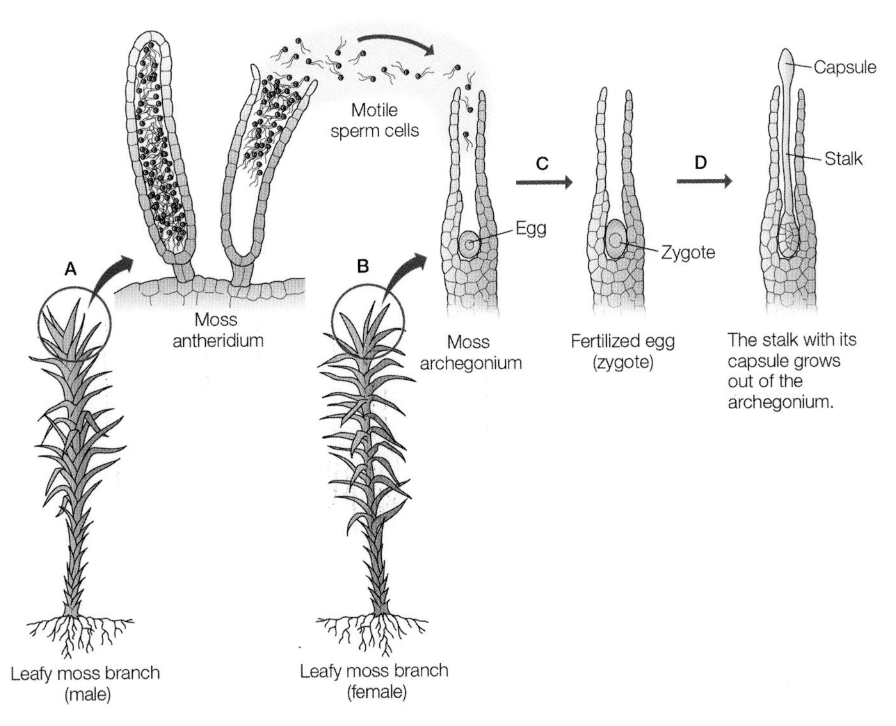

FIGURE 17.2.

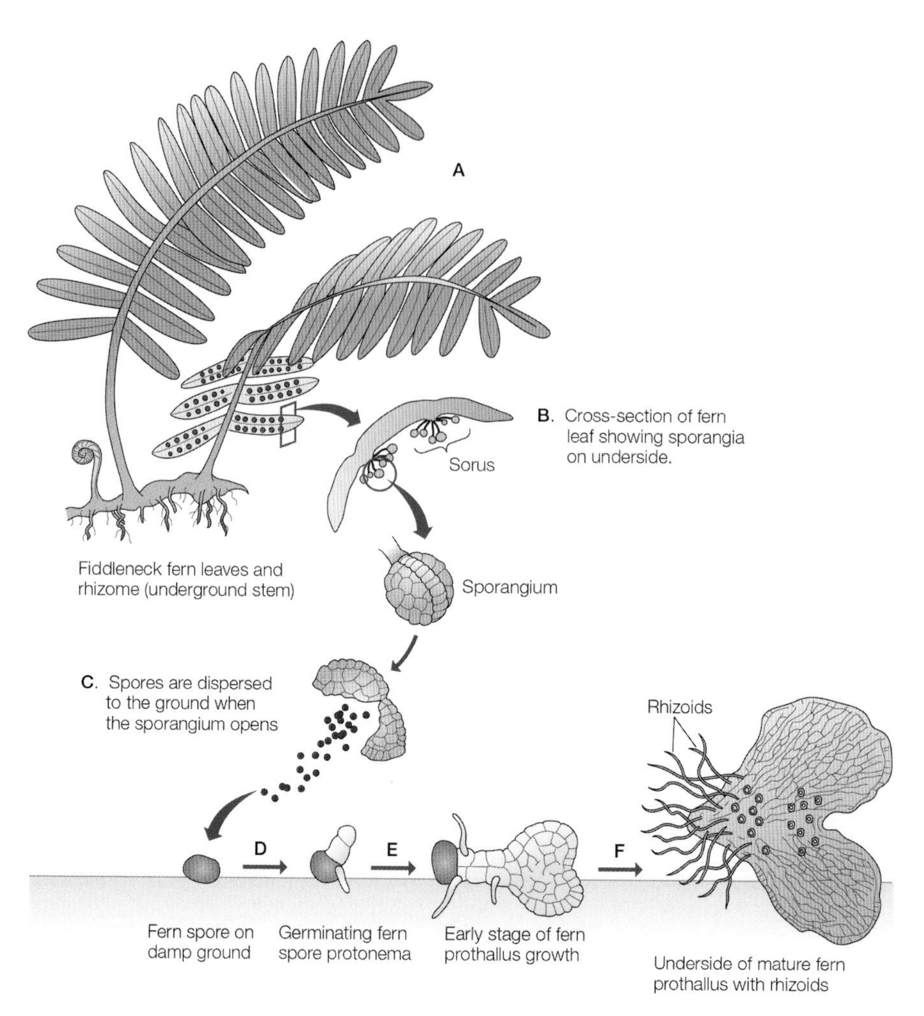
A
B. Cross-section of fern leaf showing sporangia on underside.
Sorus
Fiddleneck fern leaves and rhizome (underground stem)
Sporangium
C. Spores are dispersed to the ground when the sporangium opens
Rhizoids
D
E
F
Fern spore on damp ground
Germinating fern spore protonema
Early stage of fern prothallus growth
Underside of mature fern prothallus with rhizoids

FIGURE 17.3.

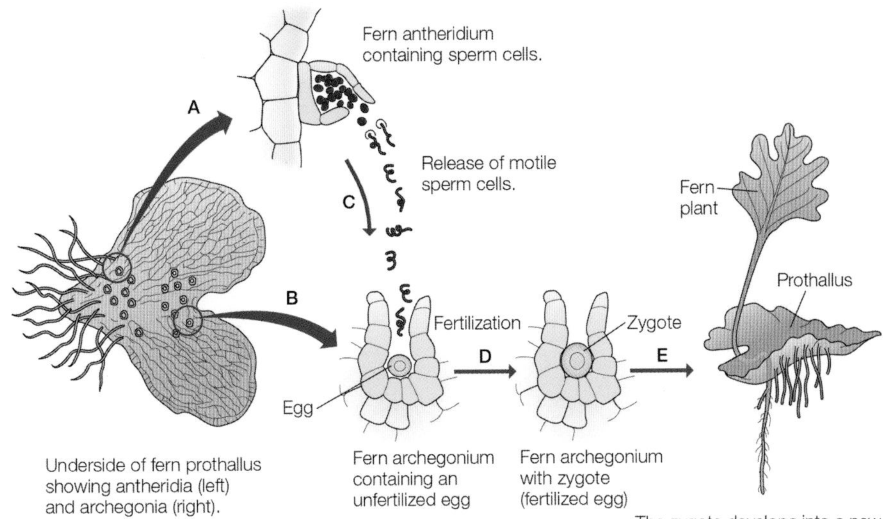
Fern antheridium
containing sperm cells.
A
Release of motile
sperm cells.
C
B
Fern
plant
Prothallus
Fertilization
Zygote
D
E
Egg
Underside of fern prothallus
showing antheridia (left)
and archegonia (right).
Fern archegonium
containing an
unfertilized egg
Fern archegonium
with zygote
(fertilized egg)
The zygote develops into a new
fern plant that grows out of the
archegonium on the prothallus.

FIGURE 17.4.

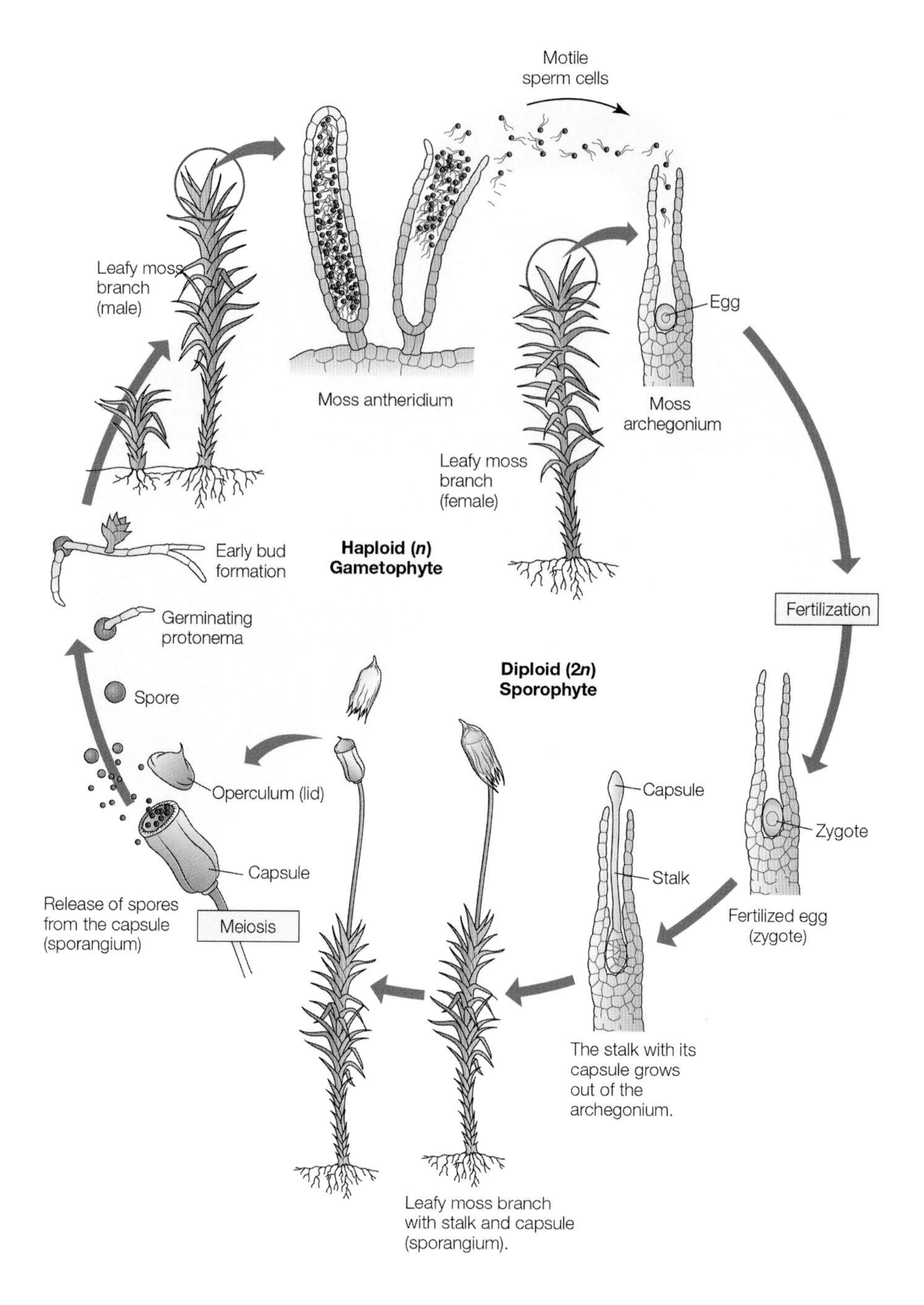

Motile
sperm cells
Leafy moss
branch
(male)
Moss antheridium
Egg
Moss
archegonium
Leafy moss
branch
(female)
Early bud
formation
Haploid (n)
Gametophyte
Fertilization
Germinating
protonema
Diploid (2n)
Sporophyte
Spore
Operculum (lid)
Capsule
Zygote
Capsule
Stalk
Release of spores
from the capsule
(sporangium)
Meiosis
Fertilized egg
(zygote)
The stalk with its
capsule grows
out of the
archegonium.
Leafy moss branch
with stalk and capsule
(sporangium).

FIGURE 18.1.

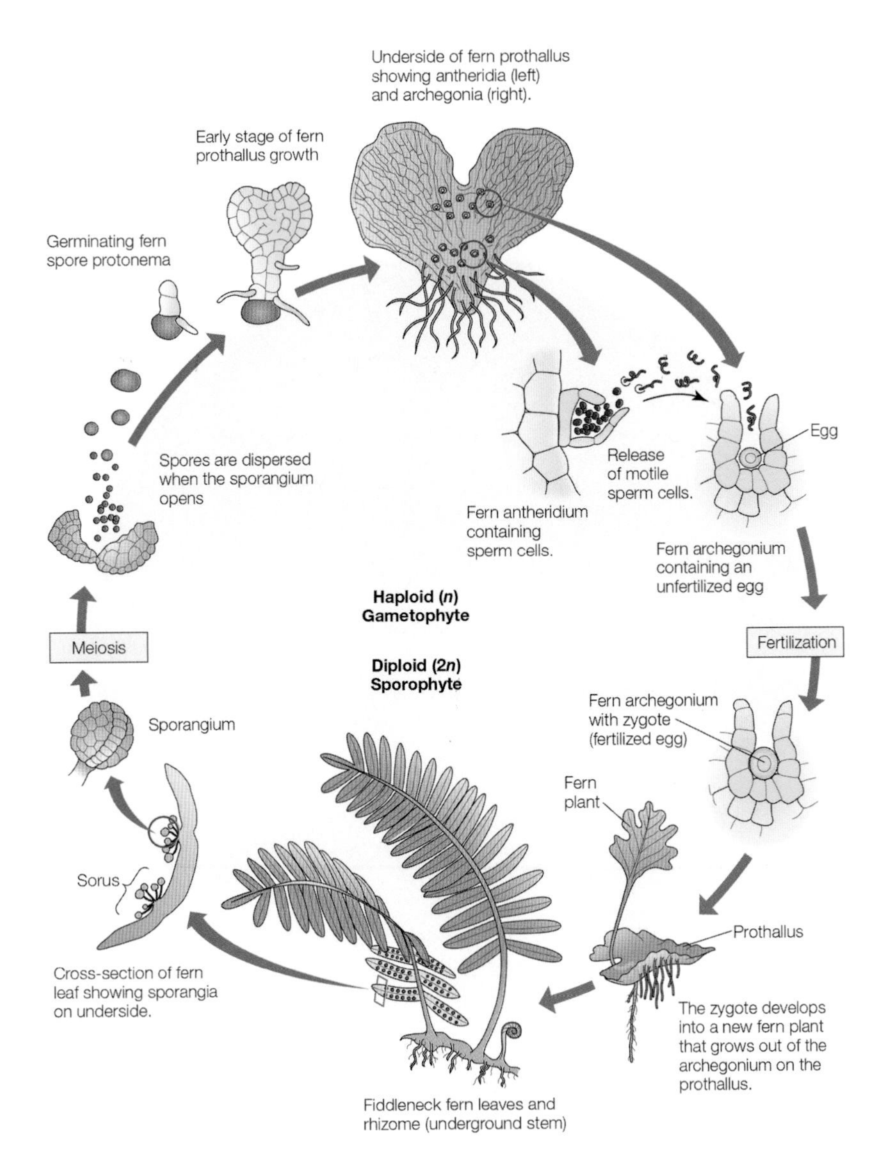
Underside of fern prothallus
showing antheridia (left)
and archegonia (right).
Early stage of fern
prothallus growth
Germinating fern
spore protonema
Egg
Release
of motile
sperm cells.
Spores are dispersed
when the sporangium
opens
Fern antheridium
containing
sperm cells.
Fern archegonium
containing an
unfertilized egg
Haploid (n)
Gametophyte
Meiosis
Fertilization
Diploid (2n)
Sporophyte
Fern archegonium
with zygote
(fertilized egg)
Sporangium
Fern
plant
Sorus
Prothallus
Cross-section of fern
leaf showing sporangia
on underside.
The zygote develops
into a new fern plant
that grows out of the
archegonium on the
prothallus.
Fiddleneck fern leaves and
rhizome (underground stem)

FIGURE 18.2.

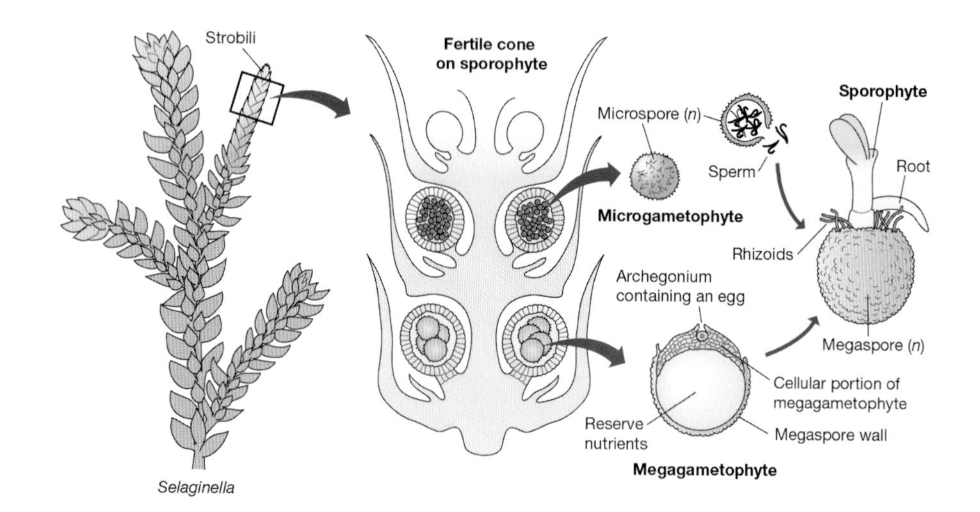

FIGURE 18.3.

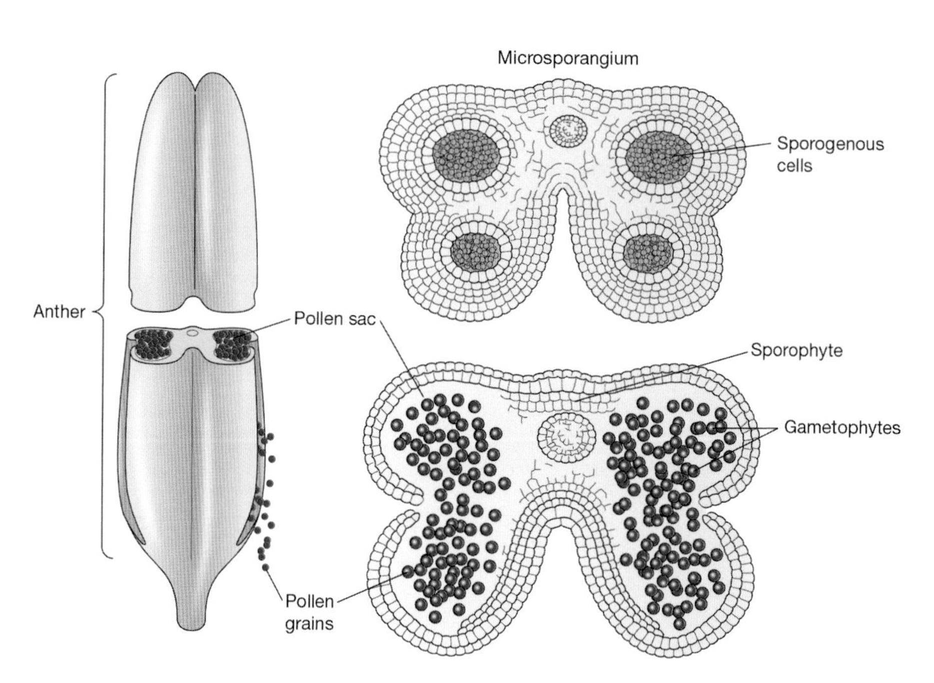

FIGURE 18.4.

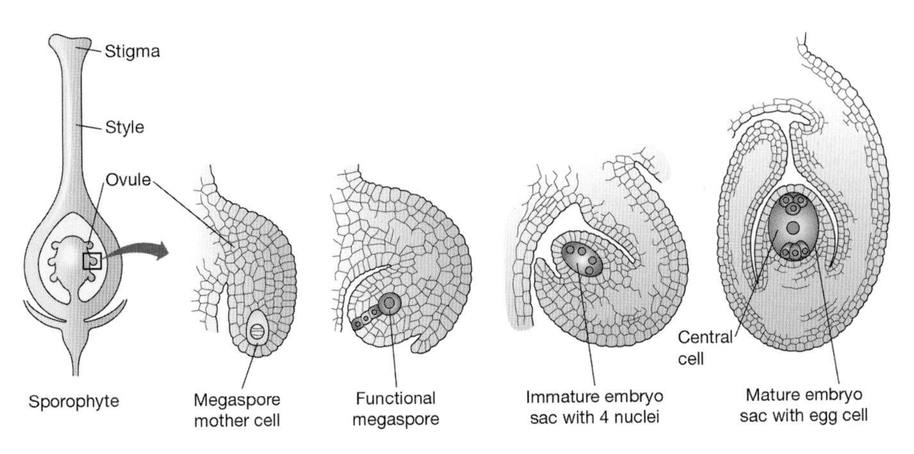

FIGURE 18.5.

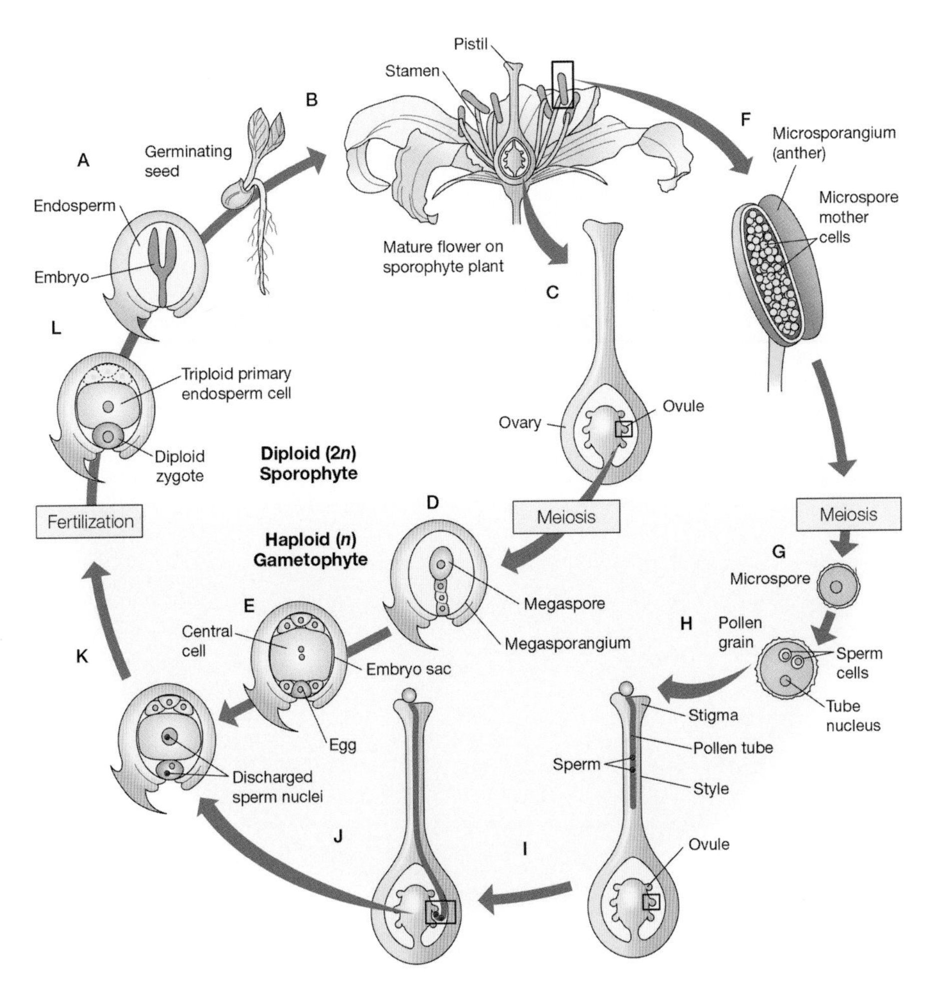

FIGURE 18.6.

painting from life.[12] The presence of such blemishes in several illustrations of the *Juliana Anicia Codex* could indicate that the practice of painting directly from nature, assumed by Singer to have died out with the Greeks, continued at least through the sixth century in the Eastern Empire.[13]

SEVEN CENTURIES OF KNOCK-OFFS AND SPIN-OFFS

The seventh-century *Codex Neapolitanus* is second in importance only to the *Juliana Anicia Codex*. The plant illustrations have been greatly simplified, and some of the leaf shapes have even been altered to suit the taste of the artist, features consistent with the herbal's having been copied rather than painted from living specimens (Figure 10.4).

In the ninth century, another Greek herbal of Dioscorides, possibly produced in Syria, contains illustrations even more stylized than those of the seventh-century *Codex Neapolitanus*, indicating that the images were again copied rather than painted from nature (Figure 10.5).

There is only one surviving illustrated copy of the Latin translation of the five books of Dioscorides, the *Dioskurides Lombardus*, also referred to as the "Old Latin Translation," which dates to the second half of the tenth century.[14] Most of the extant Latin herbals are combinations of the text of Dioscorides with that of a certain Apuleius, referred to as Apuleius Barbarus, Apuleius Platonicus, or Pseudo-Apuleius to distinguish him or her from the author of *The Golden Ass*. These texts were anthologies of medical recipes, charms, spells, and prayers and were probably compiled from Greek material around the beginning of the fifth century CE. The *Apuleian* manuscript herbals were the most widely read of the late classical herbals during the Early Middle Ages. Despite the Christian campaign against paganism that began about the middle of the fourth century CE, these works frequently included incantations to the Earth Goddess, clearly a vestige of their pagan roots.[15]

The texts of the Latin herbals of Apuleius were less accurate than those of the Greek herbals of Dioscorides, and the illustrations were even more schematic compared to the naturalistic drawings of the *Juliana Anicia Codex*, as indicated in the seventh-century examples shown in Figure 10.6.

The oldest existing Anglo-Saxon herbal dates from 1050, just sixteen years before the Battle of Hastings. It may be the first translation of a herbal (the Latin Herbal of Apuleius) in the vernacular. Two examples showing stylization to the point of iconography are shown in Figure 10.7. The addition of frames around some of the plant images emphasizes their purely decorative nature.

In the Near East, Arabic translations of the herbal of Dioscorides first appeared in Baghdad in the ninth century, while in the tenth century the Byzantine Emperor Romanus sent, as a gift to the Spanish caliph, a finely illustrated herbal of Dioscorides in Greek, which was soon translated into Arabic. Many other Herbals of Dioscorides also made their way into the Islamic Empire, and thus the whole of Arab medicine was strongly influenced by it.[16] By the twelfth century, the Norman kings had renewed contacts with Italy and Sicily, the Crusades had opened up the Near East to Constantinople, and a new "Romanesque" style of herbal evolved. These included framed compositions of historical or mythical figures as well as images of plants stylized almost beyond recognition (Figure 10.8).

(a)

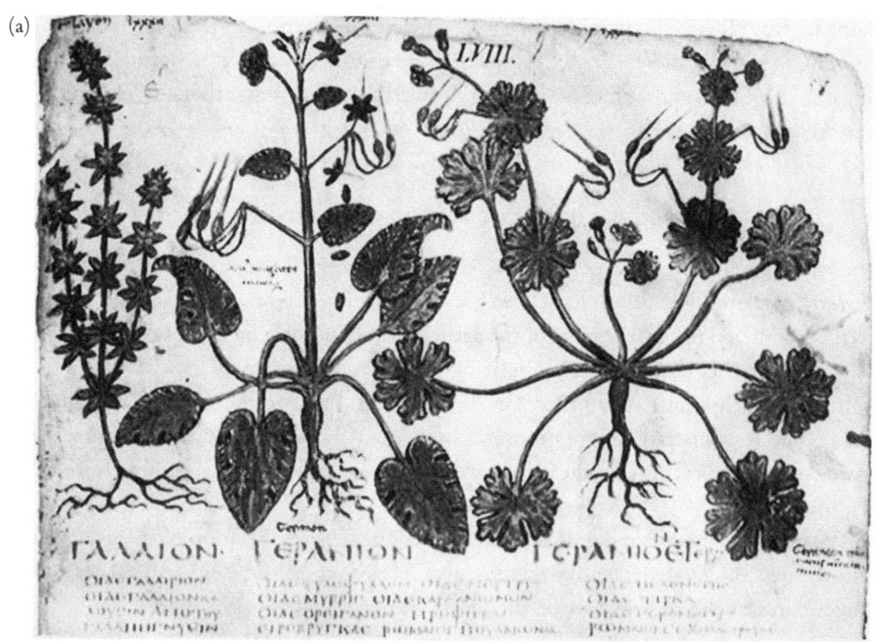

(b)

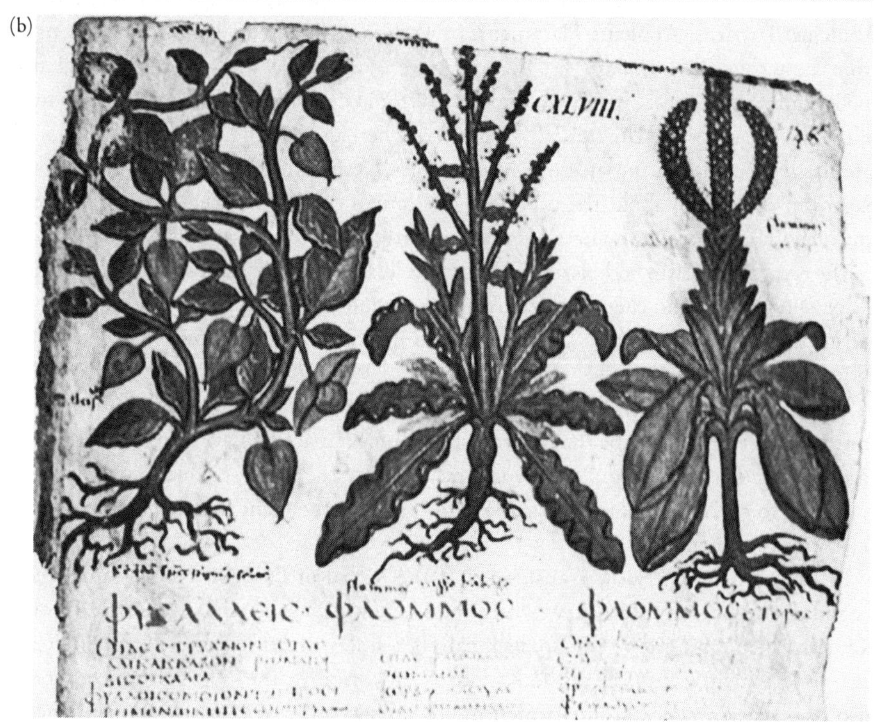

FIGURE 10.4 Illustrations from *Codex Neapolitanus,* seventh century. A. Lady's bedstraw (*Galium verum*), cranesbill (*Erodium malacoides*), and *Geranium molle. B.* Winter cherry (*Physalis alkekengi*) and two mulleins (*Verbascum* sp.).
From Blunt, W., and S. Raphael (1994), *The Illustrated Herbal*, Revised Edition; Thames and Hudson.

FIGURE 10.5 Syrian (?) Codex, ninth century. A. Grape hyacinth (*Muscari* sp.). B. Poppy (*Papaver* sp.).
Bibliotheque Nationale, Paris, MS gr. 2179. See Collins, M. (2000), *Medieval Herbals: The Illustrative Traditions*. The British Library Studies in Medieval Culture. The British Library and University of Toronto Press, p. 84.

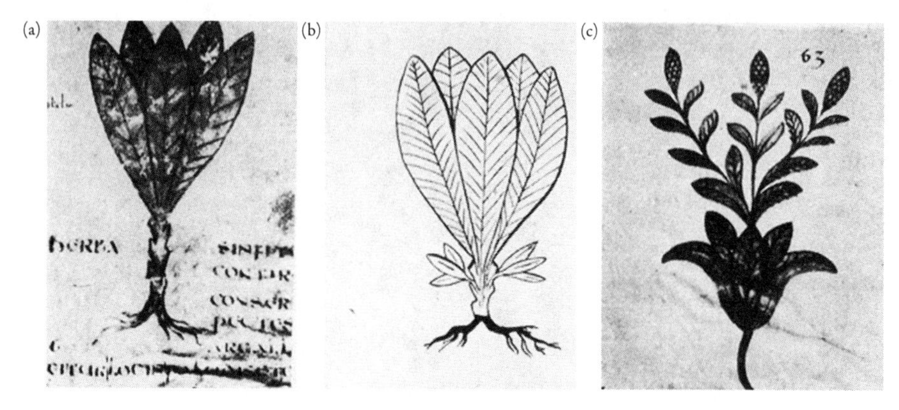

FIGURE 10.6A–C Seventh-century Latin herbals of Apuleius. Comfrey (*Simphytum* sp.).

A thirteenth-century German herbal combined elements from Dioscorides and Apuleius.[17] The German herbal is similar to the Anglo-Norman herbals in that most of the plant illustrations are highly stylized and unrecognizable (Figure 10.9).

It was not only the poor illustrations that made the medieval herbals botanically useless; the written descriptions were also badly garbled. Many of the plants discussed in the classical sources and later Arab texts did not even grow in northern Europe—a fact unknown to local physicians—and misidentification was common. According to Karen Reeds, herbalists during this period tended to regard all physical descriptions as "untrustworthy":

> The deficiencies of the plant descriptions only reinforced the common philosophical contempt of appearances and particulars of the sub-lunary world. Color, shape, texture, taste, smell—these were all accidental, untrustworthy bases of knowledge.[18]

Little wonder medieval physicians made scant use of the elaborate herbals commissioned for the aristocracy, preferring instead to depend on the accumulated folk wisdom of local herb gatherers.

Disconnected from their original purpose as practical field guides, medieval herbals were judged less by their accuracy and more by the standards of medieval decorative art as exemplified by illuminated manuscripts. The purpose of illuminations was to enliven the text and entertain the reader. An example of the complete divorce from reality can be seen in the miniature painting entitled "Spring Landscape" (Figure 10.10) from the *Carmina Burana Codex*, an early thirteenth-century collection of drinking songs, satirical poems, and short theatrical works written in Latin by the Goliards, a group of free-thinking university students. The plants in this landscape appear bizarre, alien, and surreal, suggesting that the intention of the artist was not to portray everyday reality, but to reveal the mystical spirit of springtime normally hidden from our eyes. In other words, it was a Neoplatonic spring landscape.

Compared to the fantastic vegetation in such works as *Spring Landscape*, even the most schematic illustrations in medieval herbals must have seemed like paragons of accuracy. How else to account for the otherwise inexplicable remark attributed to a sixteenth-century

(a)

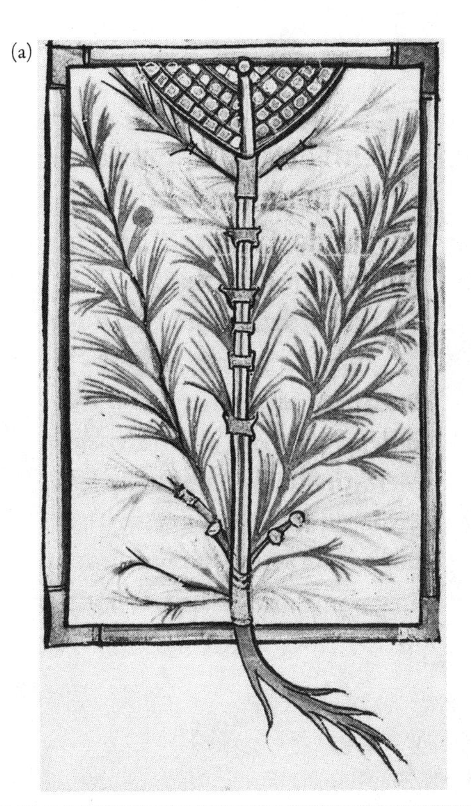

(b)

FIGURE 10.7 Plant illustrations from Anglo-Saxon Herbals. A. Henbane (*Hyascyamus* sp.) from *Apuleius Platonicus Herbarium*, England, c. 1050. B. (left) Lupin (*Coronilla* sp.), (right) an umbellifer (carrot family). from Apuleius Platonicus, England, c. 1200.
From Blunt, W., and S. Raphael (1994), *The Illustrated Herbal*, Revised Edition; Thames and Hudson, pp. 39, 42.

(a)

FIGURE 10.8 Anglo-Norman Herbals. A. "Gallitrichum"; "Immolum" (perhaps houseleek, *Sempervivum tectorum*); Homer and Mercury and "Immolum." *Herbarium of Apuleius Platonicus*, England, c. 1200. B. Asphodel (*Asphodelus* sp.); Sorrel, *Rumex acetosa*) and a Centaur holding a "Centauria," perhaps *Centaurium* sp. *Herbarium of Apuleius Platonicus*, England, c. 1200.

From Blunt, W., and S. Raphael (1994), *The Illustrated Herbal*, Revised Edition; Thames and Hudson, pp. 46–47.

German Count that the plant images in one of the Apuleian herbals were "so lifelike that it is not possible to express it more cunningly," despite the fact that this particular herbal was among the most stylized of all the medieval herbals.[19]

PRE-ISLAMIC VEGETATION GODDESSES IN THE QURAN

The Arabian Peninsula out of which Islam emerged in the seventh century is in fact a subcontinent slightly larger than India or Europe, riding its own tectonic plate. The Romans divided the Arabian Peninsula into three geographical zones, Arabia Petraea

(b)

FIGURE 10.8 Continued

(a)

FIGURE 10.9 Medieval German herbal from the thirteenth century. A. Madonna lily (*Lilium candidum*) and two other unrecognizable plants. B. Fern (upper right) plus a member of the saxifrage family.
From Blunt, W., and S. Raphael (1994), *The Illustrated Herbal*, Revised Edition; Thames and Hudson, pp. 46–47.

(b)

FIGURE 10.9 Continued

FIGURE 10.10 "Spring Landscape," miniature painting from the thirteenth-century *Carmina Burana Codex*.
Source: Wikimedia Commons.

in the north; Arabia Deserta, the vast interior desert; and Arabia Felix in the southwest. Thanks to the monsoon rains, Arabia Felix, corresponding to modern day Yemen, was much greener and more agriculturally productive than the northern and interior regions of the peninsula.

The Arabs themselves divided the peninsula into only two regions: north (*Arabia Deserta* and *Arabia Petraea*) and south (*Arabia Felix*). From the early second millennium BCE to the first century BCE, *Arabia Felix* was known as *Saba* ("Sheba" in the Bible) and its inhabitants were referred to as *Sabaeans*. Saba was the main source of commercially valuable aromatic plants, including frankincense and myrrh, and the fragrant spices cinnamon and cassia.[20] As a result, the rulers of ancient Saba were famed for their great wealth, which they used to build opulent temples decorated with gold and silver statues of their deities. The Bible records the munificent gifts given by the Queen of Sheba to King Solomon.[21] However, it was not in the wealthy south, but in the less hospitable northern region populated by Arabs that Islam was born in the seventh century and quickly swept throughout the Arabian Peninsula.

Prior to the rise of Islam, virtually all the inhabitants of Arabia, north and south, were polytheistic. The names of many local deities are known from inscriptions, although their attributes are poorly understood.[22] Both gods and goddesses were worshipped, and the most important male and female deities were often paired as divine couples. As was the practice among other polytheistic societies of the ancient world, local deities were often blended with deities from other religions (typically Greco-Roman), and the attributes of individual deities might vary from town to town or assume the attributes of other deities. For example, Herodotus reported that the Arabs of Nabataea (southern Jordan and northern Arabia) worshiped Allat (contraction of *Al-Ilat*, "the Goddess"), whose attributes were comparable to those of Aphrodite and Dionysus.

Sūra 53 and the Satanic Verses

The Ka'bah in Mecca—a large, black, cubical, masonry structure—was the most important religious shrine in northern Arabia during pre-Islamic times just as it is today, attracting pilgrims from all over the peninsula every year. By the time Muhammad's forces conquered Mecca in 630 BCE, the Ka'bah was said to house some 360 idols. Since little is known about the identities of these Meccan deities, we confine our discussion of pre-Islamic vegetation goddesses to those mentioned by name in the Quran: Allat, Al-'Uzza, and Manat. Just as the female devotees of the goddess Asherah were chastised in the Hebrew Bible (e.g., Jeremiah 44:15–28), the worshippers of Allat, Al-'Uzza, and Manat were also rebuked, both in the Quran and in traditional medieval Islamic writings, thus attesting to their importance in pre-Islamic Arabia.

The Quranic critique of the three goddesses occurs in Sūra 53 19–23, an-Najm (The Star), considered to be a revelation by the angel Gabriel:

> Have ye thought upon Al-Lat and Al-'Uzza
> And Manat, the third, the other?
> Are yours the males and His the females ?
> That indeed were an unfair division!
> They are but names which ye have named, ye and your fathers, for which Allah hath revealed no warrant. They follow but a guess and that which (they) themselves desire. And now the guidance from their Lord hath come unto them.[23]

This revelation is especially significant because Muhammad belonged to the Quraysh tribe, stewards of the sacred Ka'bah shrine where Allat, Al-'Uzza, and Manat were worshipped. As a child, Muhammad is said to have taken part in religious rites dedicated to these deities, and he understood that Meccans were devoutly loyal to them. Several early biographers of Muhammad, writing in the eighth through tenth centuries and including Ibn Ishaq, al-Wāqidī, Ibn Sa'd, and al-Tabarī, recorded the tradition that, to convert Meccans to Islam, Muhammad granted them permission to continue worshipping Allat, Al-'Uzza, and Manat so long as they accepted Allah as the supreme deity. According to this tradition, the original version of Sūra 53 contained the following lines:

Have ye thought upon Al-Lat and Al-'Uzza,
And Manat, the third, the other?
These are the exalted *gharāniq*,
whose intercession is hoped for.[24]

The term *gharāniq*, usually translated as "high flying cranes," is a metaphor for interceding angels. It was said that the Meccans were so grateful for Muhammad's words that they accepted Islam and joined him in ritual prostration at the end of the Sūrah. Later, according to tradition, Muhammad retracted these lines at the behest of the Angel Gabriel, who, in Sūra 22:52, explains that "God abrogates what Satan interpolates," heretically implying that Muhammad was tempted by Satan. To correct Muhammad's error, the Angel Gabriel substituted the current version of the Sūra (Sūra 53:21). The Scottish historian Sir William Muir coined the term "Satanic Verses" for the offending interpolation, made famous in the West by Salmon Rushdie's novel of the same name.

Modern Muslim scholars deny the authenticity of the "*gharāniq* incident." However, no one disputes the fact that several medieval Islamic historians *believed* that Meccans were reluctant to give up worshipping their three goddesses and that Muhammad tried to accommodate them.

Of the three goddesses, Manat is perhaps the easiest for modern observers to understand. As the personification of Time, Fate, and Death, her presence loomed over every waking hour of one's life. However, in the following discussion, we will focus our discussion on the agricultural goddesses Allat and Al-'Uzza, who are more complex and variable and are often confused with each other.

THE AGRICULTURAL GODDESSES ALLAT AND AL-'UZZA

Different aspects of the goddess Allat were venerated throughout the Arabian Peninsula. Like Inanna, she was associated with the planet Venus, and Herodotus equated her with Aphrodite. Depending on the region, she was also either the wife of Allah (a contraction of Al-Ilah, "the God") or his daughter. Inscriptions found in Nabataea refer to her as the "Mother of the Gods." Shams, the Sabaean sun goddess, may represent another aspect of Allat. In her function as a tutelary deity, she was depicted armed in the manner of Athena. She may also be related to the Chaldean goddess Allat and the Carthaginian goddess Allatu, both of whom were underworld deities, like the Sumerian Ereshkigal or the Greek

Persephone. The latter association implies some connection with fertility and the agricultural cycle. In this aspect, she is sometimes confused with Al-'Uzza.

Al-'Uzza is ultimately traceable to Inanna/Ishtar, the great goddess of ancient Mesopotamia. In Sasanid Iraq in the sixth century, she was worshipped by several names, including Nanai (Inanna) and Ishtar, and she was also syncretized with the goddess Anahit in Persia. As Dlibat Ishtar she was identified with the planet Venus. According to Michael Morony, the epithet applied to Dlibat Ishtar meant "passionate," which evokes the ancient Akadian epithet *ezzu* ("raging") originally applied to Ishtar. *Ezzu* then became the Arabic name Al-'Uzza for the planet Venus (az-Zuhra).[25]

Al-'Uzza was probably worshipped first by the Sabaeans of Yemen, who identified her with the planet Venus. The Nabataeans of Petra, now in Jordan, conflated her with Isis and Atargatis (the local goddess of fertility—especially the life-engendering power of water and soil), as well as other Hellenistic goddesses. Carved into the façade of the Treasury Building at Petra, dating to 131 AD, are several female figures representing different goddesses or aspects of a single goddess, Al'Uzza/Atargatis/Isis. Although they are severely worn, it is still evident that they hold cornucopias and are decorated with ears of wheat, representing agricultural abundance. At the Nabataean temple of Khirbet et-Tannur, a similar image of Al'Uzza/Atargatis/Isis appears against a background of leaves, flowers, fruit, and grain, suggesting a relationship to Ceres.[26] However, in the ninth-century *Book of Idols* by Hisham ibn Muhammad ibn al-Kalbi (discussed in the next section), Al-'Uzza, like Asherah, is characterized as a goddess of trees. Her shrines were typically located in sacred groves at the center of which stood three trees where an epiphany of the goddess descending from heaven was supposed to occur. The sacred groves of Al-'Uzza were tended by a priest and priestess.

Al-'Uzza: Goddess of Date Palms

As we have seen, the worship of date palms, or their aniconic equivalents in the form of posts or pillars, was a feature of the cult of Asherah as described in the Hebrew Bible, and the date palm continued to be revered as a sacred tree by diaspora Jews after the fall of the Second Temple in 70 CE. In the *Ethiopian Book of Enoch*, the prophet Enoch flies to Paradise and discovers that the Tree of Life is a date palm.

Arab worship of date palms can also be traced back at least as far as the first century AD, when the Greek historian Diodorus provided the following description of a religious sanctuary in a date palm oasis located in the Sinai:

> Directly after the innermost recess [of the northwest coast] is a region along the sea which is especially honored by the natives because of the advantage which accrues from it to them. It is called the Palm-grove and contains a multitude of trees of this kind which are exceedingly fruitful and contribute in an unusual degree to enjoyment and luxury. But all the country round about is lacking in springs of water and is fiery hot because it slopes to the south; accordingly, it was a natural thing that the barbarians made sacred the place which was full of trees and, lying as it did in the midst of a region utterly desolate, supplied their food. And indeed not a few springs and streams of water gush forth there, which do not yield to snow in coldness; and

these make the land on both sides of them green and altogether pleasing. Moreover, an altar is there built of hard stone and very old in years, bearing an inscription in ancient letters of an unknown tongue. The oversight of the sacred precinct is in the care of a man and a woman who hold the sacred office for life. The inhabitants of the place are long-lived and have their beds in the trees because of their fear of the wild beasts.[27]

In the tradition of the Prophets of the Hebrew Bible who railed against the planting of the sacred trees of the goddess Asherah, the Prophet Muhammad destroyed the sacred groves of Al-'Uzza. In the *Book of Idols*, ibn al-Kalbi relates that after the Prophet had captured Mecca he dispatched one of his generals to Al-'Uzza's sanctuary in the valley of Nakhlah to cut down the three sacred trees. *Nachla* is Arabic for date palm, which suggests that the sacred trees in question were date palms. When the general had cut down two of the sacred trees, the distraught priestess emerged—"an Abyssinian woman with disheveled hair and her hands placed on her shoulder[s], gnashing and grating her teeth." The general slew the priestess and cut down the third tree. Upon hearing the news, the Prophet declared, "That was Al-'Uzza. But she is no more."

Muhammad may have struck down Al-'Uzza, but the date palms and other garden plants she represented were assimilated into the Paradise Garden of the Quran and thus retained as religious symbols. Significantly, the Islamic ban on making images of "living things" applies to humans and other animals, but not to plants. Indeed, plant decorations adorn many famous mosques, including the Dome of the Rock, the mosque of Cordoba, and the Ummayyed Great Mosque of Damascus (Figure 10.11). Some of the images are naturalistic, including date palms, grapes, pomegranates, and olives, and some are more stylized. Despite the destruction of her sanctuaries in the seventh century, Al-'Uzza's symbols—date palms, other fruit trees, and garden flowers—have endured and continue to grace and beautify Islamic mosques.

MARY AND THE DATE PALM

Prior to the emergence of Islam, Christians had assimilated elements of pagan goddesses into the figure of Mary. As discussed in Chapter 11, Mary's identification with flowers and gardens reached its zenith in the High and Late Middle Ages, but, prior to this, she was associated with sacred trees—both the Tree of Knowledge of Good and Evil (in her role as the "Second Eve") and the Tree of Life, by way of Near Eastern goddesses. In the Christian literature, she is first associated with date palms in the Apocrypha (non-canonical writings) of the New Testament. The Gospel of Pseudo-Matthew, dating to around the sixth century AD, incorporates encounters with sacred trees into the tale of Mary and Joseph's flight to Egypt mentioned briefly in the Gospel of Matthew. Unlike the canonical version, however, the Gospel of Pseudo-Matthew includes miraculous feats performed by the infant Jesus along the way to Egypt, such as the taming of wild dragons. According to Pseudo-Matthew, on the third day of the journey, after traversing the desert for hours, the Holy Family arrived at a solitary date palm tree. Explaining to Joseph that she was tired and wished to rest in the shade of the date palm, Mary dismounted with the infant Jesus in her arms and sat down beneath the tree. The gospel continues:

> And as the blessed Mary was sitting there, she looked up to the foliage of the palm, and saw it full of fruit, and said to Joseph: I wish it were possible to get some of the fruit of this palm. And Joseph said to her: I wonder that thou sayest this, when thou seest how high the palm tree is; and that thou thinkest of eating of its fruit. I am thinking more of the want of water, because the skins are now empty, and we have none wherewith to refresh ourselves and our cattle.

(a)

FIGURE 10.11 Decorations from the Umayyad Great Mosque of Damascus (~710 AD). A. Cupula showing date palm and buildings bordered by ornamental plants. B. Mosque façade lavishly illustrated with fruit trees representing the Garden of Paradise.

(b)

FIGURE 10.11 Continued

Overhearing his parent's conversation, the infant Jesus looked up with a joyful countenance and said to the tree: "O tree, bend thy branches, and refresh my mother with thy fruit." Immediately, the palm "bent its top down to the very feet of the blessed Mary," allowing Joseph and Mary easily to pick the fruit. Jesus then promised the date palm a place in Paradise and further asked it to release a spring that was hidden in the earth beneath its roots: "open from thy roots a vein of water which has been hid in the earth, and let the waters flow, so that we may be satisfied from thee." Suddenly, a spring of clear, cold water began to flow from the base of the tree, and the Holy Family refreshed themselves with fresh dates and pure water before resuming their journey.

The Quran contains an alternative version of Mary's encounter with the palm tree in Sura 19:21–26. After the Annunciation, the newly pregnant and unwedded Mary, fearful of rejection by her family and friends, "retired . . . to a remote place" to bring the baby to term and give birth. Eventually, her contractions began:

> And the pains of childbirth drove her to the trunk of a palm-tree: She cried (in her anguish): "Ah! would that I had died before this! would that I had been a thing forgotten and out of sight!"
>
> But (a voice) cried to her from beneath the (palm-tree): "Grieve not! for thy Lord hath provided a rivulet beneath thee;
>
> "And shake towards thyself the trunk of the palm-tree: It will let fall fresh ripe dates upon thee.

"So eat and drink and cool (thine) eye. And if thou dost see any man, say, "I have vowed a fast to (Allah) Most Gracious, and this day will I enter into not talking with any human being'"

At length she brought the (babe) to her people, carrying him (in her arms). They said: "O Mary! truly an amazing thing hast thou brought!"

The Quranic variant of Pseudo-Matthew is highly significant for several reasons. First, the incident occurs prior to the birth of Jesus. Instead of Jesus, the active agent in the story is identified as "a voice" who "cried to her from beneath the palm tree." Although some commentators have identified the voice as Jesus inside Mary's womb, there is no reference to "womb" in the passage. At the time the Quran was composed, Arabs were still polytheistic and Al-'Uzza was worshipped as a goddess of date palms. The voice from beneath the date palm could therefore be associated with the tree itself—the roots or the soil—perhaps representing the chthonic aspect of Al-'Uzza or a similar tree goddess.

According to Sura 19, after eating the dates and drinking the water from the sacred date palm, Mary gave birth and subsequently returned to her village to show the baby to her people. Instead of rejecting her as she feared they would, her neighbors recognized that a miracle had taken place and celebrated the new arrival. By implication, Mary's eating the sacred dates and drinking the sacred water marked Jesus as a miraculous child.

The date palm persisted as an emblem of the Virgin Mary's miraculous pregnancy well into the Renaissance, as seen in Christian religious art. For example, in "The Virgin's Wedding Procession" by the Giotto, the palm leaf extending from the balcony symbolizes Mary's pregnancy (Figure 10.12).

The story of Mary and the sacred tree even made its way to northern Europe. In the process, the date palm became transformed into the more familiar cherry tree. In the "Cherry Tree Carol," Joseph and Mary happen upon a cherry orchard during their journey to Egypt:

O then bespoke Mary,
With words so meek and mild,
"Pluck me one cherry, Joseph,
For I am with child."

O then bespoke Joseph,
With answer rude and wild,
"Let him pluck thee a cherry
That brought thee now with child."

O then bespoke the baby
Within his mother's womb
"Bow down then the tallest tree
For my mother to have some."

Then bowed down the highest tree,
Unto his mother's hand.
Then she cried, "See, Joseph,
I have cherries at command."

FIGURE 10.12 “The Virgin’s Wedding Procession” by Giotto (1266/7–1337).

The carol is more light-hearted than either the early Christian or Quranic versions and seems to combine elements of both. In Pseudo-Matthew, Jesus speaks from his mother’s arms, whereas in the Quran it is a voice from below the tree. In the “Cherry Tree Carol,” Jesus does indeed speak from the womb. Many books were transmitted to the West from Islamic sources during the Late Middle Ages, and the Quran was among them. Although it may seem counterintuitive given the historically fraught relationship between Christianity and Islam, it is possible that the European Christmas carol was influenced by both the early Christian and Quranic versions of the story.

MUHAMMAD AND THE DATE GROWERS

Dates and melons were the Prophet Muhammad’s favorite foods, and date palms are mentioned twenty times in the Quran. A summary of the date palm’s unique biological attributes, drawn mainly from Islamic authors writing after the time of Muhammad, was provided by the thirteenth-century Muslim historian Zakariya Muhammad Qazwini, in his book *The Wonders of Creation and the Oddities of Existence*:

> This blessed tree is found only in countries where Islam is the prevailing religion. The Prophet said, in speaking of it, "Honor the palm, which is your paternal aunt"; and he gave it this name because it was made from the remains of the earth out of which Adam was created. The date palm bears a striking resemblance to humans, in the beauty of its erect and lofty stature, its division in two distinct sexes, male and female, and the property which is peculiar to it of being fecundated by a sort of copulation. If its head is cut off, it dies. Its flowers have an extraordinary spermatic odor, and are enclosed in a case similar to the sac in which the fetus is contained, among animals. If an accident happens to the marrow-like substance at its summit [the terminal bud] the palm dies, just as we see a man dies when his skull is severely injured. Like the members of a person, the leaves which are cut off never grow again; and the mass of fiber in which the palm is surrounded offers and analogy to the hairs which cover the human body.[28]

Qazwini uses the term "paternal aunt" for the date palm, emphasizing the female aspect of the tree because the orchard consisted primarily of fruit-bearing trees. Nevertheless, he unequivocally states that the tree is comprised of two sexes and that "a sort of copulation" is required for fruit production. Recall that Theophrastus considered a sexual explanation for pollination in date palms but rejected any direct comparisons between plants and animals (see Chapter 9).

Arab scholars also went further than either Theophrastus or Aristotle by postulating an evolutionary process to account for the *Scala Natura*. According to the *Encyclopedia of the Brethren of Purity*, also known as the *Epistles of Ikhwan al-Safa*, written in the tenth century, Allah initiated the Creation by forming the first vaporous elements, which coalesced on their own to form rocks. Next, rocks spontaneously developed into plants, and plants developed into animals, culminating in the evolution of humans. The date palm was considered a transitional form of plant because of two animal-like features: the existence of two sexes and the fact that it was killed by decapitation.

A surprisingly modern Islamic interpretation of the relationship between science and religion arose in the context of the question of date palm sexuality in one of the six major Sunni Hadith,[29] called Sahih Muslim,[30] written in the ninth century. One day, the Prophet was passing by a date orchard and noticed that the people were performing an unfamiliar task. When he asked them what they were doing, they replied that they were sprinkling the female flowers with male pollen to obtain fruit. The skeptical Prophet responded, "I do not find it to be of any use." When informed of the Prophet's words, the people immediately stopped pollinating the female trees. Later, upon being told that the yield of dates had been drastically reduced that year because the people had followed his advice not to pollinate their trees, the Prophet replied as follows:

> If there is any use of it, then they should do it, for it was just a personal opinion of mine, and do not go after my personal opinion; but when I say to you anything on behalf of Allah, then do accept it, for I do not attribute lie [sic] to Allah, the Exalted and Glorious.[31]

The moral to this parable is summarized in the title: "It Is Obligatory to Follow the Prophet (May Peace Be upon Him) in All Matters Pertaining to Religion, But One Is Free to Act on One's Own Opinion in Matters which Pertain to Technical Skill." Thus medieval Islam seems to have been unique among the three Abrahamic religions in making an explicit distinction between sacred writ and "technical skill" (science).

THE *HORTUS CONCLUSUS*: WOMEN AS WALLED GARDENS

The metaphor of the virginal young woman as a walled garden[32] had existed in the poetic tradition of the Near East since the Bronze Age. We encounter it most familiarly in the biblical "Song of Songs," as discussed in Chapter 5, in which the erotic descriptions of the *hortus conclusus*,[33] or "enclosed garden," celebrate the sensuality of love between a man and a woman, and God is never mentioned. To both Jewish and Christian theologians, such lyrical celebrations of sexuality smacked of paganism and were therefore unacceptable. As mentioned earlier, the first-century CE Jewish philosopher Philo's solution to this problem was to declare that all scripture, especially those passages dealing with romance and nuptials, must be interpreted allegorically. For Philo, the underlying meaning of the "Song of Songs" was the love of God for his Chosen People, an interpretation with which rabbis and the devout could feel comfortable.

Early Christian theologians, such as Origen, adopted the Jewish solution of interpreting the "Song of Songs" allegorically, but instead of God's love for his Chosen People, the allegory became the love of Christ for his Church. However, the metaphor of the Church as a bride in an enclosed garden (*hortus conclusus*) was apparently a difficult abstraction for ordinary people to grasp, and it required endless exegeses. Bishop Ambrose of Milan's labored attempt at an explanation probably went over the heads of the average church-goer:

> So the holy church, ignorant of wedlock, but fertile in bearing, is in chastity a virgin, yet a mother in offspring. She, a virgin, bears us her children, not by a human father, but by the Spirit. . . . She, a virgin, feeds us, not with milk of the body, but with that of the Apostle.[34]

According to Bishop Ambrose, the *hortus conclusus* clearly symbolizes virginity and chastity, *not* female sensuality as a naïve reading of the "Song of Songs" might lead one to assume. The Church was like a virgin because it was sealed off from worldly affairs ("wedlock"), yet at the same time it was "fertile in bearing." The plants of the *hortus conclusus* produced fruit by the "Spirit" alone, without sexuality or a "human father" (a gardener).

The substitution of a cold abstraction (God's love of the Church) for the flesh and blood woman in the *hortus conclusus* may have satisfied the puritanical scruples of theologians, but it robbed the poem of its vitality and popular appeal. Perhaps sensing this loss, Bishop Methodius of Olympus in his parable *The Banquet of the Ten Virgins* found a way to reinstate the feminine presence by replacing the bride of the "Song of Songs" with a group of ten nuns in a Paradise Garden belonging to the Greek goddess Arete, the personification of Virtue. Methodius wrote that the garden contained:

> different kinds of trees there, full of fresh fruits, and the fruits that were hung joyfully from their branches were of equal beauty, and there were ever-blooming meadows strewn with variegated and sweet-scented flowers.[35]

According to Elisabeth Augspach, Arete's garden is identified with the ten nuns, whereas Mary "is adorned with the fruits of virtue."[36]

The identification of the *hortus conclusus* with ten nuns proved to be a more popular metaphor than the Garden-as-Church abstraction, and it was not long before the garden was assimilated to Mary herself. As discussed in earlier chapters, many pagan goddesses were identified with flowers and fruits, and so it is not surprising that Mary, the Christian mother goddess, would also come to possess some of these attributes. For example, Saint Ephrem described Mary as the sinless and inviolate "flower unfading."[37] In the context of citing precedents in the Old Testament for events in the New Testament, Saint Jerome explicitly identified Mary with the *hortus conclusus* of the "Song of Songs", as well as with the Garden of Eden, and it was Jerome who famously described Mary as "a garden enclosed, a fountain sealed."[38] Mary can thus be thought of as a nature/agricultural goddess as well as a mother goddess.

THE VIRGIN MARY AS A NATURE/AGRICULTURAL GODDESS

The Council of Ephesus in 431 AD marked a significant shift in Church doctrine, sanctioning the cult of the Virgin as *Theotokos*, or "Mother of God."[39] The cult of the Virgin expanded most rapidly in the Eastern Church. During the sixth century, around the time of the *Juliana Anicia Codex*, the hymn writer Saint Romanos composed the Akathist ("Standing") hymns to Mary Theotokos. The third hymn in the service (here slightly abridged) makes extensive use of agricultural metaphors, reminiscent of the Sumerian love songs of Inanna and Dumuzi and the biblical "Song of Songs":

> The power of the Most High then overshadowed the Virgin for conception, and
> showed Her fruitful womb as a sweet meadow to all who wish to reap
> salvation, as they sing: Alleluia!
> Rejoice, branch of an Unfading Sprout:
> Rejoice, acquisition of Immortal Fruit!
> Rejoice, laborer that laborest for the Lover of mankind:
> Rejoice, Thou Who givest birth to the Planter of our life!
> Rejoice, grainfield yielding a rich crop of mercies:
> Rejoice, table bearing a wealth of forgiveness!
> Rejoice, Thou Who makest to bloom the garden of delight:
> Rejoice, O Bride Unwedded!

Whereas Jerome was the first to identify Mary with the *hortus conclusus*, and, almost concurrently, Ambrose equated the *hortus conclusus* with the Garden of Eden (with Mary as the "Second Eve"), in the ninth century the Frankish Benedictine monk Paschasius Radbertus compared the *hortus conclusus* to the Virgin's uterus, thus emphasizing Mary's productive yet "pristine" womb:

> Therefore enclosed garden, because the Virgin's uterus was always pristine and incorruptible in every way. It is called a garden because all of the delights of paradise flower within it, and the womb is closed from shame, where the fountain brings forth our redemption.[40]

Finally, Alcuin of York, Charlemagne's minister of education, gave the following botanical epithets to Mary: "flower of the field," "lily of the world," and "*hortus conclusus*."[41]

Berceo's "Milagros de Nuestra Señora"

Despite Mary's growing popularity, there was not enough in scripture to satisfy the public's increasing demand for stories about her life and miracles. A poem by the thirteenth-century Spanish poet and cleric Gonzalo de Berceo, "Milagros de Nuestra Señora" (Miracles of Our Lady), was written in vernacular Spanish in an attempt to disseminate Marian theology among the Castilian-speaking population of Christian Spain.

In the introduction, the poet states that "while on a pilgrimage" he came upon a flowery meadow, which turned out to be a garden, although it lacked a wall around it:

> The flowers there emitted a marvelous fragrance;
> they were refreshing to the spirit and to the body.
> From each corner sprang clear, flowing fountains,
> very cool in summer and warm in winter.
>
> There was a profusion of fine trees—
> pomegranate and fig, pear and apple,
> and many other fruits of various kinds.
> But none were spoiled or sour.[42]

Berceo goes on to explain that the meadow/orchard symbolizes the Virgin Mary, and its greenness signifies her virginity:

> This meadow was always green in purity
> for Her virginity was never stained;
> *post partum et in partu* She truly was a virgin
> undefiled, incorrupt in Her integrity.

Berceo next states that the names of the flowers are the names by which the Virgin Mary is known:

> Let us turn to the flowers that comprise the meadow,
> which make it beautiful, fair, and serene.
> The flowers are the names the book gives
> to the Virgin Mary, Mother of the Good Servant.

Although the names of the flowers are not given, Berceo lists some of the fruit, nut, and other trees with which Mary is associated, and seems to identify the Palm with the Tree of Knowledge in the Garden of Eden:

> She is called Vine, She is Grape, Almond, Pomegranate,
> replete with its grains of grace,
> Olive, Cedar, Balsam, leafy Palm,
> Rod upon which the serpent was raised.

Unlike the Garden of Eden, which belongs to a male deity, Berceo's *un*walled garden is Mary's domain. Berceo's Mary is a strong and powerful personage whose undefiled perfection requires no earthly walls for protection. Mary inhabits her garden in the same way that Zeus inhabits Mount Olympus and God resides in Heaven, and, like the male deities, she is free to come and go as she pleases. Far from being a mere passive vessel for the divine spark, Berceo's Mary Theotokos is out in the world performing miracles everywhere—a true goddess in nearly every sense of the word.

NOTES

1. Morton, A. G. (1982), *History of Botanical Science*, Academic Press. However, some historians have noted that chapter IX is inferior to Theophrastus's more theoretical works and have suggested that it may have been written by his students; see Lynn Thorndike (1924), Disputed dates, civilisation and climate, and traces of magic in the scientific treatises ascribed to Theophrastus, in *Essays on the History of Medicine* presented to Karl Sudhof, Charles Singer, and Henry Sigerist, eds., Oxford University Press.

2. Singer, C. (1927), The herbal in antiquity and its transmission to later ages, Part 1. *Journal of Hellenic Studies* 47:1–52. Mithridates's knowledge of poisons was immortalized in the famous legend in which he is said to have protected himself against being poisoned by taking sublethal doses to build up his tolerance.

3. Morton, *History of Botanical Science*.

4. Pliny (1855), *Natural History*, Book 25, ch. 4; trans. John Bostock; H. T. Riley. Taylor and Francis. http://data.perseus.org/citations/urn:cts:latinLit:phi0978.phi001.perseus-eng1:25.4.

5. Janick, J., and K. E. Hummer (2012), The 1500th Anniversary (512–2012) of the *Juliana Anicia Codex*: An illustrated Dioscoridean recension. *Chronica Horticulturae* 52: 9–15.

6. Pulteney, R. (1790), *Historical and Biographical Sketches of the Progress of Botany in England from Its Origin to the Introduction of the Linnaean System*. Vol. 1, ch. 25 ("History of the Discovery of the Sexes in Plants"), T. Cadell in the Strand., pp. 333–334.

7. Based on *Vindobona*, the Roman name for the city that later became Vienna.

8. Harrison, M. (1989), *A Temple for Byzantium: The Discovery and Excavation of Anicia Juliana's Palace-Church in Istanbul*. Harvey Miller, p. 34.

9. Collins, M. (2000), *Medieval Herbals: The Illustrative Traditions*. The British Library Studies in Medieval Culture. The British Library and University of Toronto Press, p. 46.

10. Singer, The herbal in antiquity and its transmission to later ages; Blunt, W., and S. Raphael (1994), *The Illustrated Herbal*, Revised Edition. Thames and Hudson.

11. Collins, *Medieval Herbals.*

12. Janick and Hummer, The 1500th Anniversary (512–2012) of the *Juliana Anicia Codex.*

13. Janick, J., and J. Stolarczyk (2012), Ancient Greek illustrated Dioscoridean herbals: Origins and impact of the *Juliana Anicia Codex* and the *Codex Neopolitanus. Notulae Botanicae Horti Agrobotanici Cluj-Napoca* 40: 9–17.

14. Collins, *Medieval Herbals,* p. 149.

15. In his book *The White Goddess,* Robert Graves provided a translation of the opening Latin invocation from a twelfth-century Apuleian herbal from England: "Earth, divine goddess, Mother Nature, who generatest all things and bringeth forth anew the sun which thou hast given to the nations, Guardian of sky and sea and of all gods and powers; through thy power all nature falls silent and then sinks in sleep. . . . Thou dost contain chaos infinite, yea, and winds and showers and storms; thou sendest them out when thou wilt and causest the seas to roar; thou chasest away the sun and arousest the storm. Again, when thou wilt thou sendest forth the joyous day and givest the nourishment of life with thy eternal surety . . . thou art great, queen of the gods. Goddess! I adore thee as divine." Graves, R. (1948), *The White Goddess.* Farrar Straus Giroux (original manuscript from the British Museum, MS. Harley 1585, ff 12v).

16. Singer, The herbal in antiquity and its transmission to later ages.

17. Blunt and Raphael, *The Illustrated Herbal.*

18. Reeds, K. M. (1991), *Botany in Medieval Renaissance Universities.* Garland Publishing, Inc., p. 24.

19. Cited by Reeds, *Botany in Medieval Renaissance Universities,* p. 31.

20. "True cinnamon" and "cassia," respectively, are two different species of the genus *Cinnamomum* in the Lauraceae family: *C. verum,* is native to Sri Lanka, and *C. Cassia,* is from southern China. The predominant volatile compound, cinnamic aldehyde, is the same in both species, although *C. cassia* has a stronger flavor than *C. vera.* Both spices were available in the ancient Mediterranean via commerce.

21. From 1 Kings 10:10, *The New Oxford Annotated Bible,* Third Edition.

22. Hoyland, R. G. (2001), *Arabia and the Arabs: From the Bronze Age to the Coming of Islam.* Routledge.

23. Pickethall, M., trans. (1930), *The Glorious Qur'an.* Alfred A. Knopf.

24. Ibid.

25. Morony, M. G. (1984), *Iraq After the Muslim Conquest.* Princeton University Press.

26. Bedal, L. A. (2013), *The Petra Pool-Complex: A Hellenistic Paradeisos in the Nabataean Capital (results from the Petra Lower Market survey and excavation, 1998).* Gorgias Studies in Classical and Late Antiquity 10. Gorgias Press.

27. Diodorus (1935), *The Library of History,* Book III, Section 42. Loeb Classical Library edition: (http://penelope.uchicago.edu/Thayer/E/Roman/Texts/Diodorus_Siculus/3C*.html)

28. Z. M. Qazwini. Cited by P. B. Popenoe (1924), *The Date Palm,* Field Research Projects, Coconut Grove, Fla., published in 1973.

29. *Hadith* are compilations of reports that are supposed to quote the prophet Muhammad's exact words on a given topic.

30. Sahih Muslim hadith, Book 30, ch. 35, Number 5830.

31. Hamid-Siddiqui, A. (1980), *Sahih Muslim*, Third Edition. Sh. Muhammead Ashraf, Lahore; No. 5830, p. 1259.

32. Much of the information contained in this section is taken from Elisabeth A. Augspach's book, *The Garden as Woman's Space in Twelfth- and Thirteenth-Century Literature* (2004), Edwin Mellon Press.

33. The term *hortus conclusus* is from the Vulgate, the early fifth-century Latin version of the Bible written largely by Jerome. The Vulgate ultimately became the official Latin version of the Bible used by the Roman Catholic Church.

34. Ambrose of Milan, *De Virginibus*, cited by Augspach, *The Garden as Woman's Space*, p. 30.

35. Methodius, *The Banquet of the Ten Virgins*, cited by Augspach, *The Garden as Woman's Space*, p. 32.

36. Ibid.

37. *Precationes ad Deiparam* in Opp. Graec. Lat., III, 524–537.

38. Letter of Jerome to Pammachius, 393 AD.

39. Augspach, *The Garden as Woman's Space*.

40. Paschasius Radbertus, *Espositio in Evangelium Matthaee*, cited by Augspach, *The Garden as Woman's Space*, p. 37.

41. Ibid.

42. Gonzalo de Berceo (1997), *Miracles of Our Lady*, trans. R. T. Mount and A. G. Cash. University of Kentucky Press.

11

Troubadours, Romancing the Rose, and the Rebirth of Naturalism

IN HIS LETTER to the Galatians, Saint Paul warned against the pitfall of fleshly desires:

> Live by the Spirit, I say, and do not gratify the desires of the flesh. For what the flesh desires is opposed to the Spirit, and what the Spirit desires is opposed to the flesh; for these are opposed to each other, to prevent you from doing what you want.[1]

For Medieval theologians, who regarded the human condition as a microcosm of the universe, this opposition between fleshly desires and the Spirit was played out on the cosmic scale as the clash between the transient material world, in which death triumphs, and the celestial kingdom, in which belief in Christ brings eternal life. As we saw in Chapter 10, the rejection of the material world in the Early Middle Ages coincided with the trend of the illustrated herbals away from naturalism and toward iconography and symbolism. In literature and liturgy, however, a vestige of the ancient association of women and plants was retained in the person of Mary, who took on many of the attributes of the pagan agricultural goddesses, with one important exception: their sexuality. During the medieval period, Mary was most closely identified with flowers, especially the lily and the rose. Once flowers became emblematic of Mary, devout Christians had a powerful vested interest in keeping them pure and chaste. The medieval period was therefore not an auspicious time for philosophical speculations on the role of sex in plants.

Two cultural shifts were needed before sex in plants could even be considered: a return to naturalism in the illustrated herbals and an end to the anathematization of sex in scholarly discourse. Only by looking at real plants could scholars begin to lay the foundations for the study of floral anatomy. And only in a less puritanical milieu could scholars peering at stamens feel free to hazard the heterodox opinion that stamens and pollen grains were the male sexual structures of flowers.

During the 600-year period from the eleventh to the seventeenth centuries that spanned the High and Late Middle Ages and the Renaissance, these two cultural shifts did indeed occur, preparing the ground for the two-sex model.

The greatest impetus for what has been called the "Twelfth Century Awakening" came in response to the Christian capture of the Islamic city of Toledo by Alfonso VI of Castile in 1085, which brought into Europe a sudden influx of Greek learning that had been preserved by Islamic scholars—most of which had not been translated into Latin. The response of European intellectuals was a passionate revival of classical thought and a burgeoning enthusiasm for classical literature.

THE FIRST TROUBADOUR

The first signs of a new wind blowing began in the eleventh century with the appearance of an important new literary movement. The troubadour poets appeared first in southwestern France, bordering Spain and the Mediterranean, in a region called Aquitania by the Romans. Their secular lyrics, written in the Provençal dialect, focused on the powerful themes of chivalry and courtly love, and, over the course of the next two centuries, the influence of the troubadours spread throughout Europe.

The proximity of Aquitaine to Muslim Spain placed it on the front line of cultural exchange between the Christian and Muslim worlds. Indeed, some scholars believe that the first known troubadour of Aquitaine, William IX, the duke of Aquitaine, may have been influenced by Arabic love poetry filtering in from the courts of Andalusia.[2] William IX also may have been influenced by Ovid, whose poetic works included three on the topic of romantic love: *The Art of Love, The Cure for Love,* and *The Amours*. Ovid's poems treated Love in a light-hearted, elegant manner and were widely available in southern France during the twelfth and thirteenth centuries in both Latin and Provençal.

Ovid's love poetry emphasized courtship over consummation. In one passage in *The Art of Love,* Ovid counseled a young woman not to yield too quickly to her suitor's importuning, for the more barriers that were placed in his path the more heightened his anticipation of the ecstasy that awaited him. In this hyperaroused state, the boundary between pleasure and pain becomes blurred. Ovid's idea of the heightening effect of delayed gratification was embraced by the troubadour poets, who soon took it to its logical extreme: the unending, blissful torment of unrequited love. In this way, physical love could be celebrated without transgressing the medieval Christian ideal of chastity.

William of Aquitaine's poetry uses nature imagery to express the tenderness of love, as in "For the Sweetness of Springtime":

> In the sweetness of springtime
> Forest, flower, and the birds
> Sing—each in his native chant—
> Therefore a man stands well
> When he has what he most desires . . .

He goes on to liken the aftermath of a storm to reconciliation after a lovers' quarrel:

Our love affair moves on
Like a flowering hawthorne branch
Standing above a trembling tree . . .
At night, only rain and frost;
But the next day's sunshine gleams
Through green branches and leaves.
I still remember that morning
When we pledged an end to our war,
And she gave me that great gift—
Her loving and her ring:

Having won his lady's heart, William's tone abruptly switches to locker-room braggadocio:

Oh, God, let me live long enough
To grope beneath her cloak! . . .
Let other gabbers brag about their love.
We've got the meat and the knife.

The God to whom William prays, however, is clearly not the Christian God of the medieval church. In his poem "Very Happily I Begin to Love," William's description of his lady combines Venus's power of love with the miraculous healing powers of Mary:

Every joy must lower itself
and all royalty obey
my lady, because of her kindness
and of her sweet pleasant visage;
and he will live a hundred times longer
who can partake of her love.
Because of her joy can the sick turn healthy
and because of her displeasure can a healthy man die
and a wise man turn mad . . .

The juxtaposition of erotic and spiritual elements characterizes much of the literature of courtly love and distinguishes it from classical love poetry.

CAPELLANUS'S TREATISE ON LOVE AND THE COURTLY LOVE TRADITION

The term "courtly love" was coined in 1883 by the French medieval scholar Gaston Paris to describe the particular kind of love depicted in the poem "Lancelot" by Chrétien de Troyes and in other medieval romances. Such romances typically involved situations in which the object of the lover's affections was "a married lady of high rank" for whom he could only sigh from afar.[3] His sole means of expressing his passion was through acts of chivalry and valor, which he dedicated to his idol. Courtly love poems mirror the lyrics of the troubadour poets and are consistent with the rules of love as outlined by the twelfth-century cleric Andreas Capellanus,[4] who is thought to have been a courtier of Marie of France, the daughter of Eleanor of Aquitaine. Eleanor had famously fostered the literary tradition in her court, and Marie's commissioning of Capellanus's treatise *De Amore* around 1185 is in that tradition. Nevertheless, it is difficult to tell whether Capellanus intended the work to be didactic, satirical, or merely a reflection of the times.

Typical of the courtly love tradition, the garden plays a central role in *De Amore*. Capellanus cites the example of a spurned lover who succeeds in convincing his Lady to reciprocate his love by bringing her to a beautiful garden meadow, where Venus, the Queen of Love, holds court:

> the meadows were very beautiful and more finely laid out than mortal had ever seen. The place was closed in on all sides by every kind of fruitful and fragrant trees, each bearing marvelous fruits according to it kind.

Capellanus's garden is divided by walls into concentric circles.[5] In the middle of the innermost circle is the Tree of Life, "a marvelously tall tree, bearing abundantly all sorts of fruits." A "wonderful spring of the clearest water, which . . . taste[s] of the sweetest nectar," gushes from its roots. Venus sits beside the tree on her gem-encrusted throne, attired in splendid robes and wearing a golden crown.

The inner circle of the garden is termed "Delightfulness" (Earthly Paradise), the middle circle is called "Humidity" (a kind of Purgatory), and the outer circle is "Aridity" (Hell). The "King of Love" (Cupid as Venus's husband rather than her son) arrives and sits on his throne beside Venus, enjoying a glass of red wine. His entourage soon follows, and the women take their seats according to their rank and/or conduct. Ladies of rank who accept Love sit in the area of Delightfulness, common women who accept Love take their places in the circle of Humidity, while women of any rank who reject Love must sit in the outer circle of Aridity. The latter are made to sit on chairs of thorns with the soles of their feet resting on burning ground. As the *pièce de résistance*, a man assigned to each of these love-rejecting women is tasked with periodically shaking their chairs, inflicting painful injury.

Terrified of suffering the same fate as the women of Aridity, the lover's Lady reluctantly acquiesces to his advances. Thus Capellanus combines, either naively or cynically, the sadistic theology of the era with the aestheticized ritual of courtly love. Pulling out all the stops, the lover deploys a fearful tableau in his campaign to win his Lady's favors, a technique of psychological coercion that medieval clerics had honed to a fine art, although deploying it to different ends.

MOTHER EARTH AND MOTHER NATURE

In the biblical account of the Creation, the earth is not merely a passive substrate for growing plants. Rather, it acts as God's intermediary in bringing forth vegetation:

> Then God said, "Let the earth put forth vegetation: plants yielding seed, and fruit trees of every kind on earth that bear fruit with the seed in it." And it was so.
> The earth brought forth vegetation: plants yielding seed of every kind, and trees of every kind bearing fruit with the seed in it. And God saw that it was good.[6]

The two Hebrew words used to denote "earth" in the Bible are *eretz* and *adamah,* both of which have feminine genders. *Eretz* refers to the earth which brought forth plants, while *adamah* is the earth from which God fashioned Adam. Together, *Eretz* and *Adamah* can be interpreted as a latent "Mother Earth," whose generative powers God invokes to create plants and animals.

A more explicit reference to Mother Earth in the Judeao-Christian tradition occurred in the thirteenth century, when Saint Francis of Assisi broke with centuries of medieval Church tradition by extolling the natural world. In his Latin poem, "Canticle of the Sun," composed around 1224, Francis affirmed a familial relationship with "Brother Sun," "Sister Moon," and "our sister Mother Earth." "Mother Earth" ("Matre Terra") was singled out for particular praise as the one "who feeds us and rules us":

> Be praised, my Lord, through our sister Mother Earth, who feeds us and rules us, and produces various fruits with colored flowers and herbs.

As in Genesis, it is Mother Earth, not God directly, who produces the flowers and herbs. To Francis, our organic relationship to Earth was no mere metaphor, but a deep spiritual truth. Assisi's "Canticle of the Sun" placed the Christian imprimatur on the pagan concept of an earth mother—hints of which are found in Genesis—and inaugurated a new era in the Christian attitude toward nature.

Although her name was coined in the Renaissance, Mother Nature's literary antecedent can be traced at least as far back as the twelfth century to the goddess Natura. Natura makes her literary debut in twelfth-century France in the poems of the Neoplatonist philosopher, Bernard Silvestris. Her character was further developed and elaborated by other twelfth- and thirteenth-century authors, such as Alan of Lille and Jean de Meun. E. R. Curtius, went so far as to suggest that the goddess Natura was no mere literary conceit, but a resurrection of the pagan reverence for nature:

> Natura is a cosmic power. . . . She is one of the last religious experiences of the late-pagan world. She possesses an inexhaustible vitality. . . . (Her) power over men's souls is proved by the Christian polemic against her.[7]

The name *Natura* is related to the Greek term *physis* and the Latin term *natura*, both of which are grammatically feminine. The first known use of *physis* (the essential character of a material object) occurs in Homer's *Odyssey*, fittingly, in relation to a magical plant, Moly, which Hermes presents to Odysseus to protect him against the sorceress Circe:

> So saying, Hermes gave me the herb, drawing it from the ground, and showed me its *physis*.[8]

Aristotle listed seven definitions of *physis* in *Metaphysics*, concluding that the term encompassed everything in the sublunary world. The sun, moon, stars, and planets were considered divine. According to Aristotle, everything on earth is comprised of the four elements—earth, air, fire, and water—and is characterized by change, whereas celestial objects, being perfect, never change. In describing the operations of physis, he frequently resorted to teleology and personification—"Nature does nothing in vain, nothing superfluous" —thus foreshadowing the medieval allegorical goddess Natura.[9]

Like Aristotle, Plato believed that heavenly bodies were divine beings, possessing both souls and intelligence. Plato's cosmological scheme was elaborated upon in the third century CE by Plotinus, the founder of Neoplatonism,[10] and it was the latter's exegesis of Platonic cosmology that was passed on to the Middle Ages.

Plato's view of the universe is far more mystical than Aristotle's, and there is no hint in Plato's writings of a personified nature goddess. In the *Timaeus*, Plato asserts that the "real" universe consists of eternal, ideal forms. In contrast, the material universe was generated by a Creator (the *Demiurge*), and fashioned out of formless, chaotic matter by an emanation of the Demiurge called Intellect (*Nous*). Although the material universe is modeled after eternal ideal forms, it is only a crude imitation of the "real" universe and is subject to change. The Demiurge created the *World-Soul*, which was joined to the stars by Nous. Plato's concept of Nous—representing intelligence, or the rational aspect of the Creator—suggests that the material universe, unlike abstract ideal forms, follows the rules of cause and effect and is accessible to reason. The Platonic view of the nature of the material world thus contrasts with the Christian idea of an omniscient Creator God whose continuing divine intervention defies the laws of cause and effect, rendering the material universe ultimately inaccessible to reason.

Bernard Silvestris was affiliated with the French school of Chartres, [11] where the chief goal of the scholars was to unite Neoplatonism with Christian doctrine. But they also did something revolutionary: they reinstated the feminine principle of divinity in the Creation myth, at least symbolically. In Silvestris's *Cosmographia*, "*Noys*" (Nous) is the daughter of God, and Natura is the daughter of Noys. According to Peter Dronke, no writer before Bernard had gone so far as to describe Natura as "the blessed fecundity of the womb" of the goddess Noys.[12] The poem begins with Natura complaining to her mother about the confusion and chaos of the formless universe. In response, Noys creates the heavens, the earth, and all living creatures on it—except humans. Finally, Noys teams up with Natura and Physis (who are for the first time distinguished from each other)[13] to create the first human in an earthly paradise garden. Sexuality is exalted for its role in reproduction, and the penis is singled out for particular praise as the "genial weapon" that defeats Death, while Natura performs her role by generating the semen.

God ("Tugaton," from the Greek word meaning "The Good") makes only a brief appearance in *Cosmographia* as a remote, mysterious triple-shaft of light to which Natura and her companion Urania pray for guidance. In contrast to the biblical narrative, the Creation is here brought about by a "feminine *trinitas creatrix*." As documented by Dronke, Noys, Natura, and Physis are closest in their attributes to the classical earth goddesses—Terra, Tellus, and Gaia. Bernard Silvestris, a devout Christian, regarded his fable merely as a pedagogical "wrapping" around the divine truth of Genesis— meant to illuminate, not replace, the biblical narrative. Nevertheless, his work suggests the lingering appeal of concepts associating the earth and nature with women that had been the hallmark of the pagan world.

Alan of Lille wrote two poetic works that were strongly influenced by Bernard: *De Planctu Naturae* (The Plaint of Nature) and the *Anticlaudianus.* In *De Planctu Naturae*, the goddess Natura, who functions as God's viceroy, complains to the poet that whereas all the other animals propagate themselves according to her laws, humanity is given over to lust and fruitless perversions such as extramarital sex, homosexuality, sodomy, and masturbation—symbolized by a rent in her garment. Unlike Bernard, who never provides a physical description of Natura, Alan describes Natura's attire, including her undergarments, in detail. Her crown and outer garments consist of the firmament and creatures of the air, sea, and land, whereas her most intimate apparel is represented by plants:

> I did not establish by any authority that would give certainty what fancies, shown by way of pictures, played on the upper parts of the shoes and the underclothing that lay concealed beneath the outer garments. However, . . . I am inclined to think that a smiling picture made merry there in the realms of herbs and trees. There trees, I think, were now clad in coats of russet, now tressed with leaves of green, were now bringing forth young, fragrant flower buds, now showing their age in the growing strength of their offspring.[14]

Decorated with an ever-changing panorama of herbs and trees growing, blossoming, and fruiting, Natura's undergarments are redolent of her generative powers, yet free of carnality.

THE ROMANCE OF THE ROSE

In Capellanus's treatise, Venus is identified with a tree, just as Mary is in the Apocrypha (Pseudo-Matthew and *Ecclesiaticus*), the Koran, and in Gonzalo de Berceo's *Miracles of Our Lady.*[15] In the context of the Courtly Love tradition, flowers in general—and roses in particular—came to symbolize sexual love. For example, in one of the illuminated *Tacuinum sanitatis* manuscripts from Northern Italy, which were Latin translations of an Arabic medical treatise, we find an illustration of a rose bush with a Lady seated beside her standing husband in the foreground (Figure 11.1A). The Lady hands her spouse a pair of roses from a mound of blossoms in her lap.

Although they are white roses, symbolizing modesty, the gesture is meant to suggest the intimacy of their relationship. Roses were the symbol of Venus, and, in the fifteenth-century illustration shown in Figure 11.1B a nude Venus, whose sexual power radiates from her vulva, is shown holding a rose and wearing a rose garland. Directly below her, courting couples amuse themselves to the strains of a harp within a *hortus conclusus* lined with rose bushes.[16]

A famously explicit example of the association of roses with female sexuality is the thirteenth-century poem "The Romance of the Rose." This allegorical fantasy, set in a walled garden, concerns a young man who falls in love with a rosebud. It was begun in 1237 by Guillaume de Lorris, who died after writing 4,058 lines. This version ended with the young man's passion for his rosebud still unconsummated. Subsequently, two different poets added their own endings to the poem. The first was an anonymous poet who added seventy-eight lines describing the young man's amorous romp on the lawn with his beloved rosebud. The second was Jean de Meun, a Parisian scholar, poet, and satirist, who wrote another 21,780 lines before finally allowing the impatient lover unrestricted access to his rosebud. Most scholars agree that Jean de Meun's contribution, informed both by scholastic philosophy and the comedies of Juvenal, transformed the original work—which was begun

FIGURE 11.1 Roses and secular love. A. *Roxe* (rose); *Tacuinum santitatis*. B. Venus above an enclosed garden, from the Italian *Codex Sphaera*.
A is from Bibliothèque nationale de France, Paris, MS nouv. acq. lat. 1673, fol. 83 recto. Ca. 1380–90. B is from Berlin Archiv für Kunst und Geschichte.

(b)

FIGURE 11.1 Continued

in the tradition of courtly love—into an erudite, but bawdy satire.[17] His conclusion to the poem provoked a strong negative reaction because of its pornographic aspects, so much so that an alternative version was composed by a third poet who recast the entire poem as a Christian allegory.[18]

In Guillaume de Lorris's poem, a young man falls asleep in the month of May and dreams he is walking along a river bank. He soon comes upon a beautiful garden surrounded by a high wall. After being admitted into the garden, he joins others in a dance. He then goes off to explore the garden more fully, stalked secretly by Cupid. Arriving at the very spring where Narcissus met his death, he sees the reflection of a beautiful rosebud (the virginal Lady) behind the hedge of a rosary. At this very moment, Cupid wounds him with five arrows, and the young man falls desperately in love with the rosebud.

Eventually, with the help of "Fair Welcome" (symbolizing the Lady's positive response), the lover manages to persuade "Danger" (symbolizing her ability to dominate her lover) to allow him to pass through the hedge surrounding the rosary. The lover approaches the Rose and is delighted to discover that its corolla

Had not yet spread so as to show the seed,
Which still was by the petals well concealed,
That stood up straight and with their tender folds
Hid well the grains with which the bud was filled.

In other words, the Rose was still virginal. Precisely how to interpret the hidden "seed" and the "grains" is unclear, but they clearly portray the Rose's concealed sexuality as a kind of hidden treasure. The lover then asks Fair Welcome's permission to kiss the rose, and when Venus herself appears and supports the lover's request, permission is granted:

Nor did I linger, but at once did take
A sweet and savory lipful from the Rose.
Let no man ask if then I felt delight!
My senses quickly were in perfume drowned
That purged my body from its pain, and soothed
The woes of love that had so bitter been.

But the young man's ecstasy is short-lived. "Jealousy" is aroused and imprisons both Fair Welcome and the Rose in a castle surrounded by a moat. Lorris's poem ends with the lover despairing of ever seeing his Rose again.

The finales of the two alternative additions to "The Romance of the Rose" differ starkly in tone, complexity, and subtext. In the anonymous seventy-eight line conclusion, "Dame Pity" and her allies visit the Lover and tell him how, with Cupid's help, they had stormed the castle and rescued the Rose. "Dame Beauty" presents him with the Rose, with which he is now free to do whatever he pleases without fear of "contravention":

There at our ease we took great delight;
The tender grass provided us a bed;
The rosebush petals fair made coverlets;
And all that night we nothing did but kiss
And satisfaction find in other joys.

We infer that the bud opened her petals in response to her Lover's caresses, for when Dame Beauty appears to take the Rose back to her castle, the Lover refers to her as "the blossom":

But, none the less, the blossom never closed
against me, even when 'twas snatched away.

Jean de Meun's lengthy addition uses the same plot framework as the anonymous seventy-eight line conclusion, with a multitude of philosophical and satirical digressions involving a host of new characters, the most important of which is "Dame Nature." In Jean de Meun's extension, the goddess Natura appears after the army of barons, led by Cupid and Venus, storm the castle but fail to rescue the Rose bush from the "Tower of Shame." We learn from Nature that God has subcontracted the universe to Her to rule according to Her laws. The most important law is Love, for it is the means by which species perpetuate themselves. Man alone, among all living things, defies her authority by rejecting Love. Determined to aid the Lover in his cause, Dame Nature instructs her chaplain, "Genius," to exhort Love's army to storm the castle and recover the Rose. Genius exhorts the barons to action in sexual terms, using the ancient agricultural metaphor of the plow:

Plow, barons, plow—your lineage repair;
For if you do not there'll be nothing left
To build upon. Bend well your sturdy backs
Like sails that belly to take in the wind . . .
The plow hales[19] lift with your two naked hands,
and with your arms strongly assist the beam
And strive to thrust the coulter[20] firmly home
And keep it in its proper place, to sink
More deeply in the furrow.

The barons attack and Venus shoots her arrow, setting the castle aflame. The enemies of Love are routed. The Lover mounts the ruined Ivory Tower and arrives at a shrine between two pillars. The shrine contains a narrow passageway that is blocked by a leathery barrier symbolizing the hymen. Using the staff that Dame Nature has given him, the Lover breaks through the barrier and squeezes through the passageway, where he finds the Rose:

I seized the rose tree by her tender limbs
That are more lithe than any willow bough,
And pulled her close to me with my two hands.
Most gently, that I might avoid the thorns,
I set myself to loosen that sweet bud
That scarcely without shaking could be plucked.
I did this all by sheer necessity.
Trembling and soft vibration shook her limbs;
But they were quite uninjured, for I strove
To make no wound, though I could not avoid
Breaking a trifling fissure in the skin,
Since otherwise I could have found no way
To gain the favor I so much desired.

In a radical departure from the tradition of courtly love, Jean de Meun's Lover impregnates the Rose:

This much more I'll tell you: at the end,
When I dislodged the bud, a little seed
I spilled just in the center, as I spread
The petals to admire their loveliness,
Searching the calyx to its inmost depths,
As it seemed good to me. It there remained
And scarcely could unmingle from the bud.
The consequence of all this play of mine
Was that the bud expanded and enlarged.

The Lover of the anonymous seventy-eight line conclusion of the *Romance* went no further than to shower his Rose with kisses. Jean de Meun's version, on the other hand, makes it quite clear that the constraints of courtly love have been breached. The Rose is no longer simply the symbol of the Lady; it symbolizes her vulva and womb. When the Lover finally gains access to his fair rose, he wastes no time in consummating his love, and, shortly thereafter is pleased to report the biological result. Lest anyone take offence, de Meun ends his racy narrative by invoking the age-old cop-out: "Then morning came, and from my dream at last I woke."[21]

DECONSTRUCTING "DEFLOWERING"

In "The Romance of the Rose," the Lover's "plucking" of the rose is equivalent to "deflowering" it, although this particular figure of speech does not appear in the poem. The precise origin of the verb "to deflower" (in the sexual sense) is unknown, although the idea can be traced to Roman times. Because classical authors restricted the definition of the flower to the petals, "deflowering" simply meant plucking the petals from a flower. A prime example of this usage is a poem (no. 62) by the first-century BCE Roman poet Catullus, in which a girl who loses her virginity is compared to a flower without petals:

When a flower grows secure within a fenced garden,
unknown to grazing livestock, not uprooted by any plough,
a flower which the breezes nurture,
the sun strengthens, the rain raises,
many boys, many girls desire it,
this same flower, once it has been plucked
by a slender fingernail and lost its petals (defloruit),
no boys, no girls want it, so a young girl,
as long as she remains untouched,
remains dear to her own;
but when her body has been corrupted
and she has lost her chaste flower,
remains neither pleasing to boys nor dear to girls.

It seems unlikely that Catullus's poem represents an isolated instance in late antiquity of the use of "deflower" to denote loss of virginity. The earliest known use of "deflower" in English occurred in 1393, in John Wyclif's Bible: "The lust of the gelding deflourede the {ygh}unge woman" (Ecclesiasticus 20:2).[22] The phrase *rose brechen* ("picking a rose") is the medieval German equivalent of the English word "deflower."[23]

A related metaphor is the term "flowers" used to denote menstruation in the medical literature of the Middle Ages. According to the Oxford English Dictionary, the earliest known usage of "flowers" in this sense occurred in 1400: "A woman schal in the harme blede For stoppyng of hure flowrys."[24] The English term "flowrys" (flowers) comes from the French word, *fleurs*, which was derived from the Old French word for flower: *flor* or *flour*. French scholars believe that the word *fleurs*, meaning "menses," originally arose as a corruption of the word *flueur*, meaning "flow." The use of the term "flowers" for menstruation continued up until at least the mid-eighteenth century in England.

HOW EVE'S APPLE BECAME EVIL

The domestic sweet apple, *Malus pumila*, a member of the Rose family, is the most iconic fruit in the Western world. Villages, private residences, and families are named after it. "Apple" is even assimilated into the names of other fruits and vegetables such as pineapple and star apple in English, and, in Romance languages (based on the name that derives from Pomona, the Roman goddess of fruit trees and gardens), *pomme de terre*, or apple of the earth (French for potato); *pomodoro*, or golden apple (Italian for tomato); and pomegranate, or apple with many seeds.[25]

Wherever the apple spread, it was either lionized or demonized in literature, art, music, mythology, and philosophy. In Classical antiquity, myth surrounded it with themes of immortality, sexuality, or deception—or all three. For example, Hera, goddess of marriage, received her golden apples, which conferred immortality on anyone who ate them, as a wedding present from the earth goddess Gaia. In the Judgment of Paris, the Apple of Discord awarded to Aphrodite as a prize for beauty led to the Trojan War when Paris seized Helen, the wife of Menalaus. Atalanta, a sworn virgin, was betrayed into marriage by Aphrodite's magic apples. Elsewhere, like Hera's apples, the apples of Freya (or Idun) goddess of love, lust, and beauty in Norse mythology also conferred immortality. The mythological history of apples is highly fraught.

Much of the apple's notoriety in the West arose from its identification in the Middle Ages with the forbidden fruit of the Garden of Eden. However, the identity of the "Tree of Knowledge of Good and Evil" is never given in the Bible, and the Bible's area of origin is for the most part too hot and dry for apple cultivation. Other instances in the Bible where the apple is mentioned by name, such as the "Song of Songs," are now considered to be mistranslations of the Hebrew word *tappuah*. Plausible candidates for the fruit of *tappuah* include quince and pomegranate, both of which were well-attested in the ancient Near East.

Whether or not the apples of Eden were actually quinces, how did the domesticated sweet apple come to be associated with the biblical forbidden fruit? During the Middle Ages, etymology was often used to discern the hidden meanings of words. Isidore of Seville considered it a heavenly sign that *malum*, the Latin word for apple, was identical to *malum*, the word for evil. This seems to have been one source of the sweet apple's identification with the forbidden fruit that led to the expulsion of Adam and Eve from their earthly paradise.[26] It was further alleged that an apple cut in half lengthwise resembled female genitals—a reminder of Eve's complicity in the Fall.[27]

Another hypothesis concerning the source of the apple's ill repute is based on the apple tree's supposed preference for "adulterous" affairs.[28] The apple is difficult to grow from cuttings because of the large amount of root-inhibiting tannins that are released from the bark. Because it is self-incompatible, it requires a genetically different apple variety as its pollen source to set fruit, resulting in a heterozygous embryo. Consequently, it will not breed true from seed. Medieval farmers knew nothing about the requirement for pollination for fruit and seed production, but through grafting they were able to arrive at a practical solution to the problem of propagating sweet apples. Grafting, however, was regarded by some medieval authors as the plant equivalent of sexual intercourse.

As absurd as the idea of "apple adultery" may seem, it was within the moral framework of scholastic discourse during the Middle Ages. For example, one of the questions Roger Bacon put to his students at the University of Paris was whether the production of a mutant sport or offshoot (*monstrum*) by a plant was the result of sin. This line of enquiry naturally led to the question of whether or not plants were capable of sin (Bacon argued that they were not).[29] Nevertheless, the fact that Bacon posed the question at all suggests that the possibility of plant sin was taken seriously by thirteenth-century academics.

GRAFTING AS A METAPHOR FOR INTERCOURSE

The sexual symbolism of grafting had occurred to the early horticulturalists. Several passages in the Talmud compare marriage to grafting, and there were restrictions, according to Jewish law, on the types of species that could be grafted together.[30] As noted in Chapter 5, in the Jewish Mishna, the practice of artificially pollinating date palms by tying male flowers to the female rachis was referred to as "grafting." Likewise, the practice of grafting trees was sometimes equated with sexual intercourse. The earliest written evidence for a sexual interpretation of grafting can be found in the treatise *Nabatean Agriculture* by Ibn Wahshiyya. Although Wahshiyya compiled the book in the tenth century, his sources are thought to date to around 600 AD or earlier.[31] Wahshiyya relates that, to ensure a successful graft union, the farmer must perform it ritualistically with the help of a "beautiful servant girl":

> [He] should take a beautiful (servant) girl . . . who must be of outstanding beauty. He takes her by the hand and lets her stand at the root of the tree where he wants to graft the branch. Then he prepares the branch like people do when they want to graft it and then he comes to the tree onto which he wants to graft it. The girl stands under the tree. He cuts a hole in the tree for the branch and takes off the girl's clothes and his own clothes. Then he puts the branch in its place while having intercourse with the girl, in a standing position. While having intercourse he grafts the branch to the tree, trying to do it so that he ejaculates at the same time as he grafts the branch.[32]

The sexualized tree-grafting ritual suggests a folk tradition that could be quite ancient, perhaps extending back to the tree goddesses of the Bronze and Iron Ages. According to Wahshiyya, if the girl becomes pregnant, the branch "will possess all of the tree's odor and taste." He cites as an example the grafting of a pear branch onto a lemon tree. However, he warns that for the pears to receive the color and odor of the lemon, the girl must not be forced against her will. To avoid rape, the farmer is advised to "do this with his wife whom he has married that particular year, not otherwise."

The preceding procedure was recommended for the transfer of color and odor from the stock to the scion. If, in addition, the farmer wished the fruit of the grafted branch to acquire the taste of the scion's fruit, stronger measures were needed. For example, if one wanted to graft an apple branch onto a pomegranate tree so that the apples tasted as sweet as pomegranates:

> [H]e must bring a girl to the tree where he wants to graft it and he must speak pleasantries with her until she laughs, he shall kiss her and pinch her and give her the branch so that she will graft it by her own hand. When she puts the branch in its place, he must remove her clothes from behind. While she is facing the tree and grafting the branch, he must have intercourse with her from behind and he must order her to take her time in completing the grafting—which means the planting of the branch—until he ejaculates. He should try to take care that the ejaculation and the completion of the graft should coincide.[33]

If the girl becomes pregnant, according to Wahshiyya, the grafted apple branch "will bear sweet and juicy apples."

Medieval Europeans first learned of the Nabatean grafting rituals (which may have been apocryphal) through the twelfth-century Jewish scholar Maimonides, who, in his *Guide for the Perplexed*, described them as "remarkable witchcraft" involving "disgraceful sex." Throughout the Middle Ages, a whiff of indecency clung to the practice of grafting fruit trees, the majority of which were apples. Typically, a newly emerged "lance" of first-year wood from a sweet apple tree was used as the scion to be grafted onto a wild apple (or "crabapple") stock, usually by the technique of cleft-grafting. The sixteenth-century French botanist Jean Ruel, to whose ideas about the sexuality of flowers we will return at the end of the chapter, had characterized typical grafting between species as "*miscella insitione . . . insitione adulteries*" ("mixed insertions . . . adulterous insertions").[34] Since sweet apples were normally propagated by grafting onto wild species, this may further account for the apple's association with adultery.

Indeed, the belief that grafting was the equivalent of plant sex persisted well into the seventeenth century, since no less an authority than the Renaissance scholar Francis Bacon (not to be confused with Roger Bacon, his distant relation from the Middle Ages)[35] felt the need to refute the notion. In his book, *Sylva Sylvarum* (1627), Bacon cites the ancient Greek idea that new animal species arise when "there is copulation of several kinds; and so compound creatures, as the mule, that is generated betwixt the horse and the ass" are produced. But, he adds, "The compounding or mixture of kinds in plants is not found out."

He rules out copulation in plants because "lust requireth a voluntary motion," which plants lack. If plants were able to mix sexually, it would have to be "more at command than that of living creatures." In other words, humans would have to intervene to make it happen. And if plants did indeed have sex, Bacon speculated, "it were one of the most noble experiments touching plants to find it out; for so you may have great variety of new fruits and flowers yet unknown." On the other hand, he states, "Grafting doth it not" because "it hath not the power to make a new kind. For the scion ever over-ruleth the stock." Thus, according to Francis Bacon, grafting should not be confused with plant sex.

THE ROSE AS A SYMBOL OF MARY

Ancient Greeks and Romans esteemed the rose as the most beautiful and fragrant of flowers and associated it with Aphrodite and Venus, the classical goddesses of romantic love. Venus was often depicted holding a rose or wearing a wreath of red and white roses. The association of roses with romance reached its zenith in the literature of courtly love, in which rose gardens and rose bushes became extended metaphors for the Beloved, as in "The Romance of the Rose" discussed earlier.

During the early years of the Eastern Orthodox Church, Mary assimilated many of the floral attributes of Near Eastern and Greco-Roman agricultural goddesses (see Chapter 10). The Byzantine Saint Ephrem described her as the sinless and inviolate "flower unfading."[36] One of the first to apply the rose metaphor to Mary was the poet Sedulius, whose Latin imitations of Virgil were written in the early sixth century. In his epic poem "Carmen Pascale" (Song of the Church), which summarizes the four gospels, Sedulius describes Mary as a rose that "springs forth from a thorny bush," reminiscent of the biblical epithet "lily among thorns" from the "Song of Songs" (2:2).[37]

Mary is also associated with roses through the "rosary," derived from the Latin word *rosarium*, meaning "rose garden" or "rose garland."[38] The rosary is an extended prayer addressed to Mary consisting of 150 "Hail Marys" recited in groups of ten, separated by fifteen "Our Fathers," and coupled with meditations on the mysteries of the lives of Mary and Jesus. The term is also applied to the string of beads used to count the repetitions. The story relating how the sequence of prayers to Mary came to be associated with roses is contained in a very sweet late-thirteenth-century legend known as "Aves Seen as Roses."[39]

In "Paradiso," Dante describes Mary as the "Rose in which the word of God became flesh."[40] Dante situates his beloved Beatrice within the petals of the immense celestial White Rose that represents Heavenly Paradise, of which Mary is the Queen. With this magnificent image, Dante unites the rose symbolism of the courtly love tradition with that of the Marian tradition, which began in the early eleventh century with the writings of Bernard of Clairvaux. In the sixteenth-century poem "Litany of Loreto," Mary is again referred to as the "Mystic Rose," a title still in use today. Two of the most beautiful artistic representations of Mary seated in her rose garden were painted in the fifteenth century by Stefan Lochner and Martin Schongauer of Germany (Figures 11.2A and B).

Finally, Mary's title "Mystic Rose" is illustrated in the exquisite eighteenth-century engraving by the renowned Augsburg artist Josef Sebastian Klauber (Figure 11.3).[41] Reminiscent of an Egyptian tree goddess emerging from her sacred tree (Chapter 5), Mary emerges from a huge rose blossom holding a stalk of lilies, symbolizing her virginity and purity The rose is "mystical" because it symbolizes the paradox of virgin birth and the unfathomable role of

(a)

FIGURE 11.2 Fifteenth-century paintings of Mary in her rose garden. A. Stefan Lochner, *Madonna im Rosenhag* (1448), Wallraf-Richartz-Museum. B. Martin Schongauer (1473), Cathedral of Colmar.
A is from Wallraf-Richartz-Museum. B is from Cathedral of Colmar.

(b)

FIGURE 11.2 Continued

the Holy Trinity in that event. In addition to their beauty and fragrance, the fact that roses and other flowers were thought to produce fruits and seeds asexually made them suitable emblems for the Virgin Mary. Consistent with this idea, in some illustrations, Jesus is represented as a bud growing from Mary's rosebush.

FIGURE 11.3 Mary is portrayed as the *Mystic Rose* in this engraving by Josef Sebastian Klauber (d. 1768) of Augsburg.

Permission to reprint this illustration granted by The Mary Page at: Marian Library/International Marian Research Institute, University of Dayton, Ohio.

WOMEN AND FLOWERS IN EARLY RENAISSANCE POETRY AND ART

During the early Renaissance, artists and poets turned to the themes and aesthetic styles of the classical period for inspiration. Pagan artistic associations between women, flowers, and fruits were resurrected, usually with erotic overtones.[42] For example, the Elizabethan poet Thomas Campion saw a garden in his lover's face:

There is a garden in her face
Where roses and white lilies grow;
A heav'nly paradise is that place
Wherein all pleasant fruits do flow.
There cherries grow which none may buy,
Till "Cherry ripe" themselves do cry.[43]

In "The Rapture," the seventeenth-century poet Thomas Carew uses a garden metaphor to express his amorous intentions:

Then, as the empty bee that lately bore
Into the common treasure all her store,
Flies 'bout the painted field with nimble wing,
Deflow'ring the fresh virgins of the spring,
So will I rifle all the sweets that dwell
In my delicious paradise, and swell
My bag with honey, drawn forth by the power
Of fervent kisses from each spicy flower.
I'll seize the rose-buds in their perfumed bed,
The violet knots, like curious mazes spread
O'er all the garden, taste the ripen'd cherry,
The warm firm apple, tipp'd with coral berry :
Then will I visit with a wand'ring kiss
The vale of lilies and the bower of bliss.[44]

In the realm of art, Sandro Botticelli's monumental masterpiece, *Primavera* (1482), invokes classical deities to represent spring (Figure 11.4). The central figure of the painting is Venus (not shown in Figure 11.4), but Chloris and Flora play important supporting roles. The right side of the painting shows the progression of Chloris from a Greek nymph to the Roman goddess of flowers, as described by Ovid. Zephyr (Latin Favonius), the god of the west wind, tries to make up for raping Chloris by marrying her, which will transform her into a goddess. Zephyr works his magic, and Chloris begins her metamorphosis: "As she talks, her lips breathe spring roses: 'I was Chloris, who am now called Flora.'"[45] We see Chloris with roses streaming from her mouth, and, shortly afterwards in her floral dress as the fully realized, pregnant Flora, goddess of Spring, who sprinkles roses on the ground.

FIGURE 11.4 Chloris and Flora, detail from Sandro Botticelli's *Primavera* (Spring). From the Uffizi Gallery; Wikimedia Commons.

ROGER BACON, ALBERTUS MAGNUS, AND THE REVIVAL OF DESCRIPTIVE BOTANY

As we have seen, after the death of Theophrastus, Greek natural philosophy declined precipitously. Roman agricultural writers had little patience with Greek philosophical abstractions, and Christian Europe branded as heresy any ideas that conflicted with the Bible or Church teachings. But medieval Christianity's obsession with the afterlife eventually gave way to the troubadour poets and to Saint Francis of Assisi, who turned away from the Church's fixation on death by celebrating the beauty and spirituality of God's Creation, the natural world. By the twelfth century, European scholars were actively seeking to reconnect with the philosophers, mathematicians, and physicians of ancient Greece because, paradoxically, this seemed to be the best way to move forward.

The field of medicine had undergone a small but significant advance as early as the tenth century with the establishment of an important medical school in Salerno. "Advance" in this case meant a return to Greek medical practices, facilitated by the availability of Greek medical and pharmacological texts at a nearby Benedictine monastery. Soon, new medical schools patterned after Salerno were flourishing throughout Europe. By bringing together scholars engaged in teaching medicine and pharmacology, these new medical schools made it all but inevitable that the deficiencies of the Greek texts would eventually come to light, compelling physicians to carry out direct observations of both their patients and plants to correct the errors.

Just as the faculties of the new medical schools were returning to Greek medical texts, scholastic philosophers such as Adelard of Bath in England were engaged in reviving Greek natural philosophy. Adelard of Bath was perhaps best known as the translator of Euclid's *Elements*, which led to the rediscovery of geometry. His conviction that the universe is a finite system susceptible to elucidation by logic and reason derives in large measure from his expertise in geometry. His highly influential *Quaestiones Naturales* (1107 AD) was a list of seventy-six questions and answers about the natural world written as a dialog between the philosopher and his nephew. Although Adelard stated that his answers to these questions were based on "the opinions of the Saracens," this may have been a ploy to shield himself from criticism. Modern scholars have identified two main sources for Adelard's answers: Plato's *Timaeus* and Cicero's *De Natura Deorum*, as well as other Greek writers.[46]

Quaestiones Naturales begins with six questions about plants (representing the simplest and most basic life forms) and ends with questions about the stars (the highest beings on the *Scala Natura*). Both the questions and the answers about plants are simply a rehash of Aristotelian botany (e.g. "Why do plants grow from unsown soil?" and "Why do hot plants grow from cold earth?"). Significantly, however, plants are treated as living organisms worthy of study in their own right, not merely as "simples" for medicine.

Between 1175 and 1225, a flurry of translations from Arabic into Latin made complete versions of Galen, Hippocrates, and Aristotle's zoological writings available to European scholars for the first time. A Latin translation of Nicolaus's *De Plantis* also made its appearance in Europe during this period, and because it was attributed to Aristotle, such luminaries as Roger Bacon and Albertus Magnus devoted considerable attention to it.

Roger Bacon, an English Franciscan Friar and early proponent of empirical research, argued that logic alone could never be as persuasive as a physical demonstration. As Bacon pointed out, being told that fire is hot and burns flesh is only an abstraction until we have actually experienced the heat of the flame ourselves. Bacon was a devoutly religious man who believed that experimental science would ultimately demonstrate the truth of Christianity and thus serve the best interests of the Church. He was one of the so-called *scholastics* or "schoolmen" who subscribed to the theology of medieval European universities, which sought to reconcile Aristotelian logic with the writings of the early Church Fathers. But he was also a visionary who understood the limitations of *De Plantis* and foresaw that a new science of plants would be required for the improvement of agriculture:

> Now this science [agriculture] extends to the perfect study of all vegetables, the knowledge of which is imperfectly delivered in Aristotle's treatise *De Plantis*;[47] and therefore a special and sufficient science of plants is required, which should be taught in books on agriculture.[48]

It is unfortunate that Bacon did not have access to the two major treatises by Theophrastus, *Historia Plantarum* and *De Causis Plantaraum*, and instead had to make do with the totally inadequate *De Plantis*. Sometime between 1236 and 1247, he delivered a series of lectures at the University of Paris based entirely on the questions addressed in *De Plantis*: whether plants were alive, had souls, slept, breathed, and whether they could be transformed into one another and be mutually grafted. Much space was devoted to the question of whether a grafted tree had one soul or two. If it had one soul, which was it: the scion's or the stock's? Such unanswerable questions Bacon doubtless would have dismissed as superfluous had he been able to read Theophrastus's treatises.[49]

Another thirteenth-century scholastic theologian whose botanical enquiries would have benefited greatly from a knowledge of Theophrastus was Albert of Bollstädt, better known as Albertus Magnus, the esteemed Bishop of Ratisbon and mentor of the "Angelic Doctor," Thomas Aquinas. Like Bacon, Albert revered *De Plantis* as the work of Aristotle but was keenly aware of its deficiencies, which he blamed entirely on corruptions introduced by translators.

In his seven-volume treatise *De Vegetabilibus*, Albert provided elaborate interpretations of *De Plantis* to remedy these deficiencies and, unlike Bacon's Paris lectures, Albert's interpretations in *De Vegetabilibus* were based on a significant body of his own original observations.[50] Many of his interpolations demonstrate a remarkable talent for plant morphology, which, in Agnes Arber's estimation, was "unsurpassed during the next four hundred years." [51]

Of relevance to the problem of sex in plants is the chapter titled "On the Nature and Generation of Flowers," although his understanding of the structure and function of flowers does not go much further than *De Plantis*. The flower, Albert states, "is the sign of the fruit," and the flower and fruit are of "the same substance." Then, in Book VI, he describes the stamens of several species, including their pollen. Although Pliny was the first to describe and name stamens, he failed to note the presence of pollen inside the anthers. Twelve hundred years later, Albert described the pollen grains in the anthers but failed to associate them with the fecundating "dust" of male date palms. Instead, he interpreted them as a waste product. Using the analogy of earwax in animals, which he defined as an impurity of the blood secreted by the brain, he referred to pollen as particles of "wax" released by stamens during the removal of the impurities from the plant's earthy nutriment.

Albert thus considered the function of stamens to be excretion rather than reproduction. Why didn't Albert, who must have been familiar with the ancient literature on the artificial pollination of female date palms, connect the "dust" of male date palms with the "wax" of stamens? The answer is that he was merely following Aristotle.

Although Albert adopted Aristotle's erroneous identification of pollen as "wax" and embellished it further by comparing it to earwax, to his credit, he also performed the first known chemical tests on it, such as placing it in a flame. He interpreted the fact that pollen readily ignited as a confirmation that it was composed of wax. Indeed, Albert was partially correct in his conclusion because the major component of the outer layer of pollen cell walls is a highly cross-linked lipid compound with a waxy consistency called *sporopollenin*. Albert also adopted Aristotle's notion that bees collect wax from flowers to build their hives. In reality, worker bees secrete the wax themselves in specialized glands located on their abdomens. Pollen serves as a nutritious food source for bees, not as a direct source of wax.

With *De Vegetabilibus*, medieval botany reached its apex, but in most respects it fell far short of the standard set by Theophrastus. Theophrastus did not have to contend with the philosophical disdain for "particulars" in the study of nature as Albert did. So pervasive was this disdain for "particulars" in the medieval period that Albert felt the need to apologize in *De Vegetabilibus* for his descriptions of individual plants. [52] Not until 1453, with the appearance of the Latin translations of the long-lost botanical treatises of Theophrastus, did such detrimental medieval attitudes finally die out. Remarkably, *Historia Plantarum* and *De Causis Plantarum* were still scientifically useful some 1,600 years after they were written.

THE REBIRTH OF NATURALISM IN MEDIEVAL HERBALS

The natural theology of St. Francis and the poems of the Troubadours in the thirteenth century laid the foundation for the revival of classical naturalism in art. In France and Germany, exquisite foliage motifs began appearing on the capitals of the major cathedrals. Still, the plant illustrations of the herbals remained highly stylized until the end of the fourteenth century, when professors of the newly established medical schools began to demand greater accuracy.[53]

The revival of Naturalism in the manuscript herbals took place over a period of about seventy years in the fourteenth century. The *Tractatus de Herbis*, written early in the fourteenth century, included both highly stylized illustrations copied from older herbals as well as some that were apparently drawn from real plant material. An example of the latter is the portrait of jasmine shown on the right in Figure 11.5A. Here, we see the beginnings of realism, although not on a par with the best plant portraits in the sixth-century *Juliana Anicia Codex*. The specimen is unnaturally flattened, as if the portrait had been drawn from a pressed plant with its leaves carefully spread out.

Fully naturalistic plant portraits begin to appear in the manuscript herbals at the end of the fourteenth century. The first of these was the *Herbolario Volgare* (Popular Herbal), an

(a)

(b)

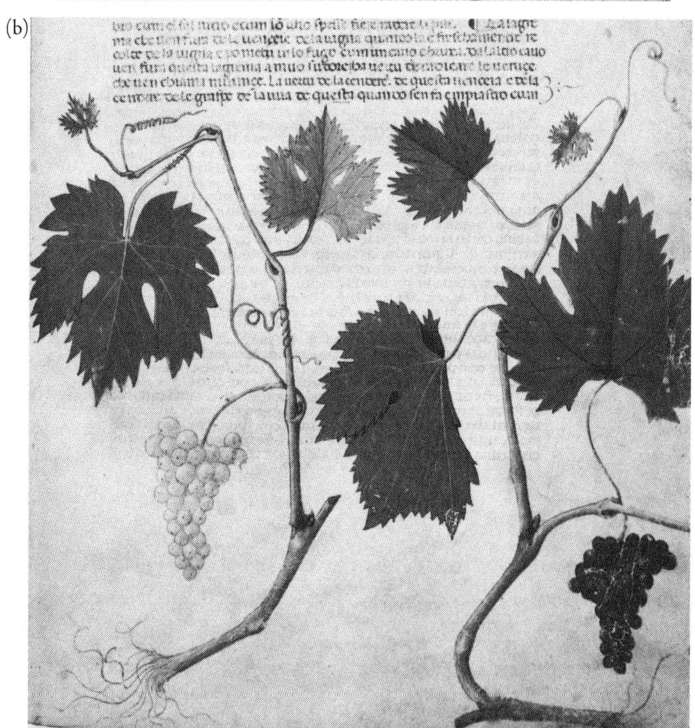

FIGURE 11.5 Early examples of Naturalism in manuscript herbals. A. *Tractatus de Herbis*, early fourteenth century (left) *Silfu* (unidentified); (right) *Sanbaco* (*Jasminum officinale*, L., Jasmine) MS. B. Grape vine (*Vitis vinifera*) *Herbolario Vulgare* (1390–1400), MS; C. Corn poppy (*Papaver rhoeas*), *Liber de Simplicibus* (1419).

A is from Egerton 747, fol. 98. B is from Egerton 2020 (27v–28r). C is from Biblioteca Nazionale Marciana, Venice, Cod. Lat. VI 59.

(c)

FIGURE 11.5 Continued

Italian translation of a much older Arabic herbal, which was presented sometime between 1390 and 1404 to the last Lord of Padua, Francesco Carrara the Younger. The anonymous artist of the Carrara Herbal broke decisively with medieval tradition by painting highly naturalistic plant portraits, such as the grape vine shown in Figure 11.5B. After more than 800 years, a medieval herbal was finally produced that equaled and even exceeded the accuracy of the *Juliana Anicia Codex*. The *Liber de Simplicibus* (Book of Simples), written by Benedetto Rinio in 1419, continued the tradition of excellence, as illustrated in Figure 11.6C. A new era of Naturalism in the manuscript herbals had dawned. However, these new naturalistic herbals were costly to produce and to copy, so they remained largely in the hands of the aristocracy, and their scientific impact on the newly established universities and medical schools was minimal.

THE PRINTED ILLUSTRATED HERBALS: FROM ICONOGRAPHY TO NATURALISM

The first Gutenberg Bibles were printed in Germany around 1455, and the new printing technique soon spread to Italy, where the first printed book with woodcut illustrations, titled *Meditationes de Vita Christi*, was published in 1467. The first printed herbal illustrated with woodcuts, *De Viribus Herbarum* by Macer Floridus, appeared in Milan in 1482. Its crude, highly stylized images were a far cry from the hand-painted naturalistic plant portraits of the manuscript herbals, and, over the next few decades, the woodcut images of the newly printed herbals, rather than improving in quality, became more and more stylized. Eventually, they became little more than iconographic symbols.[54] Various reasons have been given for this deterioration. First, there was a learning curve before European printmakers had acquired sufficient skill to reproduce the detailed botanical illustrations found in the contemporary manuscript herbals. Second, printers were businessmen who needed to make a profit on their books. Printed herbals were still a novelty with little or no competition, so there was zero economic incentive to invest in top-quality artists. Third, the early printed herbals were aimed at the general public rather than the universities and medical schools, where scientific accuracy was a high priority. The general public was content with decorative icons.

Leading artists of the fifteenth century disdained woodcut printing (with the exception of one master, Martin Schongauer), but acceptance of the new technique

(a)

FIGURE 11.6 Botanical illustrations of Albrecht Dürer. A. *Holy Family with Three Hares* (woodcut). B. *The Large Piece of Turf* (watercolor and gouache on vellum).

(b)

FIGURE 11.6 Continued

was stimulated in the sixteenth century by the extraordinary work of Albrecht Dürer, an early admirer of Schongauer's woodcuts and copper engravings. Dürer's beautifully rendered *Holy Family with Three Hares* (1497/8) set a new standard of excellence for woodcuts (Figure 11.6A). In this woodcut, Mary is seated on a turf-covered bench near a patch of weeds that crowds the single iris, symbol of Mary's fidelity. Prior to Dürer, landscapes were rendered as if they were tidy parks or ornamental gardens. Dürer's *The Large Piece of Turf* (1503) literally broke new ground with its ecologically accurate portrayal of a clump of grassland species, obviously drawn from life (Figure 11.6B). Gone are the symbolic flowers—roses, lilies, and irises—that regularly adorned medieval art. Instead, Dürer gives us only the feathery heads of grasses, perhaps at an early stage of ripening, and a shy trio of dandelions, their closed blossoms poised to reopen and release their parachuted seeds. If there is symbolism in this remarkable painting, the meaning is highly personal, reflecting the artist's direct relationship with nature.

The idea of a personal relationship with nature is in some ways the secular equivalent of the Reformation's call for a personal relationship with God. Although Dürer remained a Catholic and painted many religious subjects, he was also a humanist and quite sympathetic toward the Reformation. Perhaps the subtext of *The Large Piece of Turf* is the identification of God with even the humblest patch of earth, a religious philosophy—like that of Spinoza's in the seventeenth century—entirely compatible with a scientific view of nature. According to Karen Reeds, Dürer's *The Large Piece of Turf* provides "a missing link needed to explain the striking spurt of naturalistic botanical illustrations in German herbals, beginning with Hans Weiditz's drawings and woodcuts in Otto Brunfels's *Herbarum vivae icones*." [55]

Otto Brunfels was a German theologian and former Carthusian monk[56] who converted to Protestantism around 1420. He developed an interest in botany by way of pharmacology but soon became interested in the natural history of plants. He graduated from medical school in Basel in 1530 and, in the same year, published *Herbarum Vivae Icons*, or "Living Portraits of Plants" based mainly on his own observations of the flora of Germany. Many

FIGURE 11.7 Woodcuts of Brunfels and Fuchs. A. Pasque-flower (*Pulsatilla vulgaris*) in Brunfels's *Herbarium Vivae Eicones*, Vol. I, p. 217; B. Turkish Corn (*Zea mays*) Fuchs's *De Historia Stirpium*, p. 825.

(b)

FIGURE 11.7 Continued

of the species were not even listed in Dioscorides. *Herbarum Vivae Icons* was notable both for its text and its illustrations. The plant portraits were exquisitely rendered in woodcuts by Hans Weiditz, a prominent book illustrator who had studied with one of Dürer's close contemporaries. It is obvious from the detail and naturalism of these woodcuts that they were made from living models (Figure 11.7A). Weiditz faithfully drew the plants Brunfels set before him without idealizing them. Several of the portraits include broken stems or insect-damaged leaves.

The illustrations utilized by the Bavarian botanist Leonhart Fuchs matched Weiditz's scientific accuracy but avoided showing incidental damage, which is a property of the individual plant rather than the species as a whole. Fuch's great herbal, *De Historia Stirpum* (1542), included 512 woodcuts of wild and domesticated species, using plants growing in local fields and gardens. Among these were the first complete portraits of two exotic species from the New World, "Turkish Corn"[57] (maize) (Figure 11.7B) and red chili peppers.

ADAM ZALUZIANSKY: THE LAST OF THE SCHOLASTIC BOTANISTS

The botanical writings of the late Renaissance Bohemian scholar Adam Zaluziansky have sometimes been cited as a forerunner of the early modern sexual theory of plants.[58] Zaluziansky made the important distinction between seeds and buds, and, in his book *Methodi Herbariae Libri Tres* (1592), he stated that seeds, unlike buds, required some sort of sexual fertilization. In support of this distinction, he paraphrased Pliny's description of the impregnation of female date palms:

> If the male plant has been cut, the widowed female plants afterwards become sterile, which normally conceives by means of raised hairs, breezes, by the sight of him, and even by the dust, while leaning towards him with caressing hair. Art and human skill help this understanding of love, once the sexual act has been completed, and the male flower, down, and dust have been sprinkled on the females.

Like Pliny, Zaluziansky claimed that all plants reproduced sexually, and he went on to compare the majority of plants to the hermaphroditic fish, the sea bream, called "erythinis" by the Greeks, and to the "androgynes" (hermaphrodites) of humans. Zaluzianky's comparison of plants to hermaphroditic animals was indeed novel and seemed to imply that typical flowers contained both male and female sexual structures. For these reasons, some historians of botany have credited Zaluziansky with anticipating the actual discovery of sex in plants.

However, a closer reading of chapter XXIIII, "On the Sex of Plants," suggests that, despite his reference to pollination in date palms, he is merely restating Aristotelian doctrine regarding hermaphroditic flowers, as in the following passage:

> But all the things that are reproduced by the earth have entrusted to a most diligent Nature that plants even more so have both sexes. For some of these it is confused, for others separated. Indeed certain single [plants] have of themselves the capacity to generate something else, when the principles of maleness and femaleness have been mixed, and that is by the best plan of nature.[59]

In this passage, Zaluziansky, following Aristotle's example, distinguishes between dioecious plants composed of two sexes and hermaphroditic plants in which "maleness and femaleness have been mixed." Later, he correctly states that most plants fall into the latter category. However, Zaluziansky fails to associate maleness or femaleness in hermaphroditic flowers with specific anatomical structures. Like Aristotle, Zaluziansky believes that "maleness" is a nonmaterial "principle" of the plant that confers the ability "to do," whereas "femaleness" confers the ability "to bear." In hermaphroditic plants, these two immaterial principles are "mixed." Despite his awareness of artificial pollination in date palms, Zaluziansky makes no mention of the role of pollen in the vast majority of flowers. Thus, although his insights concerning seeds and buds are tantalizing, rather than being a forerunner of modern botany, Zaluziansky is perhaps best regarded as the last of the scholastic botanists.

JEAN RUEL'S PREGNANCY MODEL OF THE FLOWER

The sixteenth-century French botanist, Jean Ruel or Ruellius has been credited with being the first botanist after Theophrastus to undertake a general treatise on botany.[60] In *De Natura Stirpium* (1536), Ruel described plant reproduction using such anthropomorphic terms as "conception," "gestation," and "parturition." In addition, he referred to the embryonic plant inside the seed as the "fetus." To account for "conception," Ruel cited the importance of wind, which acts as the husband of the plant:

> Conception is the first thing in the order of nature, after the wind Favonius has begun to blow; then in February all vegetal things are married to it. It is the procreative spirit of the whole world. It blows from the equinoctial west, ushering in the springtime, and all nature is in lively expectancy of conceiving seed; also, this [wind] breathes the breath of life into seeds already in the ground. These are in the receptive state during a greater or less number of days, according to their different natures, remaining pregnant, some for a longer, others for a shorter period, before bringing forth. . . . This, in the case with trees, is called germination. The parturition of these is in their flowering; the flower, consisting of little disrupted wombs; and coming forth from these are the fruits, to be nourished and brought to maturity.[61]

Note that the Roman god of wind Favonius not only causes the plants to become "pregnant," but also stimulates the seeds in the ground to germinate. E. L. Greene comments that the sexual metaphors Ruel employed in this passage indicate that he regarded plant reproduction as a sexual process, even if the male fecundating factor was erroneously identified as wind. This Neoclassical conceit is not so very different from the way Christians imagined Mary's conception as pictured in the countless paintings of the Annunciation, in which a dove represents the Holy Spirit and Mary's purity is symbolized by a lily. In both cases, the process is sexual because it requires two partners: neither flowers nor Mary can conceive on their own. In the words of E. L. Greene:

> Such figures of speech could not fail to express and promulgate the idea of the femininity of all trees and shrubs in general. In the realm of animal life it was plainly otherwise. Here it was necessary to procreation that there be a conjunction of two individuals of opposite sex; for neither in the times of Pliny nor in the days of Ruel was anything known of parthenogenesis in the animal kingdom. It could

> not well be thought of as a universal condition in the plant world. A mythical personage was brought forward to balance the inequality that is manifest between animals and plants regarding sex, by supplying to the latter the male, or fecundating, element.[62]

Greene concluded that "it should be clear that men long ago held that plants are not asexual, but unisexual and feminine." Writing at the turn of the twentieth century, Greene expressed some surprise[63] that this "ancient doctrine" about plant reproduction had been overlooked by previous historians of botany:

> It has been at the cost of much time and study that I have been able to gather from the old-time botanists, and from their thoughts, as summed up and expressed by Ruel, the ancient doctrine of the physiology of plant reproduction; and as far as my reading has gone, not one of the historians has undertaken its elucidation. It is nevertheless an important topic. It is always important to the history of science, or any branch of any science, to get at the earliest views that men can be found to entertain regarding it.[64]

Indeed, as we have argued in this book, the origins of Jean Ruel's "plants-as-female" paradigm can be traced as far back as prehistoric times.

NOTES

1. Galatians 5:16–17, *The New Oxford Annotated Bible* (2007).

2. Menocal, M. R. (1990), *The Arabic Role in Medieval Literary History: A Forgotten Heritage*, second edition. University of Pennsylvania Press.

3. Porter, P. (2003), *Courtly Love in Medieval Manuscripts*. University of Toronto Press.

4. "Capellanus" means "chaplain."

5. The structure of Hell in Dante's *Inferno* can be viewed as a combination of the planar concentric circles in Capellanus's *De Amore* and the deep cavern of the underworld in Virgil's *Aeneid*. But whereas in *The Inferno* the lowest level of Hell is reserved for the worst sinners, in *The Aeneid* it corresponds to Earthly Paradise, or the Elysian Fields.

6. Genesis 1:11–12. *The New Oxford Annotated Bible*, augmented third edition (2007), M. D. Coogan, ed. Oxford University Press.

7. Curtius, E. R. (1953), *European Literature and the Latin Middle Ages*, trans. W. R. Trask, Routledge, pp. 106–107; cited by Dronke, P. (1980), Bernard Silvestris, Natura, and personification. *Journal of the Warburg and Courtauld Institutes* 43:16–31.

8. "The root was black, while the flower was as white as milk; the gods call it Moly, and mortal men cannot uproot it, but the gods can do whatever they like." *The Odyssey*, Book 10.

9. Economou, G. (2002), *The Goddess Natura in Medieval Literature*. Notre Dame Press.

10. The earliest use of the term "Neoplatonism" was in the early nineteenth century. Plotinus regarded himself as a Platonist.

11. Dronke, Bernard Silvestris, Natura, and personification.

12. Ibid. According to some writers, Bernard Silvestris and Bernard of Chartres (famous for his quotation about the "shoulders of giants") were two different individuals.

13. According to Dronke, "in the *Cosmographia*, herbs are the special concern of Physis, not Natura."

14. Alan of Lille (1980), *The Plaint of Nature*, trans. James J. Sheridan. Pontifical Institute of Mediaeval Studies, Toronto.

15. In Berceo, Mary is actually identified with the entire orchard, with each tree representing one of her miracles.

16. Winston-Allen, A. (2005), *Stories of the Rose: The Making of the Rosary in the Middle Ages,* Pennsylvania State University Press.

17. Dunn, C. W. (1962), Introduction to *The Romance of the Rose*, first edition. E. P. Dutton.

18. Winston-Allen, *Stories of the Rose.*

19. Handles.

20. According to the *Oxford English Dictionary* (OED): "The iron blade fixed in front of the share in a plough; it makes a vertical cut in the soil, which is then sliced horizontally by the share."

21. Winston-Allen cites a comparable example of overtly sexual rose symbolism in a thirteenth-century fable from Germany, which echoes the ancient Sumerian myth of Innana and the date grower (as discussed in Chapter 5). A wily farmhand spots an exhausted serving maid fast asleep. Taking advantage, he gingerly lifts her skirts, exposing her "rose garden." Before leaving, he draws a little circle with lamp black just above her "little rose bush." The circle is a reference to an obscene German song celebrating a "wreath of brown roses." Winston-Allen, *Stories of the Rose.*

22. Definition from the *Oxford English Dictionary.*

23. Winston-Allen, *Stories of the Rose.*

24. Wright, T., and J. O. Halliwell, eds. (1841–43), *Reliquiae Antiquae: Scraps From Ancient Manuscripts*, Vol I, p. 190.

25. Juniper, B. E., and D. J. Mabberley (2006), *The Story of the Apple.* Timber Press.

26. Augspach, E. A. (2004), *The Garden as Woman's Space in Twelfth- and Thirteenth-Century Literature.* Edwin Mellon Press.

27. Touissaint-Samat, M. (2008), *A History of Food*, second edition. Wiley-Blackwell.

28. Pastoureau, Michel (1993), Bonum, Malum, Pomum: Une Histoire Symbolique de la Pomme, in *L'Arbre: Histoire Naturelle et Symbolique de l'Arbre, du Bois et du Fruit au Moyen Âge.* Le Léopard D'Or.

29. Easton, S. E. (1952), *Roger Bacon*. Oxford University Press.

30. Mudge, K., J. Janick, and S. Scofield (2009), A history of grafting. *Horticultural Reviews* 35:437–492.

31. Hämeen-Anttila, J. (2006), *The Last Pagans of Iraq.* Brill, Leiden. The complete *Nabatean Agriculture* is more than a thousand pages long and written in Arabic. *The Last Pagans of Iraq*, the only English translation, includes sixty-one translated excerpts focusing mainly on the ethnographic, philosophical, and religious material. Most of the agronomic details are derived from Greco-Latin sources.

32. Ibid.

33. Ibid.

34. Ruel, Jean (1536), *De Natura Stirpium Libre Tres.* Cited by Juniper, B. E., and D. J. Mabberley, *The Story of the Apple.*

35. According to the website "Escutchions of Science" (http://www.numericana.com/arms/bacon.htm), Roger and Francis Bacon, although distantly related, belonged to the same distinguished Norman family.

36. In *Precationes ad Deiparam*, in Opp. Graec. Lat., III, 524–537.

37. Sedulius "Carmen paschale," II, 28–31, trans. Rev. See Koehler, Theodore A. (SM), *Christian Symbolism of the Rose* (http://campus.udayton.edu/mary/rosarymarkings36.html)

38. Winston-Allen, *Stories of the Rose.*

39. http://campus.udayton.edu/mary/index.html

40. *Paradiso* 23:73–74.

41. The Marian Library of the University of Dayton has in its possession rare books of the eighteenth century with engravings by the renowned Augsburg artist, Josef Sebastian Klauber (ca. 1700–68). The highly symbolic and illustrative reproductions are typical of the Baroque period.

42. Seaton, B. (1989), Towards a historical semiotics of literary flower personification. *Poetics Today* 10(4):679–701.

43. Shakespeare's Sonnet 130, which begins, "My mistress' eyes are nothing like the sun," parodies poems with lavish comparisons like Campion's. Such poems were very popular at the time.

44. In the Indian Vedic Hymn, "Ode To a Black Bee," perhaps dating to the first millennium BCE, the black bee is also portrayed as the paramour of flowers. See *Srimad Bhagavata Mahapurana* (1971), 10th Canto, Discourse XLVII; C. L. Goswandi, Gita Press, Gorakhpur, India.

45. Ovid (2004), *Fasti* V, trans. A. J. Boyle, 193 ff. Penguin Classics.

46. Burnett, C. (1998), *Adelard of Bath, Conversations with His Nephew.* Cambridge University Press.

47. Bacon was under the impression that *De Plantis* was Aristotle's lost treatise on plants.

48. Roger Bacon (1996), *Communia Naturalium* (~1268), in A. C. Crombie, ed., *The History of Science from Augustine to Galileo.* Dover.

49. Easton, S. C. (1952), *Roger Bacon and His Search for a Universal Science: A Reconsideration of the Life and Work of Roger Bacon in the Light of His Own Stated Purposes.* Columbia University Press.

50. Arber, A. (1938), *Herbals: Their Origin and Evolution 1470–1670*, second edition. Cambridge University Press.

51. Ibid.

52. Reeds, K. (1980), Albert on the Natural Philosophy of Plant Life, in J. A. Weisheipl, ed., *Albertus Magnus and the Sciences: Commemorative Essays.* Pontifical Institute of Mediaeval Studies, Toronto.

53. Reeds, K. M. (1991), *Botany in Medieval Renaissance Universities.* Garland Publishing, Inc.

54. Arber, *Herbals.*

55. Reeds, *Botany in Medieval Renaissance Universities.*

56. The Carthusian Order of enclosed monastics was founded by Saint Bruno. The name is derived from the Chatreuse Mountains in southeastern France, the site of the first Carthusian hermitage.

57. Although originally brought over from the New World by Columbus, maize reached Central Europe via the Middle East and its true origin was soon forgotten. Hence, it came to be known

variously as "Turkish corn," "Turkish wheat," "Egyptian corn," and "Syrian sorghum." Fuchs's herbal presented the first European portrait of the complete maize plant. However, earlier illustrations of maize cobs and tassels, painted between 1515 and 1517, can be found in the Villa Farnesina in Rome. See Janick, J. (2012), Fruits and nuts of the Villa Farnesina. *Arnoldia*, 70:20–27.

58. Guétrot, M. (1935), Histoire et critique de la découverte du prétendu sexe des plantes, in *Bulletin de la Société Botanique Du Centre-Ouest*, Saint-Maixent L'École. Imprimerie Garnier & Co., pp. 21–90.

59. We are indebted to Professor Gildas Hamel at UC Santa Cruz for translating ch. XXIIII of Zaluziansky's *Methodi Herbariae Libri Tres* (1592).

60. Meyer, E. H. F. (1854–57), *Geschichte der Botanik* (History of Botany). Gebrüder Bornträger, Königsberg.

61. Ruel, *De Natura Stirpium*.

62. Greene, E. L., trans. (1883), *Landmarks of Botanical History, Part II*. Stanford University Press, p. 649.

63. Greene's insight was first introduced in Chapter 1.

64. Greene, *Landmarks of Botanical History*, p. 651.

12

The Difficult Birth of the Two-Sex Model

BY THE MID-SEVENTEENTH century, the transition from natural philosophy to experimental natural philosophy, referred to as the "Scientific Revolution," was in full swing. A century earlier, Copernicus, Galileo, and Vesalius had overturned venerable Greek notions about cosmology, mechanics, and human anatomy. Then, in 1620, in his magisterial treatise *Novum Organum Scientiarum* (New Instrument of Science), Francis Bacon espoused an empirical approach to the study of nature based on observation and experiment, rather than on the received wisdom of the past.[1] Bacon's writings provided the impetus for the establishment of the Royal Society of London in 1660, the first public body devoted exclusively to "the corporate pursuit of scientific research."[2] Despite these early advances in the physical and medical sciences, however, botany remained mired in its medieval soil. Naturalists like John Ray were primarily concerned with classification and showed little interest in the mechanisms underlying plant growth and development. On such questions, Aristotle and Theophrastus still served as the ultimate authorities.

The first inkling that Greek descriptions of plant anatomy were as inadequate as their views on plant taxonomy followed swiftly in the wake of the invention of the microscope. When trained on plant tissues for the first time, the microscope revealed a multitude of new structures unknown to the Greeks, and while these discoveries helped free botanists from the influence of their classical cicerones, they also raised questions about their functions. Francis Bacon had argued that to elucidate a function, observations must be supplemented with well-designed experiments.

The discovery of sex in plants through a combination of observation and experiment was arguably the crowning achievement of seventeenth-century botany. It occurred in three successive stages corresponding to the contributions of three outstanding physician-botanists: Marcello Malpighi in Italy, Nehemiah Grew in England, and Rudolf Jacob Camerarius in Germany. Malpighi's and Grew's studies were based entirely on microscope observations and were thus descriptive in nature, whereas Camerarius was the first to conduct experiments. Malpighi's contribution is often underappreciated because, far from challenging the old plants-as-female model, he embellished it with contemporary clinical terminology. Grew, whose treatise on plant anatomy was published shortly after Malpighi's, initially adopted Malpighi's elaborate one-sex model but later added a novel feature of his own: a transsexual stamen! But Grew did not base his new hypothesis on experimental evidence. Rather, his theory was a philosophical compromise between a one-sex and a two-sex model, which could be thought of as a "one-and-a-half-sex model." It was left to the experimentalist Camerarius to sever botany's umbilical cord to classical botany once and for all by providing the first experimental evidence for a straightforward two-sex model of plant reproduction.

EARLY MICROSCOPE OBSERVATIONS ON PLANTS

Just as the telescope opened up the universe to human curiosity, the microscope revealed the formerly invisible world of the very small. By 1600, technical improvements in glass-making and lens-grinding in Holland had led to the invention of telescopes,[3] and, ten years later, Galileo employed his own telescope to observe mountains on the moon and other celestial objects. Galileo also adapted his telescope to magnify objects at close range, enabling him to observe the compound eyes of insects. His reconfigured telescope, with its two lenses, is referred to as a "compound microscope," in contrast to a "simple microscope," which consists of a single lens mounted on a stand. Galileo and his colleagues, who called themselves the *Accademia dei Lincei* (Academy of the Lynx-Eyed),[4] used the new compound microscope to examine a wide variety of materials.

In the meantime, England and Holland were heating up as centers of microscopy research. In 1628, the British physician William Harvey published his microscope observations of pulsating insect hearts, and, in 1663, Robert Hooke, who held the position of Demonstrator at the newly established Royal Society of London, used a compound microscope to view plant materials at close range, including moss "leaves" and thin sections of cork. Hooke observed that both tissues were actually comprised of hundreds of tiny, honeycomb-like chambers, which he called "cells." His book of illustrations, *Micrographia*, published in 1665, caused a considerable stir throughout Europe in the rapidly expanding scientific community.

Hooke's pioneering but somewhat haphazard microscope observations were soon taken up more systematically by the young English physician Nehemiah Grew, then only twenty-three years old, who chose to focus his efforts on plant structures. In the seventeenth century. it was quite common to couch one's scientific pursuits in religious terms, and Grew, whose father was a Puritan clergyman, explained his decision to study plants as a means of gaining insight into God's wisdom. Meanwhile, in Italy, the physician Marcello Malpighi, thirteen years Grew's senior, had already established himself as an eminent authority in the fields of anatomy, physiology, and embryology. Despite his fame, Malpighi was under constant attack from conservative elements in the Italian medical community who doubted the usefulness of applying modern scientific techniques to medicine.[5] In contrast, Grew's career was supported and encouraged by enthusiastic backers at the Royal Society of London.

Malpighi's interest in plants was sparked by a chance event in 1662, when he was a medical professor at the University of Messina. While taking a stroll in the garden of the Viscount Ruffo, Malpighi snapped off a chestnut branch and noticed the fibrous texture at the broken end. Upon returning home, he dissected the branch under a microscope and was surprised to discover that the wood was composed of tiny fibers and tubular vessels. Fascinated by the novel structures he was seeing, Malpighi spent the next ten years of his life making detailed notes and drawings based on his microscopic observations of plants.

The anatomical studies of Malpighi and Grew progressed in parallel. Malpighi's early work was first published in Bologna in 1671 under the title *Anatome Plantarum Idea*. Coincidentally, the Royal Society published Nehemiah Grew's preliminary anatomical studies, *The Anatomy of Plants Begun*, in the same year.[6] Several years later, the Royal Society published Malpighi's comprehensive treatise, *Anatome Plantarum*, in two volumes: the first in 1675 and the second in 1679. In parallel, Grew presented a series of four shorter papers to the Society between 1672 and 1676. The last of Grew's papers has since been cited in virtually every book on the history of botany because it broke with the classical tradition by presenting a new sexual theory of plants.

In 1682, Grew published a compilation of all his papers in his landmark volume, *The Anatomy of Plants*. Although he and Malpighi had worked independently, Grew, through his contacts in the Royal Society, had ready access to Malpighi's manuscripts, whereas Malpighi experienced long delays before Grew's papers arrived by post. As Grew himself diplomatically acknowledged in his preface, his own research on plants owed much to Malpighi's observations. It is also clear that Grew's hybrid version of the two-sex model was strongly influenced by Malpighi's one-sex model of the flower as a plant uterus.

MALPIGHI'S THEORY OF FLORAL MENSTRUATION

During the sixteenth century, Vesalius and other Renaissance anatomists labeled their diagrams of the human female's reproductive system using terms applied to the male genital system. Ovaries were called "testicles," Fallopian tubes (oviducts) were given the name "spermatic ducts," and the vagina was drawn to resemble a penis (see Chapter 1). By the late seventeenth century, however, the female pathway had come to be regarded as distinct from the male's, as reflected in the adoption of different terms for male and female reproductive structures.

Marcello Malpighi, like all seventeenth-century physicians, made the transition from the one-sex to the two-sex model in animals without difficulty, but he held fast to the one-sex model in plants. In fact, Malpighi went much further than the Renaissance botanist Jean Ruel had done in feminizing the flower. Whereas Ruel had used the poetic term "womb" for the ovary of flowers, Malpighi applied the gynecological term "uterus" to the same structure. In this way, Malpighi advanced the plants-as-female paradigm from a poetic metaphor to a scientific hypothesis. He went on to compare the style of flowers to the Fallopian tubes, and postulated that the sepals, petals, and stamens all contributed to seed production by ridding the sap of impurities.

Malpighi described the process of ridding the sap of waste as a type of "menstruation." This idea was not new with him. In 1631, Peter Laurenberg, a Professor of Medicine from

Rostock, Germany, published *Horticultura*, in which he stated that flowers form as the result of a "menstrual discharge":

> The florescence (*flos*)[7] of plants corresponds to the menstrual discharge of a woman (*menstruum muliebre*), for which reason these discharges are also called "florescences" (*flores* or [in German] *die Blumen*). As soon as a tree sends forth its florescences, there is hope that it will also give fruit. Nor does a plant [ever] bear fruit before it bears florescences. One that never flowers never bears fruit; this is exactly the same as with women who, once the menstrual discharge has begun to break forth, are then fit for reproduction. When the flow ceases due to [old] age, they become unfit to give birth.[8]

In Laurenberg's definition of the flower, the medieval euphemism for menstruation as "the flowers" came full circle. A seventeenth-century Italian medical text illustrates the pregnant womb in the form of an open flower, indicating that the flower/womb analogy could be applied in both directions (Figure 12.1).

Like Laurenberg, Malpighi viewed menstruation as the removal of impurities from the woman's blood, a necessary step before conception could take place. This interpretation of menstruation was at odds with the view taken by the Greeks. Both Hippocrates and Aristotle had defined menstruation as the elimination of excess fluid (humors), which women tended to accumulate because of their lack of heat and the "sponge-like" nature of their bodies.[9] Good health, according to Aristotelian and Galenic theories, depended on a proper balance of humors, but, importantly, there was nothing intrinsically unclean or impure about the humors themselves. Far from viewing menstrual blood as a waste material, Aristotle regarded it as the physical substance out of which the embryo was fashioned. Male semen provided the organizing principle that caused the embryo to coalesce from the menstrual blood, similar to the way rennet curdles milk in the making of cheese.

The Bible, on the other hand, treats menstrual blood as if were a contagion:

> And if a woman have an issue, and her issue in her flesh be blood, she shall be put apart seven days: and whosoever toucheth her shall be unclean until the even. And every thing that she lieth upon in her separation shall be unclean: every thing also that she sitteth upon shall be unclean. And whosoever toucheth her bed shall wash his clothes, and bathe himself in water, and be unclean until the even.[10]

Medieval scholars adopted the biblical view of menstruation, which was associated with God's curse directed at Eve, "I will greatly multiply thy sorrow and thy conception." Malpighi's understanding of menstruation as a means of removing impurities from the blood was thus more closely allied to biblical/medieval attitudes than to Greek humoral theories.

Malpighi's ideas about the flow of sap in plants were informed by his earlier studies on the circulatory and lymphatic systems of animals. Malpighi had provided the first visual evidence for the system of capillaries between the arteries and veins, the missing piece of the puzzle in Harvey's new theory of the circulation of the blood. In his discussion of the functions of flower parts in *Anatome Plantarum*, Malpighi combined the Aristotelian model

FIGURE 12.1 A medical illustration showing the pregnant womb in the form of a blossom. From *De Formato Foetu* (1631), Frankfurt, p. 37, plate 4: (http://www.nlm.nih.gov/dreamanatomy/da_g_I-D-4-01.html).

for the purification of sap by the "flower" with the medieval identification of menstrual blood as impurities. In the following passage, he considers two theories—either the sepals, petals, and stamens manufacture a refined sap that flows toward the uterus, or they remove impurities from the sap that rises from the roots and then transport the purified sap back toward the uterus:

> Meanwhile I have wondered if the leaves [petals] of flowers produce sap in their sacks [cells] which, flowing inside, they pour by the soft uterus and the immature seed, as I have plausibly concluded about the remaining leaves [sepals?] of the flower stalk. However, for some time I have rather believed that the role of leaves [petals, stamens, and sepals] consists in the purification of the unfit humors.[11]

Malpighi clearly favors the latter hypothesis, whereby the stamens, petals, and sepals remove impurities from the sap.[12]

According to Malpighi, once the sap has been purified by the stamens and other floral structures, it is transported rapidly back to the uterus. His description of this process is novel and strangely erotic:

> The self-contained purer fluid of the flower stalk can be brought back by a frantic discharge, so that it rushes more easily and more purely into the uterus.[13]

A modern reader could be forgiven for interpreting "frantic discharge" as a plant orgasm, especially in the context of a fluid that "rushes . . . into the uterus." However, there is no evidence that Malpighi thought of this process in sexual terms—or at least not consciously. It seems more likely that he was envisioning a pump-like mechanism that forced the purified sap from the sepals, petals, and stamens toward the uterus. Such an interpretation would be consistent with Malpighi's own research on the role of the multichambered heart in the circulation of the blood and on the contractions of the spleen, which forces lymphatic fluid into the vessels.

Malpighi's description of the "frantic discharge" of purified sap from the flower into the "uterus" is reminiscent of Descartes's theory of the physiology of love outlined in his book, *The Passions of the Soul.*[14] According to Descartes, the mere sight of a loved one forces a denser, cruder type of blood into the heart, displacing the more rarefied blood that has already passed through the heart multiple times. The less refined blood causes the heart, the seat of the soul, to send a signal to the brain compelling it to desire the love object. Although in Malpighi's model of the flower it is the refined sap from the sepals, petals, and stamens that displaces the cruder sap from the roots, the basic mechanisms both involve the explosive movement of humors. It's possible that Malpighi consciously or unconsciously associated the mechanism of fertilization in flowers with Descartes's theory of passion.

Consistent with a human sexual analogy for fertilization in plants, Malpighi compared menstruation in flowers to the events preceding "the moments of conception" in women:

> Hence it is right to consider that nature through these structures [sepals, petals, and stamens] is eliminating, as if it were mucus matter, most of the humor, which is of a

> diverse substance and unfit for the generation of seeds. Hence perhaps I apply the not unfitting derived name of *menstrual purgations*, which closely precede the moments of conception in women, just as it [conception] succeeds the discharges of flowers. [In this way] a defined portion of the sap is separated through the stamens and the leaves of the flower. . . . And since in the discharge of menstruation a certain maturity of time is required . . . so the production of flowers in plants similarly doesn't succeed immediately, but after a defined time.

According to Malpighi, "menstrual purgations" of flowers take various forms, including the secretions of glandular hairs and, most significantly, pollen, which he called "globuli." Malpighi was the first to illustrate pollen grains and their release from anthers, which he called *globulorum capsulae* (capsules of globules). Figure 12.2A shows pollen grains escaping from the apertures at the tips of an anther of "Indian wheat" or maize. In his description of the figure, Malpighi again emphasizes the dynamic nature of the process: "The excited apertures at the summits allow the globules to exit." The term "excited" ("awakened," "aroused") applied to the anther, like the term "frantic discharge," is sexually suggestive. Figure 12.2B shows pollen exiting the anther of a squash plant ("Cucurbitae"). According to Malpighi, "the capsule [anther] of globules acts as a junction which filters the staminal material in the globules." Presumably this statement refers to the role of the anther in "menstrual purgation." Thus Malpighi interprets the stamen as a female reproductive structure.

Most historians of botany have concluded that Malpighi simply missed the boat on the question of sex in plants. But neither Malpighi nor Laurenberg were true asexualists, any more than Ruel was. The Laurenberg-Malpighi anthropomorphic menstruation model of the flower feminized it to such an extent that it flung the door wide open to questions about the missing male half of the equation. Chance favored the prepared mind, and Nehemiah Grew seized the opportunity to provide the answer.

NEHEMIAH GREW'S *ANATOMY OF VEGETABLES BEGUN*

Nehemiah Grew's ideas about flowers, unlike those of Malpighi's, evolved over time. Malpighi was already a mature researcher with a long list of discoveries in anatomy, physiology, and embryology to his credit by the time he became interested in plants. His elaborate analogies comparing flowers to the human female reproductive system were based on his extensive first-hand knowledge of animal anatomy and physiology. In contrast, Grew was barely out of Pembroke College, Cambridge, when he initiated his own botanical studies in 1664.[15] These early studies consisted of anatomical drawings of plants made without the use of a microscope. His half-brother, Henry Sampson, brought Grew's investigations to the attention of Henry Oldenburg, the energetic Secretary of the Royal Society, who promptly showed them to the Bishop of Chester, John Wilkins, a leading figure in the Society. Wilkins presented them at a meeting of the Society, and the positive response led to an offer to publish the work, *The Anatomy of Vegetables Begun* (1672), under the Society's imprimatur.

Grew's discussion of floral anatomy in *The Anatomy of Vegetables Begun* could have been written by any of the classical botanists. Like Jean Ruel before him, Grew metaphorically

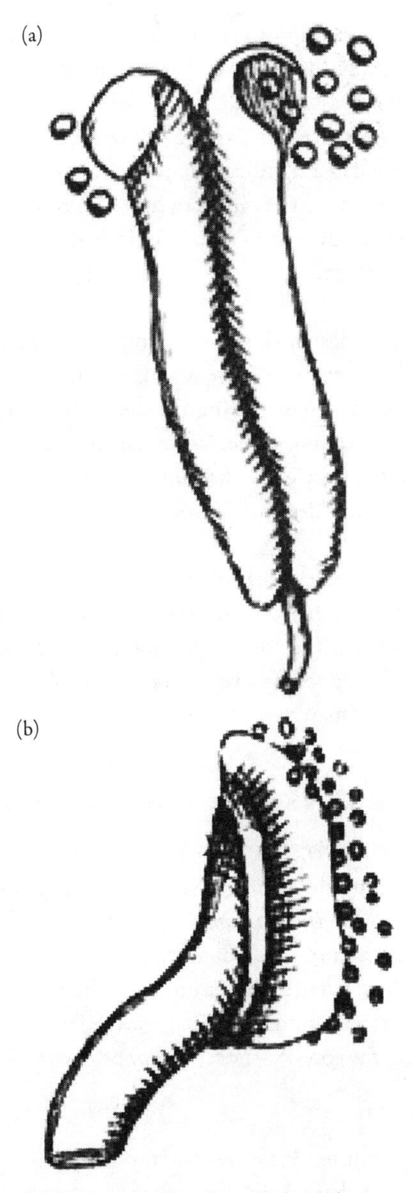

FIGURE 12.2 Illustrations of anthers and pollen from Malpighi's *Anatome Plantarum*. A. *Tritico Indico* ("Indian wheat" or corn (*Zea mays*) figure 185, table XXXI. B. Cucurbitae, figure 189, table XXXI.

likened flowers to women. For example, Grew described the calyx, which he called "the empalement," as a kind of girdle that supports and protects the internal parts of flowers:

> The Design of the *Empalement* is to be *Security* and *Bands* to the other two *Parts* of the *Flower*: To be their Security before its opening, by intercepting all extremeties of *Weather*: Afterwards to be their *Bands*, and firmly to contain all their *Parts* in their due and most decorous posture: so that a *Flower* without its *Empalement* would hang as uncouth and taudry, as a Lady without her Bodies [bodice].[16]

In this passage, Grew compares flowers to proper ladies, whose sepals act like a bodice, enabling them to maintain decorum by restraining their "Parts." The potential for "uncouth" sexuality is explicitly suggested.[17]

In 1676, five years after the preceding passage was written, Grew presented a second lecture on the flower that included a significant modification of his metaphor of the calyx as a bodice. Rather than emphasizing the calyx's tightness in restraining the internal parts of the flowers, Grew now stressed the calyx's ability to slacken during the swelling of the fruit, which he compared to pregnancy:

> they [the sepals] are aptly designed, not only to protect the *Leaves* [petals] of the *Flower* in the *Bud*; and after their *Expansion*, to keep them tite: but also, by receding, *Bredways*, one from another, and so making a greater *Circle*, gradually to give way fore the full *Growth* and safe spreading of the *Attire* [stamens]. Which, in regard it consists of *Parts* exquisitely tender, were it pinched up too close, would be killed or spoiled before it came to the *Birth*. As *Teeming Women* gradually slaken their *Laces* . . .[18]

Here we have two clear examples of the flower as a female projective system, first as the nubile young woman exhibiting her charms to best advantage, and second as a pregnant woman swelling to form a fruit. The crucial question is, why did Grew alter his metaphor? What had happened in the intervening five years to inspire Grew to change his 1671 trope for the calyx from a corset to a maternity dress?

NEHEMIAH GREW'S TRANSSEXUAL ATTIRE

"We next proceed to the *Flower*. The *Parts* whereof are most commonly three: the *Empalement*, the *Foliation*, and the *Attire*." So begins Grew's 1676 lecture on flowers, published in Book I, Chapter V of *The Anatomy of Plants,* the definitive compilation of his botanical papers published in 1682. As noted, the term "empalement" was Grew's idiosyncratic name for the calyx, whereas "foliature" was the term already in use for petals. Grew also invented the curious moniker "attire " for the frilly structures, mostly stamens, which were attached above the corolla. The term may be derived from a passage in Milton's *Lycidas*, written in 1637:[19]

> Bring the rathe [early blooming] Primrose that forsaken dies.
> The tufted Crow-toe [wild hyacinth], and the pale Gessamine [jasmine], The white Pink, and the Pansie freakt with jeat [black coal], The glowing violet.
> The Musk-rose, and the well attir'd Woodbine [honeysuckle].

In turn, Milton's metaphor "well attir'd" may have been inspired by the famous passage about lilies in the New Testament (Matthew 6:28–30):

> Consider the lilies of the field, how they grow; they toil not, neither do they spin:
> And yet I say to you, that even Solomon in all his glory was not arrayed like one of these.

> Wherefore, if God clothe the grass of the field, which today is and tomorrow is cast into the oven, shall he not much more clothe you . . .

When Grew referred to stamens specifically he used the term *seminiform attires.* Seminiform attires were composed of two structures, the *chives* (a sixteenth-century term meaning "threads") and the *semets* (summits), equivalent to the anthers. Like sixteenth-century botanists, Grew thought anthers resembled small grains of wheat, hence the term "seminiform," or seed-shaped. This in itself is suggestive, foreshadowing the discovery of the role of stamens in seed formation.

Grew used the term *florid attire* when he could not find a typical *seminiform attire* (consisting of stamens with conspicuous anthers) among the various "ornamental" structures associated with a flower—as in the case of the disk flowers of chicory, marigold, and other members of the Asteraceae or sunflower family. Because the five stamens of the disk flowers are fused, forming a sheath surrounding the ovary and style, Grew was unable to discern the individual anthers. Thus he mistakenly combined the style and stigma of disk flowers under the rubric "florid attire."

Already in 1671, however, Grew had begun to focus on the "seminiform attire," which, despite being dismissed by botanists as merely decorative, he now suspected played a more important role than either the sepals or the petals:

> The use of the Attire, how contemptibly soever we may look upon it, is certainly great. And although for our own use we value the Leaves of the Flower, or the Foliation, most; yet of all the three Parts, this in some respects is the choycest, as for whose sake and service the other two are made.[20]

Grew assumed that flowers were divinely created and that one of their functions was to delight humans. But he doubted that this was their only function:

> As for Ornament, and particularly in reference to the Semets, we may ask, If for that merely these were meant, then why should they be so made as to break open, or to contain anything within them? Since their Beauty would be as good if they were not hollow; and is better before they crack and burst open, then afterwards.[21]

Grew added the provision of "food for other animals" as another function of the attire, for he had observed "a vast number of little Animals in the Attires of all Flowers," a sign of God's love for even the smallest of his creatures:

> Go from one Flower to another, great and small, you shall meet with none untaken up with these Guests. . . . We must not think that God Almighty hath left any of the whole Family of his Creatures unprovided for; but as the Great Master, some where or other carveth out to all; and that for a great number of these little Folk, He hath stored up their Peculiar provisions in the Attires of the Flowers; each Flower thus becoming their Lodging and their Dining-Room, both in one.[22]

In this charming passage, Grew was merely echoing classical writers, but he ended his discussion with a teaser. In addition to its two "secondary" uses, Grew proposed that the Attire also had a "Primary and Private Use," beneficial to the plant itself. But on this subject Grew was not yet prepared to speculate: "[W]hat may be the Primary and Private Use of the *Attire* . . . I now determine not." [23] Grew's circumspection about the function of the Attire suggests that, in 1671, he may already have been thinking about the possible sexual role of stamens, but was waiting for confirmation of some kind before discussing it openly.

Precisely when Grew developed his sexual theory of plants is difficult to say. One of his inspirations, which he noted in his 1676 paper on flowers, was the discovery of hermaphrodism in snails. Although there has been some confusion in the literature about priority, there now seems little doubt that the discovery of hermaphrodism in snails was originally made by Grew's colleague at the Royal Society, the British naturalist John Ray, in 1660.[24] In his first book, describing new plant species growing around Cambridge, Ray noted that the deadly nightshade plant was susceptible to snails and slugs despite the plant's lethal effects on humans. He went on to add:

> In passing one may mention that they [snails] are hermaphrodite. That they alternately function as male and female by impregnating and receiving at the same time will be clear to anyone who separates them as they are having intercourse in Spring, although neither Aristotle nor any other writer on Natural History has recorded this fact.[25]

The Dutch anatomist Jan Swammerdam confirmed Ray's observations and published them in his *Historia Insectorum Generalis* (The Natural History of Insects) in 1669.

Several factors probably contributed to Grew's decision to announce his new sexual theory in 1676. The discovery of hermaphrodism in snails had, by this time, established a solid precedent that made hermaphrodism in plants seem less radical. Malpighi's first installment of *Anatome Plantarum* published in 1675 had failed even to mention the phenomenon, and Grew may have sensed that this was an opportune time to step out from under Malpighi's shadow. At the same time, he may have worried that, as a junior scientist, challenging the eminent Malpighi's one-sex model of flowers might raise some eyebrows. This may account for his private conversation with Sir Thomas Millington, the first physician to William and Mary and a prominent member of the Royal Society. He may have sought out Sir Thomas as a sounding board for his ideas. Grew's diffidence is on display in his description of their meeting:

> In discourse hereof with our Learned Sedleian Professor Sir Thomas Millington, he told me, he conceived, That the Attire doth serve as the Male, for the Generation of the Seed. I immediately reply'd, That I was of the same Opinion; and gave him some reasons for it, and answered some Objections, which might oppose them. But withall, in regard every Plant is hermaphroditic, or Male and Female, that I was also of Opinion, That it serveth for the separation of some parts, as well as the Affusion of others.[26]

Historians generally have taken a skeptical view of Millington's contribution to the sexual theory of plants. Other than Grew's comment, there are no written records of Sir Thomas's views on the subject.[27] Moreover, Grew's contemporary, the great British botanist

John Ray, gave Grew sole credit for the discovery in his *Historia Plantarum* (1686). The conversation between Grew and Sir Thomas undoubtedly took place, but it seems quite possible that Grew himself instigated it in order to enlist an influential member of the Royal Society as an ally should his sexual theory meet with any opposition from Malpighi loyalists.

In the event, Grew never actually contradicted Malpighi's menstruation model of stamens; rather, he built upon it. Echoing Malpighi, Grew stated that, initially, the "Attire" served to remove impurities from the sap. He then proceeded to compare the Attire (along with the "discharge" from the "womb") to "menses":

> And First, it seems, That the Attire serves to discharge some redundant Part of the Sap, as a Work preparatory to the Generation of the Seed. . . .
>
> Wherefore, as the *Seed-Case* is the *Womb*; so the *Attire* (which always stands upon or round about it) and those *Parts* of the *Sap* hereinto discharged; are, as it were, the *Menses* or *Flowers*, by which the *Sap* in the *Womb*, is duly qualified, for the approaching *Generation* of the *Seed*.[28]

Having fully endorsed Malpighi's menstruation model of the flower, Grew next introduced a startling innovation: the stamen began its life as the menses of the flower, but, after maturing, it functioned as the *male* sexual organ by releasing pollen:

> And as the young and early Attire before it opens, answers to the Menses in the Femal [sic]: so it is probable, that afterward when it opens or cracks, it performs the office of the Male.

In support of his theory, Grew cited the phallic appearance of seminiform Attires:

> This is hinted from the Shape of the Parts. For in the Florid Attire, the Blade does not unaptly resemble a small Penis, with the Sheath upon it, as its Praeputium. And in the Seed-like Attire, the several Thecae [sacs; anthers] are like so many little Testicles.

Indeed, Grew's interpretative drawings of "seed-like" (*seminiform*) attires do indeed have a phallic appearance, although he had to greatly exaggerate the width of the filament relative to the anther to achieve the effect (Figure 12.3).

In the case of the "florid attire," Grew regarded the sheath of fused stamens surrounding the central "blade" (his term for the style and stigma of disk flowers in members of the sunflower family) as analogous to a "praeputium" (foreskin). However, he also observed that the praeputium sometimes gave rise to "globulets" (pollen grains). This observation suggests that Grew had some inkling that the "sheath" in chicory flowers was in fact functionally related to the anthers ("thecae") of the "seminiform attires."

Completing the analogy of the attire to male genitals, Grew compared pollen grains to "Vegetable Sperme":

> And the Globulets and other small Particles upon the Blade or Penis, and in the Thecae, are as the Vegetable Sperme. Which, so soon as the Penis is exerted, or the Testicles come to break, falls down upon the Seed-Case or Womb, and so Touches it with a Prolific Virtue.

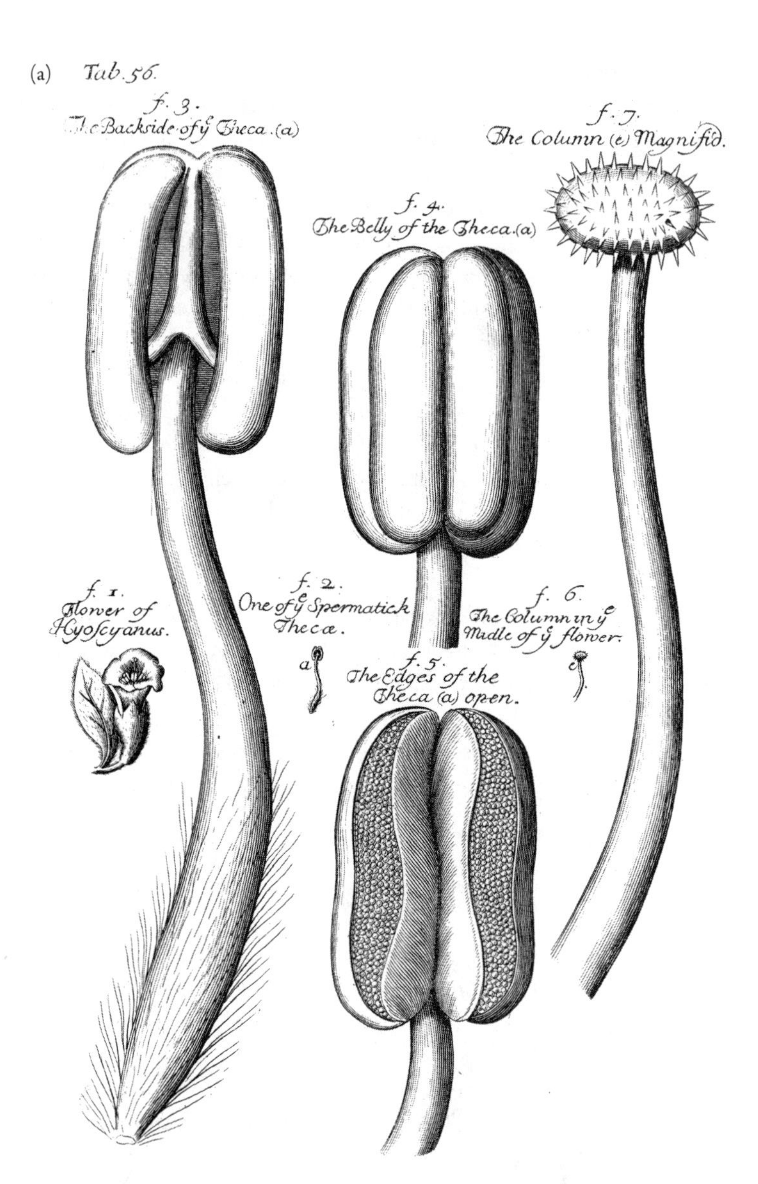

FIGURE 12.3 Grew's illustrations of "attires" (stamens). A. Seminiform; note the exaggerated thickness of the filaments. B. Florid. (Left) Disk flower of marigold consisting of an outer corolla of fused petals, an inner sheath of fused stamens, which Grew likened to a foreskin, and the stigma visible at the top of the sheath. (Right) The stigma and style of the marigold pistil removed from the sheath of stamens. Note that Grew's florid attire actually refers to the style and stigma of the disk flower. Thus, his phallic interpretation of its appearance is not only fanciful, it is based on an erroneous attribution of sex. (From Grew's *The Anatomy of Plants* [1682], figures 56 and 60.)

(b)

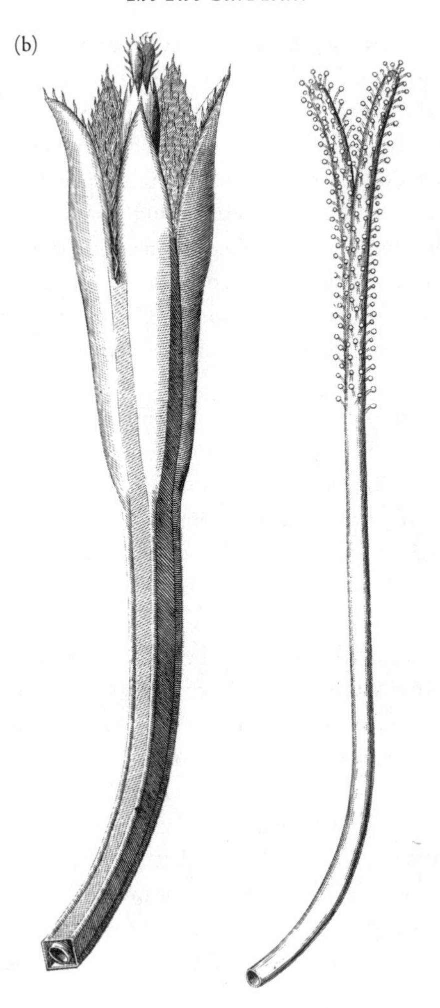

FIGURE 12.3 Continued

Grew had zero experimental evidence to support his conjecture that flowers were hermaphroditic. He argued entirely by analogy, citing hermaphrodism in snails:[29]

> That the same Plant is both Male and Female, may the rather be believed, in that Snails, and some other Animals, are such. And the Parts which imitate the Menses, and the Sperm, are not precisely the same: the former, being the External Parts of the Attire, and the Sap, which feeds them; the latter, the small Particles or moyst Powder which the External enclose.

Despite his observation about the phallic appearance of the Attire, Grew was quite evasive about its sexual identity. If the Attire initially functioned as the menses, did that mean that the sperm was the same as the menses? No, he states, the Attire ("the Parts which imitate the Menses") and the "Sperm" are "not precisely the same." Only the "small Particles or moyst Powder" are unequivocally identified by Grew as the male structure of the flower.[30]

Grew realized that his theory was incomplete. In Swammerdam's hermaphroditic snails, fertilization was brought about in the usual way among animals—by copulation—with the added feature of mutual penetration. But what went on in hermaphroditic flowers? Since copulation between two flowers was clearly impossible, the only conceivable mechanism seemed to be self-fertilization. Somehow, the "vegetable sperm" had to fertilize the young fruit from the same flower so that viable seeds could form inside it. The discovery of the pollen tube as the conduit for transmitting the sperm cells to the egg was still a long way off, so Grew devised a plausible mechanism to account for fertilization. He invoked simple gravity: the released pollen grains simply fell onto the "Uterus," thus transmitting a "Prolifick Virtue" to the ovary. To bolster his model, Grew resorted once again to animal analogies:

> And that these Particles, only by falling down on the Uterus, should communicate to it or to the sap therein, a Prolifick Virtue: it may seem the more credible, from the manner wherein Coition is made in some Animals; as by many Birds, where there is no Intromission, but only an Adosculation [impregnation by touching] of Parts: And so in many Fishes. Neither in others, doth the Penis ever enter any further than the Neck of the Womb.

But how could semen, which is a material substance, be transmitted to the interior of the Womb by mere touch? To circumvent this difficulty, Grew proposed that semen in animals, and "particles" in the case of plants, may only be the carrier of an invisible generative emanation that readily passes through the walls of the Womb:

> Nor doth perhaps the semen itself: or if it doth, it can by no means be thought, bodily or as to its gross Substance, to enter the Membranes, in which every Conception, or the Liquor intended for it, before any Coition, is involved; but only some subtle and vivifick Effluvia, to which the visible Body of the Semen, is but the Vehicle. And the like Effluvia may be easily transfused from the above said Particles into the Seed-Case or Womb of a plant.[31]

Nevertheless, Grew admitted that one should not press the animal analogy too far, for "if anyone shall require the Similitude to hold in every Thing; he would not have a Plant to resemble, but to be, an Animal."

Whatever qualms Grew may have had about the reception his lecture would receive, they appear to have been unfounded. By this time, Grew enjoyed a solid reputation, and Society members honored him by enthusiastically endorsing the publication of his papers. Grew's new sexual theory was welcomed by his colleagues as a plausible, if unproven, hypothesis about which little more could be said without experimental confirmation.

Support for the new sexual theory was slow in coming. The first published reference to Grew's hypothesis came ten years later, from his friend and fellow Society member, the British naturalist John Ray. Ray is best known as the first to develop a classification scheme, based on the concept of species, that was applicable to both plants and animals. In 1686, in the first of his three-volume treatise *Historia Plantarum*, Ray quoted Grew's description of the male sexual role of stamens approvingly. In support of Grew's theory, Ray cited the existence of species comprised of separate fruit-bearing and pollen-bearing individuals, such as date palms, willows, hops, nettles, spinach, and dog's mercury. However, he added the caveat that "this opinion concerning the use of pollen still requires confirmation." Eight years later, Ray seems to have dropped his earlier reservations. "In our opinion the pollen is equivalent to the sperm of animals," he wrote decisively, citing classical accounts of artificial pollination in date palms as evidence. Shortly afterward, Ray reviewed an account, written by the Italian botanist Paolo Boccone, of the Sicilian practice of artificial pollination in pistachio trees.[32] Ray cited the practice as further evidence of the sexual role of pollen.

It is curious that neither Grew nor Ray ever thought to test the sexual role of pollen experimentally. In the words of one historian, Grew's "clear and useful account of structure is sadly marred by guesses as to function and the propensity to put forth untested speculations."[33] But Grew, like Malpighi, was first and foremost a descriptive anatomist in the tradition of Vesalius. Since he had access to the best microscopes available, and much of what he described was new to science, it is not surprising that he dedicated the bulk of his efforts to plant anatomy.[34] Nor did any other contemporary British botanists have the time, resources, or motivation to test the new sexual theory of plants using the methodical approaches of experimental natural philosophy. Those members of the Royal Society who were in the best positions to test the new sexual theory of plants systematically showed no inclination to do so. In Tübingen, however, there was a young physician named Rudolf Jacob Camerarius who eagerly took up the challenge.

CAMERARIUS'S SCIENTIFIC INFLUENCES

Rudolf Jacob Camerer (Latinized to Camerarius) was born in 1665 into a distinguished family of doctors and apothecaries going back several generations.[35] His father was First Professor of Medicine at the University of Tübingen. Rudolf attended the University of Tübingen, earning his bachelor's degree in Philosophy and Medicine in 1679, and his master's degree three years later. He subsequently entered medical school, working under the supervision of Georg Balthasar Metzger, who was also Director of the University's Hortus Medicus. This garden had been established in 1663 by Duke Eberhard III of Württemberg, and Metzger had become its Director in 1681.

After Camerarius completed his medical degree in 1685, he traveled throughout Germany, Holland, England, France, and Italy for the next two years, returning to Tübingen in 1687, on the eve of the War of the Grand Alliance.[36] While in England, Camerarius may have paid his respects to his scientific idol, the British chemist Robert Boyle. Thanks to the Royal Society of London, the experimental movement in Britain was flourishing. Boyle had written voluminously on the limits of reason in comprehending Nature's laws, which he regarded as synonymous with God's laws.[37] Boyle, a fervent Anglican who was also strongly influenced by his sister's Puritan convictions, believed, as did many of his fellow Society members, that the experimental method was not only a tool for probing the laws of nature, but a means to decipher the divine plan and thus to grow closer to God.

If Camerarius visited Boyle in England (the record is unclear), he likely would have borne a letter of introduction from Johann Christoph Sturm (1635–1703), theologian, natural philosopher, and Professor of Mathematics at the University of Altdorf. Sturm had founded the

Collegium Curiosum sive Experimentale, a German scientific society, in 1672. Partly because Germany was not yet a unified nation, German scientific societies of the seventeenth and early eighteenth centuries tended to be smaller, more local, and less stable than their counterparts in Italy, France, and England.[38] Consequently, members of German scientific societies often sought dual memberships in the larger European societies. Sturm was a foreign member of the British Royal Society and published some of the findings of the Collegium Curiosum in the Royal Society's journal, *Philosophical Transactions*. Like Camerarius, Sturm was also a great admirer of Boyle and had written an article praising Boyle's essay, "Tractatus de Ipsa Natura" (A Treatise Concerning Nature). Like Boyle, Sturm believed that the human mind was incapable of grasping divine laws through reasoning alone. All three believed that progress in science could only be achieved through experimentation.

In his landmark *De Sexu Plantarum Epistola*,[39] Camerarius cited three authors whose views supported his own: Grew, Ray, and Sturm. Compared to Grew and Ray, Sturm is a relatively obscure figure in early modern science. Tübingen had no scientific society of its own, and, on the title page of the *Epistola*, Camerarius identified himself as a member of the "Kaiserlichen Academy (Nuremberg: Academieae Caesareo Leopold[ina] N[aturae] C[uriosorum])." Altdorf, where Sturm was on the faculty, is a suburb of Nuremburg, and Sturm was a member of the Nuremberg Academy.[40]

Although Camerarius cited Sturm along with Grew and Ray in the *Epistola*, he was somewhat equivocal about Sturm's contribution to the sexual theory:

> Since I anticipate going beyond them, I think it is worth taking the trouble to recall the reasons adduced, both my own and those of others, and in first place those that the famous Englishmen Nehemiah Grew and John Ray furnish in abundance, seeing that I have up to the present not become aware of any other authors or partisans of the affirmative case, unless it may be that which I glimpse in [the writings of] J. C. Sturm.

In other words, Sturm's support for the sexual theory was something that Camerarius thinks he may have "glimpsed" in Sturm's papers. Based on a reading of Mark Elvin's timely translations of these papers,[41] it is difficult to see anything that might be construed as anticipating the two-sex model of plants. While Sturm does state that conception in animals and plants are "exactly the same," he omits any mention of the role of pollen in plants. Quite the contrary, Sturm invokes the classical trope of the fertilizing spring breeze by stating that fertilization in plants occurs *in the soil* in response to an "animating breeze of flowers."[42] Sturm's idea of "fecundation" in the soil by an "animating breeze" is based on the notion that seeds are produced with their embryos already inside them. Sturm believed that fertilization in plants consisted of the *activation* of a pre-existing embryo after the seed had fallen to the ground. This school of thought is known as *Preformationism*, and it played an important part in the debate over sex in plants during the late seventeenth and early eighteenth centuries.

PREFORMATIONISM: OVISTS VERSUS SPERMISTS

Preformationism can be traced as far back as Genesis and the writings of Aristotle, but was first articulated as a scientific theory in the late seventeenth century by the French priest

Nicolas Malebranche and the Dutch microscopist Jan Swammerdam. Malebranche had written in 1674 that:

> We may say that all plants are in a smaller form in their germs. . . . It does not seem unreasonable to say that there are infinite trees inside one single germ, since the germ contains not only the tree but also its seed, that is to say another germ, and Nature only makes the little trees develop.[43]

Preformationists applied the ideas of the French philosopher Descartes, who posited a universe initially created by God but governed thereafter mainly by mechanical laws, to the question of embryo formation and growth in the womb. Preformationists fell into two camps: the *ovists* and the *spermists*. The ovists believed that the unfertilized egg already contained a miniaturized version of the mature organism within itself, needing only to be activated by the male semen. The first preformationists were all ovists who believed that the embryos of all the people who have ever lived on earth, and who will be be born in the future, were originally nested inside the ovaries of Eve, like an infinite series of Russian dolls.[44] The spermists, on the other hand, believed that all the preformed embryos originally resided in Adam's sperm. Spermism achieved a degree of notoriety in 1694 after the Dutch microscopist, Nicolas Hartsoeker, published a sketch of a homunculus (tiny human) curled up in the head of a sperm. Although the sketch was only meant to illustrate the *theory* of a homunculus, word spread that Hartsoeker had actually *seen* one in the microscope. Not to be outdone, several other microscopists soon claimed to have seen homunculi inside either a sperm or an egg—another example of the believing-is-seeing principle!

Ironically, it was Camerarius himself who provided the first unequivocal evidence refuting Sturm's preformationist theory of plant reproduction. As described in the *Epistola*, Camerarius, following the example of Malpighi, conducted microscope studies on the development of the embryo in the ovules of Papilionaceae (legume) flowers—most likely pea or bean—and obtained results that not only supported the importance of pollination in embryo formation, but also directly contradicted Malebranche's and Sturm's theories. Camerarius first removed the petals and stamens from the unopened floral bud at the earliest stages of its development to observe the initial "rudiment" of the immature pod prior to the release of pollen from the anthers. By holding the tiny pod up to the sun and then placing it under the microscope, Camerarius was able to see a row of tiny, greenish transparent vesicles (the unfertilized *ovules*) lining the young pod, each vesicle filled with a clear liquid. These ovules were the future seeds of the pod.[45] He then dissected flowers with intact petals and stamens that had been allowed to undergo pollination. Soon after the pollen had been released from the anthers, Camerarius could detect a new structure inside the ovule, which he described as an undifferentiated "green dot." Over time, the green dot increased in size and developed cotyledons (embryonic leaves) and a radicle (embryonic root).

The embryo, according to Camerarius, did not simply unfold into a young plantlet by enlargement, as the Preformationists claimed. As it increased in size, it underwent shape changes that transformed it from a round dot into an embryo. Camerarius was thus an early proponent of *epigenesis*, the idea that the embryo acquires its anatomical features incrementally after fertilization has taken place. Although Malpighi had made similar observations, Camerarius was the first to show that epigenesis in plants was dependent on pollination.

TÜBINGEN UNDER SIEGE: SCIENCE IN A TIME OF WAR

Upon his return to the University of Tübingen in 1687, Camerarius was awarded the honorary title of Professor Extraordinary of Medicine, and when Metzger, the Director of the Hortus Medicus, died in October of the same year, Camerarius was appointed as his replacement. Shortly afterward, he began the landmark experiments with monoecious and dioecious flowers that would culminate in the publication of *De Sexu Plantarum Epistola* in 1694, the year of Malpighi's death.

Given the frequent military incursions by French troops, it is remarkable that Camerarius was able to conduct his pollination studies at all. Forty years previously, Germany had been devastated by the Thirty Years War (1618–48). Tübingen's wealth in silver had been plundered and the university's faculty severely depleted. The end of the war brought needed relief, but a fresh conflict—the War of the Grand Alliance—broke out in 1688, just as Camerarius was beginning his experiments. Once again frequent evacuations wreaked havoc on the university, making serious scientific research all but impossible. The logistical nightmare of transporting laboratory equipment to safety whenever a French regiment encamped outside of town must have been a disheartening and frustrating experience for Camerarius and his colleagues.

Camerarius was related by marriage to the town hero credited with saving the city from pillage by the French armies.[46] Camerarius's younger sister, Agnes Susanna, was married to Johannes Osiander, a professor of theology and Chancellor of the university, and, in 1688, Osiander had successfully negotiated an agreement with French officers to spare the city in exchange for a large ransom. As a result, Tübingen and its university was spared, which may explain why Camerarius was able to carry out his pollination studies apparently without interruption for three years. By 1694, however, the war had become so disruptive that he was forced to terminate his experiments. At the end of *De Sexu Plantarum Epistola*, Camerarius complained bitterly about the misery inflicted on Tübingen during this period, which he described as "a time of war turbulence and public calamity in the fatherland."

The old Hortus Medicus, where Camerarius conducted the first controlled experiments demonstrating the sexual role of pollen, was located behind the Alte Aula (Old Auditorium) of the Tübingen campus, overlooking the tranquil, tree-shaded banks of the Neckar River. The Alte Aula is still there, but sadly, Camerarius's experimental garden has been paved over to make a parking lot.

CAMERARIUS'S HISTORIC APPEAL TO EXPERIMENT

Camerarius published his *Epistola* in 1694 as an open letter to his friend, Michael Bernhard Valentin, a professor of medicine in Giessen. Although some of his preliminary findings had already been published in the *Ephemerides Germanicae Academiae Caesalpeo Leopoldinae Naturae Curiosum*, the journal of the German scientific society, most of his experiments were summarized in the *Epistola*. It was printed as a small, palm-sized volume, 110 pages long, at Camerarius's own expense. As a private impression, the number of copies of the first and only printing was very limited, and only six copies of the *Epistola* survive. Camerarius probably sent a few copies to selected colleagues at various German universities, but international distribution was no doubt restricted by the war.[47] The only surviving copy at the University of Tübingen library was bequeathed in the nineteenth century by the estate of a Tübingen physician. A second copy, now missing, is believed to have been donated by Camerarius himself.

As discussed earlier, the British botanist John Ray, in his *Historia Plantarum* (1686), had endorsed Grew's sexual theory, noting that it would explain the presence of two separate "sexes" in various species. One of the dioecious trees not cited by Ray, but known to seventeenth-century gardeners, was the mulberry tree (*Morus nigra*). Early in his *Epistola*, Camerarius stated that plants fell into three categories based on the structure of their flowers: Class I (hermaphrodite), Class II (monoecious), and Class III (dioecious).[48] Camerarius was mulling over these categories when, in 1689, he noticed that the female mulberry tree growing in the Hortus Medicus was bearing fruit in the absence of any known male trees in the vicinity. Camerarius realized that if this were true it would contradict the sexual theory.

Boyle had written that it was the duty of the scientist to investigate the cause of all such discrepancies between the predictions of a theory and actual observation. To investigate the cause of the apparent discrepancy of the fruit-bearing female mulberry tree, Camerarius dissected the individual fruits (drupelets) under a microscope and discovered that they contained only aborted, empty seed cases—that is, seeds without embryos. He likened such sterile fruits to the "wind eggs" that chickens lay in the absence of a rooster. What at first appeared to be a discrepancy between the prediction of the sexual theory and observation, upon further investigation turned out to support the theory after all. Camerarius thus disclosed a valuable principle: the presence of fruits alone was not a reliable indicator of fertility. The true test of fertility was the production of viable seeds.

Armed with this important insight, Camerarius set about designing his first systematic experiments testing the sexual theory. According to John Ray, one of the species reputed to consist of two sexes was the common European wildflower *Mercurialis annuis* (Dog's Mercury). In May of 1691, Camerarius selected—"from an abundant growth of delicate little plants that were pushing up their shoots in a carefree fashion"—two females and transplanted them to a remote part of his garden—"far away from any other *Mercurialis*."[49] Although the isolated female plants produced abundant fruits, these fruits abruptly withered as soon as they had expanded to about half their normal size, and none produced viable seeds. Camerarius published these results in the *Ephemerides* in 1691 as a short communication titled "Wind-eggs in *Mercurialis*."

For the next three years, Camerarius relentlessly pursued his investigations into the sexual theory. In addition to *Mercurialis*, he experimented with two other dioecious species, *Spinacia* (spinach) and *Cannabis* (hemp), isolating the female plants from the males and observing whether or not they produced viable seeds. As in the case of *Mercurialis*, all the seeds produced by the isolated *Spinacia* females were sterile. In the case of *Cannabis*, however, the results were less clear-cut. Following a similar protocol to the one he had used with *Mercurialis* and *Spinacia*, he transplanted three female *Cannabis sativa* plants at an early stage of flowering to his garden, which did not contain any males. At the end of the flowering season, Camerarius was gratified to find that, as expected, the three female *Cannabis* plants were filled with sterile seeds. But his happiness was cut short by the discovery that among the mostly sterile seeds were "numerous fertile seeds"—he does not say how many—"at which," he confessed rather endearingly, "I must admit I was quite upset."

Whenever Camerarius's confidence wavered, he turned to the philosophical writings of Robert Boyle for solace. Following the distressing results with *Cannabis*, he seized upon a particular passage admonishing young scientists never to trust the outcomes of single experiments:

> Experiments on the basis of which you strongly desire to erect theories . . . should be verified repeatedly, and with the greatest of circumspection; nor should too much credence be given to those that you accomplish perfectly no more than a single time.[50]

Buoyed by Boyle's reassuring words, Camerarius determined to repeat the *Cannabis* experiment the following spring, but, when spring arrived, he was stricken with a serous illness that lasted until the summer. By the time he recovered, it was too late in the growing season to begin a new experiment. Fortunately, Camerarius discovered an isolated group of six *Cannabis* plants growing in his garden, three of which were females.[51] The plants were a little taller than he would have preferred, but none of the anthers of the three male plants had split open yet, which meant that there was still time to conduct his experiment. He quickly cut down the males and waited anxiously to see how the females would be affected. Alas, he obtained the same result as before! Although each of the female plants contained "an immense quantity of infertile seeds," there were in addition, he sadly confessed, "fertile ones that were not so few in number."

Despite the equivocal results with *Cannabis*, Camerarius persevered with other species. He conducted similar experiments with two monoecious plants: castor bean (*Ricinus*) and maize (*Zea mays*). After removing the male flowers from castor bean plants, he waited to see what would happen to the female flowers. To his immense relief, not one pistil developed into a mature fruit:

> In the second class of plants, in which the male flowers are separated from the female on the same plant, I have learned by two examples the deleterious effect produced by removing the anthers. When I removed the male flower buds of *Ricinus* before the anthers had expanded, and prevented the growth of the younger ones, but preserved the ovaries that were already formed, I never obtained perfect seeds, but observed empty vessels, which fell finally to the ground exhausted and dried up.[52]

Turning to maize, Camerarius identified the tassels at the top of the stalk as male inflorescences and the cobs below as females. Camerarius also observed the silky styles attached to each kernel and noted that they functioned to receive the pollen:

> In this cereal the protruding tassel at the end of the stalk is too well known to need a detailed description. After the wilting and drying of these tassels without producing any seeds themselves, farther down those thick cylindrical cobs are taking shape, which with their grains are covered by some leaves and protruding from each grain a long thread [silk], which spread like a tail and which receive the pollen.[53]

Camerarius tested his pollen-catching hypothesis for the silk by removing it from two of the cobs soon after they appeared, and noted that no fertile seeds formed in the ear as a result:

> I carefully cut off the stigmas ["silk"] of maize that were already dependent, in consequence of which the two cobs remained entirely without seeds, though the number of abortive seed capsules was very great.[54]

The crucial experiment, however, was to deprive an intact cob of pollen by removing the stamens and observing the effect on kernel production. In maize, it is easy to remove all the stamens of a plant at once by cutting off the entire tassel before it matures. When Camerarius performed the experiment on isolated plants, the results were dramatic:

> when the unfolding tassels were cut away before they could open, two cobs appeared that were devoid of any seed, containing a great number of empty seed-capsules instead.[55]

However, Camerarius conscientiously reported that a third cob contained eleven fertile seeds, despite the absence of tassels. Thus, as in the case of *Cannabis*, the results with maize were less than definitive. Although eliminating the male flowers resulted in sterile seeds in the vast majority of cases, the presence of a smaller number of fertile seeds in *Cannabis* and maize appeared to contradict the sexual theory. Summoning courage once again from the example and words of Robert Boyle, Camerarius, refused to despair:

> It is not, however, the case that I should lose hope even if certain experiments may have continually failed to respond to my desires, given that there is a philosopher reminding me that I have numerous comrades who have shared this same fate.[56]

Indeed, a few years later, in 1698, Camererius discovered the likely cause of the unexpected and unwanted fertile seeds in his experiments with *Cannabis* and maize, which he published in the *Ephemerides*.[57] He observed that the female plants of spinach, nettle, and other dioecious species occasionally produced either male or hermaphroditic flowers. The rare production of a few male or hermaphroditic flowers among Camerarius's "female" *Cannabis* plants could well account for the presence of the fertile seeds.[58] Similarly, hermaphroditic flowers occasionally occur in both the tassels and the ears of monoecious maize, which would allow the production of seeds even in plants whose tassels had been removed.[59] As we shall see in later chapters, eighteenth- and nineteenth-century critics of the *Epistola*, apparently unaware of his 1698 paper, continued to argue that the presence of even a small number of fertile seeds in Camerarius's emasculated *Cannabis* and maize plants constituted proof that the sexual theory was false.

Although being completely forthcoming about results that did not agree with his hypothesis, Camerarius nevertheless concluded that the sexual theory was supported by the vast majority of his data. His summary of the sexual theory was therefore the first straightforward formulation of the two-sex model of plants:

> In the vegetable kingdom, no production of seeds, the most perfect gift of nature, the general means for the maintenance of the species, takes place, unless the anthers have prepared beforehand the young plant contained in the seed. It appears, therefore, justifiable to give these anthers a nobler name and to ascribe to them the significance of male sexual organs, since they are the receptacles in which the "seed" itself, that is that "dust" which is the most subtle part of the plant, is secreted and collected, and from which it is later distributed. . . . It is equally evident that the ovary with its style represents the female sexual organ of the plant, which must provide, with all its strength, a mother's support for her new-born fetus, which she conceives and for which she cares.[60]

Camerarius was adamant that plant sex should no longer be considered a metaphor or figure of speech, but was literally true. Thus, in the case of monoecious and dioecious plants:

> They behave indeed to each other as male and female, and are otherwise not different from one another. They are thus distinguished with respect to sex, and this is not to be understood as it is ordinarily done, as a sort of comparison, analogy, or figure of speech, but is to be taken actually and literally as such.[61,62]

When citing Malpighi and Grew's contributions in his introduction, Camerarius largely ignored both the former's elaborate menstruation model and the latter's transsexual theory of the stamen. As a relatively obscure young physician from a provincial German town, Camerarius may have felt freer than his more distinguished and well-connected European colleagues to jettison the ancient one-sex model. On the other hand, Camerarius did not make a complete break with the past. He retained the old idea expressed by Laurenberg, Ray, Malpighi, and Grew that the function of the petals was to purify the sap for the nourishment of the developing seed. Thus he writes that after the embryo is conceived,

> [n]othing is left for the petals to do other than the service that is customarily assigned to them collectively, that is, to remove impurities from the sap and return it in purified form to . . . the delicate seed containers.[63]

At the end of his *Epistola,* Camerarius included a poem, written in a flowery Latin style by an anonymous poet, celebrating the new sexual theory. The English translation quoted here is taken from Patrick Blair's *Botanick Essays* (1720).[64] The poet begins his exposition with the classical trope of spring winds (Favonius/Zephyr) causing the flowers to open, but here for the first time the wind's fertilizing effect is directly associated with the transport of pollen from the stamens to the pistil ("pointal"), which is personified as a blushing bride:

When Winter's gone, and Spring succeeds,
With gentle Blasts Favonius blows,
The opening Flow'rs each Sex disclose,
    And promise future Seeds.

The Stamina with Meal abound,
And when the gentle Zephyrs blow,
They from their Double Summits throw
The Golden Dust around.

Which born by the propitious Winds,
About the Female Vessels spreads,
And round the Pointal's [pistil's] hollow Beds,
A glad Reception finds.

No anxious Thought their Love destroys,
They want no sable Night, to hide
The Blushes of the yielding Bride,
Fill'd with tumultuous Joys.[65]

As we shall see in later chapters, the two-sex model of plants inspired many versifiers, both for and against the new sexual theory.

NOTES

1. In *Novum Organum* (1620), Francis Bacon promulgated the view that philosophy should employ inductive rather than deductive reasoning; that is, general principles should be based on scientific data, not the reverse.

2. Hunter, M. (1989), *Establishing the New Science: The Experience of the Early Royal Society*. The Boydell Press, pp. 261–278.

3. The identity of the inventor of the compound microscope is uncertain. The Dutch father-son team of Hans and Zacharias Janssen, who were spectacle-makers, claimed credit for the invention, but, according to another theory, it was Galileo. Galileo's telescope had been based on several similar instruments, one of them produced by Hans Janssen, that appeared in the Netherlands in 1608. Galileo improved on the original design and, in 1610, had used his telescope at close range to examine insect parts. The term "microscope" was coined by one of Galileo's colleagues at the Accademia dei Lincei, Giovanni Faber, in 1625. It is possible that Hans Janssen modified his telescope for use as a microscope independently at around the same time.

4. The lynx, a symbol of the new science of microscopy, was reputed to have exceptionally sharp vision.

5. Piccolino, M. (1999), Marcello Malpighi and the difficult birth of modern life sciences. *Endeavour* 23:175–179.

6. Adelmann, H. (1966), *Marcello Malpighi and the Evolution of Embryology*, 5 volumes. Cornell University Press.

7. Based on the definition provided by Joachim Jung (1587–1657), the term *flos* referred to "the somewhat tender part of the plant, conspicuous by reason of its color or form or both, which adheres to the rudiment of the fruit." A "perfect" *flos* consisted of petals, stamens, and style, but not the ovary, which was equated with the fruit.

8. Elvin, M. (2015), *Transferring the Impulse of Life: The Scientific Proof of Sexual Reproduction in Plants*. Forthcoming. We are indebted to Professor Elvin, Professor Emeritus of Chinese

History at Australian National University, and Emeritus Fellow at St. Antony's College, Oxford, for providing access to his translation prior to publication.

9. Bloodletting, the treatment of choice for various ailments in the ancient world, was based on the same principle of maintaining the proper balance of the humors.

10. Leviticus 15:19–23.

11. Malpighi, *Anatome Plantarum*, p. 55. All quoted passages from Malpighi's *Anatome Plantarum* were translated from the Latin by Professor Gildas Hamil, Department of History, University of California, Santa Cruz.

12. Curiously, Malpighi also refers to "fungi" as participating in this process. Peering through his microscope, Malpighi apparently mistook the magnified glandular hairs on the styles of many flowers for tiny mushrooms and included them among the floral structures that purified the sap.

13. Malpighi, *Anatome Plantarum*, p. 56.

14. Descartes, R. (1649/1989), *The Passions of the Soul: Les Passions De l'Âme*, trans. Stephen Voss. Hackett Publishing Company.

15. Hunter, *Establishing the New Science*, pp. 261–278.

16. Grew, Nehemiah (1682), *The Anatomy of Plants*, Book I, p. 35 (first presented in 1671).

17. Many monocotyledons, including tulips, have undifferentiated "tepals" in place of sepals and petals. Later, Grew acknowledged that tulips lack sepals, but argued that the "fat and firm" petals of tulips make them "sufficient to themselves."

18. Grew, *The Anatomy of Plants*, Book II, p. 163. "On the Function of the Empalement," first presented in 1676.

19. LeFanu, W. (1990), *Nehemiah Grew: A Study and Bibliography of His Writings*. St. Paul's Bibliographies, Winchester/Ominigraphics.

20. Grew, *The Anatomy of Plants*, Book I, ch. V, p. 39.

21. Ibid.

22. Ibid.

23. Idem., p. 40.

24. Örstan, A. (2010), John Ray's hermaphrodite snails on their 350th anniversary. *Mollusc World* 23:4.

25. Ray, J. (1660), *Catalogus Plantarum circa Cantabrigiam Nascentium*. The quote is from the 1975 English translation, *Ray's Flora of Cambridgeshire* by A. H. Ewen and C. T. Prime (Wheldon & Wesley). Ray's original is available from Google Books: http://tinyurl.com/yk2j55t.

26. Grew, *The Anatomy of Plants*, Book IV, ch. V, p. 171.

27. See Pulteney, R. (1790), *Historical and Biographical Sketches of the Progress of Botany in England from Its Origin to the Introduction of the Linnaean System*. Vol. 1, ch. 25 (History of the Discovery of the Sexes in Plants), T. Cadell in the Strand, pp. 333–334. More recently, we were unable to find any mention of the sexual theory of plants in Millington's letters housed at the British Library in London.

28. Grew, *The Anatomy of Plants*, Book IV, ch. V, p. 172.

29. Ibid.

30. Grew's obscure insight anticipated by some 200 years the discovery of the phenomenon of *alternation of generations* in plant life cycles by Wilhelm Hofmeister. The stamen is part of

the sporophyte generation and is genetically distinct from the pollen grains, which belong to the gametophyte generation.

31. Grew, *The Anatomy of Plants,* Book IV, ch. V, p. 173.

32. Boccone, P. (1697), *Museo di piante rare della Sicilia, Malta, Corsica, Italia, Piemonte e Germania con figure 133 in rame, Venetiis, apud Ioannem Baptistam Zuccarum.*

33. L. C. Miall (1912), *The Early Naturalists, Their Lives and Work, 1530–1789.* Macmillan, p. 168. Cited by LeFanu, William (1990), *Nehemiah Grew: A Study and Bibliography of His Writings,* St. Paul's Bibliographies, Winchester, Omnigraphics, Inc., p. 17.

34. Hunter, *Establishing the New Science,* pp. 261–278. Hunter has pointed out that Grew's finances during the period of his research for the Royal Society were extremely tenuous because of the unreliability of voluntary support from the Society's members. Because of this, he oscillated between botany and medicine, finally opting to pursue the more lucrative profession of medicine. The need to support himself may explain why he abandoned botanical research soon after completing his anatomical studies and never tested his hypothesis concerning the sexual role of pollen.

35. R. J. Camerarius is unrelated to the famous herbalist from Nuremberg, Joachim Camerarius the Younger (1534–98).

36. The War of the Grand Alliance (1688–97), also known as "The War of the League of Augsburg" and "The Nine Years War," pitted the armies of Louis XIV against those of the principalities of Germany and their European allies.

37. Sargent, R.-M. (1995), *The Diffident Naturalist.* University of Chicago Press; Wojcik, J. W. (1997), *Robert Boyle and the Limits of Reason,* Cambridge University Press.

38. McClellan, J. E. (1985), *Science Reorganized.* Columbia University Press, p. 114.

39. The complete citation is: Academiae Caesareo Leopold. N. C. Hectoris II. Rudolphi Jacobi Camerarii, Professoris Tubingensis, Ad Thessalum, D. Mich. Bernardum Valentini, Professorem Giessensem Excellentissimum (1694), *De Sexu Plantarum Epistola,* Tubingae: Rommeius.

40. Personal communication, Dr. Gerd Brinkhus, Library of the University of Tübingen.

41. Elvin, *Transferring the Impulse of Life.*

42. From: Sturmius, M. Johann Christophorus (1687), *Diducendi alias uberius Argumenti De Plantarum Animalium; Generatione.* Altdorff, Literis Schönnerstaedtianis. p. 14–15. Translated by M. Elvin.

43. Malebranche, *The Search after Truth.*

44. Pinto-Correia, C. (1997), *The Ovary of Eve: Egg and Sperm and Preformation.* University of Chicago Press, p. 6.

45. Prévost, A.–M. (1965), Rapprochement entre l'Epistola de sexu plantarum de R. J. Camerarius (1694) et les Observations sur la structure et l'usage des principales parties des fleurs de Geoffroy le Jeune. *Compt. Rend. Séances Acad. Sci.* 261:2045–2048.

46. Based on marriage records of the Camerer family located at the Tübingen Rathaus (City Hall).

47. Sturdy, D. J. (2002), *Fractured Europe, 1600–1721.* Blackwell Publishers.

48. The terms "monoecious" and "dioecious" were introduced by Linnaeus many years later. Camerarius called them simply Class II and Class III.

49. Elvin, *Transferring the Impulse of Life.* Section 1, Excursis 4, p. 269.

50. Idem., Section 25. According to Mark Elvin, the quotation, which was slightly abridged by Camerarius, is derived from a Latin translation of four of the essays from Boyle's *Certain Physiological Essays* (Herringman, 1661).

51. *Cannabis* cultivation in Germany can be traced back to the Neolithic period. It was an important herb in the German pharmacopoeia from the medieval period to the early twentieth century.

52. von Sachs, J. (1890), *History of Botany (1530–1860)*, trans. Henry E. F. Garnsey, revised by Isaac Bayley Balfour. Clarendon Press. Slightly modified for clarity.

53. Ibid.

54. Ibid.

55. Translated by F. S. Bodenheimer (1958), *The History of Biology: An Introduction*, Dawson & Sons, London p. 285. Modified for clarity.

56. Elvin, *Transferring the Impulse of Life*. Section 25.

57. Camerarius, R. J. (1698), De spinachia & urtica androgynis. *Ephemerides Germanicae, Decuriae Tertiae, Annus quintus et sextus*, pp. 484 et seq.

58. The existence of intermediate sexual types has now been shown to be more common than once thought. In addition to hermaphroditic, monoecious, and dioecious plants, a number of other intermediate sexual states exist in angiosperms. According to Ainsworth (2000), these include "gynodioecy, in which populations are composed of female and hermaphroditic plants, androdioecy, in which populations are composed of male and hermaphroditic plants, trioecy, in which populations are composed of male, female and hermaphroditic plants, gynomonoecy, in which plants carry female and hermaphroditic flowers, andromonoecy, in which plants carry male and hermaphroditic flowers, and trimonoecy, in which plants carry male, female and hermaphroditic flowers." Ainsworth, C. (2000), Boys and girls come out to play: The molecular biology of dioecious plants. *Annals of Botany* 86:211–221.

59. Irish, E. E., and Nelson, T. (1989), Sex determination in monoecious and dioecious plants. *The Plant Cell* 1:737–744.

60. Original translation from von Sachs, *History of Botany*. With elements from with a modern translation by Mark Elvin (2015).

61. Translation from M. Leapman (2000), *The Ingenious Mr. Fairchild*. St. Martin's Press, p. 30.

62. Camerarius was well aware that the vast majority of plants were hermaphrodites, which "impregnate themselves" and "give birth from themselves to that which they have conceived," a phenomenon he regarded as "altogether extraordinary." Nevertheless, he notes, it is the smaller group of monoecious and dioecious plants that led to the discovery of sex in the hermaphrodites. Translations from Mark Elvin (2015).

63. Modified from a translation provided by Mark Elvin.

64. A more literal, modern translation by Mark Elvin (2015) provides a better idea of the scientific content of the poem:

The delicate bud, breathed open by Zephyr
sets its stamens for forthcoming seeds in position,
intertwining in marriage the separate sexes.
Then it is that the twinned anthers split,

as accustomed, apart. Scattered forth to all sides
is their "flour"—abundant, and bright yellow, pollen—
    nothing else than a fine-textured dust that, once rising,
        is borne through the air by the currents that follow.

Here let there be semen! Here masculine fluid
—the more purified fraction strained off from the sap—
    that is sprinkled nearby on the mouths of the tubes
        topping frail, minute, styles on which it is scattered

in lavish amounts! Females' sexual organs
once watered through conduits, the spouses in public
    thus couple together; nor in hearts that are floral,
        will embarrassment stir at *this* kind of love.

65. The poem continues for 12 more stanzas.

13

Plant Nuptials in the Linnaean Era

AFTER CAMERARIUS HAD so elegantly demonstrated the potential of experiments to elucidate the sexual role of pollen, one would have thought that botanists of the succeeding century would rush to follow his lead. But with a few exceptions—most notably Joseph Koelreuter—this was not to be the case. Surveying the reactions to the new sexual theory of plants, nineteenth-century plant physiologist Julius von Sachs mordantly summarized the confused responses of eighteenth-century botanists, seemingly caught off-guard by the sudden paradigm shift:

> some simply denied the new theory, many adopted it without understanding the question, others formed a perverse or distorted perception of it under the influence of reigning prejudices, while others again sought to appropriate to themselves the merit of the real discoverer.

"There were but few," Sachs concluded, "who with a right understanding of the question advanced it by new investigations."[1]

To some extent, the anemic response may have been due to the limited circulation of *De Sexu Plantarum Epistola*, so limited that only a few early eighteenth-century botanists had actually read the original. However, the main reason for the lack of experimental followup was the fact that the primary preoccupation of botanists during the eighteenth century was not experimental botany, but taxonomy, the supreme practitioner of which was the prolific Swede, Carolus Linnaeus.

There were other factors at work as well. In England, for example, the early supporters of the sexual theory were mainly gardeners and horticulturalists—practical men who recorded their observations as time permitted. Nevertheless, these amateur botanists achieved several firsts in the new field of pollination biology, including the discovery of bee pollination and the production of the first artificial plant hybrid. Some eighteenth-century British gardeners clearly understood that by demonstrating that plants could hybridize with one another, they were not only confirming the sexual theory, they were also opening the door to the creation of lucrative, new horticultural varieties.

We begin our narrative of the sexual theory during the Linnaean era with some of his important predecessors in Paris. Although French botanists, unlike their English counterparts, enjoyed considerable support from the French Academy, the environment in which they worked was fiercely competitive, especially among younger scientists, fostering a desire for quick results in order to gain fame.

PLANT SEX AT THE FRENCH ROYAL ACADEMY OF SCIENCES

Research in plant biology at the French Academy during the early eighteenth century had both descriptive and experimental components. Physiological experiments had focused primarily on the problem of the ascent of sap, inspired by Harvey's model for the circulation of the blood.[2] Significant experimental work was also conducted on reproduction in nonseed plants, including algae, fungi, bryophytes (mosses, liverworts, hornworts), and ferns. Joseph Pitton de Tournefort studied the origin of spores in ferns, bryophytes, and fungi, mistaking them for seeds. René-Antoine Ferchault de Réaumur was investigating sexual reproduction in the marine alga, *Fucus*, identifying what he called male and female "flowers" as well as "seeds." Jean Marchant was carrying out parallel studies on the liverwort species that later would be named in his honor: *Marchantia*.

Botanists at the Academy kept up with the publications of the Royal Society of London and were aware of Malpighi's anatomical description of the flower as well as Grew's sexual theory. In a 1683 article in the *Memoirs* of the Academy, Marchant further elaborated Malpighi's uterine analogy of the flower, comparing it point by point to the mammalian female reproductive system:

> the style is to the flowers what the [Fallopian] tubes of the womb are to the animals, and it contains in its membranes the siliques which take the place of the chorion and the amnios, providing the air which is necessary for perfecting the seed which links to the placenta through its umbilical cord.

Significantly, Marchant omitted any mention of Grew's comparison of the stamen to a penis. His reticence about Grew's two-sex model was shared by his colleague, Tournefort, the newly appointed Professor of Botany at the Jardin du Roi. Tournefort continued to espouse the classical interpretation of pollen as a waste product. He regarded the stamen as a mere "vessel of excretion," the vilest part of the flower, unworthy of serious study. Tournefort's stubborn opposition to the sexual theory was to dominate French botany for the next twenty years. His influence was felt outside of France as well. For example, Giulio Pontedera, the Director of the botanical garden and Professor of Botany at Padua, Italy, in his *Anthologia* of 1720, reiterated Tournefort's views on pollen as a waste product and described the male flowers of dioecious plants as useless appendages. Consistent with an overall female interpretation of flowers, he compared nectaries to breasts, which nourish the seed.

The most enduring contribution to botany of the influential and aristocratic Joseph Pitton de Tournefort was the concept of genus.[3] In his *Institutiones rei Herberai* (1700) he defined twenty-two classes of flowering plants based primarily on the structure of the corolla and 698 genera using the fruit as the main taxonomic criterion. Earlier, in 1694, he had published his important three-volume treatise, *Elements of Botany*, which hewed closely to Malpighi's uterine model of the flower.

Tournefort's opposition to the sexual theory arose from three main sources: his deference to Malpighi, his own research on spore production in the nonseed plants, and his failure to confirm the existence of sex in date palms experimentally. Tournefort believed that the spores of ferns, bryophytes, and fungi were equivalent to the seeds of flowering plants, and, because he had determined that such "seeds" (spores) were produced without any visible sexual structures, he inferred that they must have been produced asexually. If the seeds (spores) of fungi were produced without sex, he reasoned that the seeds of angiosperms and gymnosperms must also be produced asexually.

Although he rejected the sexual theory in principle, Tournefort recognized that, according to the classical literature, pollen did seem to play a role in fruit development in date palms, and he was also aware of the role of the fig wasp in the caprifig–edible fig interaction, but he adopted the views of Theophrastus regarding the asexual mechanism by which this occurred (discussed in Chapter 8). Nevertheless, shortly after being appointed Professor of Botany at the Royal Garden in 1683, he traveled to Andalusia, Spain, famous for its date palm orchards, where he attempted to carry out artificial pollination experiments. Unfortunately, the project ended in failure, for, according to Edward Lee Greene, "he made unsuccessful efforts to either establish or disprove the reputation the palms had long had with some botanists of being endowed with sexuality; but he came away knowing no more about the matter than Theophrastus had known two thousand years before."[4]

Although Tournefort remained adamantly opposed to the sexual theory, two of his students at the Jardins du Roi, Sébastien Vaillant and Claude-Joseph Geoffroy, made it the primary focus of their research. Claude-Joseph Geoffroy's older brother, Étienne-François Geoffroy, in his 1704 medical thesis on spermism versus ovism, was actually the first to present a lecture on Camerarius's pollination studies (without citing Camerarius) to the French Academy. Seven years later, Claude-Joseph presented a lecture to the Academy purportedly on his own research that was nearly identical to the experiments of Camerarius—once again, without citing Camerarius.[5] This provoked Sébastien Vaillant, a former surgeon and student of Tournefort's, who was then working on a new system of plant classification based on the sexual theory, to charge the younger Geoffroy with plagiarism in a sensational—and notorious—public lecture.[6]

SÉBASTIEN VAILLANT AND THE "INNOCENT PLEASURES" OF PLANTS

On the morning of June 10, 1717, at around 7 AM, Sébastien Vaillant strode briskly into the auditorium at the Jardin du Roi to deliver the opening lecture of the annual course in botany with the deceptively dull title *Discourse on the Structure of Flowers, Their Differences and the Function of Their Parts.* The occasion marked the opening of the Royal Garden at its magnificent new location on the banks of the Seine, and, even at this early hour, the mid-sized hall was packed with about 600 people from all levels of society, including 200 students from the medical school.[7] The opportunity to give the lecture had arisen when the regular professor, Antoine de Jussieu, who was also director of the Jardin du Roi and Vaillant's superior, had asked Vaillant, who was seventeen years his senior and whose primary job was to maintain the herbarium and serve as "Assistant Demonstrator," to lecture in his place while he traveled in Spain. To the largely self-taught Vaillant, who lacked de Jussieu's academic credentials and family connections, it was a splendid opportunity, and he intended to make the most of it.[8]

Sebastien Vaillant's rise in the scientific world had been improbable. The son of a tradesman, he had shown a precocious interest in plants. But he was also a musical prodigy who, at the age of eleven, had become organist at the Cathedral St-Maclou in Pointoise. Later, in exchange for room and board, he became organist for a nursing order of nuns, where he spent time watching the nuns treat patients at the local hospital. Borrowing books from the surgeons, he immersed himself in the medical literature. After several years of observing and studying on his own, he was accepted as an apprentice surgeon, and, by the age of nineteen, he was hired as an assistant surgeon in the Norman city of Evreux.[9]

Around 1691 Vaillant moved to Paris, hoping to be hired as a surgeon at the Hôtel-Dieu, the oldest hospital in Paris, but after arriving he attended a plant demonstration by Tournefort at the Jardin du Roi and became convinced that botany was his true calling. For the next few years, he supported himself as a surgeon in the countryside around Paris while taking every class of Tournefort's he could attend. Eventually, with Tournefort's help, he obtained the position of personal secretary to Dr. Guy-Crescent Fagon, Director of the Jardin du Roi. Shortly after Tournefort's death, he was appointed Assistant Demonstrator. Despite his unorthodox background, Vaillant had managed to transform himself into a first-rate botanist at one of the world's great botanical gardens.

Vaillant was highly critical of Tournefort's system of classification based on the corolla and fruit in *Institutiones Rei Herbariae,* and, at the time of his opening lecture, he was preparing his own *magnum opus, The Botanicum Parisiense*, in which he planned to introduce a new system of plant classification based on the sexual parts of flowers—a forerunner of the Linnaean system.

By 1717, the sexual theory of flowers was no longer novel to members of the French Academy, who had heard two previous lectures on the same subject—the thesis defense of Etienne-François Geoffroy in 1704, and a faculty research lecture by Claude-Joseph Geoffroy in 1711. Both of these lectures had excited considerable interest, but the racy version of the sexual theory Vaillant was about to present would bear little resemblance to the dry, scientific recitations of the Geoffroy brothers. As pointed out by Jacques Rousseau, society in early-eighteenth-century France was surprisingly puritanical and marked by an emphasis on good manners and refinement among the higher social classes.[10] Publicly referring to sexual organs by name and expounding on their functions was simply not done in polite company. Applying sexuality to plants, unless it was couched in the driest possible terms, bordered on the indecent. But Vaillant had entered the Academy through the back door and seemed to relish the role of interloper. Besides, he knew that the medical students in the audience, chafing at the strictures imposed by their elders, would respond enthusiastically to all things radical, especially concerning the taboo subject of sex.

Vaillant's portrait by the Dutch engraver Jacobus Houbraken shows a handsome face framed by the rich curls of a wig cascading over his shoulders. Now, attired in academic robes, he stares out with large, intelligent eyes and lips that seem to hover between a smile and a smirk. Perhaps it is the frank stare of a former surgeon whose knowledge of the human body is far more intimate than that of physician-professors. We imagine him standing behind the podium of the lecture hall clad in full academic regalia, shuffling through the pages of his lecture notes. At the appropriate moment, he glances up with a sly smile, signaling his readiness to begin. Only after the conversations have entirely ceased and all eyes are fixed expectantly upon him does he launch energetically into his lecture. "Gentleman," he begins, ignoring the fact that many elegant ladies are also in attendance,

> Since among all the parts that characterize plants the ones we call flowers are, without argument, the most essential, it seems appropriate to discuss them with you at the outset, even more so because botanists in general have provided us with rather confused ideas about them. Perhaps the language I am going to use for this purpose will seem a little novel for botany, but since it will be filled with terminology that is perfectly proper for the use of the parts that I intend to expose, I believe that it will be more comprehensible than the old fashioned terminology, which—being crammed with incorrect and ambiguous words better suited for confusing the subject than for shedding light on it—leads into error those whose imaginations are still obscured, and who have no good notion of the true functions of the majority of these structures.[11]

As most in the hall would have understood, Vaillant was referring to the "old-fashioned terminology" of his deceased mentor, Tournefort, whom he believed had made a hash of plant classification by overemphasizing the corolla and ignoring the most important parts of the flower: the sexual organs. The flower was by definition the structure that contained the sexual organs. The corolla's entire purpose, Vaillant asserted, was to cover and protect these sexual organs:

> From my definition of the true flower, one can readily understand that it should be in full bloom, because, when still a bud, the corolla not only completely surrounds the reproductive organs, but also conceals them so perfectly that one can consider the bud as a nuptial bed, since it is usually only after they have consummated their marriage that they are permitted to show themselves; or if the bud happens to open slightly before they are through, it opens completely only after they have left each other.

Here we find the first use of the term, later made famous by Linnaeus, of the closed flower bud as a curtained "nuptial bed," which conceals the bride and groom as they consummate their marriage.[12] Leaving nothing to the imagination, Vaillant describes a steamy encounter between stamen and pistil worthy of a D. H. Lawrence:

> the tension or swelling of the male organs occurs so rapidly that the lips of the bud, giving way to such impetuous energy, open with astonishing speed. At that moment, these excited organs, which seem to think only of satisfying their own violent desires, abruptly discharge in all directions, creating a tornado of dust which expands, carrying fecundity everywhere; and by a strange catastrophe they now find themselves so exhausted that at the very moment of giving life they bring upon themselves a sudden death. . . .
>
> Nor does the scene end there. As soon as this sport has ended, the lips of the flower approach each other with the same speed as they came apart, returning the bud to its original shape. One would never suspect that the flower had suffered any violence unless one had witnessed it, or unless one noticed the frail corpses of those valiant champions, which remain for some time displayed on the tip, where, like so many weathervanes, they serve as toys for the Zephyrs.

Such a consummation, of course, can only occur in hermaphroditic flowers. The marriages of monoecious flowers, Vaillant notes, cannot be consummated before their buds have opened because the two sexes reside on separate flowers. Perhaps because of their lack of erotic potential, the flowers of monoecious and dioecious plants were treated rather cursorily by Vaillant.

Vaillant's obvious analogy of the stamen's explosive release of pollen to a male orgasm may seem like an outrageous, unscientific conceit, but it is only a logical extension of Nehemiah Grew's earlier portrayal of the stamen as a penis and his comparison of pollen to sperm. Of course, Vaillant was also playing to his audience. One wonders whether the 200 or so male medical students somehow managed to maintain their decorum at this point or whether they erupted in appreciative applause. Lest there be any doubt of the truth of his description, Vaillant invited the members of the audience to witness the phenomenon for themselves by venturing out to the Jardin du Roi early in the morning to observe the stingless nettle *Parietaria* (Lichwort), whose excitable stamens react explosively when touched:[13]

> All these mechanisms can be observed easily on the Parietaire, at the shepherd's hour,[14] that is to say the dawn, the time at which the different sexes of plants ordinarily engage in their frolic. And if the flowers are unwilling to perform while being observed, one can force them to by gently prodding them with the tip of a needle; for as long as the flower has reached, as we say, a competent age, it is sufficient to pull the lips apart slightly, and the *hampes* or *filaments* of the stamen, initially bent or arched, become upright in a violent effort, immediately enabling one to discover what happens in all its particulars in this type of amorous exercise.

Vaillant concedes that the stamens of hermaphroditic flowers move much more slowly than this, and it is therefore more difficult to catch them *in flagrante*. But what they lack in speed and force they more than make up for in stamina:

> This precipitousness and vigor are a far cry from the behavior of the stamens of the flowers that bear both sexes. The vast majority of these act almost imperceptibly, but one has to assume that the slower they move, the longer the duration of their innocent pleasures.

It is tempting to interpret such passages as tongue-in-cheek theatrics, designed to shock the prudish and showcase Vaillant's rhetorical virtuosity. However, since nowhere in his lecture does Vaillant admit to any embellishment of his risqué descriptions of pollination or acknowledge that they are figures of speech, it is difficult to know where he himself drew the line between fact and fancy and to what extent he believed that plants were subject to the same passions as humans.

But these delightful and provocative revelations were not the only frissons of excitement administered to the audience in Vaillant's famous lecture. Even more scandalously, he went on to accuse his well-connected younger colleague, Claude-Joseph Geoffroy, of having stolen his, Vaillant's, work, as well as having plagiarized key results from other unnamed sources in the lecture Geoffroy had presented on the same topic six years earlier.

Altogether, it was an unforgettable occasion. Indeed, the students were so enamored with Valliant's lecture that they clamored for him to take over the course after the regular professor, Antoine de Jussieu, returned from his travels—a request to which de Jussieu reluctantly acceded.

SEX, THIEVERY, AND PRIORITY AT THE JARDIN DU ROI

When Julius von Sachs wrote that some proponents of the sexual theory of plants "sought to appropriate to themselves the merit of the true discoverer" of sex in plants, he was almost certainly referring to the case of Claude-Joseph Geoffroy. In his lecture of 1711, "Observations on the Structure and Uses of the Principal Parts of Flowers," Geoffroy broke with the asexualist views of his former mentor, Joseph Pitton de Tournefort, and embraced the new sexual theory of plants. In support of his position, he claimed to have carried out several observations and experiments—tassel removal in maize and isolation experiments with female *Mercurialis* plants—that demonstrated the necessity of pollination for seed production. As we saw in Chapter 12, both of these experiments had already been described in Camerarius's *Epistola*.

In the wake of his lecture, Geoffroy was immediately hailed by his Academy colleagues for having provided the first experimental proof of the sexual theory of plants. Sébastien Vaillant's accusation of plagiarism in his lecture six years later was a shot across the bow. First, he compared Geoffroy to the nymph Echo for appropriating certain terms to which Vaillant laid claim. This was not a credible charge because the terms in question can be traced back to classical times. However, Vaillant's second grenade, tossed off almost as an afterthought, hit closer to home:

> I return to the different sexes of plants. . . I thought it convenient to establish three types of flowers: Males, Females and Androgynes, names that a sweet and officious Echo (The Author of the *Observations on the Structure and Uses of the Main Parts of Flowers*) cared to repeat (at least the first two) in front of a Royal Assembly in order to transmit them to posterity, as well as some details which he did not report so faithfully, since he believed them to be simply a case of the fabled crow dressing itself up with the feathers of a jay. But as it would displease God if I were to take away from him these details and envy even the smallest of the beautiful facts he gleaned here

> and there from various authors to augment his observations, I abandon them to him with a light heart. The other details I take directly from pure Nature, the only book one must leaf through in order to avoid making mistakes by trying to impress people.

Note that Vaillant never actually cites the original sources of the "beautiful facts" Geoffroy is supposed to have purloined. However, Claude-Joseph Geoffroy did indeed paraphrase a key passage from Rudolph Jacob Camerarius's 1694 *Sexu Plantarum Epistola* without citing it. Other observations Geoffroy reported in support of spermism in flowers were clearly based on a 1703 paper in the Royal Society's *Philosophical Transactions* by the British botanist Samuel Morland. Morland had made microscopic observations of the hollow pistils of lilies, erroneously claiming that they provided a direct route for pollen grains to impregnate the ovule.[15] Geoffroy drew the same conclusion based on nearly identical observations of hollow lily styles in his 1711 discussion of spermism versus ovism.[16]

Questions have also been raised over whether Claude-Joseph Geoffroy actually performed the maize emasculation experiments he described or whether he plagiarized them entirely from Camerarius. In response to Vaillant's accusations, Geoffroy wrote two drafts of a letter of rebuttal in 1718 in which he bitterly defended himself against the charge of plagiarism. Focusing narrowly on the issue of terminology, he cited the 1704 medical thesis of his older brother, the physician-botanist Étienne-François Geoffroy, as the original source of the terms he had used in his 1711 lecture. Claude-Joseph then leveled the countercharge that it was Vaillant, not he, who had plagiarized his brother's thesis. But Geoffroy failed to mention that his brother's 1704 thesis also contained the same unattributed experiments of Camerarius that he himself claimed to have performed in 1711.[17] In fact, Etienne-François's lecture in Latin was so popular that it was presented a second time for the general public and subsequently translated into French. It appears that the Geoffroy brothers had both gained considerable prestige and acclaim by appropriating the same experiments of Camerarius twice in the space of seven years![18]

As the scandal raged on, Geoffroy wrote his two letters to the Academy defending himself against the charge of plagiarism. In a rare moment of candor, he conceded that the sexual theory had been around for "a long time" and that both Vaillant and he could plausibly be accused of plagiarism:

> But, in the end, this is a doctrine which preceded his by a long time, on behalf of which its authors, if they were in the same bad mood he is in, could cry "Thieves!" at us. Because let's face it, both of us are thieves in this matter, and I invite anyone to judge who is the good thief and who is the bad one.[19]

Geoffroy never names these "authors" who could cry "Thieves!" Nor does he define the difference between a "good thief" and a "bad thief." Nevertheless, the statement is the closest Geoffroy comes to admitting that he had misappropriated some of the material in his lecture, although he did not include this *mea culpa* in the version of the letter he presented to the Academy.

As a parting shot, Geoffroy expressed disgust over Vaillant's lurid descriptions of pollination, which, he stated, would be more appropriate for "Priapic festivals" than a scientific lecture. Among the passages that he found particularly objectionable was the salacious description of stamens that have explosively discharged their pollen and "find themselves so exhausted that at the very moment of giving life they bring upon themselves a sudden death":

> One needs Mr. Vaillant's imagination to bring such a tale to its conclusion. It is a free-wheeling description of a pleasure that he brings to his poetic style, not as a modest exposition as would be appropriate to a philosopher. Because at the end of the day, all this Apollonian[20] nonsense means nothing except that flower dust which is the subject here, carries a fecundity that renders the seed capable of perpetuating the species.

Such overheated language, Geoffroy asserts, is totally incompatible with scientific discourse:

> I am not saying that a Philosopher should not observe all and describe all, as long as it is done truthfully and in good taste. But it seems to me worthy of blame to concoct romanticized descriptions in which one ridiculously takes pleasure with subjects one should never represent except with a great deal of restraint and always in a serious manner. Must one endure it when a learned dissertation is given with such ill-timed badinage, and when a botanist, carried away by his imagination, uses words belonging to the loosest gallantry and gives obscene depictions for any purpose?

Claude-Joseph Geoffroy may have been guilty of plagiarizing Camerarius, but he makes a valid case for sobriety in scientific discourse. Nevertheless, the whole experience had left Geoffroy badly traumatized. The following year, he abandoned his work on the sexual theory, and, in 1715, he seized the opportunity to switch departments from Botany to Chemistry. Thereafter, he published papers on chemical and pharmacological topics only.

Precisely when Geoffroy's colleagues first became aware that there was any truth in Vaillant's charges is difficult to say. By 1728, the French botanist Henri Louis du Hamel was citing Camerarius's *du Sexe des Plantes*, along with the papers of Geoffroy and Vaillant. The first person to revive the controversy directly was Johann Georg Gmelin of the University of Tübingen, who, in 1749, republished Rudolph Jacob Camerarius's 1694 *De Sexu Plantarum Epistola*. In his commentary, Gmelin noted the striking resemblance of sections of Geoffroy's 1711 lecture to passages in the *Epistola* published seventeen years earlier.

Judged by modern standards, the Geoffroy brothers were clearly remiss in failing to cite both Camerarius and Morland, but the rules of attribution were more lax in the early eighteenth century, and the newly formed scientific societies were often competing for scientific priority over significant discoveries. Robert Merton describes the tense atmosphere that existed in scientific circles during the early modern period because of the frequent charges of plagiarism:

> [I]t is quite in keeping with the practice of the time to charge, and be charged with, plagiarism. Can you think of anyone of consequence in that energetic age who escaped unscathed, either as victim or alleged perpetrator of literary or scientific theft, and typically, as both filcher and filchee? I cannot.[21]

Among the plaintiffs and defendants were Descartes, Leibniz, Hooke, Halley, Newton, and Pascal.[22] The acrimonious exchanges between Geoffroy and Vaillant are thus symptomatic of the lax rules governing citation and intellectual property during the eighteenth century.

EARLY PLANT HYBRIDIZERS: MATHER, FAIRCHILD, AND MILLER

In 1627, Francis Bacon stated what seemed obvious at the time: since plants didn't copulate with one another, they could never produce "mixture in kinds." Copulation, he stated, arises from lust, and "lust requireth a voluntary motion." Conversely, if plants *could* copulate with one another, it followed logically that they would also be capable of lust and voluntary motion. Seen from this perspective, Vaillant's torrid description of the sexual foreplay of flowers was no mere conceit, but a straightforward application of Baconian logic.

Once sex was demonstrated in plants, the next step was to determine whether, *contra* Bacon, they were also capable of producing hybrid plants. The discovery of plant hybrids would complement Camerarius's demonstration of the requirement of pollen for seed production and provide definitive proof of sexual reproduction in plants. In this section, we focus on the work of four individuals whose casual observations and early experiments laid the foundation for future scientific studies of plant hybridization: Cotton Mather, Thomas Fairchild, Richard Bradley, and Philip Miller.

Cotton Mather and Bicolored Maize

The earliest known report of plant hybridization was Cotton Mather's observation of bicolored maize cobs in 1716, in a field "not far from the city of Boston."[23] Cotton Mather was a Harvard-educated Congregationalist minister, scholar, and prolific writer who took over as pastor of Boston's North Church after the death of his father, Increase Mather. A prey to superstition, Mather's zealous participation in the infamous witch trials of 1692 and 1693 in Salem, Massachusetts is a permanent stain on his reputation that has overshadowed his scientific activities. These trials, which he never repudiated, resulted in the execution of twenty people, mostly women. Still, he did take a rational interest in scientific questions pertaining to botany and medicine. As an example of his forays into public health, he and a physician friend led a campaign to inoculate Bostonians against smallpox, which had reached epidemic proportions by 1713. Their efforts met furious resistance from the citizenry, who regarded inoculation as unnatural and ungodly.

Like Camerarius, Cotton Mather was an ardent admirer of Robert Boyle. In 1710, he received an honorary doctorate from the University of Aberdeen, and, in 1713, he was elected to the Royal Society. He was aware of Grew's sexual theory, and he corresponded regularly with European scientists. A letter Mather wrote in 1716 to fellow Royal Society member James Petiver appears to be the earliest surviving account of plant hybridization. In the letter, Mather describes two "experiments" performed by an unnamed friend "not far from the City of Boston." The first was with maize:

> [M]y Friend planted a Row of *Indian Corn* that was Coloured Red and Blue; the rest of the Field being planted with corn of the yellow, which is the most usual colour. To

> the Windward side [i.e., direction from which the wind blows], this Red and Blue Row, so infected Three or Four whole Rows, as to communicate the same Colour unto them; and part of ye Fifth, and some of ye Sixth. But to the Lee-ward Side, no less than Seven or Eight Rows, had ye same Colour communicated unto them; and some small Impressions were made on those that were yet further off.[24]

The second "experiment," which involves hybridization between squash and gourd plants, is more problematical:

> The same Friend had his garden ever now and then Robbed of the Squashes, which were growing there. To inflict a pretty little punishment on the Thieves, he planted some Gourds among the squashes (which are in aspect very like 'em) at certain places which he distinguished with a private mark, that he might not be himself imposed upon. By this method, the Thieves were deceived & discovered, & ridiculed. But yet the honest man saved himself no squashes by this Trick; for they were so infected and Embittered by the Gourds, that there was no eating of them.

Unlike endosperm, fruit tissues are entirely maternal in origin and their genotype is that of the mother plant. For gourd pollen to cause the squash fruit to become bitter, it would have to involve some type of physiological interactions between the hybrid embryo and endosperm of the seed and the surrounding fruit tissues. Such physiological interactions do occur in some species but are not known to occur in the squash family.

If the bitterness of the squash is *not* due to physiological interactions between the fruit and the seed, it cannot have been caused by hybridization with the gourd plants. The most generous explanation for the claim that pollen from the gourd plant embittered the squash plant is that it was caused by a hybridization event that occurred in the previous generation of squash plants from which Mather's friend obtained his seeds.

Thomas Fairchild and the First Artificial Hybrid

While Sébastien Vaillant and Claude-Joseph Geoffroy were busy savaging each other over who was the worse "thief" regarding the discovery of sex in plants, a modest, unassuming British commercial florist, Thomas Fairchild, was putting Grew's sexual theory to the test by attempting to cross two different members of the genus *Dianthus*, a member of the carnation, or pink family. Fairchild operated a large garden in Hoxton from about 1692 until his death in 1729. According to author Michael Leapman, Fairchild's nursery in Hoxton was on the must-see list of gardening aficionados throughout England:

> Fairchild had gained a reputation as one of the most skillful nurserymen in England. . . . He was one of several in his trade known as "curious" gardeners, in the old-fashioned sense that they displayed curiosity about every aspect of their craft. At a time when England was experiencing a real upsurge of interest in gardens and what grew in them, more and more people were flocking to Hoxton, then a leading center of the trade, to gaze at the latest wonders on display at his and a clutch of neighboring nurseries.[25]

Chief among Fairchild's many horticultural achievements was his production of the first known artificial plant hybrid by crossing the carnation (*Dianthus caryophyllus*) with a sweet William (*Dianthus barbatus*).[26] This hybrid was propagated vegetatively for a hundred years and was proudly grown in many gardens around London. In a letter dated July 22, 1740, from botanist and Royal Society member Peter Collinson to the American botanist John Bartram, Collinson describes Fairchild's hybrid as follows:

> Where plants of a class are growing near together, they will mix and produce a mingled species. An instance we have in our gardens, raised by the late Thomas Fairchild, who had a plant from seed,[27] that was compounded of the Carnation and Sweet William. It has the leaves of the first, and its flowers double like the Carnation—the size of a Pink—but in clusters like the Sweet William. It is named a Mule—per analogy to the Mule produced from the Horse and Ass.[28]

In the eighteenth century, tampering with nature by creating "Monsters" was still generally regarded as distasteful and unnatural, even blasphemous. In Shakespeare's *A Winter's Tale*, written around 1610, Perdita expressed contempt for "streak'd gillyflowers," which she called "nature's bastards" because she believed they were created illegitimately by the gardener's art, probably by grafting.[29] In Fairchild's day, anyone attempting to create new varieties by hybridization was widely perceived as meddling with the divine plan. Members of polite society were already somewhat squeamish about biological conception in general. In a dissertation on sexual reproduction in animals and plants published in 1743, Jacob Andrew Trembley felt compelled to apologize abjectly for inflicting such a disgusting subject on his fastidious readers:

> We are wisely silent concerning the other circumstances of the formation of the fetus of an animal or plant. Already we are trying imprudently to reveal the very lofty and rather secret mysteries of nature; already, kindest readers, we have abused your patience more than is fair; it is time that we put a stop for the sake of the disgusted reader, for this reason at least, that we may repay your kindness by not abusing it any further.[30]

That such attitudes were common in the eighteenth century—along with Fairchild's natural diffidence—may help to explain his seeming reluctance to bask in the full glory of his "Mule" by describing the experiment in his own words. The only accounts we have of the origin of Fairchild's Mule are second-hand—the first by Richard Bradley in his book, *New Improvements in Planting and Gardening: Both Philosophical and Practical* (1717/1718) and the second by Patrick Blair in his *Botanik Essays* (1720). Curiously, the two accounts differ on the question of the hybrid's origin. In a passage describing the potential benefits to agriculture of the practice of artificial pollination, Bradley, a prolific writer and Fellow of the Royal Society who later became the first Professor of Botany at Cambridge, states that "a curious person may by this Knowledge, produce such rare kinds of Plants as have not yet been heard of, by making choice of two Plants for his Purpose, as near alike in their Parts, but chiefly in their Flowers and Seed-Vessels." He then goes on to cite the example of Fairchild's Mule:

> [T]he Carnation and the Sweet William are in some respects alike; the Farina of the one will impregnate the other, and the Seed so enliven'd will produce a Plant different from either, as may now be seen in the garden of Mr. Thomas Fairchild of Hoxton, a Plant neither Sweet William nor Carnation, but resembling both equally, which was raised from the Seed of a Carnation that had been impregnated with the Farina of the Sweet William.[31]

Although Bradley never actually states that Fairchild made the hybrid himself, he identifies sweet William as the pollen donor, a fact that could only be known if the hybrid was produced by hand.

Three years later, Patrick Blair, a physician, amateur botanist, and Fellow of the Royal Society, described Fairchild's hybrid again in a presentation to the Royal Society. In a report appearing in the Minutes of the Royal Society of 1720, Blair states unequivocally, and contrary to Bradley's account, that the hybrid was generated spontaneously and that Fairchild discovered it by chance:

> The other Experiment was made by Mr. Fairchild some years ago. He found a plant in his garden of a middle nature between a Sweet William & Carnation July flower (a specimen of which was produced before the Society) it grew in a bed where the seed of each of those flowers had by accident been thrown promiscuously, & he takes it to be an heterogeneous production from these two different flowers . . . these new sort of plants produce no seed, but are barren like the Mule or other Mongrel animals which are generated from different species.[32]

Blair's account seems the more credible of the two, given that Fairchild himself was present at the proceedings and even passed around a pressed specimen of his "Mule" to the assembled Fellows. On the other hand, Blair used the term "experiment," which implies a deliberate action rather than an accidental find. Assuming Fairchild created the hybrid himself, why would he allow Blair to characterize it as an accidental find?

One possible explanation for the discrepancy between Bradley's and Blair's accounts was proposed by Michael Leapman. Noting the prevailing religious prejudice against hybridization, Leapman speculated that Fairchild "may have asked Blair to let the grandees of the Royal Society believe that his discovery had been by chance rather than by design."[33]

Assuming Fairchild did, in fact, create his "Mule," it was probably not an isolated incident. According to Zirkle, five years later, Fairchild wrote in *The City Gardener* that he was continuing his researches on "the generation of plants," a euphemism often applied to hybridization at that time. But if Fairchild succeeded in making any other hybrid varieties, neither he nor his friends ever recorded them for posterity.

HERMAPHRODITIC FLOWERS AND BEE POLLINATION

All of Camerarius's experiments had been carried out with dioecious and monoecious species in which the two sexes are on separate flowers. Richard Bradley seems to have been the first to test the sexual theory in hermaphroditic flowers. His choice of tulips was felicitous because of their large, accessible stamens and pistils. Although the Dutch speculative craze known as "Tulipomania" had ended by 1639, tulips remained a ubiquitous favorite in gardens throughout Europe. Bradley had visited Amsterdam to study horticulture in 1714, but Dutch botanists had apparently not yet performed any "castration" experiments on tulip flowers. Bradley describes his own tulip experiment as follows:

> I made my first Experiment upon the Tulip, which I chose rather than any other Plant, because it seldom misses to produce Seed. Several Years ago I had the convenience of a large Garden, wherein there was a considerable Bed of Tulips in one Part, containing about four hundred Roots: In another Part of it, very remote from the former, were twelve Tulips in perfect health. At the first Opening of the twelve . . . I cautiously took out of them all their Apices [anthers], before the *Farina Fecundans* was ripe, or any ways appear'd: These Tulips being thus castrated, bore no Seed that Summer: while, on the other hand, every one of the four hundred Plants, which I had let alone, produced Seed. [34]

Bradley's pioneering demonstration of the role of pollen in a hermaphroditic flower was published in 1717, the same year as Vailliant's lecture at the Jardin du Roi.

The Scottish gardener Philip Miller was appointed head gardener of the Chelsea Physic Garden in 1722, where he remained until his retirement in 1767. In addition to greatly expanding the garden's collection, he was an active experimentalist who corresponded regularly with Bradley, Blair, and other scientific figures. In 1730, he was elected to the Royal Society. Miller was probably best known to his contemporaries as the author of *The Gardener's Dictionary*, first published in 1724. The *Dictionary*, which won high praise from Linnaeaus, went through eight editions and was translated into French, German, and Dutch.

Shortly before assuming his post at the Chelsea Physic Garden, Miller, inspired by Bradley's account of his tulip experiments, wrote a letter to the author describing his own castration experiments along with the first known report of bee pollination:

> I planted a Dozen of Tulips by themselves, and soon as they open'd, took out the Apices with a fine Pair of Nippers. . . . About two Days after, as I was sitting in my Garden, I perceiv'd, in a Bed of Tulips near me, some Bees very busy in the Middle of the Flowers; and viewing them I saw them come out with their Legs and Belly loaded with Dust, and one of them flew into a Tulip that I had castrated: Upon which I took my Microscope, and examining the Tulip he flew into, found he had left Dust enough to impregnate the Tulip . . . [for they bore good ripe seeds which afterwards grew].[35]

Aristotle had also observed bees collecting "bee bread" from flowers, but he failed to associate bees with any benefit to the plant. Here, Philip Miller, for the first time, connects the dots between bee visits and pollination—no doubt a spine-tingling revelation for the Chelsea gardener!

LINNAEUS AND THE SEARCH FOR THE "NATURAL" SYSTEM OF CLASSIFICATION

No one did more to publicize the sexual theory of plants in the latter half of the eighteenth century than the Swedish botanist Carl von Linné, better known as Linnaeus. Linnaeus achieved this feat not as an experimentalist like Camerarius, but as a taxonomist. His new sexual system of plant classification, based on the number of stamens and pistils, swept all others aside and dominated botany in Europe and the New World from around 1760 to 1800.

Linnaeus understood, better than anyone else, that the tsunami of new plant species that was inundating Europe as a byproduct of the Age of Exploration had created an urgent need for a uniform method of classifying and naming plants. He was the first botanist to break with the ancient past and dispense with Theophrastus's division of plants into the three main categories: trees, shrubs, and herbs. Yet, in other respects, Linnaeus's philosophical underpinnings were antiquated, even to his contemporaries. More than a whiff of medieval scholasticism permeated his efforts to reconcile the bewildering diversity of the Plant Kingdom with Aristotelian principles and the accounts of the Creation given in Genesis.

The son of a Lutheran vicar in the rural Swedish village of Råshult, Linnaeus was strongly influenced by seventeenth-century natural theology as articulated by the British naturalist and Puritan divine John Ray. He was barely touched by the intellectual movements of the Enlightenment, such as the rise of Deism, the new field of biblical criticism, and the growing evidence for the vastness of geological time suggested by the increasingly frequent discoveries of fossils. Yet, to Linnaeus, the lack of a universal system for classifying and naming plants posed a religious challenge as well as a scientific quandary. In a sense, Linnaeus was engaging in his own brand of higher criticism from a botanical perspective. For, if the number of species was indeed fixed as was widely believed, how did the world's vast flora squeeze into the Garden of Eden? And was this even feasible, given the diversity of habitats required to accommodate all the world's flora?

Linnaeus passionately believed that he had been called upon by God to "reform" botany.[36] Just as Luther had sought to eliminate the role of the clergy as divine intermediaries, Linnaeus aimed to bring people closer to God by taking plant identification out of the exclusive domain of professional botanists (himself excepted) and by making it accessible to the lay public. God had given Adam the sacred task of naming all of His creatures, and each name that Adam conferred was emblematic of that creature's essential being and place in the Creation. To know a plant's true name was, in a very real sense, to bring one closer to God.

Tragically, these "natural" names assigned by Adam, along with the Edenic Ur-language itself, had been lost following the Fall and the Tower of Babel episode described in Genesis. Natural historians of the seventeenth century, including Linnaeus's predecessor John Ray, believed that plant taxonomy was in a similar state of post-Babel confusion. God had created living organisms according to a "natural order" arranged in a hierarchy, with plants at the bottom and humans at the top, equivalent to Aristotle's *Scala Natura*, and Adam had assigned names that reflected the "natural" relations among the organisms. Although Ray despaired of ever recovering prelapsarian knowledge of these natural relationships, he sought to approximate it by grouping plants according to their anatomical features. He rejected the medieval practice of arranging plants according to their medicinal properties or symbolic "signatures,"[37] and he argued that Latin should be the basis for naming living things because it was the closest thing to a universal language.[38] Prior to Linnaeus, however, the Latin names of plants typically consisted of strings of descriptive terms accompanied by lists of vernacular synonyms. Such lengthy names were difficult, if not impossible, to memorize. For example, the diminutive wildflower that so enchanted Linnaeus on his youthful trip to Lapland bore the unwieldy moniker, *Nummularia major, rigidioribus et rarius crenatu foliis flore purpureo gemello*. Linnaeus rechristened it *Linnaea borealis*, and, in 1753, he published his rules for binomial nomenclature (genus and species) in his landmark treatise *Species Plantarum*.

Although Linnaeus did not invent binomial nomenclature, which was first used in plant taxonomy by the Swiss botanist Caspar Bauhin,[39] it was Linnaeus who—by dint of his encyclopedic knowledge of natural history, forceful rhetoric, and supreme self-confidence—popularized it and laid down strict rules barring any other method for the naming of plants, in response to which his adoring fans throughout the world could only say, "Amen!"[40] Binomial nomenclature was indeed a godsend for both professional naturalists and amateur botanists, who now had some hope of mastering the names of their local floras.

Over the course of his life, Linnaeus vacillated between pessimism and optimism regarding the possibility of discovering the true "natural" system of classification devised by God. Although he sometimes hinted that his sexual system might be equivalent to the natural system, he admitted in his correspondence that it was "artificial" (that is, for convenience only) rather than natural. To arrive at a natural system, he finally concluded, would require a description of every plant on earth. Only then would the natural system of classification become apparent. Put another way, Linnaeus's view of plants was analogous to Dmitri Mendeleev's view of the elements: only after they were all arranged in a periodic table according to their properties could one begin to understand the principles governing their diversity. This explains Linnaeus's lifelong obsession with collecting and describing plants from around the world and his extreme frustration, leading to severe depression, when he was prevented from doing so.

Of one thing he was absolutely certain: if anyone could discover the natural system, it was he. His supreme self-confidence coupled with a near manic enthusiasm and phenomenal productivity had much to do with his unprecedented success in convincing others to adopt the rules he established for classifying plants. His new sexual system, first published in *Systema Naturae* in 1735, gradually displaced all previous methods of plant classification.

Although he was equivocal about species, in *Philosophia Botanica* (1751) Linnaeus followed Ray and Tournefort in declaring that all genera were "natural."[41] Each genus, he believed, was created "such as it is; and for this reason it is not to be capriciously split or stuck [to another] for pleasure, or according to each man's theory."[42] But despite his aspirations to become a "second Adam," and notwithstanding his conviction that he was divinely inspired, he ultimately owned to a more modest role. He readily conceded, for example, that the classes he had designated in his sexual system of classification were artificial:

> Artificial classes are substitutes for natural ones, until the discovery is made of all the natural classes, which more genera, which have not been discovered, will reveal.[43]

If genus was the only "natural" taxon created in the beginning, how does one account for the origin of all the species on earth, let alone the even more numerous varieties? Linnaeus, a forerunner of Darwin, grappled with this problem his entire career.

THE SEXUAL SYSTEM OF CLASSIFICATION

Linnaeus's decision to base his new plant classification system on the sexual organs of flowers was not revolutionary but the culmination of an historical trend in taxonomy. Aristotle had argued that organisms should be grouped according to those traits that best represented the organism's *teleos*. By this he meant the organism's purpose according to its own self-interest, not the purpose to which any other organism might put it. It seemed obvious to Aristotle that the purpose of plants was to reproduce by generating seeds.

Following Aristotle's principles, the Renaissance botanist Andrea Cesalpino based his classification system (*De Plantis Libri*, 1583) on the "fructification," by which he meant fruits and seeds. Flowers he regarded as the "covers" of fruits. In his *Historia Plantarum* (1686), John Ray followed Cesalpino's example, but treated flowers, fruits, and seeds more or less equally.[44] Joseph Pitton de Tournefort had divided plants into classes based on the morphology of the corollas, assigning subclasses according to the position of the ovary with respect to the corolla (inferior vs. superior), but, as in Ray's scheme, no mention was made of the sexual role of pollen in his system of classification.

Once news of Camerarius's experimental confirmation of the sexual theory began to diffuse to European scientific circles, it was inevitable that sexuality would be incorporated into plant taxonomy. Camerarius had noted the distinct arrangements and numbers of stamens and pistils in different species, but he never applied these traits to a classification system. Vaillant was clearly moving in this direction when work on his *Botanicon Parisiesnsis* (The Flora of Paris and Environs) was truncated by his untimely death in 1722.[45] In his 1717 lecture, he had cited the number of stamens as an important taxonomic character and given examples of variations in numerical relationships and positions of the sexual organs in different flowers.

Linnaeus seems to have first learned of Vaillant's work through his mentor, Johan Rothman, the district medical officer and physics instructor at the Växjö Gymnasium where Linnaeus attended school. Rothman recognized Linnaeus's talents as a botanist and gave him private tutorials in botany, introducing his eager pupil to the rudiments of the new sexual theory by providing him with a surprisingly faithful synopsis of Sébastien Vaillant's risqué lecture.[46] It is possible that he became further acquainted with it through excerpts and summaries of Camerarius's *Epistola* around this time. Thus, Linnaeus's early education played an important role is guiding him toward his sexual system. According to Erikson,

> Even if Linnaeus did not remember exactly what Rothman had taught him, we may suppose that henceforth there would always be for him a subconscious link between the two matters—classification and the sexuality of plants.[47]

After a frustrating year at the University of Lund, with its meager resources, Linnaeus transferred to the more prestigious University of Uppsala in 1728, where he soon acquired another admiring mentor, Oluf Celsius, the Swedish botanist, philologist, and clergyman.[48] Oluf Celsius had recruited Linnaeus to help him collect plants for his never-published *Flora Uplandica*. It was customary for Swedish students to present their mentors with a poem in their honor on New Year's Day.[49] On New Year's Day, 1730, Linnaeus bestowed an even more heartfelt literary gift upon his patron, a brief manuscript entitled *Praeludia Sponsaliorum Plantarum* (Introduction to the Nuptials of Plants) (Figure 13.1). Suffused with the lyrical eroticism of youth (he was twenty-two at the time), Linnaeus's paean to spring and plant sexuality no doubt drew additional strength and urgency from the shortness of Scandinavian springs:

> In the springtime, when the bright sun comes to our zenith, it awakens in all bodies the life that has lain smothered during the cold winter. . . . This sun affords such joy to all living things that words cannot express it; the black-cock and the wood-grouse can be seen to mate, the fish to play, why all animals feel the sexual urge. Love even seizes the very plants, as among them both *mares* and *feminae*, even the hermaphrodites, hold their nuptials, which is what I now intend to discuss, and show from the genitalia of the plants themselves which are *mares*, which *feminae* and which hermaphrodites.

FIGURE 13.1 Frontispiece and title page of Linnaeus's *Praeludia Sponsaliorum Plantarum*. (Left) Illustration showing pollination in dioecious plants. The female plant on the left is being pollinated by the male plant on the right. (Right) Title page and illustration showing hermaphroditic flowers pollinating each other.

Linnaeus's prose style was strongly influenced by the classical poets, Ovid, Horace, and Virgil. But whereas the pagan poets had celebrated sexual liasons freely, without regard for the rules of matrimony, Linnaeus, a Lutheran minister's son, contained the erotic impulses of plants wholly within the institution of marriage. As in the poem attached to Camerarius's *Epistola*, Linnaeus, too, personified the stamens and pistils as "bridegrooms" and "brides" whose "nuptials" are consummated in a "bridal bed" of petals:

> The actual petals of the flower contribute nothing to generation, serving only as Bridal Beds, which the great Creator has so gloriously arranged, adorned with such noble Bed Curtains and perfumed with so many sweet scents, that the Bridegroom there may celebrate his *Nuptias* with his bride with all the greater solemnity. When the bed is thus prepared, it is time for the Bridegroom to embrace his beloved Bride and surrender his gifts to her: I mean, one can see how *testiculi* open and emit *pulverem genitalem*, which falls upon *tubam* and fertilizes the *ovarium*.

Regarding the origin of his sexual system of classification, Linnaeus wrote, "Before I was twenty-three, I had conceived everything."[50] Five years later, Linnaeus was ready to commit himself to print. In 1735, he journeyed to the University of Leiden, where Herman Boerhaave, Vaillant's champion and publisher, was director of the University's Botanical Garden. In Leiden, Linnaeus committed *Systema Naturae* to paper and published it in the same year.

Following well-established tradition, Linnaeus divided the natural world into three Kingdoms: Animal, Vegetable, and Mineral. In Aristotelian fashion, minerals were defined as things that grow, plants as things that grow and have life, whereas animals were defined as things that grow and have both life and feeling. Each Kingdom was further subdivided into Class, Order, Genus, and Species.

The most innovative and influential section of *Systema Naturae* was his sexual system for classifying plants, entitled "Nuptiae Plantarum" (Marriages of Plants). Plants were separated into two main types of marriages: Publicae and Clandestinae. The latter consisted of the single class, Crytogamia (Hidden Marriages), which included all nonseed plants, such as ferns, horsetails, mosses, liverworts, algae, and fungi.[51] Linnaeus's use of the terms "public" and "clandestine" referred to widespread customs regarding marriage at the time. During the late seventeenth and eighteenth centuries, arranged marriages were on the decline and marriages based on romantic love were becoming the norm. In England, for example, the absence of laws requiring a marriage license or public announcement made it relatively easy for couples to marry without their parents' knowledge or consent by eloping to Scotland, and such marriages were referred to as "clandestine."[52]

Linnaeus divided the Publicae into two categories: Monoclinia,[53] in which husbands and wives occupy one "bed" (hermaphroditic flowers), and Diclinia, in which husbands and wives have separate "beds." The Diclinia were further divided into the three classes: Monoecia, Dioecia, and Polygamia. The Polygamia included plants with hermaphroditic and unisexual flowers on the same individual. If the stamens and pistils of hermaphroditic flowers of polygamous plants represented the "husbands" and "wives," the male and unisexual flowers on the same plant became the "lovers" of the husbands and wives. The Monoclinia were subdivided into twenty classes based on the structures, positions, and number, union, and length of their stamens: Monandria, Diandria, Triandria, and the like. Each of the classes (except the Cryptogams, or nonseed plants with "hidden" sex structures), were then further subdivided into orders based mainly on the number of pistils: Monogynia, Digynia, Trigynia, and so on.

As has been noted by several scholars, Linnaeus's choice of stamens for the assignment of Class and pistils for the assignment of the subordinate category of Order had no scientific basis and probably reflected patriarchal attitudes of eighteenth-century society.[54] The inferior status of women had, after all, been laid out in Genesis, which served to justify the advantaged economic and legal positions enjoyed by men. Indeed, if Linnaeus had ranked pistils above stamens in his taxonomic hierarchy it would have been tantamount to claiming that Adam was created from Eve's rib, with all the social implications that implied. Even if he had wanted to, which is unlikely, Linnaeus was far too ambitious to raise any red flags by flying in the face of such a fundamental social prejudice, thus jeopardizing the acceptance of his sexual system.

However, it is important not to overinterpret the significance of the stamen–pistil hierarchy as a reflection of Linnaeus's patriarchal attitudes. As later immortalized in verse by Erasmus Darwin, many flowers in the Linnaean system consist of single wives with multiple husbands, a polyandrous domestic arrangement not likely to be endorsed by the beer-swilling denizens of the local pub. Moreover, Linnaeus ultimately came to question the fixity of species and proposed that the *vast majority* of plant species were created by intergeneric crosses occurring *after* the Fall. As we shall see in the next section, in the case of hybrids between different genera, Linnaeus always assigned the progeny of such a cross to the genus of the female parent. In principle, therefore, the vast majority of plant species should be grouped according to their maternal genera, the botanical equivalent of matrilineal descent. Linnaeus's views on gender as applied to plant taxonomy were thus more complex than are sometimes implied.

Linnaeus's user-friendly sexual system was greeted with rapturous applause, especially from women, and especially in England, where botanizing had attained the status of a national pastime. In the words of Londa Schiebinger,

> Well-born ladies, including the Duchess of Beaufort, Lady Margaret of Portland, and Mrs. Eleanor Glanville, led the way, collecting rare and exotic plants from around the world. The royal family (George III, Queen Charlotte, and his mother, Augusta—all botanical enthusiasts) further enhanced the popularity of botany by serving as influential patrons and enlarging the Royal Botanical Gardens at Kew, in London.
>
> It was in this new atmosphere of interest in botany, especially among the ladies of the upper classes, that Linnaeus's sexual system gained wide acclaim.[55]

In the absence of any English translations of *Systema Naturae*, English botanists introduced his principles into their texts. According to Ann Shteir,

> In England, various pioneering books introduced readers to Linnaeus's ideas about classifying and naming plants. Some were expositions of the Linnaean system, others

> applied his classifications to indigenous and regional plants of more remote areas. . . . In 1760 James Lee issued his *Introduction to Botany*, a pioneering book about the Linnaean system. It consists of translated extracts from Linnaeus's writings. . . . [It] was an early Linnaean best-seller and remained a standard introductory work for fifty years.[56]

Jean Jacques Rousseau, initially at least, was so smitten with the Linnaean sexual system that he sent a message to Linnaeus as follows: "Tell him I know of no greater man on earth." As noted by Alexandra Cook, Rousseau referred more often to Linnaeus in his *Confessions* and *Rêveries* than to any other botanist.[57] Like Linnaeus, Rousseau frequently rhapsodized about the natural world with a near religious fervor. He was a forerunner of the Romantic movement, with a particular interest in plants. In a series of eight letters addressed to a young mother who had written to him for advice on a suitable subject of scientific study for her daughter, Rousseau recommended plants and provided her with a summary of the basic concepts of botany, although he discretely omitted Linnaeus's sexual system as being unsuitable for children.[58] Published as *Lettres élémentaires sur la botanique (1771–1773)*, these letters were translated into English by the British botanist Thomas Martyn and supplemented with twenty-four additional letters "fully explaining the system of Linnaeus." It was published in 1785 as *Letters on the Elements of Botany* and "went through eight editions over the next thirty years." [59]

THE PELORIA BOMBSHELL: SPECIATION AND THE MATRILINEAL DESCENT OF PLANT GENERA

John Ray, in *Historia Plantarum* (1686), was the first to define species as a group of plants capable of propagating themselves by seed, such that their "distinguishing features" were preserved. He also recognized that species give rise to varieties, which exhibit minor variations from the parental type while retaining the species' "distinguishing features":

> Thus, no matter what variations occur in the individuals or the species, if they spring from the seed of one and the same plant, they are accidental variations and not such as to distinguish a species . . . ; one species never springs from the seed of another nor vice versa.[60]

For most of his career, Linnaeus adhered closely to Ray's definitions of species and varieties, and, like Ray, he believed in the fixity of species and rejected the ancient idea of the transmutation of one species into another. He considered varieties to be the result of "accidents" caused by local environmental conditions, such as soil or temperature. Based on the assumption that varieties did not breed true, Linnaeus excluded them from his classification scheme.[61]

Even without including varieties, the sheer magnitude of the number of species on earth, and the diversity of their habitats, meant that they could not possibly have been created en masse in the Garden of Eden. In an attempt to reconcile the biblical account of Creation with the vast number and diversity of plants, Linnaeus postulated that Paradise was a tropical island containing mountains tall enough to generate a wide range of microclimates and habitats. Initially, the island was the only landmass on earth, but at some point the sea receded, gradually exposing all the continents. By the time of Adam and Eve's expulsion,

Paradise was surrounded by land, and plant species were free to spread across the globe via seed dissemination. Linnaeus's interpolation of geography into the Creation myth made a nice just-so story, but, as he realized, it had neither biblical nor scientific support and was thus not very compelling as an explanation for the fixity of species.

Searching for a more satisfying answer, Linnaeus began to consider the implications of the sexual theory for the fixity of species concept, especially the phenomenon of hybridization. Exactly when Linnaeus first became aware of Fairchild's Mule, the artificial hybrid between carnation (*Dianthus caryophyllus*) and sweet William (*Dianthus barbatus*), is unknown. There is no record of any correspondence between the two men at the Linnaean Society in London, and, at the time Fairchild died in 1729, Linnaeus would only have been twenty-two.[62] In 1736, however, Linnaeus traveled to England and paid visits to Sir Hans Sloane and Philip Miller. Presumably, both the sexual theory and plant hybridization were prime topics of conversation. It seems likely that it was during this visit that Linnaeus first became fully informed about Fairchild's Mule and, possibly, the work of Cotton Mather and others regarding hybridization in maize, for it was shortly after this time that he began to incorporate hybridization into a revolutionary new theory of the origin of species in which the number of species increased over time by a kind of organic evolution.

The immediate catalyst for the new theory was the discovery by a student at Uppsala of an unusual, toadflax-like plant growing on one of the islands of the Stockholm archipelago.[63] The student pressed and glued the specimen to an herbarium sheet and brought it to Olof Celsius, the renowned Uppsala professor of botany, who passed it on to Linnaeus. Linnaeus initially confirmed the identification of toadflax (*Linaria vulgaris*), but upon closer inspection he realized that the flowers of the inflorescence were so unusual that he began to suspect he was the victim of a hoax. Had the student glued an inflorescence from a different plant onto the herbarium sheet in an effort to deceive him and make him look ridiculous? He therefore prevailed upon the student to bring him some living specimens and was able to confirm the presence of the aberrant flower structure on fresh material as well. Flowers of the genus *Linaria* have bilaterally symmetrical corollas with four stamens and a single nectar spur. The bizarre specimen found by the student, which Linnaeus named *Peloria* (Greek for "monstrous"), had a radially symmetrical corolla with five stamens and five nectar spurs (Figure 13.2).

The taxonomic implications of *Peloria* were profound. If *Peloria* were found to breed true, which it appeared to do, thus satisfying the definition of a species, it would overturn the fixity of species doctrine. Even worse, *Peloria*, which had two pairs of stamens of unequal length, would have to be placed in a different class from *Linaria*, which has five stamens of equal length. This would completely demolish Linnaeus's sexual system of classification because classes are supposed to be widely separated from each other. In principle, no plant belonging to one class should *ever* produce progeny belonging to another class. If species could transform willy-nilly into different classes the result would be taxonomic bedlam, and there would be no possibility of ever finding a natural classification system. In the words of Linnaeus:

> Nothing can, however, be more fantastic than that which has occurred, namely that a malformed offspring of a plant which has previously always produced irregular [bilateral] flowers now has produced regular [radial] ones. As a result of this, it does not only deviate from its mother genus but also completely from the entire class and thus is an example of something that is unparalleled in botany so that owing to the

FIGURE 13.2 The toadflax *Linaria vulgaris* with bilateral symmetry (left) and the radially symmetrical *peloria* mutant (right).
From Busch, A., and S. Zachgo (2009), Flower symmetry evolution: towards understanding the abominable mystery of angiosperm radiation. *BioEssays* 31:1181–1190.

difference in the flowers no one can recognize the plant anymore. This is certainly no less remarkable than if a cow were to give birth to a calf with a wolf's head.[64]

To avert such a taxonomic catastrophe, Linnaeus concocted an alternative hypothesis that would save his classification system. Linnaeus argued that *Peloria* must be a hybrid between *Linaria* and some unidentified, but markedly different, plant. Hybridization was infinitely preferable to "transmutation" because it left his classification system more or less intact. In his correspondence with Johann Georg Gmelin of Tűbingen, Camerarius's old mentor, Gmelin shared anecdotal evidence that *Delphinium* could give rise to new types of plants when hybridized with other genera, thus supporting Linnaeus's hypothesis.[65] A new theory of the origin of species based on hybridization began to crystallize in Linnaeus's mind. One just needed to understand how hybridization worked. Linnaeus believed his *medulla–cortex theory* provided the answer.

Nicolaus of Damascus in *De Plantis* had distinguished between the tissues of the outer cortex of plant stems and those of the medulla, or pith. Cesalpino, Malpighi, and Grew had all embraced the idea that the cortex was responsible for nutrition and the production of leaves, while the pith, or "womb," gave rise to the seed. In his treatise *Metamorphosis Plantarum* (1755), Linnaeus elaborated the medulla–cortex idea still further. According to Linnaeus, pistils arose from the medulla, indicating that the medulla is female. Stamens, on the other hand, were extensions of the cortex, so the cortex must be male. The male cortex, according to Linnaeus, functions mainly in nutrition, whereas the female medulla represents the life force of the plant, manifesting a kind of will to reproduce.[66]

During fertilization, the male pollen gives rise to the cortical tissues of the embryo, while the female ovule produces the medulla.[67] Because the female medulla gives rise to the seed and contains its essence, a genus is defined as a group of species that share the same medulla, derived from the same maternal ovule. In contrast, species are defined as groups of plants with different cortical material, having been fertilized by different fathers. Because genera are defined by the mother, the offspring of any intergeneric cross should always be assigned to the genus of the mother. And since Linnaeus regarded genus as the only natural taxon, his theory implied that *matrilineal descent was part of God's plan for plants*. One could even argue that Linnaeus's concept of the matrilineal descent of plant genera trumped his patrilineal assignment of class because he considered class to be an artificial taxon. Steeped as he was in the classical tradition, Linnaeus intuitively must have believed that it was self-evident that fruit-bearing plants are female at their "core."[68]

LINNAEUS'S NEW THEORY FOR THE ORIGIN OF SPECIES

In *Fundamenta Fructificationis* (1762), Linnaeus unveiled a new theory for the origin of species, which in many ways was a forerunner of modern evolutionary theory. The fixity of species doctrine was jettisoned. At the time of the Creation, according to Linnaeus, there was only one plant species for each natural order. In other words, there were only as many species as there were orders. This reduced the number of plant species in the Garden of Eden to a more manageable number. Note that the original species in Paradise, according to Linnaeus's new theory, represented the "natural orders," which, in his sexual system, were defined by *the number of pistils*. These orders/species began to hybridize with one another until they produced all the genera, and initially there was only one species per genus.[69] Intergeneric crosses followed, giving rise to more and more species. Each species was assigned to the genus of the mother. Finally, crosses between different species belonging to the same genus resulted in the formation of varieties. Linnaeus now regarded varieties as stable and permanent rather than the unstable effects of local environmental conditions.

Linnaeus's elaborate hybridization theory of the origin of species never gained traction. Nevertheless, the general principle it embodied—that modern species evolved from ancestral species over many generations by a mechanism involving sexual reproduction—represented a clear break with the fixity of species doctrine, making it an early forerunner of Darwinian theory. Ironically, Linnaeus's worst fears about the true nature of *Peloria* turned out to be correct. Not long after the publication of his 1744 *Peloria* dissertation, someone, perhaps another student, brought him yet another living specimen of *Peloria*. As Linnaeus inspected it he realized, with a shock, that the plant contained *both* normal *Linaria*-type flowers and *Peloria*-type flowers on the same inflorescence! According to the sexual system of classification, this new specimen would simultaneously belong to two different classes, an impossible situation! Years later Linnaeus's son remarked that "after *Peloria* had fallen short of his expectations [he] no longer wanted to hear any more said about this plant." Linnaeus clearly regarded his *Peloria* theory as his greatest blunder. In an essay published in 1745, he referred readers to his dissertation on *Peloria* for "a stupid description" of this plant's variability.[70]

Linnaeus should not have been so hard on himself. The genetics, let alone the molecular biology, of species variability was totally outside the framework of eighteenth-century natural history. We now know that *Peloria* is a variety of *Linaria vulgaris* that arises from a specific type of mutation in a single gene that controls corolla and stamen development in flowers with bilateral symmetry, such as *Linaria* and *Antirrhinum* (snapdragon).[71] The population of *Peloria* that the Uppsala student first stumbled upon was actually reproducing vegetatively rather than by seed, because such mutants tend to be sterile. The atypical *Peloria* inflorescence with the mixture of normal and mutant flowers we now recognize as an example of the phenomenon of "scattered peloria," caused by the instability of the *Peloria* mutation.

Despite the temporary setback of *Peloria*, the irrepressible Linnaeus continued to believe that the number of species was not fixed. New species had arisen since the Creation and continued to arise through hybridization. After 1745, he dropped the idea of hybridization among orders, but he was still confident that hybridization could occur between genera and that this was the source of most new species. Indeed, everywhere Linnaeus and his students looked in the field, they saw abundant examples of intermediate forms that they interpreted as hybrids. In the dissertation *Plantae Hybridae* (1751) published under the name of his student J. J. Haartman, Linnaeus listed no less than a hundred possible hybrids identified in natural settings or in botanical gardens. Among these were several putative hybrids between different genera, including *Veronica* × *Verbena*, *Saponaria* × *Gentiana*, and *Aquilegia* × *Fumaria*. These three pairs of genera also belong to different families, so their hybridization is highly unlikely. Linnaeus never tried to generate such intergeneric "hybrids" artificially.

In one case, however, Linnaeus confirmed the authenticity of putative hybrid between two species that he observed in his garden by creating the identical hybrid in an experiment. In 1760, the Academy of Sciences in St. Petersburg staged a competition for the best essay demonstrating sex in plants. Linnaeus submitted an essay entitled *Disquisitio de Sexu Plantarum* and was awarded the prize. After reviewing the history of the problem, Linnaeus described some emasculation experiments, similar to those of Camerarius and Bradley, demonstrating the requirement for pollination for viable seed production. He then cited the earlier mentioned three unverified "hybrids" as further evidence that sexual reproduction occurs between different genera. However, he also reported a fourth hybrid he observed in his garden growing between two species of goatsbeard or salsify, both members of the Aster family: *Tragopogon pratensis* and *Tragopogon porrifolius*. The following spring, he dusted the stigma of *T. porrifolium* with pollen from *T. pratense* and collected the resulting seeds. When planted, the seeds gave rise to hybrid plants bearing "purple flowers, yellow at the base," similar to the natural hybrids he had observed in his garden, and the authenticity of this hybrid is generally accepted today. "I doubt whether any experiment demonstrates the generation of plants more certainly than this," he concluded with characteristic modesty.

NOTES

1. von Sachs, J. (1906), *History of Botany (1530–1860)* trans. H. E. F. Garnsey and I. B. Balfour. Oxford at the Clarendon Press, p. 391. Originally published in German in 1875.

2. William Harvey's *The Motion of the Heart and Blood* (*De Motu Cordis*) was published in Frankfurt in 1628.

3. Tournefort was not the first to employ the term "genus." Aristotle used it in some contexts, as did Cordus (1541), Gesner (1551), and Bauhin (1623). However, as pointed out by Sachs (1875)

and by Mayr (1982), Tournefort was the first to consistently and clearly distinguish between terms for genus, species, and varieties, although genus was given the most taxonomic weight.

4. Green, E. L. (1983), *Landmarks of Botanical History,* Part II, F. N. Egerton, ed. Stanford University Press, p. 941.

5. Bernasconi, P., and L. Taiz (2006), Claude-Joseph Geoffroy's 1711 lecture on the structure and uses of flowers. *Huntia* 13:5–86.

6. Bernasconi, P., and L. Taiz (2002), Sebastian Vaillant's 1717 lecture on the structure and function of flowers. *Huntia* 11:97–128.

7. Schiebinger, L. (1993), *Nature's Body: Gender in the Making of Modern Science.* Beacon Press; Williams, R. L. (2001), *Botanophilia in Eighteenth-Century France: The Spirit of the Enlightenment.* Kluwer Academic Publishers.

8. Williams, *Botanophilia in Eighteenth-Century France*; Stafleu, F. A., and R. S. Cowan (1986), *Taxonomic Literature. A Selective Guide to Botanical Publications and Collections with Dates, Commentaries and Types.* Scheltema and Holkema.

9. Williams, *Botanophilia in Eighteenth-Century France.*

10. Rousseau, J. (1970), Sébastien Vaillant: An outstanding 18th-century botanist, in P. Smit and R. J. Ch. V. ter Laage, eds., *Essays in Biohistory.* International Association for Plant Taxonomy, pp. 195–228. (Regnum Veg. 71.)

11. All translations of Vaillant's lecture are from Bernasconi and Taiz (2002).

12. It is sometimes stated that Linnaeus conceived the nuptial bed metaphor as a nod to his conservative Lutheran upbringing, yet it is clear he only borrowed the phrase from Sébastien Vaillant, a worldly French Catholic, and that Vaillant, had lifted the conceit from the lyrical poem contained in Camerarius's 1694 *Epistola.*

13. *Parietaria* belongs to the Urticaceae family, which also includes *Elatostema* and *Urtica.* Urticaceous floral buds contain four stamens bent inward under tension. During drying, the filaments spring out explosively, scattering their pollen like tiny puffs of smoke. At the appropriate time, the explosive action can be triggered by touch. *Parietaria* may be monoecious or dioecious, and the pollen-scattering mechanism represents an ancient adaption to wind pollination (anemophily). The species described by Vaillant was monoecious.

14. Considered the most favorable time for love.

15. Morland, S. (1703), Some new observations upon the parts and use of the flower in plants. *Philosophical Transactions* 23:1474–1479.

16. Bernasconi and Taiz, Claude-Joseph Geoffroy's 1711 lecture; Prevost, A. -M. (1965), Rapprochement entre l'Epistola de sexu plantarum de R. J. Camerarius (1694) et les Observations sur la structure et l'usage des principals parties des fleurs de Geoffroy le Jeune. *Comptes rendus hebdomadaires des séances de l'Académie des sciences* 261:2045–2048.

17. Bernasconi and Taiz, Claude-Joseph Geoffroy's 1711 lecture.

18. Ibid.

19. Ibid.

20. Referring to Apollo, the sun-god of the Greeks and Romans, the patron of music and poetry. Perhaps a more apt adjective would have been "Dionysian."

21. Merton, R. K. (1965), *On The Shoulders Of Giants: A Shandean Postscript.* Free Press.

22. Mallon, T. (1989), *Stolen Words: Forays into the Origins and Ravages of Plagiarism.* Ticknor & Fields.

23. Zirkle, C. (1935), *The Beginnings of Plant Hybridization.* University of Pennsylvania Press.

24. Cited by Zirkle; note that the color of the kernel is actually derived from the underlying tissue, the *aleurone layer,* which is the outer endosperm. The genetic effect of pollen on the color of the endosperm is a special type of hybridization known as *xenia*.

25. Leapman, M. (2000), *The Ingenious Mr. Fairchild*. St. Martin's Press, p. 10.

26. Common names are typically not capitalized in modern English usage.

27. As discussed later, the actual origin of the hybrid is still unresolved.

28. Darlingtron, W., and H. Marshall (1849), *Memorials of John Bartram and Humphry Marshall*. Lindsay and Blakiston, Philadelphia. (Republished by Ulan Press, 2012).

29. In Elizabethan times, the "gardener's art" probably referred to grafting, since the role of pollination was still unknown.

30. Trembley, J. A. (1743), *Theses Physicae de Vegetatione et Generatione Plantarum*. M. -M. Bousquet & Sociorum.

31. Bradley, R. (1717), *New Improvements of Planting and Gardening Both Philosophical and Practical*.

32. Blair, P.(February 4, 1719/20). *Extracts from the Journal Book of the Royal Society*, XII (1714–1720): pp. 411–412. Cited by Zirkle.

33. Leapman, *The Ingenious Mr. Fairchild*.

34. Bradley, *New Improvements of Planting and Gardening*.

35. From Philip Miller's Letter to Richard Bradley, dated October 6, 1721, quoted in R. Bradley's *A General Treatise of Husbandry and Gardening* (1726). The crucial phrase in brackets is taken from Miller's shorter account in the 1751 edition of *Gardener's Dictionary*.

36. In this respect, he had something in common with his highly influential contemporary, the Swedish scientist-mystic Swedenborg, who believed he had been singled out by God to interpret the secrets of the universe. See Erikson, G. (1983), Linnaeus the botanist, in Tore Frängsmyr, ed., *Linnaeus, the Man and His Work*. University of California Press, pp. 63–109.

37. From the Greeks onward, herb collecting has always been surrounded by superstitions and rituals. Among these superstitions was the doctrine of signatures. Later popularized and elaborated by Paracelsus (1491–1541), a professor at the University of Basel, the "Doctrine of Signatures" held that the medicinal properties of plants can be discerned from their resemblance to the parts of the body that they are designed by God to treat. For example, the British botanist William Cole (1626–62) wrote that the walnut kernel "hath the very figure of the Brain," and thus, when properly prepared, "it comforts the brain and head mightily."

38. Harrison, P. (2009), Linnaeus as a second Adam? Taxonomy and the religious vocation. *Zygon* 44:879–893.

39. In *Pinax Theatri Botanica* (*Illustrated Exposition of Plants*), Bauhin described about 6,000 species and classified them according to their "natural affinities." Although he anticipated Linnaeus by using binomial nomenclature (genus and species) for many of the plants, he was inconsistent in its application.

40. Koerner, L. (1999), *Linnaeus: Nature and Nation*. Harvard University Press.

41. He interpreted the fact that insects often feasted upon different species of the same genus with no apparent harm as evidence that genus constitutes a "natural" grouping (*Disquisito de Sexu Plantarum* 1760).

42. Cited by Harrison, Linnaeus as a second Adam?

43. Ibid.

44. Morton, A. G. (1981), *History of Botanical Science*. Academic Press.

45. Vaillant died in Holland at the age of fifty-three, leaving the notes and plates for *Botanicon Parisiense*, on which he had worked for thirty-six years, to his friend Herman Boerhaave. Boerhaave arranged for its publication in 1727.

46. Erikson, Linnaeus the botanist, pp. 63–109.

47. Idem., p. 65

48. Not to be confused with his nephew Anders Celsius, who invented the Celsius scale.

49. A tradition sadly discontinued in modern times!

50. Broberg, G. (1990), *Brown-eyed, Nimble, Hasty, Did Everything Promptly*. Uppsala University Press.

51. Algae are no longer grouped with the land plants, and Fungi have been placed in a separate Kingdom.

52. The "Act for the Better Preventing Clandestine Marriage," also known as "The Marriage Act of 1753," was the first law in England and Wales to require a public proclamation before a wedding could take place.

53. "Clinia" from the Greek word for "bed."

54. Shteir, A. B. (1996), *Cultivating Women, Cultivating Science: Flora's Daughters and Botany in England, 1760–1860*. Johns Hopkins University Press; ##Schiebinger, L. (1993), The private lives of plants, in Schiebinger, *Nature's Body*; George, S. (2007), *Botany, Sexuality, and Women's Writing (1760–1830)*. Manchester University Press.

55. Schiebinger, *Nature's Body*, p. 28.

56. Shteir, *Cultivating Women, Cultivating Science*, pp. 17–18.

57. Cook, A. (2012), *Jean-Jacques Rousseau and Botany: The Salutary Science*. Voltaire Foundation, Oxford.

58. Indeed, the longer Rousseau studied botany, the more attracted he became to the natural systems of classifications being developed by Bernard and Antoine Laurent de Jussieu and other French botanists. See Cook, *Jean-Jacques Rousseau and Botany*.

59. Shteir, *Cultivating Women, Cultivating Science*.

60. Quoted by Mayr, E. (1982), *The Growth of Biological Thought: Diversity, Evolution, and Inheritance*. Belknap Press, p. 256.

61. For a historical perspective on the philosophical definitions of "accidents" in natural history, from Aristotle to Sebastian de Monteux and Leonhart Fuchs, see the discussion in Sachiko Kusukawa's *Picturing the Book of Nature: Image, Text, and Argument in Sixteenth-Century Human Anatomy and Medical Botany* (2012), University of Chicago Press, pp. 103—109.

62. Leapman, *The Ingenious Mr. Fairchild*.

63. Gustafsson, Å. (1979). Linnaeus' Peloria: The history of a monster. *Theoretical and Applied Genetics* 54:241–248.

64. Caroli Linnaei: Amoenitates Academicae III. Peloria 1749 (1744) 55–73, translation in Gustafsson, Linnaeus' Peloria.

65. Erikson, Linnaeus the botanist.

66. Ibid. As we shall see in Chapter 16, Goethe was to use elements of Linnaeus's medulla–cortex theory as the basis for an argument against the sexual theory of plants.

67. Linnaeus doesn't address how the medulla–cortex hypothesis would work in monoecious and dioecious plants with unisexual flowers.

68. Linnaeus also named the class of animals to which humans belong as the *Mammalia*, after the female breast, perhaps influenced by the important symbolic association between suckling and nature in religion and art; see Schiebinger, *Nature's Body.*

69. Linnaeus seemed to have believed that such crosses between orders were possible based on the mistaken impression that his distinguished colleague, the French naturalist Réaumur, had reported a successful cross between a rabbit and a hen. In reality, Réaumur had only reported a "barnyard romance" he had once observed, joking that: "It was the general wish, as well as my own, that it might have procured us chickens covered with hair, or rabbits clothed with feathers." Lindroth, S. (1983), The two faces of Linnaeus, in T. Frängsmyr, ed., *Linnaeus.* Cartron, L. et al. (2007), *Heredity Produced: At the Crossroads of Biology, Politics, and Culture, 1500–1870.* MIT Press.

70. Gustafsson, Linnaeus' Peloria.

71. Rudall P. J., and R. M. Bateman (2003), Evolutionary change in flowers and inflorescences: evidence from naturally occurring terata. *Trends in Plant Science* 876–882.

14

Behind the Green Door

LOVE AND LUST IN EIGHTEENTH-CENTURY BOTANY

IT WAS INEVITABLE that plant sex, once out of the bag, would become a favorite target of wits, wags, and scalawags, and nowhere more than in Merrie England. The Puritan Revolution had brought with it many social benefits, but humor was not one of them. The British love of satire, temporarily suppressed during the dour reign of Cromwell, erupted like a pent-up volcano after the Restoration of 1660, wilder and more iconoclastic than ever. Aiding and abetting this return to jollity, Charles II, upon returning from his sojourn in Paris, established a more permissive court atmosphere modeled on that of Louis XVI at Versailles. Theaters were reopened, signaling a return to a more hedonistic public life, and the most popular plays were sexually explicit comedies inspired by the farces of Moliere. Actresses appeared on the stage for the first time, and female playwrights, such as Aphra Behn, had their works performed to much acclaim.

Another factor fueling the popularity of satire in England was the heated rivalry in Parliament between the newly formed Whig and Tory parties. Impassioned, inflammatory rhetoric was frowned upon as dangerous to the social order. Instead, a rapier wit became the weapon of choice for skewering one's political opponents. Among the most notable literary satirists of the period were John Dryden, Alexander Pope, and Jonathan Swift. But there was also a galaxy of mostly lesser talents, many of whom worked out of Grub Street in the area known as Moorfields. Once a rare greenbelt area in the City of London, Moorfields served as a refuge for impoverished survivors of the Great Fire of London in 1666, over the strenuous objections of Charles II. By the eighteenth century, Moorfields had morphed into a full-blown slum filled with crumbling tenements, where pimps and prostitutes plied their trade amid a thriving industry of booksellers, publishers, and a mélange of scribblers ranging from the cerebral, including the young Samuel Johnson, to the mercenary, such as the notorious Edmund Curll.[1] Curll is best remembered as the publisher of numerous satirical books and pamphlets of an openly sexual nature.

The Enlightenment period could also aptly be called the Age of Eros. Erotica had become fashionable, stimulated by an influx of French erotica and pornography.[2] By the mid-seventeenth century, British connoisseurs were avidly devouring such classics as *L'Escole des Filles, L'Académie des Dames*, and *Venus dans le Cloître*. Samuel Pepys, after reading *L'Escole des Filles*, wrote in his diary that it was "a mighty lewd book, but not amiss for a sober man once to read over to inform himself in the villainy of the world."[3] Sober women were studiously informing themselves in the world's villainy as well. By the eighteenth century, freely translated English versions of French erotica had begun to engender indigenous works by English authors. John Cleland's *Fanny Hill*, published in two installments in 1748 and 1749, is generally considered the most notable of these racy narratives and the first to be written in novel form. Libertinism as a lifestyle, once the sole prerogative of male aristocrats, now extended to the once stodgy bourgeoisie.

Another French import entering the mix of cultural influences on British botany was the philosophy of mechanistic materialism, first outlined by Descartes in the seventeenth century and taken to its logical extreme by Julien Offray de la Mettrie in the mid-eighteenth century. Mettrie argued that the *Scala Naturae*, or Great Chain of Being, was, in fact, a continuum, and all creatures, from plants to humans, were governed by the same natural laws. Such reasoning encouraged analogical and metaphorical thinking about plant sexuality.

The conjunction of satire, pornography, the discovery of sex in plants, and the penchant for analogies proved irresistible to Edmund Curll and his Grub Street minions, giving rise to a new, albeit short-lived, eighteenth-century literary fashion, which we will dub "phytoerotica." Those who had read Sébastien Vaillant's 1717 lecture were incited to follow his lead. British phytoerotica fell into two main categories: bawdy verse, in which plants were stand-ins for human genitalia, and erotic, classically inspired verse, in which stamens and pistils were personified as husbands, wives, and lovers.[4] The former emphasized the sexual act itself, had little to do with plants, and was aimed at a male audience. Thomas Stretzer's poems on the "natural histories" of "*Arbor Vitae*" and "*Frutex Vulvaria*" are the type-specimens of this subgenre. In contrast, the latter type of phytoerotica emphasized romance and foreplay, served as a heuristic device for teaching the Linnaean system, and was aimed at a female audience. The supreme exemplar of this subgenre of phytoerotica is Erasmus Darwin's *The Loves of Plants*. Even before its publication in 1789, prominent opponents of the sexual system had excoriated Linnaeus for introducing lewdness into the science of botany. Yet *Systema Naturae* was quite puritanical in comparison with Erasmus Darwin's poetic rendering. Hugely popular in England, especially among women, Darwin's *The Loves of Plants* turned the Linnaean sexual system into a *cause célèbre*, with supporters and detractors lining up on either side.

THE ARGUMENT FOR HUMAN–PLANT ANALOGY: LA METTRIE'S *L'HOMME PLANT*

"*Cogito ergo sum*" Descartes famously declared: "I think; therefore I am." Actually, Descartes never wrote these words, Spinoza did—in a treatise on Descartes. In his *Meditations on First Philosophy*, Descartes wondered whether or not he could trust his senses as accurate reporters of the world around him. What if he were mad, or dreaming, or being deceived by an evil demon? What if the world as he perceived it didn't really exist? Perhaps he, himself, did not exist! Eventually, he discovered a way out of this logical trap:

> I have convinced myself that there is absolutely nothing in the world, no sky, no earth, no minds, no bodies. Does it now follow that I, too, do not exist? No. If I convinced myself of something (or thought anything at all), then I certainly existed. But there is a deceiver of supreme power and cunning who deliberately and constantly deceives me. In that case, I, too, undoubtedly exist, if he deceives me; and let him deceive me as much as he can, he will never bring it about that I am nothing, so long as I think that I am something. So, after considering everything very thoroughly, I must finally conclude that the proposition, *I am, I exist,* is necessarily true whenever it is put forward by me or conceived in my mind.[5]

Descartes had proposed a mechanical model for the universe and considered the bodies of plants and animals to be living machines accessible to science—all except the mind. The nervous system, according to Descartes, was merely a system of hollow tubes in which ghostly "animal spirits" wafted throughout the body. The mind and soul were immaterial and therefore not explainable by physical laws.

The French physician Julien Offray de La Mettrie saw no logical reason to invoke mystical animal spirits to explain mental processes. As a teenager, he had abandoned Catholicism to become a Jansenist, the French version of the Protestant reform movement. His first book, *The Natural History of the Soul*, argued that if humans had souls, then other animals and even plants must have them as well because of the uniform laws of nature. The French Parliament ordered all copies of his book burned in 1746, and Mettrie fled to the University of Leiden, where he had previously studied medicine under Hermann Boerhaave, Vaillant's former mentor.

Under the more tolerant auspices of the Dutch, Mettrie penned an even more daring, openly atheistic treatise, *The Machine Man* (1747),[6] which challenged the immateriality of the mind and represented the first modern statement of the materialist philosophy. The mind is not a *tabula rasa* at birth, as John Locke[7] had suggested—which implies that human beings are both perfectible and malleable—but rather possesses an inherited range of potential behaviors, compulsions, and limitations predetermined by inborn anatomical and neurological factors. In other words, in animals as well as people, mind and soul are inseparable from the brain and nervous system, which Mettrie speculated may possess properties "on a par with electricity." Over two hundred years before the biolinguistic theories of Noam Chomsky, Mettrie asserted that the primary mental process that separates humans from chimpanzees is the capacity for language.[8]

The arguments in *The Machine Man* were too radical even for liberal Holland, and the printer was ordered to deliver his entire stock to the Consistory of the Church of Leiden for incineration. Once more Mettrie was forced to flee, this time to Berlin where Frederick the Great of Prussia, an admirer and patron of Enlightenment philosophers, offered him protection and a pension. He immediately turned his attention to plants.

Written in a light-hearted, chatty style during his stay in Berlin between 1748 and 1751, *The Plant Man* articulated Mettrie's belief in the uniformity of life.[9] Poets had been doing this for centuries in the form of literary conceits. For example, in Sonnet 15, Shakespeare utilized a human–plant analogy to universalize the corrosive effects of time and compares his own sonnets praising his friend to freshly grafted scions, which renew the life of an aging stock:

> When I consider every thing that grows
> Holds in perfection but a little moment,
> That this huge stage presenteth nought but shows
> Whereon the stars in secret influence comment;
> When I perceive that men as plants increase,
> Cheered and cheque'd even by the self-same sky,
> Vaunt in their youthful sap, at height decrease,
> And wear their brave state out of memory;
> Then the conceit of this inconstant stay
> Sets you most rich in youth before my sight,
> Where wasteful Time debateth with Decay,
> To change your day of youth to sullied night;
> And all in war with Time for love of you,
> As he takes from you, I engraft you new.

By the eighteenth century, plant–human analogies had become respectable in scientific circles as well. Mettrie had been thunderstruck by the 1744 publication of Abraham Trembley's experiments on regeneration in Hydra.[10] It was common knowledge that entire plants could be propagated from cuttings, and it was also known that certain animals, such as starfish, crayfish, and lizards, could regenerate lost appendages. But a severed lizard tail would never regenerate a whole lizard, nor a crayfish leg regenerate a whole crayfish. According to Aristotle, cutting an animal in half invariably killed it, whereas cutting a plant in half did not, a distinction he attributed to the multiplicity of plant souls.

In the course of examining drops of lake water under a microscope, Trembley encountered an organism previously described by Leeuwenhoek, the freshwater polyp, or Hydra, which typically ranges in size from 2 to 10 mm. Because the tiny polyp was green (due to the presence of algae) he initially thought it was a plant, but as it seemed to be able to move about and even capture food with its tentacles he decided it must be an animal. Applying Aristotle's criterion, he carefully snipped the hydra in half transversely with a scissors to see if would survive. At first the two halves (head and tail) each contracted into a little green ball, but they soon re-extended to their normal shapes. Several days later, the half that contained the head regenerated its bottom half. Trembley wasn't too surprised by this since lizards also regenerated their tails. But when the tail end of the hydra regenerated a head, he was astounded. If he cut the Hydra lengthwise into quarters, each piece regenerated a complete organism, although further subdivisions resulted in death. Trembley subsequently found that when he bisected the head longitudinally, the two halves each regenerated a complete head. Using this trick, he was able to produce a seven-headed Hydra.[11]

Upon reading Trembley's report, Mettrie realized that Aristotle's blanket distinction between plants and animals was false. It was evident to him that the border between plants and animals consisted of intermediate organisms that had features of both kingdoms. Plants, animals, and even humans were all part of a continuum. Thus he felt justified in comparing men and women to dioecious plants.

In the Preface to *L'Homme Plant*, Mettrie insists that his transformation of people into plants is not merely "a story such as Ovid might have told," but is based on the discovery of the "singular analogy between the plant and animal kingdoms." We have only begun to glimpse the principle of uniformity, Mettrie asserts, so we are still not sure how far to apply it. "One should never force nature," he cautions, because nature can and does "stray from the laws she favors most." To determine where analogy is appropriate, he concludes, one should begin by comparing the parts of plants with those of animals.

Mettrie begins with the flow of sap, which he considers analogous to the circulation of the blood. "The lungs are our leaves," says Mettrie. "They do for us what leaves do for plants. . . . Plants have branches to enlarge their lungs so they can get more air." This statement was no doubt derived from *Vegetable Staticks* (1727) by Stephen Hales, an English clergyman who applied the methods of Harvey to study the ascent of sap in plants. In addition, writes Mettrie, the "Harveys of botany" have shown that plants have the same types of structures—veins, vessels, and capillaries—for the flow of sap as animals have for the flow of blood. He also asserts that heat does for plants what the heart does for animals—it drives the movement of the sap: "this fire, I say, is the heart that makes the juices circulate in the tubes of plants, which perspire like men."

Next, Mettrie compares female and male flowers of dioecious plants to the reproductive structures of men and women. He compares the nectary to a breast:

> As flowers have leaves or *petals*, we can view our arms and legs as similar parts. The nectarium, which is the reservoir of honey in certain flowers, such as the tulip and the rose, is like the breast that contains milk in the female plant of our species when the male makes it come. The breast in our species is double, and is seated at the lateral base of each *petal* or arm on the large pectoral muscle.

The pistil, says Mettrie, is analogous to the uterus, vagina, and vulva:

> One can regard the womb of a virgin or rather of a woman not yet pregnant or, if you wish, the ovary, as an unfertilized seed. The *stylus* in a woman is the vagina. The vulva or mount of Venus and the odor that exhales from the glands of these parts correspond to the *stigma*. The uterus, vagina, and vulva together form the pistil, which is what modern botanists call the female parts of the plant.

In men, on the other hand, the "stamen is rolled into a cylindrical tube, the *rod*," and "sperm is our fertilizing pollen." According to the Linnaean system, men, having only one rod, belong to the *Monandria*, while women, having only one vagina, belong to the *Monogynia*. Humankind as a whole belongs to the class *Dioeciae*. Mettrie goes on to compare the details of fertilization in plants and animals according to the limited knowledge then available. Like Vaillant, he has great fun equating the explosive release of pollen from anthers with ejaculation:

> The ejaculation of plants lasts only a second or two. But does ours last longer? I think not, although continence leads to some variations depending on how much sperm is stored in the seminal vesicles. An ejaculation is completed in a single expiration, so it has to be short. Pleasures lasting too long might kill us. Then, because of lack of air or breath, each animal would give life at the expense of its own, and would truly die of pleasure.

Although Mettrie writes whimsically, he clearly believes that his analogies between plants and people are valid. Indeed, the second chapter of *L'Homme Plant* describes the ways in which plants and animals differ from each other, and the third chapter concludes by characterizing the ladder of life as a continuum "so imperceptively graduated that nature climbs it without ever missing a step through all its diverse creations." This was a widely held view in the eighteenth century, one shared by both Ray and Linnaeus.

Whereas Mettrie approached the analogy between plants and people in a scientific spirit, the poem *The Man-Plant: or, a Scheme for Increasing and Improving the British Breed* (1752), by Vincent Miller, was clearly satirical in the tradition of Jonathan Swift.[12] The author of *The Man Plant* expresses his concern over the decline of contemporary British manhood as manifested by the recent dearth of military victories. To rectify the situation, he advocates increasing the number and vigor of British men by applying horticultural techniques that shorten the gestation time for childbirth. The female body, the poet declared, is comparable to "a flower plant, in the method of Linaeus [*sic*]": "her clothing is her *calyx*, her breasts are her *nectarium*, her womb is her *pistillus*, her vagina is her *style unicus*, her cervix is her *stigma oblongum*, her ovaries are her *pericarpium*, and her seed is her *semen*."[13] According to the poet, thirty-nine-day-old embryos would be extracted from the *pistilli* of pregnant women, placed in "bladders" containing liquid, and planted like seeds in soil-filled wicker baskets, where they would thrive and grow until harvested. Women would thus become like "those fertile fields that yield two or three crops in a season." In addition to echoing the ancient metaphor of women as "fruitful fields," *The Man-Plant* anticipates the account of test-tube babies in Aldous Huxley's *Brave New World* and the nightmarish depiction of the exploitation of the female reproductive system in the service of the state in Margaret Atwood's *The Handmaid's Tale*.

BAWDY GARDEN VERSE: *ARBOR VITAE, VULVARIA FRUTEX,* AND *THE SENSITIVE PLANT*

Even before Mettrie's *L'Homme Plant*, British wags had already begun to mine sexual analogies between plants and animals as a source of ribald verse. The lengthy anonymous poem *The Natural History of the Arbor Vitae, or the Tree of Life*, published in 1732, was penned by Thomas Stretzer, one of Edmund Curll's stable of hacks. According to a footnote in the introduction, the piece was originally read to "an honourable SOCIETY near Crane Court, in Fleet street" (a gentleman's club) who had "gone half through it before they were aware of the Deception." The author therefore addressed the published version of the poem to the "Fair Sex," which is "acknowledg'd the quickest of apprehension, especially in works of this nature:" It begins

> The TREE OF LIFE, another name
> For P-n-s, but in sense the same,
> Is a rich plant of balmy juice
> And own'd to be of sovereign use,
> Consisting of one single stem
> That's straight, as is a pistillum;

Whose top sometimes the curious say
Is like a cherry seen in May;
Or glandiform—but's found to be
More oft like nut of filberd-tree.
Quite the reverse of other fruit,
This grows, and dangles near the root,
Producing two of a nutmeg kind.
Twin-like, in one strong purse confin'd . . .
The fruits receive a strong supply
And yield a viscous balmy juice
Adapted to VULVARIA's use [14]

And so on and on and on for a great while.

In 1741, Stretzer published *The Natural History of the Frutex Vulvaria or Flowering Shrub*, an even cruder companion piece to the *Arbor Vitae*. Supposedly written by Philogenes Clitorides, "one of the missionaries of the Society of Jesuits for propagating knowledge in foreign parts," the essay was dedicated to "the two fair owners of the finest *Vulvarias* in the three Kingdoms." In the dedication, Stretzer writes that although Great Britain has always been home to the best *Vulvarias* in the "universe,"

> yet the trees that have been grafted upon them, for these two and twenty years last past, have so far degenerated, that our plants are held in the utmost contempt in all foreign countries, as fit only to be piss'd upon.

Stretzer laments that since the manly days of the Act of Habeas Corpus, the Bill of Exclusion, the exile of James II, and the conclusion of the Treaty of Utrecht, "we have never done any great feats, but seem to be damnably off our mettle,"[15] a predicament Stretzer blames on the "degeneracy of our Trees of Life":

> How much then, beauteous ladies, must the whole nation be obliged to your indefatigable endeavors to restore their vigor, by inoculating none but the finest plants upon your flowering shrubs.

Stretzer suggests that such a union should be fertile because, according to the learned herbalist "Leonard Fuckius," *Vulvaria* belongs to the same genus as the *Arbor Vitae* and is, in fact, none other than the "female *Arbor Vitae*." As to the natural history of *Frutex Vulvaria*, it is "a flat low shrub, which always grows in a moist, warm valley, at the foot of a little hill, etc." Stretzer continues for some time in a similar vein, a testimony to the existence of a receptive audience for this sort of work.

The poem *Mimosa: or, The Sensitive Plant* (1779) by James Perry belongs to the same subgenre of phytoerotica as *Arbor Vitae*, in which the plant as a whole serves as a phallic symbol.[16] Dedicated to the botanist Joseph Banks, who traveled to the South Pacific with Captain Cook on the *Endeavour* between 1768 and 1771, the poem satirizes Banks's purported sexual exploits in Tahiti, although the bulk of its verses feature salacious gossip about the sex lives of prominent aristocrats.[17] *Mimosa pudica* ("modest mimosa") was a problematical phallic symbol, however, because its behavior in response to touch—folding its leaflets and collapsing its petioles—was the opposite way a proper phallic symbol ought to behave. Undeterred, Perry soldiered on verse after verse, straining and stretching his metaphor to the breaking point, while propping up his narrative with copious scientific and gossipy footnotes. The poem is addressed to a "Kitt Frederick, Duchess of Queensberry, Elect," although it seems likely that the name was thinly disguised, like the other names cited in the poem.[18] The poem opens with an invocation to the Duchess:

O Thou! Who hast, so often proved
The virtues of the plant, beloved,—
That from the touch recedes.—
Assist, its magic to display;
For thou hast felt it every way,
And know'st how it suc-ceeds.

In Perry's conceit, the Mimosa with its leaflets open is like an erect phallus. Upon being touched by the Duchess's "lovely hand," the erect *Mimosa* reaches its climax and undergoes "a momentary death":

Yet, not until its magic power
Titilates in every pore,
And takes away the breath;
Miller and many more relate
Its virtual qualities create
A momentary death.
Oft have you felt upon the touch,
The force of contact to be such,
"As made it poor indeed."
The pendicles, alas give way,
It shews each emblem of decay;
And droops the sickly head.

The poem soldiers on for another forty-nine verses with variations on the same sexual joke, raising the question why such works were so popular. The fact that they were recited at all-male clubs offers a partial explanation, and the liberal quaffing of ale and spirits no doubt lowered the bar for hilarity. Increasing talk of women's rights may account for the appeal of the misogynistic aspects, whereas class resentment no doubt added further zest as the sex lives, both "natural" and "unnatural," of notable aristocrats were paraded before the gathering. At the end of the poem, a natural meritocracy based on the size and vigor of one's "plant" was proposed:

Search all nature's wide domains,
And you will find she still maintains
Each thing, within its station.
Lavish and frugal in degree,

And blessing all, yet still doth she
Delight in due gradation.

As in her other plants and flowers,
So here, she proves her mighty powers;
And makes them great and small:
Some more erect, and some more strong,
Some whose motions are more long;
And some who sooner fall.

Comparable sexual satires were also written about nonbotanical subjects, including new theories about human reproduction and electricity.[19] Scientific advances were a common subject of eighteenth-century satirical verse in Britain, whereas the Catholic Church was the main target of French erotica. In England, scientists were beginning to rival the clergy as purveyors of truth, but their privileged status also made them targets for ridicule. Poems like *The Sensitive Plant*, in which botany was mocked, took scientists down a peg and may have helped ease anxieties about unsettling new discoveries, such as plant sex.

Hoping to stem the tide of such ribald satires, a small but vocal minority of botanists reacted by rejecting both the sexual theory of plants and its taxonomic offshoot, the Linnaean sexual system, damning them both as scientifically untenable and highly offensive. The two opposing camps came to be known as the "sexualists" and the "asexualists."

SIEGESBECK'S ATTACKS ON THE SEXUAL SYSTEM AND THE SEXUAL THEORY

"What man," J. G. Siegesbeck railed in his 1737 diatribe against the Linnaean sexual system, "will ever believe that God Almighty should have introduced such confusion, or rather such shameful whoredom, for the propagation of the reign of plants. Who will instruct young students in such a voluptuous system without scandal?"[20] Siegesbeck is generally regarded as a middling botanist, and if it were not for his over-the-top rant against the Linnaean classification system his name would receive scant attention in the annals of botany. Yet his harsh criticisms of the sexual system caused Linnaeus sleepless nights and provoked a cascade of vindictive, tit-for-tat exchanges between the two botanists that illustrates what high stakes the combatants believed they were playing for.

Johann Georg Siegesbeck, a Prussian, had studied botany at the University of Helmstedt under the distinguished physician-botanist, Lorenz Heister. Heister had established a botanical garden at Helmstedt and, with Siegesbeck's help, had built it into one of the largest in Germany. In 1735, Siegesbeck was appointed director at the Apothecary Garden of St. Petersburg and continued as director for seven years, after which he left to become Professor of Botany at the Russian Academy of Science. During his tenure as director of the Apothecary Garden, he organized many collecting expeditions throughout Siberia and brought back a huge number of species, making it one of the richest plant collections in Europe and attracting the attention of Linnaeus, who was particularly interested in the Siberian flora.[21]

A possible clue to Siegesbeck's hatred of the Linnaean sexual system lies in a previous dispute between Linnaeus and Siegesbeck's mentor, Lorenz Heister. Heister had corresponded amicably with the younger Linnaeus and had given him encouragement. In 1732, Heister and his seventeen-year-old son, Friedrich, published a dissertation touting a new method for assigning plant genera based on leaf morphology. Linnaeus apparently regarded Heister's method as archaic and idiosyncratic, although the subject never came up in their correspondence. In the first edition of *Systema Naturae* (1735), however, he listed all the botanists who had used fructification as the basis for classification, singling out Heister as the lone exception. In the same year, he published a short history of botany in which he pronounced Heister's method a mere "draft," and, in a table of taxonomists, he grouped Heister among the "*heterodoxi*." Heister was deeply offended, and he struck back by publishing a paper rejecting Linnaeus's sexual system as completely worthless.[22]

Siegesbeck remained loyal to Heister, and it seems likely that he nursed a grudge against Linnaeus for insulting his former mentor. This would explain why, after a series of friendly letters between Siegesbeck and Linnaeus that also involved exchanges of plant material, Siegesbeck suddenly launched his Jeremiad against the sexual system. As farcical as the events now seem, Linnaeus suffered serious mental anguish over the dust-up and never forgave his Prussian nemesis. He found it particularly galling that Siegesbeck had attacked *Systema Naturae* mainly on theological grounds, a domain that Linnaeus, a minister's son who believed that his reform of botany was sanctioned by God, considered to be his unique calling.

In none of the preliminary letters he had exchanged with Linnaeus did Siegesbeck ever hint that an attack was imminent, but Linnaeus got wind of it through his Russian contact, Johann Amman, Professor of Botany in St. Petersburg.[23] Amman warned Linnaeus that Siegesbeck was preparing an extremely harsh critique of the sexual system. Linneaus, who was then living in the Netherlands and working on a description of the gardens of his wealthy patron, George Clifford, decided on a pre-emptive strike. In keeping with his taxonomic rule that there should be an organic link between a plant and the botanist after whom it was named, Linnaeus quickly gave the name *Siegesbeckia orientalis* to a small weedy composite with an unpleasant odor and published it in *Hortus Cliffortianus* in the summer of 1737.[24]

In December 1737, Siegesbeck published his anticipated blast, innocuously titled *A clear evaluation: Linnaeus's recently published sexual system of plants, and his method for organizing the superstructure of botany*.[25] Siegesbeck began by arguing that according to the Bible, God created plants on the third day, and he did not start with seeds. Rather he created plants fully formed, with their flowers, fruits, and seeds present from the start. There is no mention in the Bible of the existence of two sexes in plants or of sex being required for seed formation. The seed is an integral part of the vegetative plant—a vegetative propagule. And although it was true that Ray, Camerarius, and Vaillant had demonstrated something *resembling* sex in plants, this still did not justify describing stamens and pistils as male and female organs. Alternative explanations had not been ruled out. In other words, Siegesbeck challenged the legitimacy of the sexual theory upon which Linnaeus had based his entire system of classification.

However, Siegesbeck's primary argument against the Linnaean system was its shocking immorality:

> For what strange, discordant orders and classes, totally contrary to Nature, is it not necessary to subordinate in such a Method because of this fictitious matrimony of plants, e.g. when eight, nine, ten, twelve, even twenty or more husbands are found here in the same [bridal] chamber together with one woman.[26]

Would the Creator, Siegesbeck asked rhetorically, allow twenty or more men (stamens) to share a single wife (pistil) or to sanction the practice of keeping adjacent female flowers as their concubines? Ignoring the many instances of polygamy in the Bible, Siegesbeck then answered his own question with the thunderous quote at the beginning of this section: God would never allow such "a shameful whoredom!"

Amman sent Linnaeus a copy of Siegesbeck's polemic in January of 1738. At first Linnaeus refused to dignify it with a response. There was no point arguing with him, he said, because Siegesbeck did not understand scientific evidence. Writing to Albrecht von Haller, he swore that Siegesbeck would "never provoke an angry word from him, though he poured thousands on Linnaeus' blessed head." But when he returned to Sweden from Holland in 1738 he was dismayed to discover that he had become the laughingstock of Stockholm. His own countrymen believed that his sexual system had been annihilated by Siegesbeck—on theological grounds! He was treated like a pariah and even had trouble finding anyone willing to work for him as a servant. Clearly, he had underestimated the role that religion would play in the acceptance of his system.

Vowing never to engage Siegesbeck personally, he approached Carl-Fredrik Mennander, the future archbishop of Uppsala, and Johan Browallius, the future archbishop of Åbo, both of whom were physicists as well as theologians, to ask them if they would be willing to defend him against Siegesbeck's charges. Linnaeus was hoping an appeal to their Swedish national pride would induce them to neutralize a particularly nasty German critic. In 1739, Browallius answered the call. He published a response to Siegesbeck in which he stated that there was no fundamental conflict between the account of the Creation in Genesis and Linnaeus's sexual system of classification. Morality, stated Browallius, refers to the laws given by God to humanity. In nature, there are other laws, also sanctioned by God, which cannot be classified as either moral or immoral. Plants reproduced just as Linnaeus described, and there was nothing lewd or obscene about it. Just as monogamy is appropriate for people, polygamy is often called for in plants and animals. For example, both quadrupeds and birds are often polygynous, otherwise farmers would have to keep as many cocks as hens. It could also be said that bees practice polyandry, since one Queen is served by many males. Siegesbeck's criticism of Linnaeus was thus absurd and must be completely rejected.[27] In the end, Browallius's paper had its intended effect, rehabilitating Linnaeus's image in the eyes of his fellow Swedes. In return, Linnaeus named the genus *Browallia*, sometimes called the amethyst flower or sapphire flower because of the beauty of its blossoms, in his honor.

But Siegesbeck would not give up so easily. In 1741, he published yet another sharp-tongued critique of Linnaeus entitled *Vaniloquentiae botanicae specimen* (Vainglorious Botanical Specimen), in which he stated that when Linnaeus calls seeds ovaries and talks of matrimony, happiness, and the loves of plants, he, Siegesbeck, can't tell whether Linnaeus is competing with botanists or with poets and orators. Linnaeus's sexual system of classification should properly be called "the *lascivious* system." Polygamy may be sanctioned in the Old Testament, but prostitution is not!

The antagonism between Linnaeus and Siegesbeck blossomed into an intense mutual hatred during the 1740s. The situation was further exacerbated by a famous incident involving a packet of *Siegesbeckia* seeds, which sundered their relationship forever. There are two versions of the story, so we will present both of them. According to one version, Linnaeus had come across a stray packet of *Siegesbeckia orientalis* seeds and, in a fit of pique, relabeled it *Cuculus ingratus* (Ungrateful cuckoo) and put it aside. In 1744, Count Sten Carl Bielke of

Stockholm, together with one of Linnaeus's earliest pupils, Pehr Kalm, journeyed to Russia to collect plant material for Linneaus's collection and the Count's experimental garden in Uppsala. By this time, Linneaus's Russian ally Johann Amman had died, and Siegesbeck had succeeded him as Professor of Botany at the Russian Academy of Sciences. As was customary, Bielke and Kalm both brought seed packets from their own collections with them to exchange for Russian seeds.[28] However, as chance would have it, one of the seed packets that Count Bielke had traded for Russian material in St. Petersburg had been the packet of *Siegesbeckia* that Linnaeus had relabeled *Cuculus ingratus*. The latter eventually found its way to Siegesbeck, who planted a few of the seeds to see what the new species with the funny name looked like. As soon as the plants had grown he recognized it as *Siegesbeckia orientalis* and was understandably furious.

According to the second version of the story, which seems more plausible, it was Linnaeus himself who sent the relabeled seeds directly to Siegesbeck. It is the sort of reckless impulse that Linnaeus was prone to when he was sufficiently provoked. Although Bielke and Kalm did their best to mediate between the two, Linnaeus never received another plant from St. Petersburg as long as Siegesbeck was alive. When Bielke wrote to Linnaeus begging him to apologize to Siegesbeck so as not to jeopardize future plant exchanges with the Russians, Linnaeus adamantly refused, referring to Siegesbeck as *Ingratissimus cuculus et nebulo* ("The most ungrateful cuckoo and a wretch"), perhaps the only time he ever deviated from binomial nomenclature.[29] "I must say with Pilate," he wrote, 'What is written is written' . . . I will never forgive him his roguery." Bielke's fears turned out to be groundless because Linnaeus had many admirers at the Russian Academy, including Johann Georg Gmelin and Grigorii Demidov, and he continued to receive generous collections of plant material sent to him by other Russian botanists.

There was one last episode in the Linnaeus–Siegesbeck Punch and Judy show. By 1745, Russian botanists had discovered a strange new plant in Siberia with small composite flowers. When they dissected the individual florets of the new species they were unable to find any stamens. Siegesbeck was ecstatic. He believed that he had finally found the evidence disproving the sexual theory that he had been looking for, thus invalidating Linnaeus's sexual system of classification. The plant was hastily named *Anandria siegesbeckioides* in Siegesbeck's honor.

Gmelin informed Linnaeus about the discovery of *Anandria* ("without males"), but Linneaus was highly skeptical. However, he could not immediately comment on the finding because Siegesbeck was keeping a tight lid on his stock of *Anandria* seeds, and Linnaeus was the last person on earth to whom he would ever send them. Luckily for Linnaeus, Count Bielke visited St. Petersburg in 1745 and managed to wangle some *Anandria* seeds. Linnaeus received them with mounting excitement, sensing that a decisive victory was at hand. Although the plants refused to flower at first, eventually they obliged, and when Linnaeus examined the florets under a microscope, sure enough, he found the stamens! They were small and difficult to detect, but they were unquestionably present and full of pollen. Based on his sexual system, he identified the now-misnamed *Anandia* as a new member of the genus *Tussilago*. In *Dissertatio Botanica de Anandria* (1745), Linnaeus announced his discovery, once again depriving Siegesbeck of his longed-for triumph over the sexual theory and the sexual system.

In spite of his victory, Linnaeus continued in such an extreme state of animosity that he was unable to muster a shred of sympathy, even when Siegesbeck's son committed suicide a short time later. Instead, he regarded the tragedy as divine retribution for Siegesbeck's treatment of himself. Prudently, he refrained from saying so publicly.[30]

If Linnaeus learned anything from his feud with Siegesbeck, it was to tread more carefully in matters of religion. Before he published two of his more philosophical works, *De Curiositate Naturali* in 1748 and *Oeconomia Naturae* in 1749, he asked three prominent Catholic theologians from Venice—one from the Franciscan order—to review the manuscripts for any taint of heresy or blasphemy. None could find anything that would conflict with Christian morality or teaching.[31]

Despite Linnaeus's caution, in 1759 Pope Clement XIII banned all of his books from Vatican territories and ordered all copies of his works to be burned. The immediate cause was Linnaeus's taxonomic category of Primates, which placed humans in the same group as monkeys. While he was at it, the Pope also banned *Systema Naturae* for good measure. However, in 1774, Pope Clement XIV, Pope Clement XIII's successor, reversed the ban on the *Systema Naturae*, and Linnaeus rejoiced at the news that botanists were now being invited to present lectures on his method at the Vatican. The new Pope's approval of Linnaeus's sexual system was a dramatic sign that, by 1774, the Catholic Church had become reconciled to the fact that plants had sex.

AVOIDING THE "S"-WORD IN BRITISH BOTANY TEXTS

It took a few decades for the general public to become fully aware of Linnaeus's *Systema Naturae* (1735), which they at first learned by means of summaries in popular books on botany and eventually through translations into English and other languages. By the 1760s, Linnaeus had become an international celebrity, hailed as the codifier of the new rules of plant names and classification. Women, especially, were attracted to the new method of binomial nomenclature and the user-friendly sexual system for identifying plants. [32] Increasing numbers of botany books based on the Linnaean sexual system were being written by and for women, setting Linnaeus's anthropomorphized vocabulary for describing plant sex on a collision course with contemporary social mores. Discussing the mechanics of "monogamous nuptials" was awkward enough, let alone the more usual case of polygamy.

Those authors who rejected the sexual theory itself dodged the issue entirely by retaining Tournefort's outdated asexual system of plant classification. Others, like William Withering, who accepted both the sexual theory of plants and the Linnaean system of classification, nevertheless found it necessary to avoid any mention of the word sex, or even the distinction between male and female floral structures. In the Introduction to his book, *The Botanical Arrangement of All the Vegetables Naturally Growing in Great Britain* (1776), Withering stated that while it "is natural to ask the uses of these different parts—[a] full reply to such a question would lead us to a long disquisition, curious in itself, but quite improper in this place." In other words, a full discussion of the roles of stamens and pistils would inevitably lead to a discussion of sex, which would be "improper" in a book intended for young women. As Withering explained in his preface, "from an apprehension that botany in an English dress would become a favorite amusement with the ladies, . . . it was thought proper to drop the sexual distinctions in the titles to the Classes and Orders." So far as the uses of floral structures were concerned, it was sufficient to state that "the production of a perfect seed is the obvious use of the flower."

Rather than use the terminology for floral structures established by Tournefort, Ray, and Linnaeus, Withering reverted to those of Grew: *empalement* for the calyx, *chives* for stamens, tips for anthers, *pointals* for pistils, *summits* for stigmas, and *seed vessels* for ovaries. It has been suggested that this was an effort to avoid sexual connotations,[33] but this would not account for the use of *empalement* for calyx, a structure that has no sexual function. Perhaps Withering also believed that Anglicized terms would be more congenial to an English audience than Latinate terms.[34] Despite Withering's fastidious avoidance of any mention of sexuality, his description of pollination was reasonably accurate as the process was then understood:

> The fine dust, or meal (farina) that is in the tips is thrown upon the summit of the pointal: This summit is moist and the moisture acting upon the particles of dust occasions them to explode and discharge a very subtle vapour. This vapour passing through the minute tubes of the pointal arrives at the embryo seeds in the seed bud and fertilizes them. The seeds of many plants have been observed to become, to all appearance, perfect without this communication; but these seeds are incapable of vegetation.

Like Grew, Withering was a preformationist and an ovist. According to Withering, the ovary contained an "embryo seed" prior to pollination, and pollen merely served to activate the embryo by means of a "subtle vapour." Ovism was, of course, more compatible with an asexualist presentation than was spermism, which required the physical transfer of the germ to the female.

By the end of the century, authors of botany books were still struggling with the dilemma of how to present the Linnaean system to women. This was especially true of female authors, even though they accepted the sexual theory itself. For example, in her book *An Introduction to Botany, in a Series of Familiar Letters* (1796), Patricia Wakefield wrote:

> Each anther is a kind of box, which opens when it is ripe, and throws out a yellow dust that has a strong smell; this is termed pollen or farina, and is the substance of which bees are supposed to make their wax. The progress of the seed to maturity is deserving of the most curious attention. First the calyx opens, then the corolla expands and discovers the stamens, which generally form a circle within the petals, surrounding the pointal [pistil]. The pollen or dust, which bursts from the anthers, is absorbed by the pointal, and passing through the style, reaches the germ, and vivifies the seed, which without this process would be imperfect and barren.

Wakefield's account of fertilization includes everything then known about pollination except the word "sex." She goes on to cite the experimental evidence in support of the sexual theory while maintaining female modesty by steadfastly avoiding the "S" word. Also absent from her account are any analogies, poetic or otherwise, between floral and human sexual organs—a style of thinking and writing that, in any case, had been dropped by this time.

ANNA SEWARD, ERASMUS DARWIN, AND "THE LOVES OF PLANTS"

William Withering's censorship of Linnaean botany nettled his friend and mentor, Erasmus Darwin, Charles Darwin's grandfather. Darwin—an inventor, poet, and physician of some note—had recommended Withering for membership in the Lunar Society as a replacement for Dr. William Small, who had died in 1775. The Lunar Society of Birmingham was an informal dinner club and learned society that included intellectuals, physicians, industrialists, and natural philosophers and counted Benjamin Franklin among its illustrious members. The primary focus of the group was the interface between science and technology.

Erasmus Darwin regarded William Withering's scrupulous avoidance of all mention of sex in a book that purported to describe the Linnaean sexual system as a prudish bowdlerization. It also gave him the idea that there might be a market for an uncensored popular presentation of the Linnaean sexual system. Around 1777, he installed a botanic garden in a field about a mile from his home in Lichfield, arranging the plants according to the Linnaean system. When the landscaping was completed, he invited the poet Anna Seward, the daughter of long-time friends from the cathedral close community in Lichfield, to visit the new garden.[35] However, on the day of her visit, he was called away on a medical emergency, so she visited the garden alone, bringing her notebook along. While seated among the flowers she wrote a short poem inspired by Darwin's handiwork and showed it to him when next she saw him. He was so taken by it that he suggested that she should use it as the basis for a larger poetic work on the Linnaean System, an idea that he himself had been mulling over for some time. In her memoir, she relates that Darwin told her that her poem "suggests metamorphosis of the Ovidian kind, though reversed. Ovid made men and women into flowers, plants, and trees. You should make flowers, plants, and trees into men and women. I . . . will write the notes, which must be scientific; you shall write the verse."[36] But Anna declined the invitation on the grounds that the subject was "not strictly proper for a female pen."[37]

Although Seward cited feminine modesty as the reason for declining Darwin's invitation, she was no shrinking violet. She is assumed to be the author of the anonymous poem, "The Backwardness of Spring Accounted For," which commemorated the publication of the first English translation of *Systema Naturae* in 1783. The translation had been the first of several translations of Linnaeus's works by the Lichfield Botanical Society, founded by Erasmus Darwin. "The Backwardness of Spring Accounted For" describes the anarchic state of plants before the arrival of Linnaeus, with classes and sexes cavorting willy-nilly.[38] By suggesting that Linnaeus was on their side, the side of law and order, Seward's portrait of the pre-Linnaean plant kingdom as a licentious and ungovernable mob was designed to appeal to English conservatives, increasingly disturbed by the ominous rumblings coming out of Paris. Before Linnaeus, says Seward's Mother Nature, vegetation was running amok in the grip of a "leveling spirit":

Vegetation of course was o'er run with disorder
From the wood & the wall to the bank & the border
Her wisest oeconomy strangely distorted
And her government cou'd not be longer supported
"Here rank & high titles," say she, "have no merit
"And my weeds are brought up with a leveling spirit . . .

". . . No wonder we see such a grinning of Corols [corollas]
"Amidst this confusion of manners and morals . . ."

The malevolent forces of anarchy and free love had even infected Mother Nature's flower arrangements:

"By the legs shall a right perfect flower & his bride
"With mules and hermaphrodites daily be ty'd
"Can a marriage made public & marriage clandestine
"The same common bed with strict decency rest in . . .

"Ye wives with ten husbands, say will they content ye
"Whilst a neighbor lies by you with no less than twenty
"Ye husbands in wives tho' not stinted to few
"Don't envy the Flower that has concubines too
"Ye ladies with eunuch, no is it not hard
"That your virtue should seem to require such a guard
"While your gay-painted cowslip may gad where she will
Yet her husband good creature suspecteth no ill
"Patricians stand forth and say what lady's bosom
"Makes amends for your joining a plebian blossom . . ."

In "The Backwardness of Spring Accounted For," Seward paints a daringly racy spectacle of sexual promiscuity among the flowers in which social hierarchies are ignored, which no doubt explains why Seward published the poem anonymously. Never fear, Mother Nature exhorts, "some scholars at Lichfield" are "compiling my classification":

"These great legislators will shortly prescribe
"The laws rules & habits of every tribe
"Thus their manners no longer each other will shock
"What is wrong for the rose may be right in a dock . . ."

Seward cleverly turns on its head the asexualist argument that the Linnaean sexual system is both immoral and a threat to the social order. Linnaeus did not invent plant sex, she asserts; he simply brought it under control and made it conform to the natural laws appropriate for each tribe. Thanks to Linnaeus, proclaims Mother Nature, plants from different classes will now restrict themselves to their proper taxonomic sphere and will no longer shock each other by attempting to cross class barriers. According to Mother Nature, everyone who values stable government has reason to commend the Linnaean system for bringing harmony and order to the plant kingdom:

"Rejoice then my children the hour is at hand
"When Botanical knowledge shall govern the land . . ."

Erasmus Darwin eventually wrote the extended poem on the Linnaean classification system that he had previously urged Anna Sward to write. "The Loves of Plants" was published in 1789, as the French Revolution was getting underway. Fearful that it might damage his reputation as a physician, Darwin published the book anonymously. At the same time, he had high hopes for commercial success, even if it meant shocking a few pious prudes.

Darwin's epic poem was written in heroic couplets—rhymed couplets in iambic pentameter—which had been used to such powerful effect in John Dryden's translation of Virgil's *Aeniad* (1697) and Oliver Goldsmith's *Deserted Village* (1770). Drawing on the pastoral tradition, Darwin clothed the stamens and pistils of Linnaeus's twenty-four classes in the garb of shepherds and shepherdesses, knights and ladies, and other courtly inhabitants of Arcadia so popular in eighteenth-century art and literature. Canto I begins with a ringing invocation:

> Descend, ye hovering Sylphs! Aerial choirs,
> And sweep with little hands your silver lyres;
> With fairy footsteps print your grassy rings,
> Ye Gnomes! Accordant to the tinkling strings;
> While in soft notes I tune to oaten reed
> Gay hopes, and amorous sorrows of the mead.—
> From giant oaks, that wave their branches dark,
> To the dwarf moss, that clings upon the bark,
> What beaux and beauties crowd the gaudy groves,
> And woo and win their vegetable loves . . .

Darwin presents eighty-three species representing examples of each of the Orders and Classes. Classes end with *andria* (monandria, diandria, triandria, etc.) and Orders end in *gynia* (monogynia, digynia, etc.). He begins with one of the few monogamous, and therefore "virtuous," angiosperm flowers belonging to the class and order *monandria monogynia*—*Canna*, or Indian cane:

> First the tall CANNA lifts his curled brow
> Erect to heaven, and plights his nuptial vow;
> The virtuous pair, in milder regions born,
> Dread the rude blast of Autumn's icy morn;
> Round the chill fair he folds his crimson vest,
> And clasps his timorous beauty to his breast.

Each verse is accompanied by a footnote translating the ornate metaphors into dry, unadorned prose. For example, the note for Canna reads:

> Canna. Cane, or Indian Reed. One male and one female inhabit each flower. It is brought from between the tropics to our hot-houses, and bears a beautiful crimson flower; the seeds are used as shot by the Indians, and are strung for prayer-beads in some catholic countries.

However, the modern interpretation of the monogamous relationship in the *Canna indica* flower (canna lily) is actually far more complex than Darwin could possibly have imagined. [39]

Erasmus Darwin's fertile imagination finds richer material to work with in the other Linnaean categories. For example, the water-starwort *Callitriche* is given as an example of *monandria digynia*.

Thy love, CALLITRICHE, *two* virgins share,
Smit with thy starry eye and radiant hair;—
On the green margin sits the youth, and laves
His floating train of tresses in the waves;
Sees his fair features paint the streams that pass,
And bends for ever o'er the watery glass.

Once again, the situation is a bit more complicated than Darwin suspected. Rather than two pistils, *Callitriche* only has one, consisting of a single ovary with two styles. What would Darwin have made of this situation? A starry-eyed youth smitten by conjoined twins?

The following three stanzas serve to illustrate Darwin's flair for the romantic. The personified stamens and pistils for four genera are described—*Collinsonia*, a mint; *Melissa,* balm plant; *Meadia*, American cowslip; and *Gloriosa*, a lily:

Two brother swains, of COLLIN'S gentle name,
The same their features, and their forms the same,
With rival love for fair COLLINIA sigh,
Knit the dark brow, and roll the unsteady eye.
With sweet concern the pitying beauty mourns,
And soothes with smiles the jealous pair by turns.

Two knights before thy fragrant altar bend,
Adored MELISSA! And *two* squires attend.
MEADIA'S soft chains *five* suppliant beaux confess,
And hand in hand the laughing belle address;
Alike to all, she bows with wanton air,
Rolls her dark eye, and waves her golden hair.

When the young Hours amid her tangled hair
Wove the fresh rose-bud, and the lily fair,
Proud GLORIOSA led *three* chosen swains,
The blushing captives of her virgin chains.—
—When Time's rude hand a bark of wrinkles spread
Round her weak limbs, and silvered o'er her head,
Three other youths her riper years engage,
The flattered victims of her wily age.[40]

Darwin bases his interpretation of the bending of pistils and stamens in *Gloriosa* and other species on Linnaeus, who assumed that most flowers self-pollinated and was unaware of the major role of insects in cross-pollination. Stamen- and pistil-bending was thus interpreted as copulation. In reality, however, the *Gloriosa* pistils typically bend *away from* the stamens, a phenomenon which maximizes the chances of out-crossing.

Janet Browne has noted that, in the majority of cases, Darwin focused on the actions of the females in his verses.[41] In each of the scenarios, it is the behaviors of the females that tend to drive the dramatic action. In contrast, the males "were not given the same attention or depth of characterization, even in some cases being sketched solely in terms of almost empty labels such as 'swain' or 'beau.'" In the realm of plant sexual reproduction, it is the feminine principle that governs the outcome, consistent with Linnaeus's claim that the feminine "medulla" provides the impetus or "will" to produce seed, whereas the masculine "cortex" only determines the external appearance of the plant. Both Linnaeus and Erasmus Darwin granted female sexuality a dominant role in the "oeconomy" of plants, which may help to explain why asexualist critics, all of whom seem to have been men, were so morally outraged. This is not to suggest that either Linnaeus or Darwin were proto-feminists—far from it. As discussed by Londa Scheibinger and Janet Browne, both Linnaeus and Darwin shared the same patriarchal attitudes as their contemporaries, although Darwin was clearly the more radical of the two in other areas, such as religion and politics.[42] Both men were romantics, and romantics by definition are sentimentally attached to the past. Thus it is not surprising that they would both invoke the classical poets while metaphorically emphasizing the feminine aspect of plants.

"THE ANTHERS AND STIGMA ARE REAL ANIMALS"

Although tongue-in-cheek, Erasmus Darwin's personification of plants was more than just a literary conceit. Like many other botanists in the eighteenth century, Darwin believed that plants could experience a variety sensations: including hunger, fatigue, embarrassment, desire, and so on, although not as keenly as animals do. And he was not alone in these beliefs. Writing in 1787, René-Louis Desfontaines, Professor of Botany at the Jardins des Plantes, described the quivering of sexually aroused stamens in response to "the action of the pistil itself, which incites each stamen to orgasm, similar in a sense to the familiar orgasm that occurs in the sexual parts of animals." Concurrently, according to Desfontaines, the style expands slightly, "as if the law requiring a certain modesty in females were common to all organized beings."[43] His reasoning here is unclear, to say the least, but perhaps Desfontaines interpreted expansion of the style as the plant equivalent of blushing. Along the same lines, the British botanist Benjamin Cooke, FRS, in 1748, claimed that the styles of white-seeded maize plants consciously blushed crimson upon contacting pollen from a red-seeded variety:

> On the Manner of Impregnation of the Seeds in Mayze—I can add this, that if the seed and whole Species of Mayze be planted about two Yards Distance from each other, there will be a Mixture of red and white Grains in the Ears of each Plant, and you may with Pleasure observe the Filament in the white Plant, which hath been struck with the red Farina, discovering its alien Commerce by a conscious Blush.[44]

If Cooke actually observed the coloration of the silk as he describes, it is, of course, not due to embarrassment but to the presence of the red pigment, anthocyanin, which can be triggered by a variety of external conditions. In his poem "Zoonomia" (1794–1801) Darwin writes, "The anthers and stigma are real animals. . . . They are affected with the passion of love and furnished with powers of reproducing their species." As he wrote in his philosophical poem *The Temple of Nature* (1803), flowers reveled in the joy of sex:

Hence on green leaves the sexual Pleasures dwell,
And Loves and Beauties crowd the blossom's bell;
The wakeful Anther in his silken bed
O'er the pleas'd Stigma bows his waxen head;
With meeting lips, and mingling smiles, they sup
Ambrosial dew-drops from the nectar'd cup;
Or buoy'd in air the plumy Lover springs,
And seeks his panting bride on Hymen-wings.[45]

In "The Loves of Plants," Darwin cited the dioecious aquatic plant *Vallisneria*, or eelgrass, as an example of entire male flowers exhibiting animal-like mating behavior: "The male flowers of *Vallisneria* come even closer to apparent animality, for they detach themselves from the plant and float on the surface of the waters to meet their females." An illustration of this phenomenon is shown in Figure 14.1A,B. Gazing at the pile up of tiny male flowers on the (relatively) large floating female flower, it is easy to understand why Darwin and other eighteenth-century botanists thought they were witnessing active mounting of the female flower by the male flowers, although, what actually happens is somewhat different.

The leaves of *Vallisneria* grow submerged in lakes or streams, while the larger female flowers, borne singly within a spathe, rise to the surface by the uncoiling of their long spiral stalks. The tiny submerged male flowers grow crowded together on a spadix enclosed by a spathe tethered by short stalks. At maturity, the male spathes open and the small, unopened flowers detach from their spadix. Upon reaching the surface, they slowly open and are moved along the water surface by wind and currents. When they come near the female flower, they aggregate there within the cup-like depression created by the weight of the female flower on the water surface. In his 1917 paper, R. B. Wylie remarks that "as many as 50 staminate flowers may be caught in a single depression, thus forming conspicuous patches on the surface of the water." [46] Under these conditions, the staminate flowers slide down and upend onto the female flower, transferring their pollen to the two sticky stigmas (Figure 14.1B).

Given his assumption that plants had sensations like animals, Darwin was especially intrigued by plants exhibiting thigmonasty, the ability to move in response to touch. In stark contrast to James Perry's phallocentric interpretation in *The Sensitive Plant*, Darwin personified *Mimosa* as a chaste "eastern" maiden:

Weak with nice sense, the chaste Mimosa stands,
From each rude touch withdraws her timid hands,
Shuts her sweet eye-lids to approaching night,
And hails with freshen'd charms the rising light,
Veiled with gay decency and modest pride,
Slow to the mosque she moves, the eastern bride.

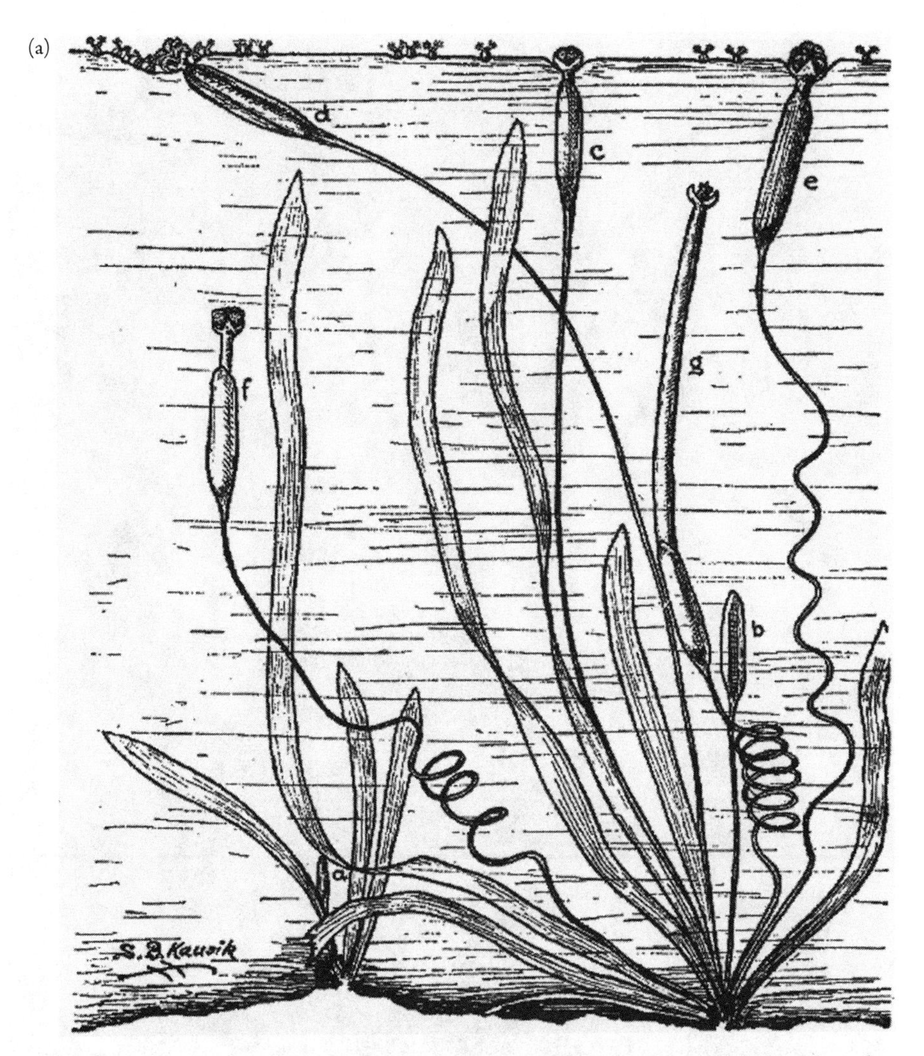

FIGURE 14.1 A. Illustration of *Vallisneria spiralis* L. showing female flowers on long stalks and freely floating male flowers, some of which are accumulating in the depression in the water's surface caused by the female flower. B. Tiny male *Vallisneria* flowers accumulating on top of a single large female flower in a depression in the water surface caused by surface tension. Pollen then germinates on the sticky stigmatic surfaces.

A is from Kausik, S. B. (1939), Pollination and its influences on the behavior of the pistillate flower in *Vallisneria spiralis. American Journal of Botany* 26:207–211.

B is from Wylie, R. B. (1917), The pollination of *Vallisneria spiralis. Botanical Gazette* 63:135–145.

FIGURE 14.1 Continued

Continuing the feminine theme in botany, Darwin represents the insectivorous sundew plant (*Drosera*) as the imperial "Queen of the marsh," treading the "rush-fringed banks" in robes of "glossy silk":

A zone of diamonds trembles round her brows;
Bright shines the silver halo as she turns;
And as she steps the living lustre burns.

In his personifications of plants as women, Darwin presented women—whether timorous, predatory, or regal—as romantic, dynamic individuals. It is easy to understand why "The Loves of Plants" was such a hit with the ladies in the late eighteenth century. As we shall see in the next chapter, the supposed negative impact of the sexual system on the morals of young women became a rallying cry of eighteenth-century asexualists, even as Joseph Koelreuter's hybridization studies were establishing the validity of the sexual theory of plants beyond any shadow of a doubt.

NOTES

1. Peakman, J. (2007), *Mighty Lewd Books: The Development of Pornography in Eighteenth-Century England*. Palgrave Macmillan.

2. The term *erotica* is derived from *eros*, the Greek word for love, whereas the term *pornography* is derived from the Greek word for prostitute. The two terms first came into use during the nineteenth century, but their meaning overlapped considerably. Peakman (2007) defines pornography as "graphic descriptions of sexual organs and/or action . . . written with the prime intention of sexually exciting the reader." Peakman considers pornography to be subcategory of erotica, which she defines as any literary treatment of sex. Others define pornography as sexual exploitation, in contrast to erotica, which implies mutual pleasure, and opinions differ as to where to draw the line. Neither term was in use during the eighteenth century, and the distinction between the two was not made.

3. Cited by Peakman, *Mighty Lewd Books*, p. 17.

4. A third category, "geo-erotica," which we omit from this discussion because it is not strictly botanical, has also been termed "sexual utopias" by Peakman. In geo-erotica entire landscapes are identified with the female body: mountains equal breasts, caves equal vaginas, which lie supine and sexually available to its male discoverers. The most famous example is Thomas Stretzer's *A New Description of Merryland. Containing a Topographical, Geographical and Natural History of that Country* (1740). A projection of the ancient idea of Mother Earth, geo-erotica has been linked to British imperialism, in which colonization is imagined as a kind of rape; see Lewes, D. (1993), Nowhere: Pornography, empire, and utopia. *Utopian Studies* 4:66–73.

5. Cottingham, J., R. Stoothoff, and D. Murdoch, eds. and trans. (1984), *The Philosophical Writings of Descartes*. Cambridge University Press, Vol. II, 16–17.

6. de la Mettrie, Julien Offray (1747–48/1994), *Man a Machine and Man a Plant*, trans. R. A. Watson and M. Rybalka. Hackett Publishing Co., Inc.

7. The English philosopher and physician John Locke (1632–1704) was a significant figure of the Enlightenment period. Considered a liberal, he was the author of "An Essay Concerning Human Understanding" and other important and influential works.

8. See Introduction by Justin Lieber in de la Mettrie, *Man a Machine and Man a Plant.*

9. de la Mettrie, *Man a Machine and Man a Plant.*

10. Hydra is a simple, freshwater invertebrate, up to 30 millimeters (1.2 inches) long, consisting of a cylindrical tube with a mouth surrounded by a ring of stinging tentacles. It is named after the mythical many-headed snake whose heads grew back as quickly as they were severed.

11. Trembley, A. (1744), *Mémoires pour servir à l'histoire d'un genre de polypes d'eau douce.* Gebr. Verbeek, Leiden.

12. Miller, V. (1752/2010), *The Man-Plant: or, a Scheme for Increasing and Improving the British Breed.* Gale ECCO.

13. Harvey, K. (2004), *Reading Sex in the Eighteenth Century.* Cambridge University Press.

14. Quoted by Leapman, M. (2000) *The Ingenious Mr. Fairchild.* St. Martin's Press.

15. Although Britain suffered no defeats in the intervening period, there were no splendid victories either. But in 1739 and 1741, Britain suffered humiliating defeats while attempting to seize Spanish colonies in the Caribbean. The poem was republished in 1741 with a new dedication advising women to grow only strong and beautiful plants and eschew variety.

16. *Mimosa pudica*, or the "sensitive plant," is a leguminous herb native to South and Central America, which is widely grown as a botanical curiosity because its compound leaves are sensitive to touch. Upon contact, the leaflets fold rapidly and the petiole bends downward, a response thought to protect the leaves from herbivores. The leaves begin to recover after a few minutes.

17. Fara, P. (2003), *Sex, Botany and Empire: The Story of Carl Linnaeus and Joseph Banks.* Icon Books.

18. Perhaps it is a veiled reference to Catherine [Kitty] Douglas, Duchess of Queensberry and Dover (1701–77), who had died two years earlier. She was a patron of the playwright John Gay and was reputed to be the most beautiful woman in Europe in the eighteenth century.

19. Peakman, *Mighty Lewd Books.*

20. Siegesbeck, J. G. (1737), *Botanosophiae verioris brevis sciographia in usum discentium adornata: accedit ob argumenti analogiam, epicrisis in clar. Linnaei nuperrime evulagtum systema plantarum sexuale, et huic superstructam methodum botanicam*; Academia scientiarum (Petropolis). Typis Academiae, Petropoli (St. Peterburg).

21. Shetler, S. G. (1969), The herbarium: past, present and future. *Proceedings of the Biological Society of Washington* 82:687–758.

22. Ann-Mari Jönsson from the Linguistics and Philology Department of the University of Uppsala has given a fascinating account of the vituperative battle between Siegesbeck and Linnaeus, and much of the material in this section is taken from her symposium presentation, available online at http://www.phil-hum-ren.uni-muenchen.de/GermLat/Acta/Jonsson.htm. Jönsson, A. -M. (2002), The reception of Linnæus's works in Germany with particular reference to his conflict with Siegesbeck, in E. Kessler and H. C. Kuhn, eds., September 2001 Munich conference "Germania latina—latinitas teutonica." Humanistische Bibliothek/Reihe I (Abhandlungen), Vol. 54.

23. Rowell, M. (1980), Linnaeus and botanists in eighteenth-century Russia. *Taxon* 29:19–20.

24. Jönsson, The reception of Linnæus's works in Germany.

25. Loosely translated from "*Epicrisis in clar. Linnæi nuperrime evulgatum systema plantarum sexuale, et huic superstructam methodum botanicam.*"

26. Ibid. p. 49, cited by Jönsson (2002).

27. Ibid.

28. According to Ann-Mari Jönsson, the two Swedes returned to Uppsala with a "rich collection of dried plants, seeds and more than 200 sorts of herbs" from Siberia.

29. Jönsson, The reception of Linnæus's works in Germany.

30. Linnaeus considered divine justice to be an integral part of the "oeconomy of nature," and, starting in the early 1740s he collected historical, political, and personal anecdotes that demonstrated either God's compensation for good deeds or His harsh punishment for bad deeds. These private notes in the form of maxims never meant for publication were intended solely as a moral guide for his son, Carl Linnaeus the Younger. They were eventually published posthumously as *Nemesis Divina*, and an English translation appeared in 1968.

31. Jönsson, The reception of Linnæus's works in Germany.

32. Shteir, A. B. (1996). *Cultivating Women, Cultivating Science: Flora's Daughters and Botany in England, 1760–1860*. Johns Hopkins University Press.

33. George, S. (2007). *Botany, Sexuality, and Women's Writing (1760–1830)*. Manchester University Press.

34. Shteir, *Cultivating Women, Cultivating Science.*

35. A cathedral close is a series of buildings associated with a cathedral, sometimes forming a square surrounding a courtyard, or close.

36. Seward, A. (1804), *Memoirs of the Life of Dr. Darwin.*

37. King-Hele, D. (1977) *Doctor of Revolution: The Life and Genius of Erasmus Darwin*. Faber; George, S. (2005). "Not strictly proper for a female pen": Eighteenth-century poetry and the sexuality of botany." *Comparative Critical Studies* 2:191–210.

38. George, *Botany, Sexuality, and Women's Writing.*

39. What appears to be a red perianth is actually a collection of highly modified petaloid stamens. Only one of the petaloid stamens is fertile, and it faces a single petaloid style, which possesses two stigmas, one at the apex and one on the side just below the apex. Before the flower bud opens, the anther transfers pollen to the laterally located stigma, a process known as "secondary pollen presentation." The stamen then dies. Although there are two stigmas, pollen can grow to form a pollen tube only on the apical stigma. The lateral stigma functions solely to present pollen to visiting bees.

40. Darwin explains his gothic account of *Gloriosa* in a footnote: "The petals of this beautiful flower with three of the stamens, which are first mature, stand up in apparent disorder; and the pistil bends at nearly right angle to insert its stigma amongst them. In a few days, as these decline, the other three stamens bend over and approach the pistil."

41. Browne, J. (1989), Botany for gentlemen: Erasmus Darwin and "The Loves of the Plants." *Isis* 80:592–621.

42. Schiebinger, L. (1993), The private lives of plants, in L. Schiebinger, ed., *Nature's Body: Gender in the Making of Modern Science*. Beacon Press; Browne, Botany for gentlemen.

43. Desfontaines (1787), Observations sur l'irritabilité des organs sexuels d'un grand nombre de plants. *Mémoires de l'Académie des Sciences*, 473. Cited by Delaporte, F. (1982), *Nature's Second Kingdom*. MIT Press.

44. Cooke, B. (1749), The mixture of the farina of apple-trees,—of the mayze or Indian Corn. *Philosophical Transactions of the Royal Society of London* 46:205–207.

45. Darwin, E. (1803), *Temple of Nature*, 2, pp. 263–270. J. Johnson.

46. Ibid.

And then that Spaniard of the rose, itself
Hot-hooded and dark-blooded, rescued the rose
From nature, each time he saw it, making it,
As he saw it, exist in his own especial eye.
—WALLACE STEVENS, *Esthétique du Mal*

15

Wars of the Roses

IDEOLOGY VERSUS EXPERIMENT

THE EIGHTEENTH-CENTURY ENLIGHTENMENT was a giddy time for botanists, gardeners, and flower lovers of all stripes. So many tantalizing questions! Did plants have sex or didn't they? If they did, did they also have feelings and passions, or were they numb to all sensations and emotions? Did modest corn silks actually blush on making contact with a wind-blown pollen grain? Were flowers miniature nuptial beds upon which frenzied marriage rites were enacted? Did aroused stamens literally vibrate with anticipation in the presence of a wet, receptive stigma? Or were all such claims lewd slanders against the virginal purity of flowers, intended to corrupt female morals? The jury was still out, but whichever side one took, the heated debate made gardens—those erstwhile havens designed for sensory pleasure as well as for pious reflection—even more fascinating places in which to spend one's time.

Of course, much of the debate missed the point. The only *scientific* question at stake was whether plants reproduced sexually, not whether they had feelings. But for many opponents of the sexual theory, sexuality itself, with or without feelings, equated with carnality—the sinful indulgence of the flesh. As long as flowers were thought of as purely feminine, the subject of concupiscence did not arise, and many found the new two-sex model disgusting and abhorrent. J. G. Siegesbeck's vitriolic outburst against the Linnaean sexual system came closer to physical revulsion than to mere scientific disagreement. There is a disturbing scene in the 1992 movie *The Crying Game* in which an IRA foot soldier discovers—at an extremely awkward moment in the bedroom—that his new girlfriend has a penis. She is, in fact, a pre-op transgender female, and, at the moment of discovery, Fergus reflexively vomits.

One suspects that Siegesbeck and other horrified asexualists were responding viscerally to the dissonant news of the presence of male sexual organs in flowers, which they had always conceived of as parthenogenically female. By the eighteenth century, the flower, a traditional symbol of virgin purity, was so thoroughly feminized in art, literature, religion, and fashion that the sexualist claim that most flowers were hermaphroditic struck some as beyond perverse. Although the majority of botanists embraced the new paradigm, a small but vocal minority vigorously opposed it on moral grounds. In their desperate efforts to discourage the use of the Linnaean sexual system, eighteenth-century asexualists cited a raft of philosophical, religious, and pedagogical reasons as justification for rejecting it.

During the latter half of the eighteenth century, opposition to the sexual theory grew more intense. Ironically, the same period also saw remarkable progress in two crucial areas of floral biology: hybridization and insect pollination. The breakthrough studies on plant hybridization by Joseph Gottlieb Koelreuter, together with the brilliant insights into floral ecology by Christian Konrad Sprengel, provided the most convincing evidence yet for the sexual theory. Yet despite this growing mountain of evidence, late eighteenth-century asexualists maintained their ideological purity.

THE IDEOLOGY OF ASEXUALISM

As late as the eighteenth century, Aristotelian hierarchical beliefs about organisms, according to which plants were too low on the *Scala Natura* to be capable of sex, still persisted. To some, the sexual theory of plants challenged not only traditional hierarchies in nature, but traditional hierarchies in society as well. Conservative asexualists, especially those who felt threatened by Enlightenment political theories, generally objected to *any* idea that undermined the *status quo* in either nature or society.

Religion also played a role in the debate. In his book, *Nature's Second Kingdom*, François Delaporte pointed out that, during the eighteenth century, the study of botany was closely allied with religion.[1] One of the French *philosophes* who made the connection explicit was the Marquis de Condorcet, a French mathematician, philosopher, and friend of Thomas Jefferson. For Condorcet, the close relationship between botany and religion formed the basis of Natural Theology:

> In reading the history of the sciences, some people have formed the belief that some scientists are more disposed to piety and others less, depending on what sort of knowledge they cultivated; and botanists, these people think, deserve to be placed in the front rank.[2]

Botanists, states Condorcet, were "more disposed to piety" because they studied the plant kingdom, which "seems to call up more forcefully the idea of a first cause, to tell us more about its boons, and to incline our soul more naturally to gratitude." Eden was, after all, a garden in which Adam and Eve would have lived out their sexually inchoate, trouble-free existences in the benign presence of God, if they had only refrained from eating the forbidden fruit. Having sinned, they must suffer, in Hamlet's words, the "heart-ache and the thousand natural shocks that flesh is heir to." Plants, Condorcet claims, have the unique ability to evoke memories of that lost Eden, when humanity was still innocent and under God's protection. No doubt the belief in the asexuality of plants contributed to the perceived compatibility of plants and religion, reinforced by hundreds of years of Christian depictions of the Holy Virgin with flowers.

If plants were associated with innocence and piety, a certain repugnance was felt toward animals and their bodily functions, especially their sexuality, which was associated with original sin. The rejection of carnality was something upon which both sexualists and asexualists could agree. To asexualists, the disgust they felt toward carnality was based on morality. Sexualists, on the other hand, were more strongly influenced by aesthetic considerations than by morality. For example, in *Sponsalia Plantarum* Linnaeus wrote that:

> Just as the genital parts of all animals have a strong and repulsive odor during the season of rut, so do flowers, or the genital parts of plants, exhale an odor, which, though quite varied in different plants, is most of the time very sweet. This is why man himself imagines he is drinking in nectar with his nostrils.

Linnaeus makes it clear that while sex may be repellant in animals, it is delightful in plants. According to Linnaeus, the sexual organs, which are "considered as almost shameful in the Animal Kingdom, are almost always hidden by nature." On the other hand, Linnaeus notes, "[I]t is agreeable to recall that the genital organs of plants are exposed to the view of all in the Plant Kingdom." In other words, plant genitals cannot be shameful, otherwise God would not have displayed them for all to see.

Rousseau expressed a similar sentiment when he remarked that whereas sexual union in animals "revealed itself to me only in a hideous and disgusting form,"[3] the opposite was true of sex in plants:

> There is no rarer rapture or ecstasy than that which I felt each time I observed the structure and organization of a plant and the interplay of the sexual parts.[4]

Like Linnaeus, Rousseau based his contrasting responses to sex in animals versus plants on the pleasure principle.

During the late seventeenth and early eighteenth centuries, materialistic philosophy gradually undermined the medieval duality between the debased body and the exalted soul. As the moral status of the body improved, there emerged a new emphasis on the senses. British philosophers John Locke and David Hume challenged Descartes by asserting that the evidence of the senses was a better proof of one's existence than reason alone. French philosophers, such as Étienne Bonnot de Condillac, Charles Bonnet, and Claude Adrien Helvétius, argued for the establishment of "a new authority of experience" based on seeing, hearing, smelling, tasting, and touching—an epistemological theory known as "sensationism."[5] Linnaeus and Rousseau responded to sexuality in animals versus plants in a sensationist context. If plant sexuality were shameful or immoral, God would not have made flowers so pleasing to the senses.

In contrast, Rousseau considered the study of human anatomy particularly repulsive, as he reflected after visiting an anatomical amphitheater at a medical school:

> What a frightful paraphernalia an anatomical amphitheater contains, with its stinking cadavers, fresh, livid, and oozing, blood, disgusting intestines, terrifying skeletons,

> and pestilential vapors? By my word, it is not to such a place that Jean-Jacques will go looking for amusement.[6]

Even minerals could evoke disgust because they were extracted from the very bowels of the earth. This left the study of plants alone as a suitable vocation for sensitive souls, provided, Rousseau warned, that one avoided the pharmacological uses of plants, which forced one to dwell on disease and ugliness:

> These medicinal notions are hardly apt to make the study of botany agreeable. They cause the dappled meadows and bright flowers to fade, suck the cool damp out of the hedged fields, and make the green glade and the forest shade insipid and disgusting.[7]

In short, anything that recalled the body's physical needs and frailties was a reminder of humanity's debased state. Plants, being clean, pure, fragrant, and aesthetically appealing turned the mind away from images of corruption and death to reveries of sensual pleasure and joy:

> The sweet fragrances, the lively colors, the most elegant shapes seem to vie with one another for the right to hold our attention. One need only love pleasure to abandon oneself to such sweet sensations.[8]

Such qualities of plants, Rousseau believed, made botany the subject most likely to lead one closer to God.[9]

Rousseau greatly admired Linnaeus and determined to master his sexual system of classification. He also believed that the same aesthetic qualities that made the study of botany so conducive to piety also made it the most suitable subject of study for women—provided they did not take it "too seriously." According to Rousseau, women, being closer to nature than men, had a natural affinity for botany. However, this same closeness to nature steered women's minds toward disorder and irrationality. The study of botany, he argued, could counteract these irrational tendencies by introducing order and discipline into their lives. However, Rousseau cautioned that the "search for abstract and speculative truths, for principles and axioms of science . . . is beyond a woman's grasp."[10] The sexual theory of plants was one of the "speculative truths" that Rousseau thought women should avoid. In his *Letters on the Elements of Botany: Addressed to a Lady*, Rousseau referred to stamens and pistils only obliquely as "essential parts" in the course of warning against the dangers of finding double-flowered blossoms in the garden:

> Whenever you find them double, do not meddle with them, they are disfigured; or, if you please, dressed after our fashion: nature will no longer be found among them; she refuses to reproduce any thing from monsters thus mutilated: for if the more brilliant parts of the flower, namely the corolla, be multiplied, it is at the expense of the more essential parts [reproductive organs], which disappear under this addition of brilliancy.[11]

Indeed, some eighteenth-century women writers reinforced this patronizing attitude in their own literary work. In her poem "To a Lady with Some Painted Flowers" Anna Laetitia Barbould compared women to decorative garden flowers:

> Flowers to the fair: to you these flowers I bring.
> And strive to greet you with an earlier spring.
> *Flowers* SWEET, *and gay*, and DELICATE LIKE YOU;
> Emblems of innocence, and beauty, too.
> With flowers the Graces bind their yellow hair,
> And flowery wreaths consenting lovers wear.
> Flowers, the sole luxury which nature knew,
> In Eden's pure and guiltless garden grew . . .
>
> . . . Gay without toil, and lovely without art,
> They spring to CHEER the sense, and GLAD the heart.
> Nor blush, my fair, to own you copy these:
> *Your* BEST, *your* SWEETEST *empire is* — TO PLEASE.

According to Barbauld, women, like flowers, are "delicate," "innocent," and "sweet;" like the lilies of the field they are "gay without toil" and their sole purpose is "to please." The type of woman the poem addresses is young, virginal, and privileged. It is certainly not addressed to ordinary middle- or working-class women.

Mary Wollstonecraft, author of *A Vindication of the Rights of Woman* (1792), forcefully attacked such hyperfeminine floral metaphors for women as degrading and inimical to the struggle for equality.[12] To her, the age-old identification of women with flowers had devolved into decadence, and the ideals of strength and productivity had been eclipsed by ideals of weakness and nonfunctionality. Women, she wrote, were classed with mere "smiling flowers that adorn the land" or "sweet flowers that smile in the wake of man." She did not, however, disown the use of all floristic metaphors. She made use of them herself:

> The conduct and manners of women," she stated, "evidently prove that their minds are not in a healthy state; for like flowers which are planted in too rich a soil, strength and usefulness are sacrificed to beauty; and the flaunting leaves, after having pleased a fastidious eye, fade disregarded on the stalk, long before the season when they ought to have arrived at maturity."[13]

This "barren blooming" she attributed primarily to "a false system of education."

PROMINENT EIGHTEENTH-CENTURY OPPONENTS OF THE SEXUAL THEORY

Many of the eighteenth-century botanists who continued to oppose the sexual theory did so, in large part, because of their opposition to the Linnaean sexual system of classification. In Italy, Giulio Pontedera, Professor of Botany and Director of the botanical garden in Padua, rejected the sexual theory in his book *Anthologia* (1720), hewing to the classical explanation of pollen as a waste product. In direct contradiction to the results of Camerarius, he regarded the male flowers of dioecious species as of no use to the flower.

In 1752, the German botanist Hans Möller, an asexualist critic of Bradley's emasculation experiments with tulips, wrote in the *Hamburg Magazine* that since anthers functioned in excretion, which was required for purifying the sap, it was not the absence of pollination that prevented seed production when the anthers were removed, but the presence of impurities that could no longer be removed from the sap.[14] The question of how such a requirement for waste excretion could possibly account for Camerarius's results with dioecious plants, Möller fails to address. To circumvent the problem of dioecious species, some asexualists resorted to the equivocation that excretion was more important for some plants than for others.

One of the chinks in the armor of the sexual theory upon which asexualists pinned their hopes was the existence of those few apparent exceptions, faithfully reported by Camerarius (see Chapter 12), to the general rule that pollination was required for the production of viable seed. According to Julius von Sachs, the importance of the exceptions was exaggerated in an abridged version of Camerarius's results published later by his friend, Michael Bernhard Valentin, Professor of Medicine at the Justus Liebig University in Giessen, to whom Camerarius had dedicated *De Sexu Plantarum*. The asexualists were also unaware that Camerarius had later shown that many dioecious species occasionally produce hermaphroditic flowers, which could account for the small number of fertile seeds he found in some of his isolated female plants.

Having read only Valintin's abridged version of *De Sexu Plantarum*, Charles Alston, Professor of Botany at the University of Edinburgh, therefore gained the erroneous impression that the experimental support for the sexual theory was weak. In 1754, Alston presented a dissertation in which he cited his own attempts to repeat Camerarius's isolation experiments with dioecious spinach, hemp, and dog's mercury. Despite separating male and female plants by seemingly insurmountable distances, he nevertheless reported obtaining "good seeds" that germinated and produced normal male and female plants. Alston considered it highly improbable that the wind could have been involved, given the wind direction and the various barriers that insulated the plants. He also stated that he examined the female plants and could find no stamens. "It therefore follows," he concluded, "that the liquor of the apices is not necessary for the fructification of plants." As for the ancient claim that female date palms require the dust from the male to bear fruit, he cited a contradictory report by a gentleman named Labat:

> But Labat directly contradicts this doctrine by a fact. "We have," says he, "a date tree beside our monastery in Martinico, which carries ripe fruit though single, whether it is a male or female I know not, but this I know for certain, that there was not another kind within two leagues [six miles] of it."

Far from being a fructifying agent, Alston insisted, pollen was a waste product, and "Nature has arranged for this dust to be thrown away as far as possible, for it is useless, if not injurious to the style."[15] The reason for Alston's failure to repeat Camerarius's results is unclear. Perhaps experimental errors or errors in record keeping led him to obtain the results he obviously desired. His ready acceptance of Labat's anecdotal report suggests that he was motivated to disprove the Linnaean system.

The same can be said for the botanical experiments of Lazzaro Spallanzini, the Italian Catholic priest and experimental biologist who made important contributions to the study of animal reproduction. Spallanzini was the foremost champion of the ovist camp of preformationism in the eighteenth century. He was also the first to demonstrate that physical contact with semen was required for the fertilization of an egg, in this case, a frog egg. However, he concluded that it was the seminal fluid rather than the sperm that effected fertilization.

Around 1777, Spallanzini attempted to test the sexual theory of plants but obtained mostly negative results. In the cases of basil and dog's mercury, he confirmed that the "dust" was necessary for seed production, but he was unable to duplicate these results in pumpkin, watermelon, hemp, or spinach. Julius von Sachs attributed Spallanzini's failures to carelessness. According to Sachs, there was gossip at the time that Spallanzini's assistant had actually performed the experiments. Alternatively, Spallanzini's error could have resulted either from the presence of a few hermaphroditic flowers or from the rare occurrence of parthenogenesis, termed *apomixis*, among some of his plants. Apomixis occurs in about 0.1% of angiosperms, in more than 400 species belonging to 40 different families. During apomixis, an embryo forms abnormally from a cell in the ovule other than the egg cell. For example, during reproductive development in the monoecious vine pumpkin (*Curcurbita pepo*), the plant undergoes a transition from male to female flowers. According to geneticist G. Van Nigtevecht, this process culminates in the formation of a parthenogenic female flower.[16] Such parthenogenic flowers can form fruits with viable seeds in the absence of fertilization, which may have been the source of Spallanzini's error. If Spallanzini was misled by apomixis, his confusion about the requirement for pollination would be understandable.[17]

On the other hand, François Delaporte favors the theory that Spallanzini may have unconsciously misinterpreted his results to make them consistent with his ovist convictions, which were strengthened by the recent discovery of parthenogenesis in aphids.[18] If he had read Camerarius, however, he would have discovered that the young German botanist had demonstrated the absence of preformed embryos in unfertilized ovules of bean plants.

Another prominent asexualist was William Smellie, a Scottish printer, editor, translator, Professor of Natural History at Edinburgh University, and friend of Robert Burns. He was also the author and compiler of the first edition of the *Encyclopedia Britannica*, which was issued in installments between 1768 and 1771. In a discourse on the sexes of plants, Smellie stated that as a young university student he had uncritically accepted the sexual theory, having been misled by the "alluring seductions of analogical reasoning." But, he states, after "perusing Linnaeus and many other works on the subject" he was "astonished to find that this theory was supported neither by facts nor arguments which could produce conviction in even the most prejudiced minds." Instead, "its principal support is derived from the many beautiful analogies which subsist between plants and animals," including the notion that plant eggs must be fertilized by pollen to produce seeds. But, says Smellie, fertilization is impossible in plants because the seeds have already acquired "bulk and solidity long before the pollen, or supposed fecundating dust, is thrown out of its capsules." Without providing any evidence, Smellie asserts that seeds mature and develop hard seed coats *before* pollen is shed from the anthers, making it impossible for pollen to make physical contact with the egg or ovum. Since pollen shed from the anther cannot fertilize the ovum of its own flower, it must therefore drift "promiscuously abroad" impregnating other species. If this happened:

> the whole vegetable kingdom in a few years would be utterly confounded; Instead of a regulated succession of marked species, the earth would be covered with monstrous productions, which no botanist could either recognize or unravel.

While it is true that pollen is sometimes shed from the anthers before the stigmas become receptive, we now know that this is a mechanism that ensures outcrossing within a species and that hybridization between different plant species is rare in nature. Smellie dismisses the idea that wind and insects could serve as pollination vectors, commenting:

> is there any thing, in northern climates at least, more desultory and capricious than the direction and motion of the winds? Can we form a conception of any thing more casual and uncertain than the wayward paths of insects?

The very notion, Smellie continues, that something as important as the fertility of the "whole vegetable kingdom" should be left to chance is "repugnant to every idea of sound philosophy." Besides, he concludes triumphantly, "the reverse has been proved by Dr. Alston, Camerarius, and Tournefort." This is neither the first nor the last time that Camerarius's own exceptions would be cited as if they were the rule in his pollination studies. As we shall see in the next chapter, German "nature philosophers" would repeat the same mistake, extending the debate on plant sex into the nineteenth century. Smellie reprinted his *Encyclopedia Britannica* article on plant sex in 1790 in his book *The Philosophy of Nature*. Although it was generally ignored in England, it was well-received in the United States.[19]

Opposition to the Linnaean sexual system in England reached its zenith late in the eighteenth century. As the gathering storm of the French Revolution stoked fears of foreign invasion and social unrest, it fueled a backlash against the liberalism of the first half of the century, which had seen a modest expansion of women's rights.[20] Some self-appointed male guardians of female morality viewed the popularity of the Linnaean sexual system among women as a stalking horse for a variety of social ills: libertinism, radical Jacobinism, feminism, and anarchy. The Cornish clergyman and poet Richard Polwhele went so far as to blame Mary Wollstonecraft for the popularity of the Linnaean system among women. In 1795, Wollstonecraft had defended the French Revolution in response to a critical pamphlet by the moderate conservative statesman Edmund Burke.[21] Adding to her notoriety, while living in Paris, she had had an affair with the American adventurer Gilbert Imlay and had borne a child out of wedlock. To Richard Polwhele, she was wickedness incarnate. Largely due to her influence, botany had become little more than lascivious sex education in disguise. Those "unblushing" female botanists who were "eager for illicit knowledge" were clearly "disciples of Miss W," fumed Polwhele. "Botany has lately become a fashionable amusement with the ladies," he noted, "[b]ut how the study of the sexual system of plants can accord with female modesty, I am not able to comprehend." Worst of all, Polwhele complained, "I have, several times, seen boys and girls botanizing together."[22]

In his satirical poem "The Unsex'd Females" (1798), Polwhele lampooned female botanists for taking an unseemly interest in plant sex:

With bliss botanic as their bosoms heave,
Still pluck forbidden fruit, with mother Eve,
For puberty in signing florets pant,
Or point the prostitution of a plant;
Dissect its organ of unhallow'd lust,
And fondly gaze the titillating dust;
With liberty's sublimer views expand,
And o'er the wreck of kingdoms sternly stand;
And, frantic, midst the democratic storm,
Pursue, Philosophy! thy phantom-form.

Once again, Polwhele lays the blame for this shocking state of affairs squarely on the shoulders of Mary Wollstonecraft:

See Wollstonecraft, whom no decorum checks,
Arise, the intrepid champion of her sex;
O'er humbled man assert the sovereign claim,
And slight the timid blush of virgin fame.

Thanks to Mary Wollstonecraft, the women of England had lost their winsome "weakness" and "reserve," daring to have rights and ambitions of their own:

No more by weakness winning fond regard;
Nor eyes, that sparkle from their blushes, roll,
Nor catch the languors of the sick'ning soul,
Nor the quick flutter, nor the coy reserve,
But nobly boast the firm gymnastic nerve;
Nor more affect with Delicacy's fan
To hide the emotion from congenial man;
To the bold heights where glory beams, aspire,
Blend mental energy with Passion's fire,
Surpass their rivals in the powers of mind
And vindicate the Rights of womankind.

According to Polwhele, women who used the Linnaean sexual system of plant classification had abandoned womanhood—they had become "unsex'd"—and it was all Mary Wollstonecraft's fault!

Despite such attacks, or perhaps because of them, late eighteenth-century asexualists became increasingly isolated as increasing numbers of botanists embraced the Linnaean sexual system—and by extension, the sexual theory of plants. One senses a certain siege mentality in the writings of the British asexualists. As early as 1761, seven years before

William Smellie published his *Encyclopedia Britannica* article debunking the sexual theory, definitive proof of its validity had finally been provided by Joseph Gottlieb Koelreuter. Indeed, Koelreuter is the unsung hero of the sexual theory, even though his work had surprisingly little impact on the contemporary debate.

KOELREUTER'S PRODUCTIVE JOURNEY FROM TÜBINGEN TO KARLSRUHE

Part of the confusion surrounding the role of sex in plants during the eighteenth century was a consequence of the limited circulation of Camerarius's *Epistola*. Even those botanists who were aware of Camerarius's work seemed to have read only abstracts or summaries that overemphasized the exceptions in his experiments rather than the main findings. It was the German physician and botanist Joseph Gottlieb Koelreuter, a fellow Tübingener, who set the record straight. In his 1761 paper, *Preliminary Report of Some Experiments and Observations Concerning Sex in Plants*, he gave Camerarius full credit for placing the sexual theory of plants on a scientific footing:

> Rudolph Jacob Camerarius is indisputably the first who proved the sex of plants through his own experiments instituted from this point of view. He, my fellow countryman, it is whom the learned world has principally to thank for this great truth, which is so general, and of such great influence upon the physical and economic sciences.

As persuasive as Camerarius's results were, however, they did not constitute "indisputable" proof of sex in plants. One could argue, for example, that removal of the maize tassel (male inflorescence) injured the plant and prevented normal seed development. On the other hand, Camerarius's results with dioecious species such as spinach, in which female plants were isolated from male plants, could not be explained away as the result of damage to the female plant. In these cases, asexualists focused exclusively on the few exceptions reported by Camerarius and ignored his more general findings.

The most compelling argument in favor of the sexual theory was plant hybridization. As we saw in Chapter 13, preliminary evidence for plant hybridization was reported in the early eighteenth century by Cotton Mather and Thomas Fairchild, but these observations were never followed-up. Of the four plant hybrids claimed by Linnaeus, three were putative intergeneric hybrids that were never verified experimentally, whereas the fourth apparent interspecific hybrid was never repeated. Since neither Mather, Fairchild, nor Linnaeus provided unequivocal proof of plant hybridization, the skepticism of the asexualists could still be justified on scientific grounds. It was left to Joseph Koelreuter to eliminate this last potential objection to the sexual theory by providing the first rigorous demonstration of plant hybridization.

Joseph Gottlieb Koelreuter was born on April 27, 1733, in Sulz, a little town on the Neckar River in southern Germany, about twenty miles southwest of Tübingen. At the age of fifteen he entered the University of Tübingen to study medicine, where he came in contact with the naturalist and explorer J. G. Gmelin, who had just returned from St. Petersburg.

Gmelin had been corresponding with Linnaeus about plant hybrids, and was preparing a republication of Camerarius's *Epistola*. In his 1749 inaugural lecture, Gmelin discussed the possible origin of new species of plants and animals, either by hybridization or spontaneous processes, and he underscored the need for further research in this area. Under Gmelin's influence, Koelreuter began to take an interest in the sexual theory of plants, especially the question of plant hybrids, and, in 1752, while still a medical student, he published a summary of all the research conducted on the subject of plant sex since Camerarius. Gmelin must have been impressed. In 1755, the year Koelreuter graduated, Gmelin used his connections to procure a position for his star pupil as custodian of the natural history collections at the Imperial Academy of Sciences in St. Petersburg.[23]

As discussed in Chapter 13, Linnaeus had won the prize offered by the Imperial Academy for the best essay in support of the sexual theory. Koelreuter, who was then in St. Petersburg, had always regarded Linnaeus as something of a dilettante compared to Camerarius. Hoping to best the celebrated Swede at his own game, the young Koelreuter had quietly begun his own experiments on plant hybridization, but he was unable to complete his analysis in time to submit an essay for the contest. He was thus forced to watch from the sidelines as Linnaeus garnered all the glory. Along with his essay, Linnaeus had sent the Academy seeds of his *Tragopogon* hybrid, and Koelreuter was assigned the task of confirming their identity. Based on his own preliminary results with tobacco hybrids, he had already formed the opinion that true hybrids are always sterile and intermediate in form between their parents. Applying this criterion, he concluded that Linnaeus's *Tragopogon* hybrid was actually "only half a hybrid" because it was neither sterile nor intermediate in appearance. (We will return to the meaning of "half a hybrid" later in the chapter.)

Frustrated by his failure to submit an essay, Koelreuter redoubled his efforts, determined to amass an overwhelming body of evidence that would dispel any lingering doubts about the sexual theory. He pursued his hybridization experiments at St. Petersburg in 1760 and 1761, after which he returned to Germany. While journeying from town to town, he managed to conduct experiments during brief stopovers: in Leipzig and Berlin in 1761, in Sulz in 1762, and in Calw in 1763.[24] It was during this period of gypsy-like freedom that he published three of his six landmark papers on plant hybridization.

In 1763, Koelreuter had the good fortune to come to the attention of the Princess Caroline Louise, wife of Charles Frederick, Margrave of Baden. Princess Caroline was a brilliant, multitalented woman, fluent in five languages, and well-versed in the arts and sciences. She corresponded with both Voltaire and Linnaeus and made her palace at Karlsruhe into one of Germany's premier cultural centers.[25]

An early admirer of Linnaeus, Princess Caroline's knowledge of botany was judged to be the equal of any professor's. At her Karlsruhe Schloßgarten she cultivated both native and exotic species, and she had paintings and copper engravings made of each specimen for a planned catalogue of her collection, with the plants arranged according to the Linnaean system. It was for this purpose that, in 1764, she appointed Koelreuter Director of the Palace Gardens at Karlsruhe, as well as Professor of Natural History in charge of her extensive natural science collections. Prior to his arrival, Koelreuter published the fourth paper in the series on the sexual theory, and, with Princess Caroline Louise's encouragement, he quickly set up experimental plots in the Palace Garden, doubtless anticipating a brilliant future ahead of him.

Koelreuter's joy was short-lived, however. No sooner had his experimental plants begun to grow than the head gardener began to complain. Juxtaposed with the scrupulously designed, meticulously manicured formal gardens of the palace, Koelreuter's purely functional experimental plots must have stood out like the sorest of sore thumbs. The head gardener clearly regarded himself as an *artiste*, and Koelreuter's mélange of vegetables sprouting untidily in the midst of his carefully crafted horticultural masterpiece were as welcome as a large warren of rabbits. Over the next two years, Koelreuter's life was made utterly miserable by the head-gardener-from-hell, whose acts of petty sabotage, although unrecorded, we can easily imagine: valuable specimens "accidentally" pruned, weeded, trampled, or allowed to wither unwatered and unloved.

Koelreuter eventually threw in the trowel. He abandoned all his hybridization experiments, even terminating the ones he was conducting at his own home! Thus, except for one posthumous paper, all his publications dealing with hybridization were published before 1777.[26] Princess Caroline's untimely death in Paris in 1783 left Koelreuter without a sponsor, and, following another dust-up with the feisty gardener he was relieved of his post as Director of the Palace Garden, although he managed to retain his position as Professor of Natural History overseeing the Princess's natural history collection.

KOELREUTER'S HYBRIDIZATION EXPERIMENTS

Koelreuter was a workaholic and a perfectionist whose scientific output, in terms of both quantity and quality, dwarfed that of all previous botanists working on the sexual theory. Between the years 1760 and 1765 Koelreuter reported the results of 140 crosses involving thirteen genera and fifty-four species.[27] Most of his experiments were conducted with three genera: *Nicotiana, Dianthus*, and *Verbascum* (mullein).

Koelreuter was also the first to study insect pollination systematically. His patience and meticulousness are legendary. In an experiment to test the efficiency of insect pollination, he divided 310 hibiscus flowers into two equal groups. One group he allowed to be pollinated by bees. The other group he hand-pollinated using a small paintbrush. He found that the bee-pollinated flowers produced 10,886 seeds, while the hand-pollinated flowers yielded 11,237 seeds. He attributed the slight advantage of the hand-pollinated blossoms to a few days of rainy weather, which prevented the bees from visiting the flowers. Fascinated by numbers, he counted the number of pollen grains produced by flowers of different species: a *Hibiscus* flower produced 4,863 pollen grains, in contrast to *Mirabilis* (four o'clocks) flowers, which produced an average of 307 pollen grains. In another experiment, Koelreuter watched a *Hibiscus* flower from dawn until dark, shooing away any insect that dared to approached it. As expected, the flower withered and dropped without producing a single seed. In another experiment, Koelreuter tested Swammerdam's claim that nectar must undergo fermentation in the crops of bees to become honey. Using a glass capillary, he collected nectar from hundreds of orange blossoms into a vial and allowed the liquid to partially evaporate. To his delight he found that the resulting viscous fluid tasted exactly like honey, disproving Swammerdam's theory.

Koelreuter's first success in creating an artificial plant hybrid came in 1760, while he was still in St. Petersburg. As he wrote in his 1761 paper,

> after many experiments instituted in vain with many kinds of plants, I have finally . . . in the case of *Nicotiana paniculata* and *Nicotiana rustica*, gotten so far that I have fertilized with the pollen-dust of the former, the ovary of the other, obtained perfect seeds, and from these, still in the same year, have raised young plants.

Although the hybrid tobacco plants grew and flowered normally, the pollen produced by the stamens was dried and shriveled in appearance and there was less of it. Since Koelreuter did not believe Linnaeus's *Tragopogon* hybrid was a true hybrid, and he was unaware that Fairchild had already created the first plant hybrid decades earlier, he described his own tobacco hybrid as the "first botanical mule":

> The fertility of this new plant appeared to me, therefore, extremely questionable, and the results confirmed my suspicion completely; for among the almost innumerable quantity of flowers there was not one to be found which had borne even a single seed, even though they had been immediately covered with a large quantity of their own pollen dust; while on the other hand, with the two natural species, every capsule is accustomed to bear four to five hundred seeds. This plant is thus in the real sense a true, and so far as is known to me, the first botanical mule which has been produced by art.

Fairchild and Linnaeus may have produced their artificial hybrids earlier, but Koelreuter deserves scientific priority because of his thorough documentation, quantitative analysis, and exhaustive follow-up experiments. For example, the floral organs of *Nicotiana rustica* are shorter than those of *Nicotiana paniculata*. Koelreuter's measurements of the floral organs of the hybrid flower showed that they were intermediate between the two parents (Figure 15.1). This was consistent with Koelreuter's *materialist* conceptual framework regarding the mechanism of inheritance, as discussed next.

KOELREUTER'S THEORY OF UNIFORM LIQUID ESSENCES

Although Koelreuter was refreshingly modern in his scientific methodology, his conceptual framework was deeply rooted in eighteenth-century philosophy. Not even a hint of evolutionary thinking is to be found in his writings. Whether he was discussing the geographical distribution of species or the elegant adaptations between flowers and insects, it was always in terms of a well-designed world and the wisdom of the Creator. A believer in the fixity of species, he viewed the sterility of hybrids to be a necessary barrier to prevent "incredible confusion" in nature. Paradoxically, Koelreuter's adherence to the fixity of species concept helped him to avoid Linnaeus's mistake of assuming that intergeneric crosses occurred with some regularity in nature. *Contra* Linnaeus, Koelreuter insisted that species hybrids were very rare in nature and that intergeneric hybrids were prohibited. Most hybridization, he argued, occurred in gardens, where related species that normally grow geographically isolated from each other were cultivated in close proximity. This, he explained, was the plan of the Creator, who, in his wisdom, had distributed closely related species around the world in order to prevent them from hybridizing.

(a)

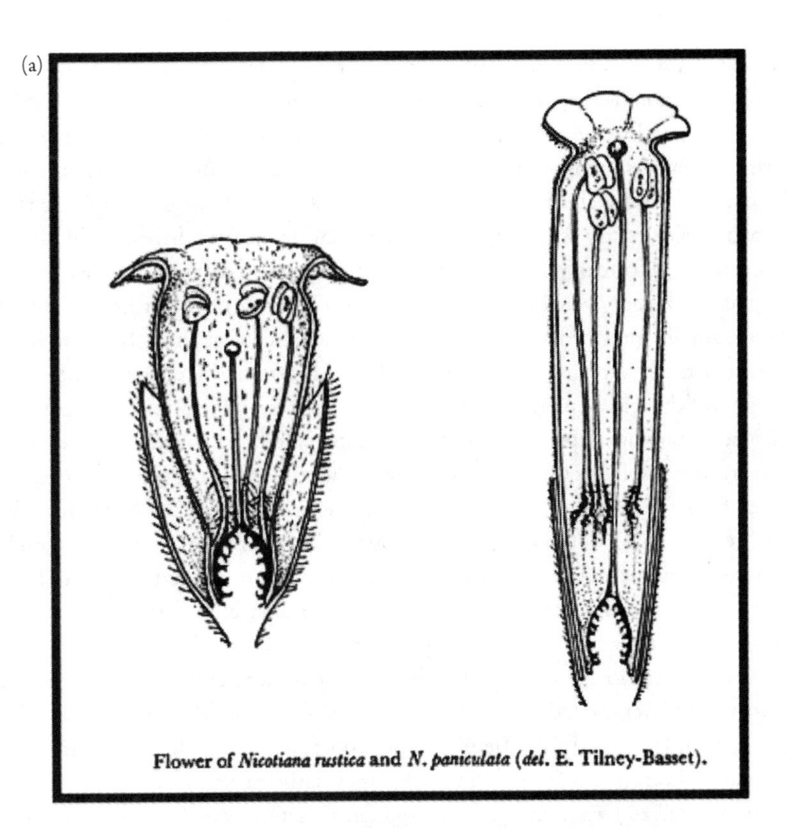

Flower of *Nicotiana rustica* and *N. paniculata* (*del.* E. Tilney-Basset).

(b)

TABLE 1. KOELREUTER'S COMPARISON OF THIRTEEN FLORAL CHARACTERISTICS

Characteristic	*N. rustica*	Hybrid	*N. paniculata*
	inches	*inches*	*inches*
Length of:			
flower	7	$9\frac{3}{8}$	$13\frac{1}{8}$
longest sepals	$5\frac{1}{2}$	$5\frac{1}{2}$	$3\frac{5}{8}$
flower projecting above sepals	$1\frac{1}{2}$	4	$9\frac{1}{2}$
corolla tube	$6\frac{5}{8}$	9	$12\frac{5}{8}$
narrow basal portion of corolla tube	$1\frac{1}{8}$	$2\frac{1}{4}$	3
filaments	4	$5\frac{1}{2}$	$8\frac{1}{8}$
style	$4\frac{1}{2}$	7	$11\frac{1}{4}$
ovary	$1\frac{1}{8}$	$1\frac{1}{8}$	$1\frac{1}{8}$
Diameter of:			
corolla tube at its mouth	$2\frac{3}{4}$	$2\frac{1}{4}$	$1\frac{3}{8}$
corolla tube in the throat	$3\frac{1}{2}$	3	$2\frac{1}{4}$
ovary	$1\frac{1}{8}$	1	$\frac{3}{4}$
Maximum width of:			
flower	$7\frac{1}{8}$	$5\frac{1}{8}$	$4\frac{1}{8}$
flower when over	$2\frac{1}{8}$	2	$1\frac{1}{4}$

FIGURE 15.1 Koelreuter's cross between *Nicotiana rustica* (left) and *Nicotiana paniculata* (right). A. Comparison of the flowers of *N. rustica* (left) and *N. paniculata* (right). B. Lengths or widths in inches of the floral parts of the parental stock versus the hybrid.

From: Olby, R. C. (1966), Joseph Koelreuter, 1733–1806, in R. C. Olby, ed., *Late Eighteenth Century European Scientists*. Pergamon Press, Oxford, pp. 50–51.

Koelreuter cited a second barrier to hybridization: the species-specificity of pollen fertilization, a phenomenon he discovered serendipitously during a futile attempt to produce hybrid plants fertilized by a mixture of pollen grains from different species. The design of the experiment was inspired by the canard, widely accepted during the eighteenth and nineteenth centuries, that a child could have multiple biological fathers.[28] Applying this principle to plants, Koelreuter assumed that it was possible to generate hybrids of multiple species by applying a mixture of pollen grains to the stigma. It was on this basis that he had declared Linnaeus's *Tragopogon* hybrid "only half a hybrid" because of its fertility and, in his opinion, its marginally intermediate appearance. He speculated that the hybrid had been generated by accidentally applying a mixture of pollen grains from *both* of the parental species to the stigma of the female parent, so that it more closely resembled the female parent.

However, Koelreuter was unable to demonstrate multiple paternity in plants when he tested his theory experimentally. Every time he dabbed a pollen mixture onto the stigma of a flower, only the pollen from the *conspecific* (same species) plant fertilized the flower. Similarly, if he placed a mixture of all-alien pollen grains on the stigma, only the pollen from the most closely related species successfully fertilized the flower. Koelreuter therefore concluded that God had arranged things so that flowers always favored conspecific pollen or, failing that, closely related pollen, thus preventing the random mixing of species. Based on his own findings, Koelreuter dismissed Linnaeus's claims regarding the abundance of intergeneric crosses in nature as "the premature births of an over-excited imagination," as indeed they probably were.

Unlike Nehemiah Grew and Sebastien Vaillant, who described the fertilizing effect of pollen in immaterial terms as a "prolifick breath" or "volatile spirit," Koelreuter believed that fertilization occurred by a mixing of male and female semen, which induced a chemical reaction:

> Since it was produced from those two simple forces, it has also an intermediate, composite, force. Just as from the mixture of an acid and alkaline salts a third, namely an intermediate salt arises.

He was, however, aware that the process of sexual reproduction was far more complex than a simple chemical reaction. All of the information about an organism and its component parts had to be passed on to the next generation via the semen. Knowing nothing about genes or DNA, Koelreuter postulated that male and female semen were each homogenous liquids that nevertheless encoded the instructions for making another individual of the same species. In plants, the oily substance inside pollen grains represented the male semen and the sticky fluid on the surfaces of stigmas represented the female semen. Koelreuter hypothesized that after the male semen from the pollen grain mixed with the female semen on the surface of the stigma, the mixture was absorbed by the style and transported to the ovary where it produced the seed.

Based on his theory of the mixing of fluids, it was logical for Koelreuter to predict that during hybridization the *uniform essences* of the two parents would be blended to form an organism that was precisely intermediate in type. As noted earlier, Koelreuter's first artificial hybrid between *Nicotiana paniculata* and *Nicotiana rustica* approximated this criterion, but some of his later hybrids did not, which puzzled him greatly.

To explain why some of the traits of the hybrids resembled those of one parent more than the other, Koelreuter postulated incomplete blending of semen, resulting in pockets of pure male or female semen that produced traits closely resembling those of the parent. However erroneous, the "incomplete blending" theory was an ingenious solution to the problem, consistent with the facts as they were then understood. In a way, it anticipated Gregor Mendel's concept of the "particulate" nature of inheritance in peas, the difference being that Mendel's interpretation allowed predictions to be made about the progeny of a cross, whereas Koelreuter's "accidental" model did not.

A detailed discussion of Koelreuter's numerous contributions to the sexual theory is beyond the scope of this discussion, but it is worth noting a few other crossing experiments because of their importance to the history of genetics.[29] For example, Koelreuter carried out what we now refer to as a *reciprocal cross* to demonstrate that the hybrids between *Nicotiana paniculata* and *Nicotiana rustica* were identical regardless of which parents served as the male or female. This experiment effectively demolished the doctrine of preformationism—both its ovist and spermist versions. It also disproved Linnaeus's idea that the female contributed the medulla (pith) of the plant while the male contributed the cortex.

Although most of Koelreuter's hybrids were sterile, to his surprise a few were self-fertile. However, even the hybrids with sterile pollen, such as *N. paniculata* × *N. rustica*, were fertile if crossed to one of its parents, provided the parent served as the pollen donor. This phenomenon allowed him to perform what is now referred to as a "backcross experiment" between, for example, the *N. paniculata* × *N. rustica* hybrid and either *N. paniculata* or *N. rustica* as the pollen donor species. Using the language of alchemy applied to metals, Koelreuter's claimed that repeated backcrosses caused the "transmutation" of the hybrid back to the parental type.

Koelreuter made good use of those few hybrids he produced that were self-fertile by crossing them with each other. According to his theory of inheritance based on the blending of uniform liquids, the initial hybrid (referred to in modern parlance as the "first filial" or "F_1" generation) should contain a uniform mixture of the two essences of the parental species. If an F_1 plant is crossed with another F_1 plant (i.e., selfed), the next generation is called the "second filial" or "F_2 generation." Based on Koelreuter's uniform liquid theory of inheritance, mixing two identical homogenous mixtures should produce the same homogenous mixture. But this is not what Koelreuter observed when he selfed two hybrids. The F_2 generation was much more variable than the F_1 generation, and, remarkably, some of the progeny even reverted to their parental types. This was equivalent to mixing two cans of identical green paint and having the mixture spontaneously separate into layers of yellow and blue! Koelreuter was completely baffled:

> This much is . . . quite clear, that matters must be rather uneven or disorderly in the selfing of hybrids; indeed, it seems as if it would occasionally lead to the production of monstrosities.

Koelreuter went on to note the tendency of the F_2 generation to restore "the original natural [parental] form and fertility." Koelreuter had unwittingly recreated the phenomenon of the "degeneration" of olive and other cultivated fruit trees into wild forms when propagated by seed, which had been recorded since ancient times. One hundred years later, Mendel, who adopted a particulate rather than a fluid model for the hereditary material, based his first two laws of inheritance, the Law of Segregation and the Law of Independent Assortment, on similar experimental results in peas.

KOELREUTER'S OTHER DISCOVERIES

In the course of his crossing experiments, Koelreuter also discovered the phenomenon of "hybrid vigor," or heterosis. For example, he noted that hybrid *Dianthus* plants had "increased vegetative power" compared to their parental stock, even though they were self-sterile. He was quick to see the potential of the phenomenon for agricultural improvement and thought it would be most beneficial for the breeding of trees:

> I would wish that I or somebody else would be so lucky someday to produce a species hybrid of trees which, with respect to the use of its lumber, would have a large influence on the economy. Among other good properties such trees might perhaps also have the one that they would reach their full size in one half the time of normal trees.[30]

Arguably, Koelreuter's most influential work was on insect pollination and the adaptations of flowers that facilitate interactions with insects. Since the time of Aristotle, it was known that flowers provided food for bees, which feasted on nectar and pollen. In addition, bees converted nectar to honey, thus indirectly providing nourishment for people. Philip Miller had noted the role of bees in pollination, but it was Koelreuter who advanced the fundamental ecological argument that the primary function of the floral nectaries was to attract insect pollinators:

> I was astonished when I first made this discovery, that the reproduction of plants is a matter of chance, a lucky accident. But astonishment changed to admiration of the method—at first sight accidental, but in fact certain—used by the wise Creator to ensure their reproduction. To be sure, all the movements of these little servants of nature show that, when they visit the flowers, they have nothing less in mind than the execution of such an important task. But what does this matter? It is enough, that without knowing it they undertake this all-important job for their own benefit as well as for that of the plants. Their needed sustenance, little drops of sweet liquid, lies hidden inside of these flowers. It requires effort and work to collect it; and in their various movements, it happens that they brush off onto the stigmas the pollen which they so easily had amassed on the hairs of their bodies to which it so readily clings.[31]

Unlike Camerarius, Koelreuter received many accolades in his lifetime.[32] Three plant genera were named in his honor, and he was elected to numerous scientific academies and societies. Albrecht von Haller, who was a preformationist, had become aware of Koelreuter's publications even before they were published and quickly spread the news to other prominent naturalists. Both the asexualist Spallanzini and the sexualist Erasmus Darwin cited Koelreuter, and the German botanist Johann Hedwig, the first to describe male and female reproductive structures in nonseed plants (cryptogams), declared in 1798 that Koelreuter had demonstrated "beyond all doubt" that "propagation by sexual union takes place in the plant world also." Koelreuter's studies on insect pollination strongly influenced both Christian Konrad Sprengel and Charles Darwin. Indeed, after Koelreuter, the asexualist movement could no longer claim any scientific legitimacy.

Thirty years later, Sprengel published his own great book on floral ecology, *The Discovery of the Secret of Nature in the Form and Fertilization of Flowers*, in which he generously acknowledged his indebtedness to Koelreuter. Koelreuter was still alive at the time and may have had the pleasure of learning about Sprengel's work through a visit from Sprengel's doctor, E. L. Heim,[33] who had helped Sprengel through a personal crisis.

SPRENGEL'S TRANSFORMATION: FROM TROUBLED HEADMASTER TO APOSTLE OF FLOWERS

The more Christian Konrad Sprengel stared at the minutiae of floral structure, the more amazed at God's infinite wisdom he became. [34] Miraculously, his former resentment, bitterness, and hypochondria melted away, and he felt a strange, long-repressed passion for life, a spiritual re-awakening. Thus, Sprengel described his own "road to Damascus" experience following a trying psychological period in this life.

Born in 1750 in Brandenburg, Prussia, Sprengel was the youngest of fifteen children—both siblings and half-siblings. Both his father and grandfather were Lutheran Archdeacons.[35] His father, Ernst Victor Sprengel, who was sixty-four when Sprengel was born, was Archdeacon at St. Gotthard Church. In keeping with the family tradition, several of Sprengel's brothers from both sets of siblings pursued similar religious careers: first studying theology at the University of Halle, next becoming secondary school teachers, finally receiving pastoral appointments in nearby towns.

At first, Christian Konrad appeared to be destined for the same career path as his brothers. In 1770, he matriculated at the University of Halle, studying theology and classical philology. Upon graduating in 1775, he taught at two different secondary schools in Berlin, and, in 1780, having performed his duties satisfactorily at these schools, he was offered the position of headmaster of the Große Schule (a Lutheran public school) in Spandau, a promotion he readily accepted.

The Lutheran Große Schule at Spandau was a secondary school, which, in Prussia at the time, meant that it was voluntary, rather than mandatory. Because poor families could not afford to support their children beyond a primary school education, the students at the Große Schule were almost all the sons of the most affluent and influential families of Spandau, who relentlessly sought privileges for their offspring well beyond those of ordinary students. Over the years, incessant parental demands for special treatment had turned the job of headmaster at the Große Schule into a living nightmare. Indeed, the position of headmaster had become vacant because the previous headmaster had resigned in a huff after an infuriating encounter with a particularly aggressive parent—and "nothing," his supervisor wrote, "could induce him to stay." Nevertheless, buoyed by youthful enthusiasm, Sprengel took hold of the reins of the Große Schule and charged ahead. The subjects he taught included German, Latin, Greek, French, theology, elementary mathematics, and natural history.

As a Lutheran public school, the Große Schule was administered by both the town council and the church. Any educational changes Sprengel wished to make thus had to be approved by both. In 1781, a month after his inaugural lecture, Sprengel wrote a forceful letter to both the Lutheran Superintendant and the town council in which he proposed major organizational changes to the teaching plan. Instead of large classes made up of students at two to four levels, which were extremely difficult to teach, Sprengel proposed substituting four smaller classes, each composed of students at the same level. To make room for the additional classes Sprengel proposed doing away with the morning worship class, substituting instead a short prayer before each of the lesson periods.[36]

This was a rather daring proposal. One can easily imagine the grumbling it provoked from the conservative town councilors, let alone from the Lutheran School Superintendant, Pastor Daniel Friedrich Schulze, who were unaccustomed to dealing with headmasters with ideas of their own. No doubt Sprengel's self-confidence derived from his strong family credentials, and, in the end, his superiors reluctantly allowed him to implement the two reforms he requested. "We put up with both of them," Pastor Schulze wrote grudgingly. The incident provides a window on a surprising aspect of Sprengel's personality not apparent from the sketchy information we have about his early years, but which is essential to understand his subsequent boldness as an innovator in floral ecology despite the lack of any formal training in botany.

Whatever hopes he may have had that his reforms would make his job as Headmaster any easier were soon dashed. Struggling to maintain student discipline and bedeviled by the same parental interference that had driven away his predecessor, he began to show signs of stress. In 1782, Pastor Schulze reported that Sprengel was reprimanded on numerous occasions for being "cruel when disciplining the students and treating his lessons arbitrarily." The Superintendent went on to list several alleged examples of Sprengel's harsh punishments, which he seemed to mete out regardless of the parents' status. For example, one day he made the mayor's son stand during an entire lesson and later struck the boy with a stick near his eyes, causing sufficient damage to require medical attention. The nine-year-old son of a town councilor was allegedly beaten so badly that the boy could not lie down for several nights. A church official's son was pursued and beaten black and blue for stealing plums from Sprengel's tree. A shoemaker's son, who had not understood what Sprengel had said in class and had asked another student, was struck about the head with a stick until he bled. He also lit into another shoemaker's son for snatching a paper from a boy on the street, causing bruises and contusions on the boy's back and shoulders. Finally, Schulze reported that Sprengel administered thirty-one strokes to another student for playing a prank at school.

Nothing of what is known of Sprengel's early years prepares us for such abusive behavior. Indeed, we have only Schulze's word that the incidents actually occurred. It is possible they were based on hearsay or blown out of proportion by Schulze, who may have been biased against Sprengel. The fact that Sprengel managed to keep his job despite Schulze's allegations may indicate that such beatings, even if true, were not considered beyond the pale. However, according to Zepernick and Meretz, Sprengel was in "a particularly poor mental and spiritual state" during 1782, and in addition was suffering from serious eyestrain and a host of other physical ailments, real or imagined.[37]

Increasingly desperate and fearful of losing his position, Sprengel sought the advice of Dr. Ernst Ludwig Heim, the distinguished district medical officer in Spandau. The choice of Heim, who was only three years older than Sprengel, could not have been better. Like Sprengel, he had attended the University of Halle, and the two may even have known each other during their college days. They hit it off immediately, and Heim soon became a role model and father figure to Sprengel.

It didn't take Heim long to determine that Sprengel's problems at the school had a psychological component. To combat depression and alleviate his physical symptoms, Heim urged Sprengel to seek comfort and relaxation in the study of nature. Heim himself was an amateur botanist of some renown who had given botany lessons to no less a personage than the young Alexander von Humboldt. Thus it was natural for Heim to encourage Sprengel to take an interest in botany—a time-honored anodyne for the spiritually troubled—and Heim even tutored Sprengel in basic botany. As early as 1782, Pastor Schulze reported that Sprengel had begun taking nature walks for his health. The following year, his walks blossomed into extended trips. Heims's prescription was working. By keeping his mind firmly focused on nature, Sprengel managed to avoid any further allegations of abuse during all of 1783. But in 1784, Heim unexpectedly closed his Spandau office and moved to Berlin. The sudden departure of his physician, friend, and mentor may have precipitated a relapse, for shortly afterward Schulze recorded that Sprengel was in trouble again for giving a physician's son twenty strokes and banishing him to the school detention hall, ostensibly for not knowing a Greek vocabulary word. Once again he was ordered to the town hall, where he promptly expressed contrition and promised to mend his ways. This time he really meant it, but not in the way his supervisors imagined. He had finally discovered a way to cope with misbehaving and underperforming students: he ceased to care. From then on, according to Schulze:

> the headmaster went to the other extreme. He showed indifference towards whether the children learned anything or not and hardly made an effort. He grumbled and let everything go.[38]

It was a decisive moment in his career. From 1784 until his early retirement in 1793 at the age of forty-three, he did no more than the minimum required to discharge his duties as headmaster. Over the objections of his supervisors he took the unprecedented step of cancelling all his private lessons (which he was not contractually obligated to give) even though it meant a loss of income. Never again would he allow himself to become unhinged by petty conflicts with students, parents, or supervisors. Henceforth, he would devote all his creative energy to unraveling the secrets of nature and, by so doing, become closer to God. He became an apostle of flowers.

SPRENGEL'S LANDMARK CONTRIBUTIONS TO FLORAL ECOLOGY

As an amateur botanist working in isolation from other scientists, Sprengel wrote *The Discovery of the Secrets of Nature in the Structure and Fertilization of Flowers*[39] for the edification and amusement of the general public rather than for professional botanists. In it, he described himself as a philosophical botanist—one who employs observation and reason to determine the structure–function relationships in plants, specifically flowers. During the classical period, the definition of "flower" included only the "showy" parts—the sepals, petals, and stamens. The pistil was considered the immature fruit, separate from the flower proper. In addition to functioning to protect the fruit, the flower was thought to refine the sap, a necessary step for seed formation. This conception of flowers reached its apogee in the seventeenth century in the uterine model of Malpighi.

Based on the writings of Camerarius, Linnaeus, and Koelreuter, Sprengel took it as a given that pistils and stamens were sexual structures. As we saw earlier, Koelreuter's discovery of insect pollination led him to conclude that the primary function of floral nectaries is to attract insect pollinators. Although insect pollination appeared at first glance to be a random process, it is made purposeful by the insects' continual search for nectar. In his 1761 paper *Vorläufige Nachricht* (Preliminary Report), Koelreuter had presciently predicted that "(t)he decline of the insect [species] would inevitably be followed by the decline of the plant species."[40]

Sprengel only learned about Koelreuter's work on nectaries and insect pollination some time after beginning his own studies.[41] Like Koelreuter, he came to regard insect-mediated pollination as a divinely ordained arrangement to ensure the perpetuation of plant species. His own brilliant entry into the field of pollination biology literally began with a splash—the splash of raindrops on petals. He had been examining the petals of a type of geranium (cranesbill) when he suddenly became curious about the many fine hairs inside the margins of the petals near the base, just above the nectaries, and "wondered what purpose they might serve." The introduction of his book begins with this insight:

> When I carefully examined the flower of the wood cranesbill (*Geranium sylvaticum*) in the summer of 1787, I discovered that the lower part of its corolla was furnished with fine, soft hairs on the inside of the margins. Convinced that the wise creator of nature had not created even a single tiny hair without definite purpose, I wondered what purpose the hairs might serve. And it soon came to my mind that if one assumes that the five nectar droplets which are secured by the same number of glands are intended as food for certain insects, one would at the same time not think it unlikely that provision had been made for this nectar not to be spoiled by rain and that these hairs had been fitted to achieve this purpose.[42]

Sprengel found similar small hairs located at nearly the same location in other species of *Geranium*. He therefore concluded that:

> Since the flower is upright and tolerably large, drops of rain must fall into it when it rains. But no drop of rain can reach one of the drops of nectar and mix with it, because it is stopped by the hairs, which are just above the nectaries, just as a bead of sweat which has run down a man's forehead is caught by his eyebrows and eyelashes and is prevented from running into his eye. An insect is not hindered by these hairs from getting at the drops of nectar.[43]

In the flowers of every species he examined, Sprengel found comparable floral structures that protected nectar droplets from rain, from which he inferred a functional relationship:

> The longer I continued this investigation, the more I saw that flowers that produce nectar drops are so contrived that insects can easily reach it, but that the rain cannot spoil it; but I gathered from this that it is for the sake of the insects that these flowers secrete their nectar, and that it is secured against rain so that they may be able to enjoy it pure and unspoiled.[44]

The following year (1788) Sprengel made another important discovery: the positions of colored spots on the petals of certain flowers often served as guides directing insects to the nectaries:

> If the corolla has a particular color in particular spots . . . it is for the sake of the insects that it is so colored; and if the particular color of a part of the corolla serves to show an insect which has lighted on the flower the direct path to the nectar, the general color of the corolla has been given to it, in order that insects flying about in search of their food may see the flowers that are provided with such a corolla from a long distance, and know them for receptacles of nectar.[45]

To Sprengel, such floral features designed specifically for insect pollinators provided proof of God's existence because it showed that even the tiniest, most inconspicuous structures of flowers had been created with a specific purpose in mind. At a time when the study of natural history was still hampered by creationist ideas, Sprengel's application of teleology to virtually every visible structure of the flower opened the door to a series of important discoveries in floral ecology.

Sprengel's *magnum opus* was more than four hundred pages long and richly illustrated with more than a thousand minutely detailed anatomical figures crammed into twenty-five copperplates. These included analyses of 461 individual species of flowering plants arranged in Linnaean taxonomic sequence. Each species was discussed according to its pollen vector (insect vs. wind) and the floral mechanism for attracting insects and facilitating pollination. Since no one had ever done this before, Sprengel was in the enviable position of making one discovery after another, laying the foundation for the new field of floral ecology. Stefan Vogel at the University of Mainz has listed nine major findings in floral ecology and fifteen additional insights into other important phenomena attributable to Sprengel.[46]

Arguably, the most significant question Sprengel addressed was the source of the pollen that was ultimately deposited on the stigma by the pollinator. If, for example, a bee simply transferred pollen from a flower's anther onto the stigma of the same flower, the bee would simply be facilitating self-pollination. If, on the other hand, the pollen that the bee deposited on the stigma was derived from a different flower, the bee would be carrying out cross-pollination or "out-crossing." Sprengel realized that, in the case of dioecious species, outcrossing was the rule, and it was also common in monoecious species such as maize. However, it had long been assumed by Sprengel and his contemporaries that all hermaphroditic flowers, which make up more than 85% of flowering plants, were self-pollinating. Self-pollination seemed to Sprengel, as it did to Koelreuter, the logical mechanism, designed by the wise Creator to prevent mixing and to assure the fixity of species. Thus he was surprised to discover features of flower morphology that seemed to prevent self-pollination, such as *dichogamy*, in which the stamens and pistils of the same flower mature at different times, thereby preventing self-pollination.[47] Sprengel drew the counterintuitive conclusion that "nature does not seem to allow any flower to be fertilized by its own pollen." Insect-mediated pollination, according to Sprengel, causes outcrossing, not self-pollination.

Even Sprengel was troubled by such a radical idea, which flew in the face of the conventional wisdom that self-pollination was necessary to maintain the fixity of species.

This apparent paradox may partially explain the rather cold reception that Sprengel's book received from his botanical peers. It was known, for example, that in some taxa self-pollination did, indeed, occur, as in the case of certain kinds of orchids, peas, and composites. These examples seemed to show that nature did not bar self-pollination as Sprengel had asserted. However, we now know that self-pollination is either a fail-safe mechanism that ensures fertilization under conditions that adversely affect insect pollination or a relatively recent evolutionary branch from an insect-pollinated ancestor.[48] As will be discussed in Chapter 18, the true significance of insect pollination would not be grasped until Darwin placed it in an evolutionary context in his landmark study of outcrossing in 1876.[49]

In addition to their qualms about outcrossing, Sprengel's more pious contemporaries also objected to what they considered to be a crassly materialistic interpretation of floral ecology. Such critics refused to accept that God's divine plan could be reduced to a series of banausic transactions between flowers and insects, for God's motives are, and always shall be, mysterious and unknowable. But what his skeptical contemporaries found most objectionable was Sprengel's less than flattering portraits of some of God's creatures, as when he reported examples of "floral deception"—such as a rewardless flower (*Orchis*) with fake nectaries—and the "stupid insects" who were fooled by the ruse, or when he described "larcenous" insects, which stole nectar without performing pollination services, or insects too clumsy to properly navigate floral structures to the nectaries.[50] To pious botanists, the all-wise and all-knowing Creator would never incorporate fakery, stupidity, incompetence, venality, or criminality into His works, and to say otherwise was tantamount to blasphemy. But Sprengel, who had observed these traits in abundance at the Große Schule, thought he knew better.

After publishing his book in 1793, Sprengel promptly resigned his position at the Große Schule, much to the relief of his superiors. Sixty years later, a school historian wrote that "[d]uring the directorship of Sprengel, an irascible and obstinate man, the school began to lapse into ruin"—no doubt an exaggeration.[51] Soon after retiring at the age of forty-three, supported by a modest pension, Sprengel moved to Berlin, where his friend Heim had his practice, and took up residence in an attic apartment. There he happily pursued his scholarly interests free at last from the demands of teaching and administration. Although he had planned to write a second volume of his flower book, he later dropped the idea due to a lack of interest in the first volume. In 1811, he published a practical treatise entitled *The Usefulness of Bees and the Necessity of Bee-Keeping, Viewed from a New Perspective*. In it he advised farmers to place bee-hives in clover and alfalfa fields to increase seed set. His only other botanical activity was to lead natural history field trips, open to the public for a small fee, in the tradition of Linnaeus. After his death in 1817, those who joined him on these excursions fondly remembered him for his encyclopedic knowledge of plants as well as his piquant sense of humor, a side of his personality that his colleagues at the Große Schule rarely, if ever, saw.

Although Koelreuter and Sprengel, alone in the eighteenth century, carried out research on the sexual theory in the tradition of Camerarius, their contributions were so compelling that any modern reader would have thought that the debate over plant sexuality was effectively over. In 1761, Koelreuter had expressed the hope that even the most stubborn doubter of the truth of the sexuality of plants would be completely convinced. If, contrary to all conjecture, there should be someone who, after a rigid examination, maintained the contrary, it would astonish me greatly as though I heard someone maintain in the middle of a clear day that it was night.

But Koelreuter badly underestimated the depth and tenacity of the asexualist view of flowers, especially among his fellow Germans, who, in the nineteenth century, fanned the lingering sparks of asexualism into a new blaze.

NOTES

1. Delaporte, F. (1982), *Nature's Second Kingdom*, trans. Arthur Goldhammer. MIT Press.
2. Marquis de Condorcet (1799), Eloge de M. Duhamel, in *Eloges des Academiciens*, Vol. III, p. 319.
3. *The Confessions of Jean-Jacques Rousseau*, Vol. XXIII, Book I. Quoted by Delaporte, F. (1982), p. 141.
4. Rousseau, Jean-Jacques (1793), Les Rêveries du promeneur solitaire (promenade V), Quoted in: Delaporte, F. (1982), *Nature's Second Kingdom*, trans. Arthur Goldhammer, p. 141.
5. O'Neal, J. C. (1996), *The Authority of Experience: Sensationist Theory in the French Enlightenment*. Pennsylvania State University Press.
6. Rousseau, Les Rêveries du promeneur solitaire (promenade VII), Quoted in Delaporte, Nature's Second Kingdom, p. 137.
7. Ibid.
8. Idem., p. 140.
9. Cook, A. (2012), *Jean-Jacques Rousseau and Botany: The Salutary Science*. Voltaire Foundation, Oxford.
10. Rousseau, Jean-Jacques (1762), *Émile, Or Treatise on Education.*
11. Rousseau, Jean-Jacques (1807), *Letters on the Elements of Botany*, trans. T. Martyn. John White. Mary Wollstonecraft deplored this kind of squeamishness. She wrote, "Children very early see cats with their kittens, birds with their young ones, &c. Why then are they not to be told that their mothers carry and nourish them in the same way? As there would be no mystery they would never think of the subject more." Wollstonecraft, M. (1996), *A Vindication of the Rights of Woman*, Dover, pp. 53–54. Originally published in 1792 by J. Johnson.
12. Mary Wollstonecraft (1759–97) was a prolific writer. As an early feminist, political theorist, and novelist, she was regarded as a radical during her lifetime. She is perhaps best known for *A Vindication of the Rights of Woman* (1792). She was the wife of William Godwin, author of *Political Justice* and the mother of Mary Shelley, the author of *Frankenstein.*
13. Wollstonecraft, *A Vindication of the Rights of Woman.*
14. Möller, Hans (1752), Fortsetzung der muthmasslichen Gendanken vom Bluhmenstaube. *Hamburgerisches Magazin,* Hamburg and Leipzig III:427. Cited by F. Delaporte (1982).
15. Alston, C. (1754), *A Dissertation on the Sexes of Plants. Essays and Observations, Physical and Literary.* Read before the Philosophical Society in Edinburgh. Translated from French posthumously in 1771, pp. 228–318; See also Alston, C. (1754), *A Dissertation on Botany*, translated from the Latin. B. Dod. (Originally published as *Tyrocinium Botanicum Edinburgense.*)

16. Van Nigtevecht, G. (1967), *Genetic Studies in Dioecious Melandrium*. Springer, p. 336.

17. Farley, J. (1982), *Gametes & Spores: Ideas About Sexual Reproduction 1750–1914*. Johns Hopkins University Press, p. 39.

18. Delaporte, F. (1982), *Nature's Second Kingdom*. MIT Press, p. 119; Pinto-Correia, C. (1997), *The Ovary of Eve: Egg and Sperm and Preformation*. University of Chicago Press, p. 209.

19. Ewan, J. (1982), Smellie's "Philosophy of Natural History." *Taxon* 31:462–466.

20. George, S. (2007), *Botany, Sexuality, and Women's Writing (1760–1830)*. Manchester University Press.

21. Wollstonecraft, M. (1795), *An Historical and Moral View of the Origin and Progress of the French Revolution and the Effect it Has Produced in Europe*. J. Johnson. Accessed online July 12, 2015, http://oll.libertyfund.org/titles/226

22. George, *Botany, Sexuality, and Women's Writing*, p. 121.

23. Morton, A. G. (1981), *History of Botanical Science*. Academic Press.

24. Mayr, E. (1986), Joseph Gottlieb Kolreuter's contributions to biology. *Osiris* 2:135–176. See also Olby, R. C. (1966), *Origins of Mendelism*. Constable, pp. 37–38; and Roberts, H. F. (1929), *Plant Hybridization Before Mendel*. Princeton University Press.

25. Among the frequent guests at Princess Caroline's salon were Johann Wolfgang von Goethe, Christoph Willibald Gluck, Johann Gottfried von Herder, Friedrich Gottlieb Klopstock, and other luminaries. A member of the Copenhagen Academy of Arts, she was an accomplished pastel artist as well as a member of the harpsichordist court orchestra. She was particularly interested in the sciences, including botany, zoology, physics, medicine, mineralogy, geology, and chemistry. Her vast mineral collection, some of which she collected in the field herself, formed the basis of the future *National Collection of Natural History* in Karlsruhe.

26. Mayr, Joseph Gottlieb Kolreuter's contributions to biology.

27. Ibid.

28. Fraternal twins can indeed be conceived by two fathers, a rare phenomenon called heteropaternal superfecundation, which occurs when two (or more) of a woman's eggs are fertilized by different men within the same ovulation period, as occurs frequently in dogs and cats.

29. For a comprehensive discussion, see Mayr, Joseph Gottlieb Kolreuter's contributions to biology.

30. In fact, the most successful application of hybrid vigor to agriculture has been not trees, but maize, with more than 95% of corn acreage planted in hybrid varieties. Other crops improved by hybrid vigor include sorghum, rice, sugar beet, onion, spinach, sunflowers, broccoli, and cannabis.

31. Koelreuter, J. G. (1961), *Vorläufige Nachricht von einigen das Geschlecht der Pflanzen betreffenden Versuchen und Beobach-tungen* [Preliminary Report of Some Experiments and Observations on the Sexuality of Plants]. An English translation by Margaret Mayr and Ernst Mayr is available in the library of the Department of the History of Science at Harvard University.

32. Ibid.

33. Hagen, H. A. (1884), Christian Conrad Sprengel. *Nature* 29:572–573.

34. Most of the biographical information for this section is from Zepernick, B., and W. Meretz (2001), Christian Konrad Sprengel's life in relation to his family and his time. On the occasion of

his 250th birthday. *Willdenowia* 31:141–152, the Botanischer Garten und Botanisches Museum, Berlin-Dahlem; and Vogel, S. (1996), Christian Konrad Sprengel's theory of the flower: The Cradle of Floral Ecology, in D. G. Lloyd and S. C. H. Barrett, eds. (1996), *Floral Biology: Studies on Floral Evolution in Animal-Pollinated Plants.* Chapman and Hall, pp. 44–62.

35. The second pastor for a protestant city church.

36. Zepernick and Meretz, Christian Konrad Sprengel's life.

37. Ibid.

38. Quoted by Zepernick and Meretz.

39. *Das entdeckte Geheimnis der Natur im Bau und in der Befruchtung der Blumen* (Berlin 1793).

40. This prediction has particular relevance for today, in light of the continuing concern over bee colony collapse disorder (CCD) and other environmental threats to insect pollinators.

41. Vogel, Christian Konrad Sprengel's theory of the flower.

42. Sprengel, C. K. (1793), *Discovery of the Secret of Nature in the Structure and Fertilization of Flowers*, trans. Peter Haase, in D. G. Lloyd and S. C. H. Barrett, eds. (1996), *Floral Biology: Studies on Floral Evolution in Animal-Pollinated Plants.* Chapman and Hall.

43. Ibid.

44. Ibid.

45. Ibid.

46. For example, Sprengel found that most flowers are insect-pollinated and contain features that attract insects, such as nectaries, petal pigmentation, and fragrance. He showed that wind-pollinated flowers are generally nectarless and inconspicuous and produce dry, powdery pollen, in contrast to insect-pollinated flowers, whose pollen tends to stick together in clumps. He also noted that insect pollinators may be either generalists or specialists. Vogel, S. (1996), Christian Konrad Sprengel's theory of the flower.

47. Another floral feature that favors outcrossing is *heterostyly*, in which a species population includes individuals with two or three types of flowers that differ in the lengths of their stamens and pistils. However, it was Darwin, not Sprengel, who in 1862 published the first study of heterostyly in *Primula* (primroses).

48. Vogel, Christian Konrad Sprengel's theory of the flower.

49. Darwin, C. R. (1876), *The Effects of Cross- and Self-Fertilisation in the Vegetable Kingdom.* John Murray.

50. Insect larceny, or "nectar robbing," refers to the harvesting of nectar from the flower without performing pollination services, typically by drilling a hole at the base of the corolla, thus avoiding contact with the anthers. Examples include carpenter bees, bumblebees, wasps, and ants, as well as some birds and mammals.

51. Wichler (1936), cited by Lloyd and Barrett, *Floral Biology.*

Form is mobile, emerging, passing. The study of form is the study of morphology. The study of metamorphosis is the key to all signs of nature.
—J. W. GOETHE (1749–1832)[1]

[F]rom first principles, systematics can be characterized as a dividing discipline, whereas plant morphology is a unifying discipline.
—D. R. KAPLAN (1938–2007)[2]

16

Idealism and Asexualism in the Age of Goethe

WHILE STROLLING IN the Grand-ducal Botanical Garden at the University of Jena, Franz Joseph Schelver, the newly appointed Curator of the garden, confided to his senior colleague, Johann Wolfgang Goethe, that he had long doubted the "theory ascribing two sexes to plants" and that he was now convinced that it was no longer tenable. The year was 1804. Goethe—poet, dramatist, and accomplished amateur botanist whose treatise on *The Metamorphosis of Plants*, published fourteen years earlier, had laid the foundation for the field of plant morphology—had used his influence to hire Schelver and to transfer his appointment from the Medical School to the Philosophy Department. His motivation was twofold. First, he hoped to establish botany as a discipline in its own right apart from medicine. Second, Jena was then the center of *Naturphilosophie*, a new philosophical movement in Germany that had arisen as a reaction against the Cartesian mechanical model of the universe. Taking a page from Plato, the "nature philosophers" granted equal weight to reason and the senses on the one hand, and to the subjective imagination on the other. Goethe believed that if scientists could only learn to use their imaginations as poets do, they could leapfrog over much dreary data-collecting and accelerate the process of scientific discovery. Goethe, then fifty-five years old, felt that Schelver's ideas were in harmony with his own.

As they wandered through the garden, Schelver asserted that the sexual theory was incompatible with the very nature of vegetative life and that the production of fruits and seeds was strictly a vegetative process. Contrary to the claims of Camerarius and Koelreuter, the pistil did not require any external agents, such as pollen, to form seeds. Schelver based his new asexual theory on Goethe's own theory of metamorphosis, an argument Goethe found difficult to resist.[3]

We'll defer our discussion of Goethe's theory of plant metamorphosis until later in the chapter. For now it's sufficient to note that Goethe's conversation with Schelver in the garden was a major turning point in Goethe's thinking, as he related in an essay written sixteen years later:

> Now the doubts that had been raised from time to time about the sexual system came to my mind, and the ideas that I myself had had on the subject came to life again. The new viewpoint supported certain views of Nature that now seemed clearer and more significant to me; and accustomed as I had been to preserve complete flexibility in my application of metamorphosis, I likewise found this viewpoint not uncomfortable, although at the same time I could not immediately relinquish the other [i.e., the sexual theory].[4]

The last sentence of this paragraph has generated a debate as to whether Goethe did in fact change his mind and become an asexualist.[5] In 1804, at least, he still felt constrained by the opinions of most academic botanists for whom the sexual theory had become axiomatic. In addition, Goethe's deep disappointment over the muted reception of his *Metamorphosis of Plants* inclined him to be cautious about associating himself with Schelver's heterodox theory, lest he do irreparable harm to his scientific reputation:

> No one familiar with the situation of our botanical science at that time will blame me for imploring Schelver not to let his thoughts get abroad. It was to be foreseen that his theory would get a most unfriendly reception and that the theory of metamorphosis, which as it was had found no acceptance, would be banished from the boundaries of science for a long time to come. Our own academic standing also made such secrecy advisable.[6]

By 1820, however, when Goethe was writing these words, asexualism was no longer being anathematized. One of Schelver's former students, the physician August Henschell, had just published a 600-page treatise, *On the Sexuality of Plants*, challenging the validity of Koelreuter's hybridization experiments, and his arguments were gaining traction. Goethe was jubilant on Schelver's behalf:

> Now his brilliant theory takes on substance through Henschell's momentous study; it is earnestly demanding its place in science, although one cannot yet foretell how that place will be found. However, interest in the theory is already astir. Reviewers, instead of preaching and scolding against it as before, confess that they have been converted, and now we can only wait to see how things will develop further.[7]

At this late date, Goethe's enthusiasm for the philosophical arguments of Schelver and Henschell is difficult to fathom in the face of the abundance of experimental evidence in support of the sexual theory marshaled by Camerarius and Koelreuter.[8] Goethe makes scant mention of Koelreuter's experiments in his own writings, although Henschell discusses them explicitly. Ten years earlier, Goethe, in his *Zur Farbenlehre* (Theory of Colors), had openly challenged the light experiments of no less a luminary than Sir Isaac Newton, so his failure to take on the lesser-known Koelreuter can hardly be ascribed to timidity. More likely it reflected Goethe's genuine ambivalence toward the sexual theory.

The resurgence of asexualism in Germany during the first half of the nineteenth century coincided with the emergence of Naturphilosophie, an intellectual movement associated with Romanticism that blended neoclassicism, rational philosophy, and spirituality in novel ways. Inspired by the writings of Spinoza and Immanuel Kant, nature philosophers attracted many disciples in Germany, and their ideas spread throughout Europe and the Americas. Because it formed the intellectual milieu in which Goethe and other botanists were working, some consideration of Naturphilosophie is required for an understanding of the origin and appeal of asexualist thinking in the early nineteenth century. For example, although Goethe frequently defended empiricism against subjective idealism, his enthusiasm for experimentation was tempered by moral concerns about "torturing" Mother Nature to force her to reveal her secrets. Much as ethicists today have questioned the value of information obtained from prisoners of war by torture, nature philosophers challenged the validity of experimental results obtained by "torturing" nature. To Goethe and his companions, nature was sacred. Only by treating her with the utmost respect and reverence could her secrets be grasped. On the other hand, Naturphilosophie was never monolithic, and Goethe was often at odds with different schools of thought within the movement. Whether or not Goethe ever fully embraced asexualism in plants, he was clearly drawn to it, and the underlying causes of this attraction have never been adequately explored.

Many eighteenth-century asexualists had been horrified by the licentiousness of the Linnaean system, ostensibly because of the popularity of botany among women, and this was also a concern of nature philosophers, especially Goethe. In the aftermath of the French Revolution, the political climate underwent a shift in attitudes regarding women's social status. Arguments for biological determinism based on traditional gender stereotypes re-emerged with a vengeance, and women's biological role as mothers was cited as justification for laws restricting their civil rights. Schelver, Lorenz Oken, and Georg Wilhelm Friedrich Hegel all espoused the view that women were by nature passive and plant-like. Goethe himself made abundant use of floral metaphors when referring to women in his poetry, and in his essay on "The Spiral Tendency in Plants," written toward the end of his life, he attributed the reproductive capacity of plants to the feminine principle.

As has been noted by many scholars, Goethe's development as a dramatist and poet is intimately connected with his personal life. In parallel with his shifting views about plant sexuality, Goethe's attitude toward his own sexuality seems to have undergone transformations as well—from a Werther-like passion for unobtainable women in his younger years, to a relatively stable married life in his middle years, to the more detached spirituality of his later years. Throughout his long career, Goethe's views about plant sex, like his plays and poetry, also appear to have been influenced by his personal life.

THE ROMANTIC RESPONSE TO HOLBACH

A watershed event in France that provided an early impetus to German Romanticism was the publication of Baron d'Holbach's *The System of Nature*.

During the seventeenth and eighteenth centuries, Descartes and Newton had transformed the scientific landscape by conceiving of the universe as a gigantic, clock-like machine. The mechanical universe had been created and put in motion by God, but, after that it was capable of running on its own according to Newton's laws, although God was still able to influence the outcome of events. Christian denominations each staked out their territory based on the extent to which God intervened in human affairs. Calvinists believed that the entire history of the universe, including human history, was predetermined by God from the outset. Deists made generous allowances for free will, with the occasional miracle at critical junctures. Most Protestants and Catholics shared the view that God regularly intervened in human affairs, differing mainly in the importance of the Church as divine mediator. In the atheist version, God was dispensed with entirely and the universe functioned on its own for all eternity according to natural laws. Just such an uncompromisingly mechanistic vision of the universe was put forward by Baron d'Holbach, a French-German philosopher and encyclopedist. To no one's surprise, it was denounced by clergy of all denominations, but especially by the Catholic clergy.

Holbach published his *The System of Nature* in 1770, under the name of one of his enemies, Jean-Baptiste de Mirabaud, a pious religious philosopher and former Secretary of the Académie Française. Mirabaud had died ten years earlier at the age of eight-five and was thus unable to defend himself.[9] Sardonically written in the guise of Mirabaud's deathbed testimony, Holbach exhorted Parisians to cease their slavish dependence on a nonexistent deity in the vain hope of procuring "a happiness nature refuses to grant." Instead of embracing a life of willful ignorance and superstition, one should submit to one's fate as Nature unfolds and seek solace through an understanding of her laws:

> [L]et them study that nature, let them learn her laws, and contemplate the energy and the unchanging fixity with which she acts; let them apply their discoveries to their own felicity, and submit in silence to laws from which nothing can withdraw them; let them consent to ignore the causes, surrounded as they are by an impenetrable veil; let them undergo without a murmur the decrees of universal force.[10]

In a deterministic, mechanical universe, "free will" is a pernicious illusion invented to justify human suffering by blaming the victim:

> The system of man's liberty seems only to have been invented in order to put him in a position to offend his God, and so to justify God in all the evil that he inflicted on man, for having used the freedom which was so disastrously conferred upon him.[11]

The impact of *The System of Nature* on the rigid theocratic society that was eighteenth-century France was electrifying. The Catholic Church threatened to withhold its monetary largesse from the crown unless the king immediately ordered all copies of Holbach's book confiscated and burned. Many celebrated intellectuals felt it prudent to dissociate themselves from the heretical tract by publishing high-minded refutations. According to Goethe, however, the reaction among students in Germany was more muted, as he describes in his autobiography:

> We had neither impulse nor tendency to be illumined and advanced in a philosophical manner; on religious subjects we thought we had sufficiently enlightened ourselves, and therefore the violent contest of the French philosophers with the priesthood was

tolerably indifferent to us. Prohibited books condemned to the flames, which then made a great noise, produced no effect upon us.[12]

Reading between the lines, Goethe seems to imply that young German intellectuals had already advanced beyond the traditional view of an anthropomorphic, personal God to a more abstract and pantheistic conception of the divine. Thus Goethe and his friends regarded Holbach's book as grim and repellant, but perfectly harmless:

> We did not understand how such a book could be dangerous. It appeared to us so dark, so Cimmerian,[13] so deathlike, that we found it difficult to endure its presence, and shuddered at it as at a specter.[14]

Because of the book's title, they had hoped to learn more about nature from it but were greatly disappointed:

> Not one of us had read the book through, for we found ourselves deceived in the expectation with which we had opened it. A system of nature was announced, and therefore we hoped to learn really something of nature, our idol. . . . But how hollow and empty did we feel in this melancholy, atheistical half-night, in which the earth vanished with all its images, the heaven with all its stars. There was to be matter in motion from all eternity, and by this motion, right, left, in every direction, without anything further, it was to produce the infinite phenomena of existence.[15]

They did not deny the operation of natural laws, nor did they consider themselves immune to such laws. Nevertheless, they felt they possessed something like free will:

> [W]e nevertheless felt within us something that appeared like perfect freedom of will, and again something which endeavored to counterbalance this freedom.[16]

In keeping with the spirit of the Reformation, Goethe and his peers rejected the idea of the duality of the material and spiritual planes. It was the notion of this duality, they reasoned, that allowed sophists like Holbach to dispense with the spiritual plane entirely. Goethe and his friends preferred a pantheistic theology along the lines of Spinoza's, in which God and nature were unified. In such a system, God is also a necessity:

> "All was to be of necessity," so said the book, "and therefore there was no God." "But could there not be a God by necessity too?" we asked.[17]

Goethe and his friends dismissed Holbach's treatise as the misanthropic ravings of a senile old man:

> We laughed him out; for we thought we had observed that nothing in the world that is loveable and good is in fact appreciated by old people. "Old churches have dark windows; to know how cherries and berries taste, we must ask children and sparrows." These were our gibes and maxims; and thus that book, as the very quintessence of senility, appeared to us as unsavory, nay, absurd.[18]

Holbach had eliminated God the Creator, leaving in His place a soulless machine controlled by natural laws based on the equations of Newton and Boyle—meager compensation for the rich brocade of universal certainties woven by religion. Atheism also posed heuristic problems for late eighteenth-century biologists. Faced with a paucity of data, naturalists routinely cited God's wisdom in lieu of a physical mechanism for the baffling complexity and seeming purposefulness of living organisms. Because Newton's and Boyle's equations failed to describe life processes, atheism would have deprived naturalists of their fall-back explanation.

At the same time, young seminarians in German universities were busily expanding the Protestant Reformation by rejecting religious fundamentalism. The American and French revolutions were greatly admired for they seemed to demonstrate the power of reason and idealism to overturn antiquated systems of thought and governance. Some young scholars were attracted to the field of comparative religion, which exposed them to the historical antecedents of Judaeo-Christian traditions. Before long, the biblical account of the Creation was being downgraded from revealed truth to the status of myth. Such a radical departure from fundamentalist thinking set German intellectuals apart from most of their European colleagues. With the enthusiasm of converts, German philosophers took up the challenge of providing an alternative explanation for the Creation in more rational and historical terms.

The problem was how to account for the manifold forms and patterns of living organisms and the reciprocal interactions among their parts if not by the intervention of a Creator God. With God now banned from eighteenth-century scientific discourse, a suitable substitute (short of atheism) had to be found. Naturphilosophie, a diverse amalgam of pantheistic notions and physical laws, fit the bill nicely. A consensus emerged that the best way to discover the rules governing the growth and development of a living organism was to describe its history. But if God is not directly involved at each stage of an organism's development, where did the apparent purposefulness of development come from? This was one of the thorniest questions that Immanuel Kant posed in his *Critique of Pure Reason* (1781), and, over the next few decades, Kantian ideas were enthusiastically elaborated by his youthful disciples into a confusing web of convoluted and often contradictory schemes.

In effect, the powers of the old Creator God were disbursed to two new pantheistic forces borrowed from Spinoza and eastern religions, "subjective consciousness" and "universal consciousness," with some favoring the former and others favoring the latter. However, neither of these two supposed forces quite explained the "purposefulness" of nature. The missing piece of the puzzle was, of course, Charles Darwin's theory of evolution published in 1859. The frustrating historical predicament that Goethe and his fellow Romantics faced as they tried to make sense of the natural world without resorting to a traditional Creator God was the absence of the concept of natural selection.

IMMANUEL KANT: ON EMPIRICISM, *BILDUNGSTRIEB*, AND ARCHETYPES

Eighteenth-century empiricism rested on the assumption that the senses were accurate reporters of external reality. *Induction* was the process by which the mind used reason to infer truths about external objects from the raw, unfiltered data of the senses. By this process, an accurate and complete view of the natural world could be obtained.

The Scottish philosopher David Hume was among the first to question the infallibility of the mind as a reliable interpreter of external reality. Contrary to Descartes and the rationalists, Hume argued, in Treatise of Human Nature (1739–40) and other works, that the mind was incapable of pure rationality and that "[r]eason is . . . the slave of the passions." Because of the mind's irrational tendencies, Hume cautioned that subjective bias could never be eliminated from scientific studies. Thus, we can never fully understand anything outside ourselves. Even our ideas about ourselves are cobbled together from fragments of memories, which are subjective and therefore unreliable. Hume's skepticism served to undermine the Enlightenment's faith in reason and science as the only paths to truth.

In *Critique of Pure Reason* (1781), Immanuel Kant, attempted to rescue science from Hume's doubts by defining a limited set of properties or patterns of the phenomenal world that the mind was capable of comprehending. Such "categories of understanding" included cause and effect, space and form, time, and quantity. Outside of these categories, according to Kant, lay the "noumenal" sphere about which nothing could be known. But within the framework of the categories of understanding, the results of science—at least physics and chemistry, which obeyed mechanical laws—could be shielded from Hume's skepticism.

Kant was less sanguine about the ability of scientists to comprehend living organisms, which did not seem to obey mechanical laws, and, for this reason, he did not believe that biology could ever become a true science. Living organisms were thus qualitatively different from machines. They could grow, develop, and reproduce without any external input other than food.[19] According to Kant,

> [a]n organized being is . . . not a mere machine: for the latter has only the power of motion, while the former has a formative power [*Bildungstrieb*], of a kind that it imparts to material not possessed of it (it organizes these materials). Hence such a propagative power of formation cannot be explained merely through the ability of motion that a machine has.[20]

Kant had borrowed the term *Bildungstrieb* from a young biologist at the University of Göttingen, Johann Friedrich Blumenbach. Blumenbach had begun his career as a preformationist but had switched to epigenesis after observing the results of surgical experiments with the fresh water polyp *Hydra*, similar to those Abraham Trembley had performed decades earlier. Struck by the ability of the polyp to regenerate lost parts, Blumenbach concluded that:

> there exists in all living creatures, from men to maggots and from cedar trees to mold, a particular inborn, life-long active drive. This drive initially bestows on creatures their form, then preserves it, and, if they become injured, where possible restores their form. . . . I give it the name of *Bildungstrieb*.[21]

Other authors had invoked similar vaguely defined "inborn drives," such as *vis plastica* and *vis essentialis*, but these terms were understood to refer to occult forces. Blumenbach's *Bildungstrieb* gained greater currency for two reasons: first, because it was adopted by Kant, and second, because both Blumenbach and Kant insisted that *Bildungstrieb* was not an occult, spiritual force, but was in some way associated with the organic matter it directed.

On the other hand, Kant scoffed at the idea that *Bildungstrieb* could ever be explained by mechanical principles:

> It is quite certain that we cannot become sufficiently acquainted with organized creatures and their hidden potentialities by aid of purely mechanical natural principles, much less can we explain them; and this is so certain, that we very boldly assert that it is absurd for man even to conceive such an idea, and to hope that a Newton may one day arise even to make the production of a blade of grass comprehensible, according to natural laws ordained by no intentions, such an insight we must absolutely deny to man.[22]

Although Kant was adamant that scientists would never comprehend the nature of *Bildungstrieb*, he believed that they could at least chip away at the problem by studying those aspects of *Bildungstrieb* that fell within the categories of understanding.[23] For Kant, the utility of the term *Bildungstrieb* lay mainly in its heuristic value. It permitted biologists to proceed with their investigations of mysterious phenomena *as if* the organism were acting purposively, guided by some unseen intelligence.

During the Enlightenment, the psychological gulf between artists and scientists had grown wider. Humanists felt the pain of this schism more acutely than scientists did, and one of the chief goals of the Romantic movement was to bring about a reunification of aesthetics and the natural sciences. In *Critique of Judgement*, Kant attempted to provide the philosophical foundation for such a reunification by comparing the "purposive character of beautiful objects and organic nature."[24] The primary "purpose" of art was to be beautiful.[25] But the creation of a beautiful art object could not be reduced to a set of instructions; otherwise, anyone could be an artist. How, then, does the artist know how to create beautiful objects? And how does the observer gazing at a work of art recognize it as beautiful?

According to Kant, the act of perceiving a work of art sets the imagination free to resonate with the harmony of the artwork, thus evoking a feeling of pleasure. If the feeling exceeds a certain threshold, the object is judged to be beautiful. In general, artists, composers, and poets are people who have an intuitive grasp of the forms and patterns that create harmonious feelings in others, a deep-seated awareness of what Richards calls the "subjectively-felt laws of harmony."[26] "Genius," according to Kant, "is the talent that gives the rule to art." The art critic may attempt to identify elements that make a work of art beautiful, but ultimately the attempt fails because beauty is an ineffable quality grounded in consciousness.

Living organisms, like works of art, exhibit beautiful forms and patterns produced by harmoniously interacting elements, which, while ineffable, *seem* to be created by the *Bildungstrieb*, or formative power. The only way biologists can study living organisms, says Kant, is teleologically—that is, as though they had been created according to a blueprint or "archetype," just as the Renaissance sculptor, Michelangelo, is reputed to have said that his job was to liberate the forms imprisoned in marble:

> In every block of marble I see a statue as plain as though it stood before me, shaped and perfect in attitude and action. I have only to hew away the rough walls that imprison the lovely apparition to reveal it to the other eyes as mine see it.[27]

In this way biologists can strive to discover the hidden rules that govern the harmony of living organisms, but, as with art, their attempts to fully comprehend living organisms will ultimately fail, just as Michelangelo's hewing of the excess marble from his masterpiece fails to account for the masterpiece itself. Thus, Kant believed that biology, like art, could never become a true science.

FICHTE'S SUBJECTIVE IDEALISM AND SCHELLING'S NATURE PHILOSOPHY

Nature philosophers were in general agreement that the senses were not transparent windows through which information passed directly to the reasoning faculties of the brain. Instead, the senses were filters that allowed only certain types of information to be recognized, corresponding to Kant's categories of understanding. An object's "noumenal" properties, which Kant referred to as *das Ding an sich* ("the thing in itself"), would forever remain unknowable.

As for free will, Kant stated in the introduction to *Critique of Judgement* that natural laws seem to have been specifically designed to accommodate human freedom of action and moral choice in an otherwise determinate universe.[28] This was exactly what Goethe and his young friends, who valued their personal freedom above all else, wanted to hear. But Kant agreed with Hume that self-knowledge, without which free will is useless, is also limited because of its dependence on unreliable memories. Kant called the knowable self the "empirical ego." But he proposed that there was another self, the "transcendental ego," which was united to nature and therefore unknowable.

One of Kant's young disciples, Johann Gottlieb Fichte, took the subjective element in Kant's philosophy to its logical extreme. According to Fichte, because all our ideas about the natural world are based on mental constructs, there is no way to demonstrate that external reality actually exists. Everything we know about the external world (which Fichte called the "not-I") emanates from our own minds, and Fichte went so far as to claim that the "absolute ego," roughly equivalent to Kant's transcendental ego, creates its own reality, which we mistake for the "real world."

Goethe and his close friend, the poet, dramatist, and historian Friedrich Schiller,[29] both considered Fichte's rejection of external reality absurd. When a disgruntled university student threw a stone through Fichte's office window at Jena, Goethe remarked to Schiller that it was a "most unpleasant way to become convinced of the existence of the not-I." [30]

It was Friedrich Wilhelm Joseph Schelling, the boyish "philosopher king" of the Romantics, who coined the term "Naturphilosophie" in his book *Ideas for a Philosophy of Nature* (1797). An ardent admirer of both Fichte and Goethe, Schelling set himself the goal of finding the golden mean between Fichte's subjective idealism and Goethe's relatively hard-minded objectivism. But rather than achieving the synthesis he hoped for, he oscillated between these two opposing perspectives throughout his career. As a student, he was a pure Fichtean, but upon joining the faculty at the University of Jena he came under Goethe's influence. His enthusiasm for subjective idealism was further dampened by a skeptical reviewer of one of his books, who wrote: "Is not the [subjective] idealist fortunate that he is able to consider as his own the divine works of Plato, Sophocles, and all the other great minds?"[31]

Schelling's formulation of Naturphilosophie owed much to the writings of the Dutch philosopher, Baruch Spinoza. Spinoza was born into a Jewish émigré community in Amsterdam populated by families who had fled Portugal to escape forced conversion and the Inquisition. Although an excellent student, he abandoned his rabbinical studies at the age of seventeen to work in his family's export business. By the time he was twenty-three, his reputation as a heretical thinker was so widespread that he incurred the wrath not only of the Talmud Torah congregation, but also of the local Calvinist clergy. According to Stephen Nadler, Spinoza was probably expressing those ideas he would later incorporate into his philosophical writings, which negated beliefs fundamental to Jewish identity, such as the existence of a providential God, the claim that Jews received their Laws directly from God (and are therefore bound by them), and the immortality of the soul.[32] He was soon expelled from the congregation and forced to leave Amsterdam, after which he spent the remainder of his years as a private scholar in nearby towns, earning his living as a lens grinder and instrument maker.

The aspect of Spinoza's philosophy that most appealed to Schelling and the other Romantics was his contention that God and nature are one. No longer does God stand apart from nature, immune from nature's laws. The unity, God/Nature, obeys physical laws and proceeds in a temporal chain of cause and effect. And since humans are part of God/Nature, human consciousness is an aspect of God/Nature as well. Thus, in one stroke, Spinoza eliminated the dualism between the spiritual and the material, between mind and body.

This was the holistic argument that Schelling was seeking in order to extricate Naturphilosophie from the logical absurdities of Fichte's subjective idealism. On the other hand, Schelling rejected Spinoza's belief in mechanistic determinism. Like Holbach, Spinoza had argued that free will was an illusion and that humans could no more choose to do what they do than a newborn "chooses" to suck its mother's breast. Such a view ran counter to the Romantics' insistence on the existence of free will. While conceding that most human activities may be instinctual, Schelling asserted that it was nevertheless possible to experience free will through the study and practice of Naturphilosophie, which enabled the mind to free itself from the bondage of determinism. Taking his cue from Kant, however, Schelling argued that great artists had no need of free will because of their ability to perceive pre-existing, universal ideals of beauty and translate them into material form.[33]

GOETHE'S EARLY CAREER

Goethe was born in 1749, in Frankfurt am Maim, a Free Imperial City of the Holy Roman Empire, to a relatively affluent family that owed its prosperity to inherited wealth.[34] His father was trained as a lawyer and had purchased the honorific title of Imperial Councillor, which carried prestige but involved few official duties. His mother came from a distinguished line of lawyers and city officials. Goethe attributed his imagination and creative gifts to his mother. From his father, who had made the grand tour of Europe in his youth, he absorbed a deep desire to study classical art and to experience the exotic landscapes of Italy.

As a boy, Goethe aspired to become a dramatist, but his father insisted that he follow a more practical path. After completing his training as a lawyer, Goethe had the good fortune to meet the young Duke Karl August of Weimar, who was sufficiently impressed by

Goethe's wide-ranging intellect and engaging personality to make him his privy counselor. After seven years in the Weimar court, Goethe was so highly esteemed by Karl August that he was granted noble status, allowing him to insert "von" before his surname. In a letter to a mutual acquaintance, the philosopher Gottfried Herder provided a jaw-dropping list of Goethe's official duties at Weimar:

> He is . . . privy councilor, president of the chamber, president of the war council, overseer of buildings and mines, and also director of *plaisirs*, court poet, orchestrator of beautiful festivals, of court operas, of ballets, of masks, of writing, of art, etc., director of the drawing academy, in which during the winter he holds lectures on osteology, and himself, principally, the first actor, dancer, and, in short, the factotum of the house of Weimar. . . . He has become a baron.[35]

The only portfolio Herder seems to have omitted was minister of agriculture. To keep up with his duties, Goethe became a voracious reader of philosophical, scientific, and technical works in addition to scholarly writings on music and art. The more administrative and cultural responsibilities he took on, the greater the scope and depth of his readings.

However, Goethe was first and foremost a prolific literary genius. He first gained attention in 1773 as the author of the play *Götz von Berlichingen*, which was loosely based on the memoirs of a sixteenth-century German Imperial Knight who lost one arm in battle and had it replaced by a prosthetic constructed of iron. Goethe's *Götz* epitomized the Romantic hero, a free spirit who died tragically rebelling against the stultifying and deceitful laws of society. If *Götz* brought Goethe notoriety, his next work, the sentimental romantic novel *The Sorrows of Young Werther* (1774), brought him instant celebrity and spawned an international cult of lovelorn, suicidal youth. However, his greatest masterpieces remain his epic, two-part drama *Faust* and his piercingly beautiful lyric poetry, which inspired lieder by Mozart, Schubert, Beethoven, Schumann, Brahms, Liszt, and Mendelssohn.

Goethe's primary scientific focus was on living organisms, especially plants. Over his lifetime he investigated many subjects in the natural sciences, but his work on plant morphology was arguably his most lasting and original contribution. He was especially interested in the morphology of leaves and flowers, and he was familiar with the researches of Camerarius, Koelreuter, and Sprengel on pollination, although he rarely, if ever, cited their work. As noted earlier, his reticence may have been due in part to persistent doubts he entertained about the sexual theory, doubts that had their origin in nature philosophy.

The key reservations that Goethe and many nature philosophers continued to have about the sexual theory were twofold: first, its failure to integrate the process of sexual reproduction with the entire life cycle of the plant; second, its overemphasis (according to their lights) on experimentation rather than on the philosophical principles guiding development. Goethe viewed the process of "metamorphosis" as a continuum in which development proceeded according to a set of rules. These rules were philosophical in nature rather than mechanistic, having a closer affinity to Platonic idealism than to the laws of chemistry or physics. Because nature philosophers had for the most part abandoned traditional Christianity in favor of a pantheistic view of the universe, no model of plant reproduction would be complete in their eyes without a spiritual component. Although Goethe as a young man readily accepted the canonical version of the sexual theory, over the years he reinterpreted it until it became so philosophical and abstract that it all but lost its original connection to sex. How can we explain this shift in Goethe's thinking? Part of the answer lies in Goethe's evolving attitude toward sex in general.

GOETHE'S "*WEIBERLIEBE*"

Goethe's botanical writings, especially his views about the sexual theory of plants, were influenced to an unusual degree by his life experience, especially his love life. Like his alter ego, Faust, Goethe's curiosity about the natural world was boundless, as was his lifelong attraction to women. Indeed, he seemed to have derived much of his poetic inspiration from his many passionate, yet chaste, affairs, which Schiller later referred to as "the *Weiberliebe* (women-love) that plagues him."[36] In his poetry, women are strongly identified with nature, and conversely, nature is personified as female. Thus, to understand Goethe's ideas about sex in plants, one must view them, at least in part, through the lens of his attitudes toward women.

Goethe's close relationship with his sister Cornelia may have served as a template for his many platonic relationships with women, especially those before his two-year sojourn in Italy. Cornelia was born a year after Goethe in 1750. The two siblings became inseparable companions, playing, studying, and reading together. They frequently entertained the family by staging plays of their own composition. In his autobiography, Goethe stated that upon reaching adolescence they both experienced "the amazement of the awakening of sensuous drives" that were held in check only by "the holy dread of the close relationship." Several years later, Cornelia consoled her forlorn brother after his first of a series of romantic rejections. Goethe wrote that at such moments brother and sister felt the pain of the barrier between them most keenly, as they "regarded themselves utterly unhappy, the more so since in this particular case the confidants could not transform themselves into lovers."[37] When Cornelia married one of Goethe's friends, the lawyer Georg Schlosser, Goethe reacted negatively: "This [news] rather took me aback . . . and now I first noticed that I was really jealous with regard to my sister."[38] He always believed that Cornelia's marriage to Schlosser was an unhappy one, and he remained convinced that she was better suited to a spiritual calling than to marriage: "I liked to imagine her, when I sometimes engaged in fancies about her destiny, not as a wife, but as an abbess, the mother superior of a noble convent." When Cornelia died in childbirth at the age of twenty-six, Goethe must have reflected bitterly that had his sister entered such an imagined convent, she would not have died so young and deprived him of her company.

Until his trip to Italy between 1786 and 1788, when most biographers agree he finally lost his virginity at the age of thirty-seven, Goethe's courtships fell into two main categories: romantic relationships which he, himself, terminated before they became physical, and Werther-like obsessions with married or engaged women who were physically unavailable. His longest and most ardent romance with a married woman began when he was twenty-six years old and lasted ten years, until his departure for Italy. The object of his desire was the Baroness Charlotte von Stein, a pious, attractive, and intelligent woman. She had heard reports that he was "the most handsome, liveliest, most original, fieriest, stormiest, softest, most seductive, and for the heart of a woman, the most dangerous man" she would ever meet.[39] And Goethe did not disappoint. He was "maddeningly impetuous and foolish with her, trampling on her sensitivities, and shocking her with his ribald jokes and vulgarities."[40] Her constant efforts to reform him provoked sheepish apologies but had no lasting effect. She was flattered by the constant outpouring of his passionate love letters, as exemplified by the following passage:

> Why should I plague you! Lovely creature! Why do I deceive myself and plague you, and so on—We can be nothing to one another and are too much to one another.—Believe me when I speak as clear as crystal to you; you are so close to me in all things.—But since I see things only as they are, that makes me crazy. Good night angel and good morning. I do not wish to see you again—Except—You know everything—I have my heart—Anything I could say is quite stupid. —I will look at you just as a man watches the stars—think about that.[41]

As with Goethe's earlier soul-mates, Goethe conflated the Baroness with Cornelia, writing a poem to her that included the lines:

> Oh, you were in ancient times
> my sister or my wife.[42]

Their friendship took on a sacramental nature, and, in 1780, she even sent him a ring to seal their spiritual bond. However, toward the end of their decade-long relationship, Goethe grew restive; he began to yearn for something more than a spiritual bond: "I should wish that there was some oath or sacrament that bound me obviously and legally to you," he confessed. But despite his importuning, the Baroness had no intention of abandoning her marriage and position, or risking her children's security. For his part, Goethe would never have jeopardized their friendship in a clumsy attempt at seduction.

In the meantime, Goethe was developing a keen interest in morphology, the study of the principles of form in biological development, a path that would eventually lead him to the question of sex in plants.

GOETHE AND THE PREMAXILLARY BONE

Goethe's foray into biology began in 1780 after attending a series of anatomical demonstrations given by Justus Christian Loder, the distinguished professor of medicine at Jena. Soon he was studying comparative anatomy in Loder's laboratory. The fact that all vertebrates shared many features in common was widely understood at the time. The question was: Why? The Darwinian answer is, of course, that vertebrates are all descended from a common ancestor, but the concept of phylogenetic trees was not yet known. Goethe favored the Neoplatonic idea that Nature, directed by God, had formed all vertebrate animals based on a single plan or archetype. He wrote, "nature proceeds from ideas, just as man follows an idea in all he undertakes."[43]

Goethe was especially proud of his discovery of the presence of the premaxillary bone in the upper jaws of humans. The premaxillary bone is that segment of the upper jaw that normally bears the incisors. The anatomist Johann Friedrich Blumenbach, had claimed that although the premaxillary bone was present in mammals and apes, it was absent in humans, as judged by the lack of visible sutures demarcating the premaxillary zone from adjacent maxillary tissue. Much was made of this apparent difference between humans and apes because it supported the biblical Creation narrative. However, several earlier studies had demonstrated the presence of suture lines around the premaxillary bone in the skulls of fetuses and children.[44] These lines gradually went away during development due to bone fusion. Goethe tried to find evidence of the premaxillary bone in a wide variety of vertebrates, including humans, by looking for residual signs of the premaxillary sutures. He was able to detect remnants of the premaxillary sutures in every vertebrate skull he examined. The latter observation convinced him that the premaxillary bone was part of the vertebrate body plan, shared by humans and apes, proving that they were both based on the same blueprint or archetype.[45]

GOETHE'S *ITALIENISCHE REISE*

In spite of his success with his work on the premaxillary bone, Goethe was growing increasingly impatient with other aspects of his life. His mounting physical frustration over his relationship with Frau von Stein coincided with a low ebb in his poetic output. He began to wonder whether his talents, like his sex life, were withering on the vine. The crushing weight of administrative responsibilities at the Weimar court was also beginning to take a toll. Today, it would be fashionable to describe Goethe's emotional state as a midlife crisis. A plan was forming in his mind to escape the barren, chilly confines of the Weimar court and to reinvent himself in the hot, sensuous landscapes of Italy.

Goethe planned his escape carefully, and, at 3:00 AM on September 2, five days after his thirty-seventh birthday a passenger identifying himself as Filippo Möller, carrying only a leather trunk and a knapsack, boarded the mail coach in Carlsbad and began the long journey to Italy. His choice of an Italian first name suggests that Goethe had already begun to reinvent himself. Six days later, he crossed the Brenner Pass into Italy, where he hoped to spend a rejuvenating two-year sabbatical pursuing his three primary passions: art, botany, and women.

Along the road to Bolzano, Goethe recorded an observation that indicates that several years before writing his *Metamorphosis* he accepted the sexual theory of plants. He noted that the foothills he was passing through were covered with vineyards and that maize was growing tall between the rows of grape vines: "The fibrous male flowers had not yet been cut off, for this is not done until some time after fertilization has taken place."[46] Although he doesn't mention pollination specifically, he calls the pollen-bearing inflorescences at the top of the stalks "male."

During his first year in Italy, Goethe befriended a group of ex-patriot German artists and intellectuals in Rome who had formed an art colony. He immediately revealed his true identity to his compatriots and was given a place of high honor among them. Although not described in Italienische Reise, he also pursued many alluring women, but his sex life failed to improve as quickly as he had hoped. In a letter to Karl August on February 3, 1787, Goethe complained that although "public girls of pleasure" were readily available, he was afraid of catching the "French disease."[47] Moreover, all of the respectable women to whom he was attracted were either married or in search of a husband. Nowhere could he discover the respectable, yet sexually pliant, woman of his imagination. Karl August, who tended to throw caution to the winds in such matters, wrote back that he should follow his own example and become more daring. On February 16, 1788, Goethe wrote back triumphantly:

> Your good advice . . . seems to have worked, for I can already mention several delightful excursions. It is certain that you, as a doctor *longe experientissimus*, are perfectly correct, that an appropriate movement of this kind refreshes the mind and provides a wonderful equilibrium of the body.

The identity of the woman to whom Goethe owed his physical emancipation has never been conclusively established. In his *Roman Elegies* (Stanza 21), written either at the end of his trip or soon after returning, he writes of a brown-skinned woman with flowing black hair named Faustine. However, another candidate for the lover in the *Elegies* is the twenty-three-year-old Christiane Vulpius, who became his mistress shortly after he returned to Weimar, and who eventually became his wife and the mother of his children.

GOETHE AND THE *URPFLANZE*

Shortly before embarking on his sabbatical in Italy, Goethe had begun reading Linnaeus under the tutelage of the botanist August Johann Batsch, author of *Botany for Women and Lovers of Plants* (1795). At first, Goethe was awed by Linnaeus's encyclopedic knowledge of plants. In a burst of enthusiasm, he rapidly mastered the Linnaean sexual system and spent many happy hours identifying the local flora. However, he soon wearied of counting stamens and pistils, chafing at the arbitrary and static nature of an artificial system, which he termed "a rigid way of thinking." In the meantime, Batsch had introduced him to the extensive French literature on "natural" systems of classification in which plants were arranged into families progressing from the simple to the complex. Natural classification systems were more compatible with Goethe's taste for embryology and development, which, as we have seen, he had so successfully deployed in his study of the human premaxillary bone.

Whereas Linnaeus's goal had been to make sense out of the bewildering array of plant forms by dividing them up into species, Goethe's objective as a morphologist was to search for the common features among all plants in order to discover the fundamental laws of development that differentiated plants from animals. How is it, he asked himself, that I can immediately recognize a plant as a plant regardless of species? This was the same question that both Aristotle and Theophrastus had attempted to answer centuries earlier. Aristotle's answer had been that plants were analogous to upside-down animals, with their "mouths" in the soil. Theophrastus had insisted that plants were entirely different from animals and could only be understood by analyzing their parts.

Echoing Theophrastus, Goethe also regarded plants as distinct life forms, not variants of animals. But in what ways did plants and animals differ, and what were the laws that governed these differences? Goethe turned, as he had in his study of the premaxillary bone, to the only conceptual framework available to him, Neoplatonic idealism, hoping that he could discover the ideal form of plants. The idea of an *Urpflanze*, or primordial plant from which all other plants are derived, had occurred to him prior to his trip to Italy, but he had regarded it more as an abstract ideal than as a real plant. However, while walking in the Public Gardens of Palermo, admiring the luxuriant growth of the many exotic species with their unfamiliar forms, he wondered whether the Urpflanzen might not be a real plant after all:

> Here where, instead of being grown in pots or under glass as they are with us, plants are allowed to grow freely in the open fresh air and fulfill their natural destiny, they become more intelligible. Seeing such a variety of new and renewed forms, my old fancy suddenly came to mind: Among this multitude might I not discover the Primal Plant? There certainly must be one. Otherwise, how could I recognize that this or that form was a plant if all were not built upon the same basic model? [48]

But although Goethe collected plant specimens from every garden in Italy that he visited, he never found the Urpflanzen he sought. Upon his return to Germany in 1788, he explained the concept of the Urpflanze to his new friend Friedrich Schiller and drew a picture of what he thought it would look like. "That's not an observation, that's an idea," Schiller observed dismissively. "Well," Goethe replied defensively, "I am quite happy that I have ideas without knowing of them, and that I can even see them!"[49]

Schiller was, of course, correct. What Goethe had conceptualized was not a real plant but a schematic drawing of the sort found today in every introductory biology textbook illustrating the fundamental arrangement of plant organs. Nevertheless, such heuristic diagrams can provide important biological insights. For example, Goethe astutely observed that the vegetative body of plants is comprised of repeating anatomical units consisting of an internode, leaf, and axillary bud. The modern term for this repeating morphological unit is "phytomer."

Goethe's most famous statement on plant morphology was his conjecture that "Alles ist Blatt" ("All is leaf"), which came to him suddenly in the Public Gardens of Palermo:

> [I]t came to me in a flash that in the organ of the plant which we are accustomed to call the *leaf* lies the true Proteus who can hide or reveal itself in all vegetal forms. From first to last, the plant is nothing but *leaf*, which is so inseparable from the future germ that one cannot think of one without the other. [50]

Goethe interpreted the various lateral appendages on plant stems, including floral organs, as successive modifications ("metamorphoses") of the idealized primordial leaf. That floral organs actually do represent modified leaves has now been confirmed at the molecular level, and Goethe is justifiably celebrated for his original insight, which helped pave the way for the early acceptance of Darwinian evolution.[51]

Inspired by his experiences in Italy, Goethe began work on his famous treatise outlining his theory of the metamorphosis of plants, which was published in 1790.

METAMORPHOSIS AND PLANT SEXUALITY

"Metamorphosis," as discussed in Chapter 7, is a Greek word made famous by Ovid in his poems about the transformation of various mythic characters, mostly female, into plants.

The term was first used in biology in the seventeenth century to describe the seemingly sudden transformation of insects from one form to another. For example, in 1669, the Dutch biologist Jan Swammerdam used a microscope to study the stages of insect development and demonstrated that these transformations were gradual and continuous. Linnaeus applied the term to plant development in his *Metamorphosis Plantarum* (1755), which outlined his medulla–cortex theory for the formation of pistils and stamens (see Chapter 13).[52] Linnaeus's *Metamorphosis Plantarum* is the likely inspiration for Goethe's German title, *Versuch die Metamorphose der Pflanzen zu erklären* (Attempt to Explain the Metamorphosis of Plants).

In *Metamorphosis of Plants* (1790), Goethe compares plant development to the ascension of a ladder that culminates in sexual reproduction. Plant metamorphosis, he states,

> may be observed at work step by step from the first seed leaves to the final development of the fruit. By transmutation of one form into another, it ascends as though on the rungs of an imaginary ladder to the climax of Nature, reproduction through two sexes.[53]

At this early stage of his own transformation into a botanist, Goethe expressed no doubts about the sexual theory as he now interpreted it, although, as we shall see, his interpretation was unorthodox. Like Aristotle's *Scala Natura*, Goethe's developmental ladder is a hierarchical system from the least perfect (vegetative) structures to the most perfect (sexual) structures. In contrast to the *Scala Natura*, which places plants at the bottom, Goethe considers plants on a par with animals, at least in the realm of sexuality. For esoteric reasons that will be explained later, Goethe even considered hermaphroditic flowers to be superior to animals as exemplars of sexual reproduction.

Goethe states at the outset that he will restrict his discussion of metamorphosis to annual plants, which "advance continuously from seed to fructification." Trees and other perennials that produce seasonal fruits are excluded from his discussion. In addition, he excludes all plants that required outside agents for pollination. Curiously, in view of Sprengel's pioneering work on this subject, Goethe regarded examples of insect- or wind-mediated pollination as "excrescences," which are "abnormal" and "restricted" in nature. As discussed later, neither wind nor insect pollination accorded well with his theory of "anastomosis," or re-fusion, which formed the basis of his view that plants are, by definition, self-sufficient organisms.

Starting with the cotyledons, or seed leaves, Goethe traces the variations in leaf form and size from node to node as the stem elongates. Goethe explains that the "cruder fluids" of the sap are drawn off and replaced by "purer ones," enabling the production of the flower. Encouraged by the eighteenth-century discovery of spermatozoa in animals, Goethe hazards the guess that petals may harbor male sperm, as suggested by their color and scent. Such a proposal, he states, is consistent with the "close relationship of petals and staminal organs," as demonstrated by the presence of intermediate forms in some flowers and the transformation of stamens into petals in double-flowers, a phenomenon that results in sterility.

According to Goethe, leaf transmutation during metamorphosis follows a pattern of expansion and contraction of the modified leaf. Considering Goethe's aversion to the mechanics of Newton and Boyle, his theory of leaf contraction and expansion is surprisingly mechanical. Goethe invokes the role of spring-like "spiral vessels" to explain the contraction and expansion of all leaf-like organs.[54]

While leaf transmutation may be driven by the action of spring-like spiral vessels, the final step in the process, self-pollination, is a spiritual rather than a mechanical process. From the highly contracted state of the anther, the tiny pollen grain emerges. No longer constrained by its spiral vessels, the pollen grain expands freely by the force of its sap and "seeks out the female parts." Goethe was a proponent of Blumenbach's concept of *Bildungstrieb*—the idea that the development of all creatures, including plants, was guided by an inborn force or "drive." It was therefore crucial to his theory of metamorphosis that pollen grains play an active role in attaching themselves to the pistil. This accounts for his peremptory dismissal of insects and wind as pollen vectors. Presumably, by bending toward the pistil, the stamen brings the pollen close enough to the stigma for the pollen grain to move toward it by its own power. Once attached to the female stigmas, the pollen grain "suffuses them with its influence," thereby bringing about fructification. According to Goethe, the union of the two sexes can be thought of as "anastomosis [rejoining] on a spiritual level," during which the plant briefly achieves sexual perfection. Ascribing this spiritual process to the vagaries of wind or insects would, in Goethe's view, make a mockery of Nature's divine plan.

The idea of "anastomosis" is key to understanding Goethe's conception of plant sex at this relatively early stage in his botanical career. Anastomosis is defined as the rejoining of two streams or branches that have previously been united. In the vegetative plant, Goethe notes that the "veins" (vascular strands) of stems and leaves branch and fuse at various points, creating a vascular network. He explains that something analogous happens during sexual reproduction in the flower: the two sexes are initially united in the vegetative plant, separate during the metamorphosis of the stamen and pistil, and then reunite during pollination in a kind of "spiritual" anastomosis:

> We see a fully formed pollen emerge from [the anthers]. . . . Now released, it seeks out the female parts that the same effect of nature brings to meet it; it attaches itself to these parts, and suffuses them with its influence. Thus we are inclined to say that the union of the two genders is anastomosis on a spiritual level; we do so in the belief that, at least for a moment, this brings the concepts of growth and reproduction closer together.

According to Goethe, spiritual anastomosis involves purification of the sap: "The fine matter developed in the anthers looks like a powder, but these tiny grains of pollen are just vessels containing a highly refined juice."

Nine years later, Goethe published a poem of the same title ("The Metamorphosis of Plants"), which was intended as a pedagogical device to introduce his theory to a general audience, especially to women. In it Goethe describes the course of plant metamorphosis to his lover (Christiane Vulpius), culminating in the formation of fruit and seed. First the leaves contract to form the stamen and pistil (only one of each for the sake of propriety), after which they recombine (pollination is not mentioned) in a mystical union that promptly gives rise to "unnumbered germs . . . concealed in the womb":

> Yes, the leaf with its hues feeleth the hand all divine,
> And on a sudden contracteth itself; the tenderest figures
>
> Twofold as yet, hasten on, destined to blend into one.
> Lovingly now the beauteous pairs are standing together,
>
> Gather'd in countless array, there where the altar is raised.
> Hymen hovereth o'er them, and scents delicious and mighty
>
> Stream forth their fragrance so sweet, all things enliv'ning around.
> Presently, parcell'd out, unnumber'd germs are seen swelling,
>
> Sweetly conceald in the womb, where is made perfect the fruit.
> Here doth Nature close the ring of her forces eternal.[55]

Goethe and his contemporaries understood sexual union as the reconciling of opposing (polar) forces. In the case of plant growth, the outward expansion of the sap is opposed by the contraction of the spiral vessels. In the vegetative plant, maleness and femaleness co-exist in the Aristotelian sense of being "combined," but are at the same time in opposition to one another. Later in life, Goethe was to speculate that maleness promoted vertical growth, whereas femaleness promoted spiral growth. During flower formation the opposing sexes separate into "beauteous pairs"—stamens and pistils. Later, when the two sexes spontaneously "blend into one," they depolarize and become a harmonious whole. Goethe considers this moment of consummated self-love, which presumably ends soon after conception, as a state of spiritual perfection—or, in the words of the twentieth-century Bohemian-Austrian poet Rainer Maria Rilke, "Narcissus fulfilled."[56]

Soon after *Metamorphosis* was published in 1790, Goethe learned that his idea that floral organs are transformed leaves had been published forty-one years earlier by Caspar Friedrich Wolff in his dissertation *Theoria Generationis* (1759). Wolff, a staunch Aristotelian, had been conducting microscopic studies in animals and plants to determine which theory of embryogenesis—preformationism or epigenesis—was correct. His microscope observations clearly supported epigenesis in both animals and plants, and his continuous observations of plant growth and development led him to propose, like Goethe, that floral organs arose as modifications of leaves.[57] Goethe contrasted Wolff's interpretation of plant morphogenesis with his own in his essay "My Discovery of a Worthy Forerunner," published in 1817. Whereas Goethe viewed metamorphosis as an alternating process of leaf expansion and contraction, Wolff regarded the transition from leaves to flowers as degenerative, involving only contraction:

> [Wolff] saw the same organ always contracting, getting smaller. The fact that this contraction alternates with expansion, he did not see. He saw it decreased in volume without noticing that it was at the same time perfecting itself, and he absurdly attributed to degeneration this path toward perfection.[58]

Thirty years later, Goethe was to incorporate Wolff's degenerative model into his final interpretation of pollination.

METAMORPHOSIS AND OLD AGE

In his posthumously published collection of aphorisms, *Maxims and Reflections*, Goethe laid out the four stages of a man's life in philosophical terms:

> Every stage of life corresponds to a certain philosophy. A child appears as a realist; for it is as certain of the existence of pears and apples as it is of its own being. A young man, caught up in the storm of his inner passions, has to pay attention to himself, look and feel ahead; he is transformed into an idealist. A grown man, on the other hand, has every reason to be a skeptic; he is well advised to doubt whether the means he has chosen to achieve his purpose can really be right. Before action and in the course of action he has every reason to keep his mind flexible so that he will not have to grieve later on about a wrong choice. An old man, however, will always avow mysticism. He sees that so much seems to depend on chance: unreason succeeds, reason fails, fortune and misfortune unexpectedly come to the same thing in the end; this is how things are, how they were, and old age comes to rest in him who is, who was, and ever will be.[59]

Goethe's youthful phase lasted much longer than it did for most of his peers, but he returned from his Italian adventures a changed man. He not only set up a household with his devoted, twenty-three-year-old mistress, Christiane Vulpius, he also shed many of his most burdensome official duties, giving him more time to pursue his artistic and scientific interests. In the same year, he met Friedrich Schiller, and the two began a close friendship. Disenchanted with the subjectivism of their fellow Romantics, they set about to educate the German public on the superiority of classical culture. They co-founded the movement of Weimar Classicism, with the goal of synthesizing classical, Enlightenment, and Romantic aesthetic values.

Romanticism had grown out of the *Sturm und Drang* (Storm and Stress) movement that flourished in Germany from the 1760s to the 1780s. Although Schiller always considered Goethe one of the founders of Romanticism, Goethe resisted the label, preferring to identify with eighteenth-century rationalism. It was only after Schiller's death, while writing his own autobiography, *Dichtung und Wahrheit* (Poetry and Truth), that Goethe finally accepted his friend's assessment—that he had, indeed, been a Romantic and had played a major role in defining the movement.

The outbreak of the French Revolution in 1789 both intrigued and disturbed Goethe. Although he had little sympathy with the French aristocrats who escaped across Germany's border, he also took a dim view of the ability of ordinary citizens to rule themselves, a view that was increasingly reinforced in 1793 with the beginning of the Reign of Terror. Goethe believed in gradual rather than catastrophic change, whether it was the formation of the earth, the metamorphosis of plants, or the evolution of human society. He welcomed Napoleon's rise to power at the end of the century, viewing him as a savior who had led France out of anarchy.

During the decade between 1795 and 1805, he was involved with several major projects: promoting Weimar Classicism, writing Part I of *Faust* and other literary projects, and developing his color theory. He was also plagued by chronic bouts of illness and depression. Many of his closest friends died, including his mentor Herder in 1803, and, most traumatic of all, Schiller in 1805. The Dowager Duchess Amalia, of whom he was very fond, died in 1807, followed by his mother in 1809. Such losses inevitably led him to reflect on his own mortality, and, in 1809, he began work on his autobiography.

Seeking escape from the Napoleonic turmoil around him, Goethe turned for comfort to the work of the fourteenth-century Persian poet Hafez, a new German translation of which had appeared in 1814. In his poem "Hegira," published in the book *West-East Divan* (1819), Goethe reflects on the military and political upheavals raging around him:

North and West and South are breaking,
Thrones are bursting, kingdom's shaking:
Flee, then, to the essential East,
Where on patriarch's air you'll feast![60]

In reality, Goethe fled west, to his native Rhineland, stopping off at various universities along the way to chat with renowned scholars about arcane topics. In this whirlwind tour of academia, he found the intellectual stimulation he needed to revive his creative energy, and, in the poetry of Hafez, he found the model he was seeking for the synthesis of the sensual and the spiritual. In his celebrated poem "Blessed Longing," written in 1814, Goethe writes as an old man driven like a moth by a "new desire/ to a higher union," a union he associates with a transcendent "death by fire":

Tell it only to the wise,
For the crowd at once will jeer:
That which is alive I praise,
That which longs for death by fire.

Cooled by passionate love at night,
Procreated, procreating,
You have known the alien feeling
In the calm of candlelight;

Gloom-embraced will lie no more,
By the flickering shades obscured,
But are seized by new desire,
To a higher union lured,

Then no distance holds you fast;
Winged, enchanted, on you fly;
Light your longing, and at last,
Moth, you meet the flame and die.

Never prompted to that quest:
Die and dare rebirth!
You remain a dreary guest
On our gloomy earth.[61]

Goethe's wife, Christiane, who had begun to suffer from a kidney disorder around this time, died after a long and painful illness. His writings immediately afterward suggest heartfelt distress, bereavement, and loneliness, but in fact, at the age of sixty-five, Goethe was experiencing a creative renewal reminiscent of his Italian journey.

Even prior to Christiane's death Goethe continued to seek inspiration from the company of women. His intense, although chaste, relationship with the former actress and singer Marianne Jung, who was married to a wealthy Frankfurt banker, lasted from 1814 to 1816. According to John R. Williams, in the poems of *West-East Divan*, the two "played out the roles of the legendary lovers. . . . Hatem and Suleika, exchanging mutual gifts of oriental knick-knacks, playing exotic charades, corresponding in a private code."[62]

In 1821, at the age of seventy-two, he indulged himself in one last futile attempt to rekindle his love life. While vacationing at Marienbad, Goethe became besotted with a seventeen-year-old girl, Ulrike von Levetow. It ended badly, but his disappointment in love once again bore him poetic fruit in the form of "The Marienbad Elegy," written in 1823 and the second of three poems in his *Trilogy of Passion*, which ends on a note of despair worthy of Werther:

> Leave me here now, my life's companions true!
> Leave me alone on rock, in moor and heath;
> But courage! open lies the world to you,
> The glorious heavens above, the earth beneath;
> Observe, investigate, with searching eyes,
> And nature will disclose her mysteries.
> To me is all, I to myself am lost . . .
> [The immortals] urged me to those lips, with rapture crowned,
> Deserted me, and hurled me to the ground.

In "The Marienbad Elegy," Goethe reprised the theme of hopeless love that he had mined so successfully in his earlier lyric poetry. Yet, Goethe's state of mind had actually undergone a radical change since his youth, having embraced eastern mysticism with its detachment, serenity, and resignation. In a letter to his friend Zelter, written a few years earlier in 1820, Goethe sees death in terms of two alternative worlds, the real versus the ideal or symbolic:

> Unconditional submission to the unscrutable will of God, viewing with serenity the ever-circling and spiral recurrence of the earth's restless bustle, love, affection, suspended between two worlds, all that is real refined, dissolving into symbol.[63]

One of Goethe's signal achievements during his last decade was the completion of Part II of *Faust* in 1831, the year before his death at age eighty-two. In contrast to Part I, with its lusty account of Faust's carnal obsessions, Part II is highly abstract, metaphysical, and in the end, mystical. Whereas Part I is focused on the seduction and abandonment of the girl, Gretchen, Part II presents Faust as a medieval knight who falls in love with Helen of Troy, the ideal of beauty and love. The play ends with a paean to "The Eternal Feminine" spoken by the "Mystic Chorus":

> All that is transient
> Is but a likeness;
> Here the ineffable
> Becomes known;
> Here the inexpressible
> Turns into deeds;
> The Eternal Feminine
> Draws us on.[64]

David Duke quotes Goethe as stating in 1828 that he regarded Part I as a youthful and naïve effort compared to the lofty idealism of Part II.[65] Although Part II never enjoyed the popularity of Part I, it provides a useful window into Goethe's philosophical metamorphosis in old age. This transformation from *weiberliebe* and passionate striving to the cares and disappointments of old age, attended by mystical hopes for transcendence in death, can also be seen in Goethe's ambiguous, but changing definition of pollination from a sexual, biological function to an asexual, mystical final stage of life.

GOETHE AND THE NINETEENTH-CENTURY ASEXUALISTS

Back in 1804, during their leisurely stroll through the University's Botanical Garden at Jena, Goethe was gratified that Schelver had based his arguments against the sexual theory on his own theory of metamorphosis. Schelver pointed out that, according to the theory of metamorphosis, plant development proceeds stepwise from seedling emergence to flower, fruit, and seed production on its own, using its inherent "force and power." Goethe described the conversation sixteen years later, when he was seventy-one, in his essay *Pollination, Volatilization, and Exudation*:

> Schelver proceeds literally from the concept of healthy and regulated metamorphosis, which holds that plant life, rooted in the earth, struggling upward toward light and air, is forever raising itself by its own bootstraps and developing step by step, scattering about even the last seed by its own force and power. The sexual system, on the other hand, requires for this final act an external agent, which is conceived as the counterpart of the flower itself, exerting influence and being influenced—with, beside, or even apart from the flower.[66]

Schelver reminded Goethe that, according to his own theory of metamorphosis, plant development involves a process of gradual refinement, leaving behind the "material" and "base" so that only the "higher, incorporeal, and better" remains:

> Schelver pursues the tranquil course of metamorphosis which, in advancing, is refined to such a degree that it gradually leaves behind all that is material, insignificant and base, permitting what is higher, incorporeal, and better to emerge in greater freedom.

Why, then, should not this latter type of pollination also be a liberation from burdensome matter, allowing the inherent abundance in the heart of the plant, through the energy of primary force, to proceed toward endless propagation? [67]

Rather than having anything to do with sex, pollination, according to Schelver, is the "liberation from burdensome matter." Once the base, material components have been removed from the sap and discarded, the pistil spontaneously forms the seed and fruit without the aid of any external agent. Goethe quotes Schelver as saying that the actual stimulus for seed production occurs underground "in the earth":

The highest level of vegetative life is the formation of a basis for future reproduction in which only the pistil is participating . . ., fertilization of this basis is taking place in the earth through the water, air and temperature.[68]

Looking back on his conversation with Schelver from the vantage point of 1820, Goethe admits that at first he was "taken aback" by Schelver's "heretical" views. But the fact that Schelver had based his pollination theory on his own theory of metamorphosis was a powerful argument in its favor:

In my nature studies I had religiously accepted the dogma of sexuality in plants and was, therefore, taken aback now to hear a concept directly opposed to my own. Yet I could not consider the new theory wholly heretical, for from the account given by the ingenious [Linnaeus] I could draw the conclusion that [Schelver's] pollination theory was a natural consequence of the theory of metamorphosis which seemed to be significant to me.[69]

Goethe portrays himself as being intrigued, but not entirely convinced, by Schelver's theory. Three years after his conversation with Schelver, however, he makes statements about the nature of seed formation that seem to place him squarely in the camp of the asexualists. Early in the essay "Formation and Transformation," Goethe defines the science of morphology as the study of the process, or "*Bildung*," that gives rise to plants.[70] All living organisms, he states, are complex aggregates of living and independent parts. The "perfection" of an organism is a function of its complexity, and organisms differ according to their complexity. The simpler the organism, the greater its imperfection, with "simplicity" being defined as a similarity of its parts:

The more imperfect a creature is, the more do these parts appear identical or similar to each other and the more do they resemble the whole. . . . That a plant or even a tree, though it appears to be an individual, consists purely of detached parts resembling both each other and the whole—of this there can be no doubt.[71]

Plants, according to Goethe's definition, are therefore less perfect than animals, as is demonstrated by the ease with which a mother plant can be cloned, which Goethe states is the same process as propagation by seeds:

How many plants are indeed propagated by slips! The bud of the least complex variety of fruit tree puts forth a shoot that in turn produces a number of identical buds; and *it*

> *is precisely in this manner that propagation by seeds takes place.* Such propagation is the development of countless *identical* individuals from the womb of the mother plant. (emphasis added)[72]

Here, Goethe reiterates the scholastic interpretation of asexual seed reproduction by comparing seeds to buds, both of which are said to produce clones from "the womb [pith] of the mother plant." By equating seed and bud formation, Goethe resurrects the Aristotelian doctrine that seed production occurs by something resembling female parthenogenesis. Whereas in *The Metamorphosis of Plants* he seemed to accept the sexual model of seed production, by 1807 he is describing seed production in parthenogenic terms.

What role did gender politics play in the reactionary asexualist movement that became associated with Nature Philosophy? In his three volume *Critique of the Theory of Plant Sexuality* published in 1812, Schelver argued that because plants were dependent for their development on the external environment, such as light and soil, they lacked self-sufficiency and were therefore too passive (feminine) to embody "male potency," which represents a "contradiction" of the female:

> Only animal life has sex. It issues from an internal stimulus, it contains contradiction between man and woman. Animal life exists through the power of difference within itself. Plant (i.e. vegetative) life does not contain the stimulus of its own development; if it is not excited by an external stimulus, it remains in a primordial state. It does not govern itself; showing internal passivity, it is only potentially able to develop. It does not embody male potency and never attains it.[73]

According to Schelver, plants play the role of the "prolific wife," while Nature (soil, water, air, and temperature) serves as their "husband":

> The life of plants is that of the always prolific and embracing wife. Nature, its husband, is the general external stimulus of development.[74]

Other early nineteenth-century German philosophers and natural scientists, such as A. W. Henschell, and Georg Wilhelm Friedrich Hegel, were also challenging the sexual theory of plants. Hegel's use of the female–plant analogy was blatantly political. In 1820, Hegel published a treatise on legal and moral philosophy, *The Philosophy of Right*, in which he grounded his arguments against equal rights for women on a plant/woman metaphor:

> The difference between men and women is like that between animals and plants. Men correspond to animals, while women correspond to plants because their development is more placid and the principle that underlies it is the rather vague unity of feeling.[75]

Regarding the sexual theory of plants, Hegel acknowledged that some plants produced seeds sexually, as in the case of dioecious species, but since sex was unnecessary and redundant in plants, he considered such cases to be exceptions representing intermediate stages between plants and animals. Further evidence that sexuality is very weak in plants can be seen in the absence of sexual dimorphism in plants. Thus, Hegel concluded that the "seed which is produced in a fruit is a superfluity."

The only nineteenth-century asexualist to criticize Koelreuter's experiments in any detail was Schelver's student, A. W. Henschell. Early in Henschell's 600-page treatise *On the Sexuality of Plants*, he stated that the sexual theory had attained the status of conventional wisdom. He then undertook a detailed analysis of Koelreuter's hybridization experiments and offered an alternative explanation for his results. Koelreuter, Schelver argued, had based his evidence for the sexual theory on the observation that different species or varieties of plants could form hybrids that had intermediate phenotypes between those of the two parents. However, Henschell pointed out, unlike hybrids in the animal kingdom, which are sterile, many of Koelreuter's hybrids, especially those between *Dianthus* species, displayed variable degrees of fertility, raising the question of whether or not Koelreuter's putative hybrids were true hybrids:

> Have plant hybrids the essential qualities of animal hybrids? Infertility with its own kind and with other species is reckoned as the principal quality of animal hybrids. Only Koelreuter speaks of hybrids which did not entirely possess this quality. Some fruit formation took place on hybrids which had been dusted with paternal or maternal pollen or with their own pollen.

Henschell offered an alternative explanation for the differences in phenotype Koelreuter observed between parents and offspring in such apparent hybrids. Because he grew his plants in cramped pots, Henschell explained, the roots were probably damaged, which would cause the plant to develop abnormally. Henschell also noted that Koelreuter routinely "castrated" the "maternal" parent in his crosses, which would further injure the plant and lead to abnormal growth. Finally, the stigmas of the "maternal" parent were dusted with alien pollen, which was contrary to Nature and therefore likely to sicken the plant. All of these disturbances would be expected to reduce fertility and bring about the production of "monsters," which could easily account for the phenotypic deviations observed between parents and offspring.[76]

Henschell's concluding argument was based on what seemed like common sense: since pollen was only a tiny part of the plant, it couldn't possibly transmit the whole form of the plant to the embryo. Instead Henschell endorsed the theory of his mentor, Schelver: the release of pollen represents the progressive removal of base matter from the spiritual essence of the plant.

DEATH AND POLLINATION: GOETHE'S FINAL EQUIVOCATION

As we have seen, Goethe equated seed production with vegetative propagation (bud formation), although he never explicitly repudiated the sexual theory. Thus Goethe seems to have hedged his bets by representing himself both as an advocate of the sexual theory and as a skeptic. Although he may have accepted the sexual theory as scientifically valid, he felt powerfully attracted to Schelver's alternative theory of pollination as spiritual refinement. Schelver had redefined pollination as the "liberation from burdensome matter," which allows "the inherent abundance in the heart [pith/medulla] of the plant, through the energy of primary force, to proceed toward endless propagation." This mystical definition of pollination was derived from Goethe's discussion in *Metamorphosis of Plants* of the refinement of the sap required for "spiritual anastomosis." It also bore a resemblance to Goethe's description of death in his 1820 letter to Zelter as the transition from one world to another, in which "all that is real" is refined and dissolved into "symbol."

According to Goethe's 1820 account of his conversation with Schelver, Schelver explicitly associated pollination with death by equating it with sporulation by pathogenic fungi. For example, Schelver confused fungal spores with pollen on infected barberry leaves:

> We know that the blossoming barberry bush diffuses a strange odor that may cause wheat fields in the vicinity to become unproductive. An unusual quality may be hidden within this plant, as we may indeed infer from the sensitiveness of its anthers. It diffuses its pollen insufficiently during the flowering period; thus we may later finds bits of pollen emerging from leaves which may even develop in the manner of calyx and corolla to form the most magnificent cryptogam.

Goethe also cites wheat rust disease as an example of pollination as a symptom of disease:

> Grain rust furnishes an example of delayed pollination terminating in nothingness. Through what irregularity of growth does a plant sink into a condition where, instead of joyfully and vigorously developing numerous progeny, it tarries on a lower step and finally executes the pollination act perniciously?

Somewhat bizarrely, Goethe even cites a moldering housefly as an example of pollination:

> In autumn we can observe that flies fasten themselves on windows, remain motionless for a time, and then gradually eject a white pollen. The chief source of this natural phenomenon seems to be situated at the point where the midriff is joined to the hind part. . . . The pollination takes place gradually and continues for some time after the creature's death.

Such a macabre interpretation of pollination reflects a darker, more "Cimmerian" view of nature, which Goethe and his youthful companions had once attributed to old people who could no longer remember "how berries and cherries taste." Indeed, the association of sexuality with contagion and death was always a powerful undercurrent in German Romanticism, but in Goethe's case this aspect of Naturphilosophie only surfaced in his later years.[77]

If Goethe never explicitly rejected the sexual theory, he qualified it and redefined it so broadly as to make it meaningless. However, he was clear and unequivocal on one point: Schelver's new pollination theory was ever so much more suitable for the instruction of "young persons and ladies" than the standard version:

> For the instruction of young persons and ladies this new pollination theory will be extremely welcome and suitable. In the past the teacher of botany has been placed in

> a most embarrassing position, and when innocent young souls took textbook in hand to advance their studies in private, they were unable to conceal their outraged moral feelings. Eternal nuptials going on and on, with the monogamy basic to our morals, laws, and religion disintegrating into loose concupiscence—these must remain forever intolerable to the pure-minded!

No doubt referring to Sebastien Vailliant, Erasmus Darwin, and others, Goethe channels the eighteenth-century critics when he scolds the sexualists for their licentiousness:

> one might point an accusing finger at the naturalists when they take as ribald a delight in Mother Nature as in the goddess Baubo[78] herself—just because they have discovered a few little weaknesses in the good mother. Indeed, we recall having seen arabesques in which the sexual relations within a flower calyx were represented in the manner of the ancients, in an extremely graphic way.

By characterizing plant sexuality as one of the "little weaknesses in the good mother," Goethe here seems once again to grudgingly accept the basic tenets of the sexual theory, contrary to some of his earlier pronouncements. Indeed, a year before his death in 1832, Goethe wrote that the reception of Schelver's theory had been extremely negative and had "degenerated into abuse," in part because Schelver had "rashly overrated" the work of his pupil, Henschell, only to distance himself from it shortly afterward. According to Goethe, Schelver's mistake was to base his argument on the "impossibility of hermaphrodism in the individual." If Schelver had, instead, based his theory entirely on Goethe's theory of spiritual anastomosis:

> the theory of sexuality in plants would have been rescued, purged, and strengthened. Wind and insects would have been abandoned, amply compensated for by metamorphosis.

Goethe's final word on plant sexuality retained his original rejection of wind- and insect-mediated pollination because it violated his concept of plant self-sufficiency. Because he restricted his idea of the sexual theory to self-pollination, he could see no difference between reproduction by seed versus vegetative propagation by cuttings or buds. Perhaps Goethe's idea of an "asexual sexuality" in plants had a psychological counterpart in his close attachment to his sister and his many platonic relationships with women. What Goethe and his peers found most appealing about the theory of metamorphosis was the neoclassical idea of the spiritual essence of the plant being liberated from its material prison. As Goethe grew older, he began to associate pollination with death and spiritual transcendence.

It has often been pointed out that Goethe's lifetime spanned the period between the last witch-burning in Europe and the advent of steam locomotives. Tremendous changes were afoot in society—and in botany as well. During the next few decades, nearly all of the misconceptions of the nineteenth-century asexualists would be clarified once and for all.

NOTES

1. Cited by Karl Richter: Morpholgie und Stilwandel: Ein Beitrag zu Goethes Lyrik, in Frityz Martini, Walter Muller-Seidel, Bernhard Zeller, eds., *Jahrbuch der Deutschen Schillergesellschaft* (1977). Alfred Kroner, no. 21, p. 199.

2. Kaplan, D. R. (2001), The science of plant morphology: definition, history, and role in modern biology. *American Journal of Botany* 88(10):1711–1741.

3. Goethe, J. W. (1820/1952), Pollination, volatilization, and exudation, in Goethe's Botanical Writings, trans. Bertha Mueller. University of Hawaii Press.

4. Ibid.

5. Coen, E., and R. Carpenter (1993). The metamorphosis of flowers. *Plant Cell* 5:1175–1181; Lönnig, W. -E. (1994), Goethe, sex, and flower genes. *Plant Cell* 6:574–577.

6. Goethe, Pollination, volatilization, and exudation.

7. Ibid.

8. The great nineteenth-century plant physiologist Julius von Sachs expressed his astonishment at the persistence of asexualism after Koelreuter: "Those who have read the writings of Camerarius and Koelreuter carefully find it difficult to believe that after their time doubts were still entertained, not about the manner in which the process of fertilization are accomplished, but about the actual existence of sex in plants. And yet those doubts were expressed repeatedly during the succeeding sixty years in various quarters and with the greatest confidence." von Sachs, J. (1890), *History of Botany (1530–1860)*, trans. Henry E. F. Garnsey, revised by Isaac Bayley Balfour. Clarendon Press, p. 422.

9. Holbach himself was only forty-seven at the time and was to live nineteen more years. Voltaire, Mirabaud's friend, was outraged at the deception: "Alas! our good Mirabaud was not capable of writing a single page of the book of our redoubtable adversary." Voltaire (1764), *Dictionnaire philosophique*, article *Causes finales*, accessed at https://en.wikipedia.org/wiki/Jean-Baptiste_de_Mirabaud]

10. Morley, J. (1884/2013). *Diderot and the Encyclopedists*. Forgotten Books, pp. 350–351.

11. Ibid., p. 357.

12. von Goethe, J. W. (1872), *The Autobiography of Goethe: Truth and Poetry: From My Own Life*, vol. 1, trans. John Oxernford. Bell and Daldy, p. 424.

13. In Greek mythology, a tribe of people who lived near the land of the dead in mist and darkness.

14. von Goethe, J. W. (1883), *The Autobiography of Goethe: Truth and Fiction: Relating to Life*, vol. 2, trans. John Oxernford. Estes and Lauriat, p. 86.

15. Ibid.

16. Ibid.

17. Ibid.

18. Ibid.

19. A few prominent naturalists, such as Charles Bonnet, Albrecht von Haller, and George Cuvier persisted in their belief in preformationism, consistent with a mechanical interpretation of embryogenesis.

20. Kant, I. (2002), *Kritik der Urteilskraft*, in *Werke*, 5:486 (A 288–289, B 292–293), trans. R. J. Richards, *The Romantic Conception of Life: Science and Philosophy in the Age of Goethe*. University of Chicago Press.

21. Cited in Richards, *The Romantic Conception of Life*, p. 218.

22. Kant, I. (1790/1908), *Critique of Judgement*, in *Biology*, trans. E. B. Wilson, Columbia University Press, p. 9.

23. Richards, *The Romantic Conception of Life*.

24. Ibid.

25. Kant's premise that beauty is the "purpose" of art was discarded by many in the twentieth century. In her book, *Venus in Exile: The Rejection of Beauty in 20th-Century Art* (2001), Wendy Steiner quotes the Peruvian poet Mario Vargas Llosa as follows: "Contemporary aesthetics has established the beauty of ugliness, reclaiming for art everything in human experience that artistic representation had previously rejected."

26. Richards, *The Romantic Conception of Life*.

27. Source unknown.

28. Richards, *The Romantic Conception of Life*.

29. The younger Schiller (1759–1805), who was trained as a doctor but took up writing instead, had admired Goethe and Rousseau as a student. Goethe and Schiller became acquainted in 1789, but became close friends around 1794 after Goethe's return from Italy. Both lived in Weimar and were the seminal figures of the literary movement known as Weimar Classicism.

30. Richards, *The Romantic Conception of Life*.

31. Eschenmayer, K. (1801), Spontaneität = Weltseele oder über das höchste Princip der Naturphilosophie. *Zeitschrift für Speculative Physik* 2:1–68. (Cited in Richards, *The Romantic Conception of Life*.)

32. Nadler, S. (1999), *Spinoza: A Life*. Cambridge University Press.

33. Schelling, Friedrich Wilhelm Joseph (1797), *On the World Soul*.

34. Williams, J. R. (1998), *The Life of Goethe: A Critical Biography*. Blackwell, p. 5.

35. Johann Gottfried Herder to Johann Georg Hamann (July 11, 1782). Cited by Richards, *The Romantic Conception of Life*, pp. 355–356.

36. See Richards, *The Romantic Conception of Life*, p. 326. Schiller, for reasons of his own, apparently did not appreciate the positive inspiration, as well as the angst, Goethe derived from his many relationships with women.

37. Dichtung und Wahrheit (pt. 2 bk6), in Sämtliche Werke 16: 250–254. Cited by Richards, *The Romantic Conception of Life*, p. 331.

38. Cited by Stelzig, Eugene L. (2000), *The Romantic Subject in Autobiography: Rousseau and Goethe*. University of Virginia Press, p. 203.

39. Cited by Richards, *The Romantic Conception of Life*, p. 358.

40. Ibid.

41. Johann Wolfgang von Goethe to Frau Charlotte von Stein, May 1, 1776, in Mandelkow, Robert, ed. (1988), *Goethes Briefe* (Hamburger Ausgabe), third edition, 4 vols. C. H. Beck, I: 213.

42. Cited by Stelzig, *The Romantic Subject in Autobiography*, p. 205.

43. Goethe, J. W., Zur Morphologie. Verfolg (WA vi, 348); cf. Bildungstrieb (WA vii, 7I–72); cited by Wells, G. A. (1967), Goethe and the intermaxillary bone. *British Journal for the History of Science* 3: 348–361.

44. Wells, Goethe and the intermaxillary bone.

45. von Goethe, J. W. (1784), Über den Zwischenkiefer des Menschen und der Tiere. Handschriftlich, mit Tafeln, M_rz 1784; ohne Tafeln 1820 zur Morphologie, Band I Heft 2: "Dem Menschen wie den Tieren ist ein Zwischenknochen der obern Kinnlade zuzuschreiben";

1831 mit Tafeln in den "Verhandlungen der Kaiserlich Leopoldinisch-Carolinischen Akademie der Naturforscher." Cited by Barteczko, K., and M. Jacob (2004), A re-evaluation of the premaxillary bone in humans. *Anatomy and Embryology* 207:417–437.

46. von Goethe, J. W. (1962), *Italian Journey*, trans. W. H. Auden and Elizabeth Mayer. Penguin Books, p. 37.

47. Syphilis. According to *Wikipedia*, "The first written records of an outbreak of syphilis in Europe occurred in 1494 or 1495 in Naples, Italy, during a French invasion (Italian War of 1494–98)." It was assumed that the disease was spread by French troops.

48. von Goethe, *Italian Journey* (April 17, 1787).

49. Cited by Richards, *The Romantic Conception of Life*, p. 424.

50. von Goethe, *Italian Journey* (April 17, 1787).

51. Frequently overlooked, however, is the fact that Goethe also stated that roots and stems were modified leaves. Although some underground structures, such as seed cotyledons and bulbs, are indeed modified leaves, both stems and roots predated the appearance of leaves in the fossil record and therefore cannot be considered modified leaves.

52. Portman, A. (1987), Goethe and the concept of metamorphosis, in F. Amrine, F. J. Zucker, and H. Wheeler, eds., *Goethe and the Sciences: A Re-Apprasial*. Springer, pp. 133–145.

53. von Goethe, J. W. (1790/1952), *The Metamorphosis of Plants,* in *Goethe's Botanical Writings*, trans. Bertha Mueller. University of Hawaii Press, pp. 31–78. Goethe may be referring here to a passage in Plato's *Symposium* in which the philosopher and priestess Diotima of Mantinea describes the "ladder of love," in which the lover progresses from physical beauty to spiritual beauty, finally ascending to ideal beauty: "Starting from individual beauties, the quest for the universal beauty must find him ever mounting the heavenly ladder, stepping from rung to rung—that is, from one to two, and from two to every lovely body, from bodily beauty to the beauty of institutions, from institutions to learning, and from learning in general to the special lore that pertains to nothing but the beautiful itself—until at last he comes to know what beauty is."

54. The cell walls of the water-conducting vessel elements of the xylem are often reinforced by such spiral wall thickenings, which can pop out of the tracheary elements during preparation of plant tissue for microscopy. Hedwig was the first to suggest that that these stiff coils ("spiral vessels") functioned as elastic springs that could contract or expand. Goethe adapted Hedwig's model to his theory of metamorphosis. Normally, the spiral vessels elongate, but as the sap becomes purified the spiral vessels somehow contract. The contraction of the spiral vessels overcomes the expansive force of the sap, resulting in short, wide vessel elements instead of elongated, narrow ones. Unable to stretch, the shortened vessel elements then fuse with one another ("anastomose"), preventing the future stamen from expanding into a petal.

55. von Goethe (1853), *The Poems of Goethe*, trans. Sir Edgar Alfred Bowring. The Henneberry Company.

56. Goethe's theory of anastomosis may have been the inspiration for one of Rainer Maria Rilke's more enigmatic poems, "Dirait-On," from the song cycle, *Les Roses* (1927), in which "self-caressing" is likened to "Narcissus fulfilled" (trans. Morten Lauridsen):

Abandon surrounding abandon,
Tenderness touching tenderness . . .
Your oneness endlessly
Caresses itself, so they say;

Self-caressing
Through its own clear reflection.
Thus you invent the theme
of Narcissus fulfilled.

57. Aulie, R. P. (1961), Caspar Friedrich Wolff and his Theoria Generationis, 1759. *Journal of the History of Medicine and Allied Science* 16(2):124–144.

58. von Goethe, J. W. (1817), *Natural Science in General; Morphology in Particular,* vol. i, no. 1. Cited by Mueller, Bertha (1790/1952), *Goethe's Botanical Writings*, trans. Bertha Mueller. University of Hawaii Press, pp. 176–181.

59. von Goethe, J. W., Maxims and Reflections: An English translation of *Maximen und Reflexionen,* accessed at http://wolfenmann.com/goethe-maxims-and-reflections-full-text.html

60. Opening lines of "Hegira" by Goethe, written in 1814. In Spender, S. (1958), *Great Writings of Goethe*, trans. Michael Hamburger. Mentor Books, p. 262.

61. From *West-East Divan*, trans. Michael Hamburger.

62. Williams, J. R. (2001), *The Life of Goethe*, Blackwell, p. 44.

63. Idem., p. 48.

64. Translated by the authors.

65. von Goethe, J. W., *Faust, Part Two*, trans. David Duke. Oxford University Press.

66. von Goethe. Pollination, volatilization, and exudation, pp. 105–114.

67. Ibid.

68. Ibid.

69. Ibid.

70. von Goethe J. W. (1807/1952), Formation and transformation, in *Goethe's Botanical Writings*, trans. Bertha Mueller. University of Hawaii Press, pp. 21–29.

71. Ibid.

72. Ibid.

73. Schelver, F. J. (1812), *Kritik der Lehre von den Geschlechtern der Pflanze.* Braun, Heidelberg. Translated in Žarský, V., and J. Tupý (1995), A missed anniversary: 300 years after Rudolf Jacob Camerarius' "De Sexu Plantarum Epistola." *Sexual Plant Reproduction* 8:375–376.

74. Cited by Žarský and Tupý, A missed anniversary.

75. In Snow, D. E. (1996), *Schelling and the End of Idealism: The Horizons of Feeling.* SUNY Press, p. 107.

76. Goethe had used a similar argument in his *Theory of Colors* (1810), in which he dismissed Newton's results because the light had been forced to pass through a prism, which was equivalent to "torturing" Nature.

77. Krell, D. F. (1998), *Contagion: Sexuality, Disease, and Death in German Idealism and Romanticism*. Indiana University Press.

78. In some versions of the Demeter/Persephone myth, the goddess Baubo is an old woman whom Demeter encounters while, disguised as an old woman herself, she searches for news about Persephone's whereabouts. Baubo tries to cheer her up with ribald jokes and finally makes her laugh by suddenly lifting her skirt and exposing her genitals. See Foley, H. P. (1994), *The Homeric Hymn to Demeter.* Princeton University Press, p. 46.

17

Sex and the Single Cryptogam

GOETHE AND THE nature philosophers had gained many adherents to their Neoplatonic view of plant morphology, but, for the most part, German scientists took a jaundiced view of Goethe's idealism and considered his *Metamorphosis* inferior to Caspar Friedrich Wolff's *Theoria Generationis* (1759).

Foremost among the detractors of Naturphilosophie was Matthias Jacob Schleiden, Professor of Botany at the University of Jena. Born in Hamburg, in 1804, Schleiden obtained his doctorate in law in 1826 and promptly set up practice in his hometown. However, the anxiety of dealing with the legal system was apparently too much for him, and he suffered from chronic, debilitating depression. In 1831, following a failed suicide attempt using a pistol, which left a noticeable scar on his forehead, he quit law and moved to Göttingen to study medicine at the university. At Göttingen, as at other universities, a thorough grounding in botany remained foundational to the study of medicine, and there he acquired a taste for it.[1] Impatient to indulge his new passion, Schleiden moved to Berlin without completing his medical degree, to study with Johann Horkel, a physician and plant physiologist. Horkel, a member of the Prussian Academy of Sciences who was also Schleiden's uncle, seems to have exerted the greatest influence on him.[2]

By 1840, Schleiden was so well thought of in botanical circles that he was offered the chair of botany at Jena, Goethe's former university. Possibly the administration hoped that Schleiden, an ardent opponent of Naturphilosophie, would help to counteract whatever damage Goethe had done to the credibility of the sciences at Jena. However, some faculty members objected to Schleiden's appointment because his doctorate was in law rather than medicine. To appease his critics, Schleiden stayed on in Berlin long enough to obtain doctorates in both philosophy and medicine, after which he made the move to Jena.[3] But before he could get settled into his new post, he suffered a second mental breakdown and made another abortive attempt on his life. This time, after a period of convalescence, he regained his momentum and went on to make significant contributions to plant biology.

Schleiden is chiefly remembered as the co-founder of the cell theory in biology, along with zoologist Theodore Schwann. In opposition to Kant, he believed that biology was equal to chemistry and physics as a science, and he was contemptuous of all vitalist theories. In *Principles of Scientific Botany: Botany as an Inductive Science* (1849), regarded as the first modern textbook of plant biology, Schleiden blamed Goethe and the nature philosophers for promoting the notion that "a poetic treatment of nature could be placed on a level with, or even preferred to, strictly scientific approaches":

> The unfortunate seed which Goethe sowed, sprang up with sad rapidity; and . . . we owe it to him that, in Botany, whims of the imagination have taken the place of earnest and acute scientific investigation. In that unbounded region every individual's imagination had naturally equal right; there was a total want of any scientific principle which should undertake the decision between differing opinions of any method.[4]

Schleiden blamed Goethe and the nature philosophers for botany's backwardness as a science compared to zoology. No doubt Schleiden's words were music to the ears of the science faculties at Jena and other German universities, who had chafed in silence during the mercurial reign of the great genius. Yet despite Schleiden's reputation as a champion of the scientific method, he has the dubious distinction of being the last botanist to dispute the sexual theory. In this regard he was, ironically, Goethe's scientific heir.

To fully appreciate the rationale behind Schleiden's radical reinterpretation of pollination, it is necessary to return to the immediate aftermath of August Henschell's provocative book, *On the Sexuality of Plants,* which challenged the findings of Koelreuter.

HYBRIDIZE, WIN A PRIZE: THE VINDICATION OF KOELREUTER

Goethe had lived long enough to see Henschell's critique of Koelreuter's experiments refuted by A. F. Wiegmann, a physician from Braunschweig. In 1822, the Royal Prussian Academy of Sciences offered a prize for a resolution of the question of whether plants could form hybrids. Since hybrid formation was central to the sexual theory, the unspoken aim of the prize was to obtain unequivocal evidence for sex in plants.

As a boy of sixteen, Wiegmann had crossed different species of geraniums and had succeeded in producing two different hybrids. Later, he continued his observations of plants while maintaining a medical practice, but he didn't initiate serious hybridization studies until the Prussian prize was announced. Four years later, in 1826, the Prussian Academy awarded him the prize for confirming Koelreuter's results, although they granted him only half of the prize money because he had not, in their collective opinion, conclusively demonstrated that his putative hybrids were actual hybrids. However, to do so would have been a tall order, given the absence of any methods for quantifying maternal and paternal contributions to hybrid progeny. Gregor Mendel's studies of inheritance in peas were still several decades away and, in any event, did not become widely known until the turn of the century.

What Wiegmann *had* accomplished, however, made him a worthy recipient of the prize. Henschell had claimed that Koelreuter's methodology was flawed because the plants he used were grown in pots, which cramped their roots. As a result, Henschell asserted, monstrous offspring had been produced, which Koelreuter had misinterpreted as hybrid plants. Koelreuter had probably grown his plants in pots for convenience, since he had to emasculate the flowers to prevent self-pollination. To circumvent Henschell's criticism, Wiegmann, then in his early fifties, grew all his plants in the open ground—despite, he complained piteously in his report, "weak sight, a trembling hand, and painful bending and kneeling."[5] However, his perseverance paid off. At the end of four years, he had performed thirty-six crosses and confirmed Koelreuter's major findings. Because the plants were grown in open soil, the argument could no longer be made that Koelreuter's hybrids had actually been "monsters" caused by pinched roots.

Henschell had also claimed that the fertility of many of Koelreuter's interspecies hybrids were proof that they were not true hybrids, since it was well-documented that hybrids in the animal kingdom were sterile. However, Wiegmann obtained the same results that Koelreuter did:

> I have found his observations well-founded, that the plants produced from seed from one capsule of hybrid plants, often differ from one another with respect to fertility, and especially in the structure of certain parts, now approximating more to the father, now to the mother.[6]

Having demonstrated that he could repeat virtually all of Koelreuter's results in plants grown under field conditions, there could no longer be any doubt that plants could form hybrids in accordance with the sexual theory. It seems likely that Goethe, a voracious reader, had kept abreast of the results of the Prussian Academy's contest and was aware of Wiegmann's prize-winning study. In 1831, the same year that Wiegmann's essay was published, he acknowledged Henschell's—and by association Schelver's—stinging defeat.

Two years later, the Dutch Academy of Sciences offered another prize to answer additional questions about plant hybridization raised by Wiegmann's studies: "What does experience teach regarding the production of new species and varieties through artificial fertilization of flowers of the one with the pollen of the other, and what economic and ornamental plants can be produced and multiplied this way?" No doubt the thriving Dutch tulip industry, eager to apply the new technique of cross-pollination to the creation of profitable new tulip varieties, was behind the new prize. In any case, the question of plant hybridization had now moved beyond whether it occurred to how it could be exploited commercially.

To the surprise of no one, the winner of the Dutch prize was the German physician Carl Friedrich von Gaertner, the celebrated son of the noted botanist Joseph Gaertner, whose extensive work on plant hybridization was already well-known to members of the Dutch Academy. The research for which Gaertner was awarded the prize in 1837 had been conducted over the previous twenty-five years and involved nearly 10,000 crossing experiments among 700 species belonging to 80 different genera. By the end of his career, Gaertner had obtained some 350 different hybrid plants, far more than Koelreuter and Wiegmann combined. These results were ultimately published in 1849, in an updated edition of his book, *Experiments and Observations on the Production of Hybrids in the Plant Kingdom*.[7]

Gaertner's primary contribution to botany rests not on his demonstration of sex in plants, which had already been well-established, but on the sheer volume of his hybridization studies. Darwin cited Gaertner thirty times in his brief section on "Hybridism" in the first edition of *Origin of Species* (1859), and Mendel cited him seventeen times in his historic paper, "Experiments on Plant Hybrids," published in 1866. Mendel also singled Gaertner out as a worthy predecessor:

> Numerous careful observers . . . have devoted a part of their lives to [plant hybridization] with tireless persistence. Gaertner, especially, in his work, *Hybrid Production in the Plant Kingdom*, has recorded very estimable observations.[8]

Coming from someone of Mendel's monumental stature, this is very high praise indeed!

THE CRYPTOGAM CONUNDRUM

There remained one last impediment to the universal acceptance of the sexual theory: the apparent absence of sex in the group of plants Linnaeus had named the cryptogams. The term "Cryptogamia," which literally means "hidden marriage," includes all the members of the plant kingdom that do not produce seeds (the "nonseed plants"): algae, bryophytes (mosses, liverworts, and hornworts), ferns, horsetails, and club mosses.[9] In this section, we'll briefly examine the discovery of sex in the cryptogams from a historical perspective. Such an approach is essential to understand the genesis of the ideas that ultimately led to the Theory of Alternation of Generations, which revealed the unity of the life cycles of the cryptogams and the seed plants and changed forever the way we think about flowers.

The angiosperms and gymnosperms—which make up the vast majority of the plant kingdom—are comprised of what Linnaeus called the *Nuptiae Publicae Plantarum*, or plants with "public marriages." Today, this group is referred to either as the *phanerogams* ("visible marriages") adopted around 1814, or the *spermatophytes* ("seed plants"), which came into use around 1897.[10]

Naming the nonseed plants "cryptogams" was an act of faith on Linnaeus's part because it assumed that this group of plants, like the seed plants, reproduced sexually, even though the sexual structures had not yet been identified. For years, botanists had searched for sexual structures in the cryptogams without success. Linnaeus's struggle to understand the sexual life cycle of mosses is a case in point. He had observed that in the spring the leafy structures of mosses, which he equated with the leafy stems of flowering plants, instead of producing flowers, produced a small capsule at the end of a slender stalk, which, when mature, released a fine powder. Today, we call the small capsule that sits atop the green, leafy stage of moss a *sporangium* (literally, "spore vessel") and the powder it contains, *spores* (Figure 17.1).

Linnaeus initially identified the moss capsule as an anther and the powder as pollen. However, he soon changed his mind after observing that the powder of moss capsules behaved nothing like the pollen from anthers. Instead of landing on a stigma, it fell to the ground, and each powder particle germinated and formed another little moss plant (see Figure 17.1). Although Linnaeus could not discern any of the microscopic intermediate structures that formed during the transition of the spore to a leafy moss plant, the ability of the moss "powder" to give rise to another moss plant suggested to Linnaeus that it must represent seeds.

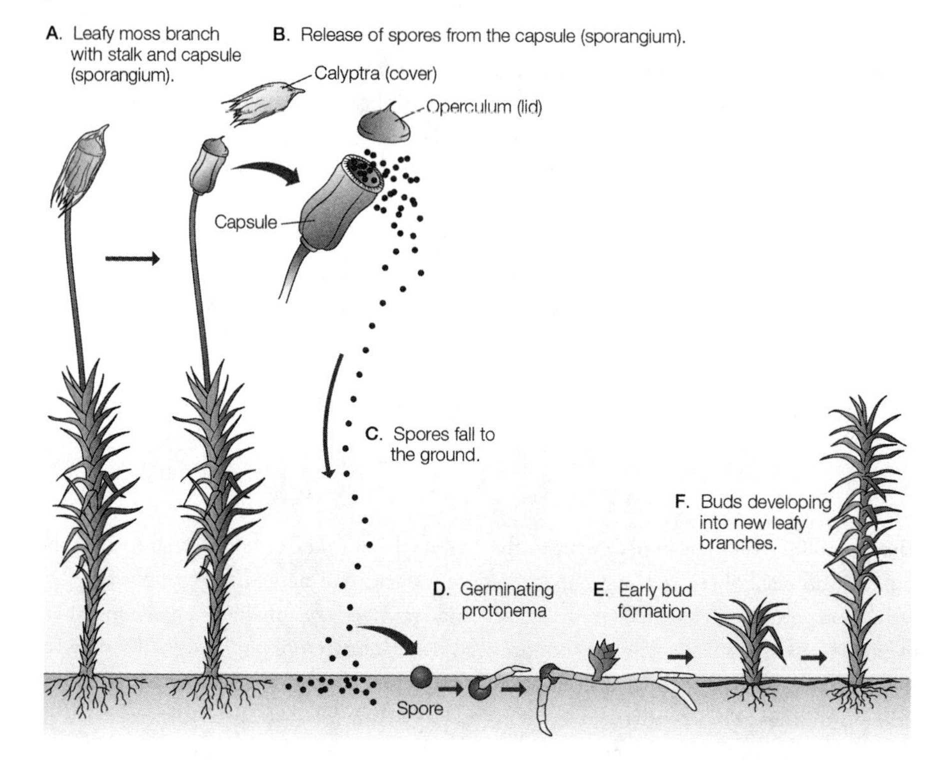

FIGURE 17.1 Stages of asexual reproduction via spores in the life cycle of a typical moss. A. In the spring, a spore-containing capsule (*sporangium*) attached to an elongated stalk grows out of the tips of the leafy structures of mosses. B. At maturity, the outer covering of the capsule, the *calyptra*, is shed, and a circular lid (the *operculum*) pops open. C. The released spores fall to the ground. D. The spore germinates on the soil, forming a multicellular filament, the *protonema*. E. Certain cells of the *protonema* develop tiny buds. F. The buds grow up into new leafy structures. In addition, root-like *rhizoids* grow below the buds, anchoring the moss plant to the ground. See the insert for color figures of the life cycles.

But this interpretation also had its problems. As a champion of the sexual theory, Linnaeus believed that seeds in flowering plants were produced by flowers. If the tiny particles of moss powder were indeed seeds, then they should have been formed by a process resembling pollination. But, try as he might, Linnaeus could find neither stamens nor pistils anywhere on the leafy structures of mosses. Convinced that sex was universal and that male and female sexual organs must be present, he contented himself with placing mosses in the *Cryptogamia* group. Meanwhile, French botanists correctly identified the powder produced by mosses, ferns, and other cryptogams as asexual *spores*. In their natural systems of classification, French botanists regarded the cryptogams generally as asexual. This was a direct challenge to Linnaeus's idea that sex was universal in the plant kingdom. Thus, by the early nineteenth century, the debate over plant sex had begun to shift from angiosperms to cryptogams.

The main barrier to discovering the sexual organs of the cryptogams was that they were invisible to the naked eye, and botanists were hampered by the poor quality of their microscopes. As microscopes gained in resolving power, botanists began to train their improved

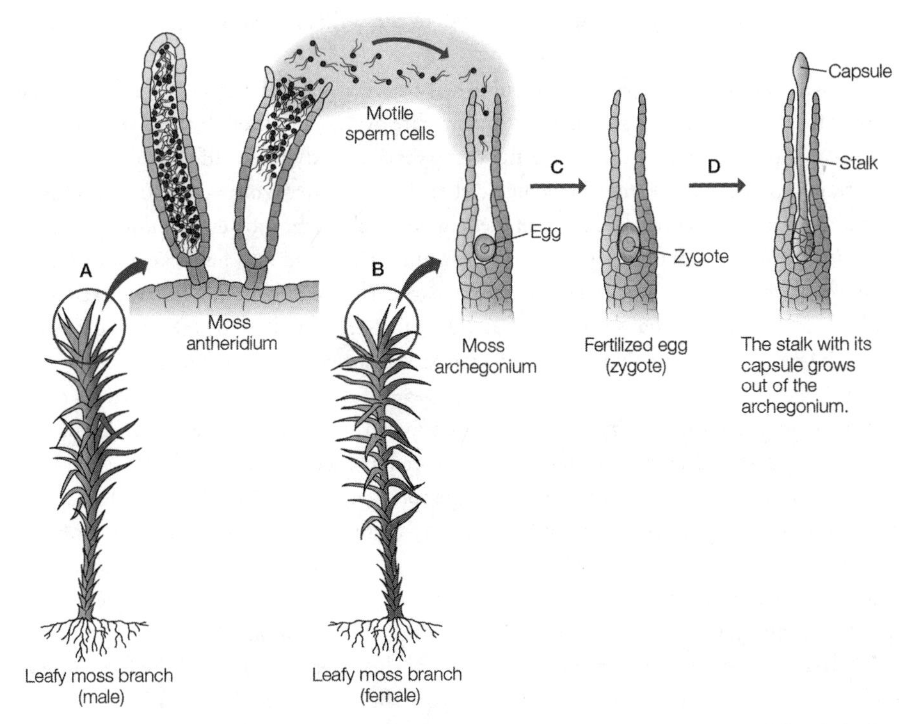

FIGURE 17.2 Stages of sexual reproduction in the life cycle of a moss. A,B. Male and female leafy structures bear *antheridia* and *archegonia* at their tips. C. The motile sperm cells produced by the *antheridia* are carried in rain droplets to the neck of the *archegonia*, where they swim down the narrow canal and fertilize the egg. D. The resulting zygote develops into a new spore capsule with stalk and grows out of the archegonium.

instruments on the various stages of cryptogam life cycles, with the result that many new structures never before seen were discovered. An important advance was made by Johann Hedwig, a physician who became Professor of Botany and Director of the Botanical Garden at the University of Leipzig. While studying the germination and growth of moss spores, Hedwig observed that the first product of germination was a green, algal-like filament, which today we call the *protonema* (see Figure 17.1).[11] Soon, tiny buds emerged from the filament, and these buds grew vertically to form the green, leafy structures we associate with moss (see Figure 17.1).

Hedwig also identified two types of sexual structures at the tips of some of the leafy structures (Figure 17.2). One type had bulbous ends, which he assumed were miniature "stamens." The other type was vase-shaped with a long neck, which he compared to tiny "pistils." Peering through a microscope at the tiny "stamens," he was able to observe the release of what he called "corpuscles." Hedwig hypothesized that these "corpuscles" were the pollen-like structures that entered the necks of the moss "pistils" and fertilized them. In today's terminology, Hedwig's tiny "stamens" on the tips of male moss branches are called *antheridia* (singular, *antheridium*), while the miniature "pistils" on the tips of female branches are termed *archegonia* (singular, *archegonium*) (see Figure 7.2).[12] The fertilized egg,

Hedwig correctly surmised, grew and gave rise to the stalk and capsule that later emerged from the archegonium.[13]

Hedwig's pioneering observations, published between 1782 and 1784,[14] confirmed that mosses did, indeed, have a sexual stage in their life cycle, even though these sexual structures were invisible to the naked eye. After many years of studying the life histories of a wide selection of bryophytes, Hedwig concluded that the two other groups of bryophytes (liverworts and hornworts) both had the same general life cycle as the mosses, including "hidden" male and female sexual organs present on the diminutive structure that developed directly from the spore. Linnaeus, who had died four years earlier, would no doubt have exclaimed with vexation, "*Vad var det jag sa*?" ("What did I tell you?").[15] Following Hedwig's discovery of sex in bryophytes, more than sixty years were to elapse before the basic features of the fern life cycle were elucidated.

It had long been known that fern spores are typically produced on the underside of fern leaves (Figure 17.3). The sporangia of fern leaves are bunched together in clusters called *sori* (singular, *sorus*). At maturity, the tiny sporangia spring open and fling their powdery spores some distance away onto the ground where they germinate to produce a short multicellular filament, also called a *protonema*. When the *protonema* reaches a certain length, the growing tip begins to expand laterally resulting in a small, flat, structure about the size of a quarter. In modern terminology this structure is called a *prothallus*.

At first, the *prothallus* was thought to represent the direct precursor of the fern embryo and was therefore referred to as a "pro-embryo." Then, in 1844, the Swiss botanist Karl von Nägeli made the startling discovery of *antheridia* (microscopic male sexual organs) on the underside of the so-called "proembryo" (Figure 17.4). These antheridia, similar to those of mosses, released "spiral filaments," which were in fact motile sperm cells. This made absolutely no sense to botanists at the time. Unlike the diminutive mosses, ferns were comparable in size to seed plants. In seed plants, sexual reproduction was assumed to be the exclusive function of the adult vegetative plant. It was as if the sexually mature stage of human beings was discovered to be not the fully grown adult, but the embryonic fetus! Botanists began to suspect that there was something very unusual about the so-called "pro-embryo" of ferns. If it was not an early stage of embryo development, what was it? Although Nägeli had detected antheridia on the undersides of these weird "pro-embryos," he was unable to find any archegonia, so the question of the sexuality of these baffling little plants remained unresolved.

The problem languished for several years, until 1848, when the Polish Count Michael Jerome Lesczyc-Sumińsky, a young, amateur botanist and illustrator working in Berlin, took a closer look under the microscope. Voila! He discovered the missing archegonia clustered just above the antheridia on the underside of the so-called "pro-embryo" (see Figure 17.4). The following year, Sumińsky also reported that he saw "spiral filaments" (sperm cells) exit the antheridia, enter the canal-like neck of the archegonium, and ultimately fuse with a large cell at the bottom of the canal, which he called the "ovule" or egg.[16] Subsequently, the fertilized egg divided to form the true embryo, which emerged from what we now call the prothallus and grew into the mature fern plant (see Figure 17.4).

Although Sumińsky's astute observations helped to clarify the fern life cycle, they were peremptorily dismissed by dyed-in-the-wool asexualists such as Schleiden, who regarded Sumińsky as a rank amateur. He declared that a

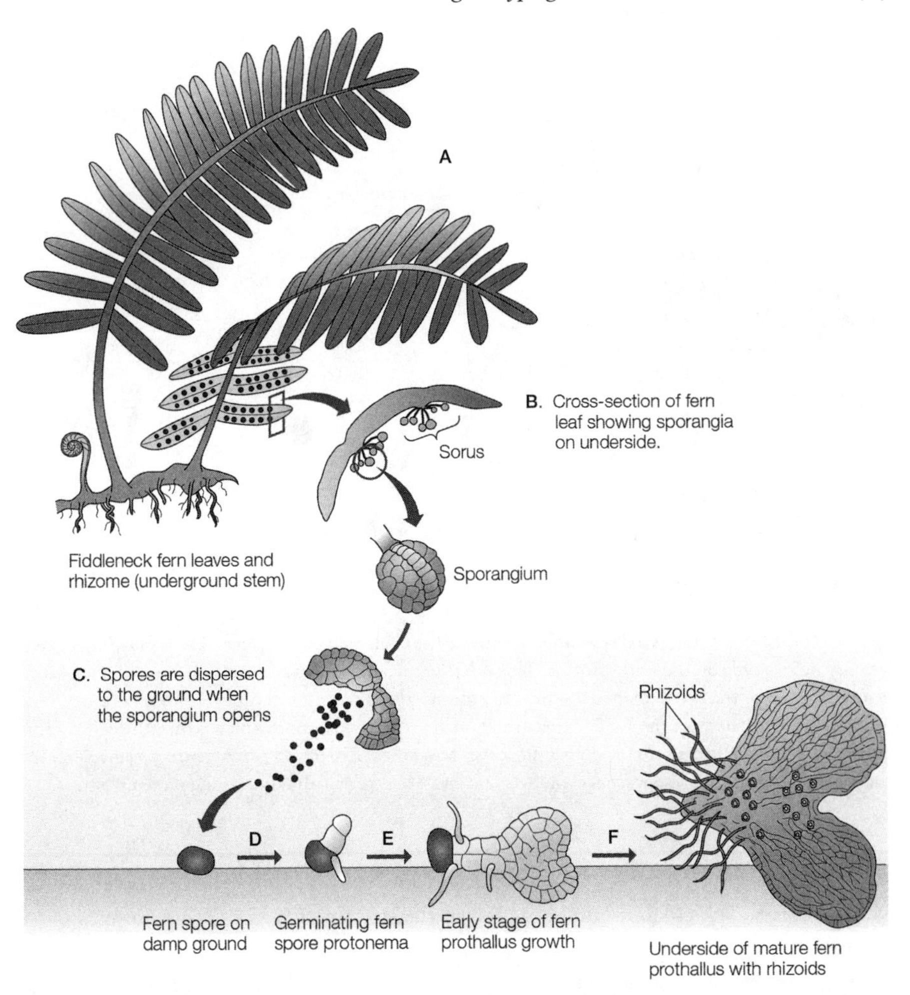

FIGURE 17.3 Stages of asexual reproduction in the life cycle of a fiddlehead fern. A. Mature fern fronds (leaves) grow out from underground stems called *rhizomes*. B. Sporangia are grouped in clusters called *sori* (singular, *sorus*) on the underside of the mature fern frond. C. At maturity, the sporangia break open and disperse their spores to the ground.[30] D. The spores germinate on the ground, forming a short, multicellular filament called a *protonema*. E. The cells at the tip of the *protonema* begin dividing laterally to form a small, flattened, heart-shaped structure called a *prothallus*. F. The prothallus expands and forms root-like rhizoids and sexual structures (not labeled) on the underside of the prothallus.

> [l]ively imagination, most likely accompanied by a faulty microscope and incorrect preparation of samples, have led Sumiński to the curious belief that the mobile spiral threads enter the fern's germinal organs.[17]

To Schleiden, the take-home message from the "cryptogams" was *not* that they possessed sexual organs, which he denied, but that they reproduced *asexually* via spores. According to Schleiden, asexual spores, not seeds, were the primary units of propagation—not only in the cryptogams, but in seed plants as well!

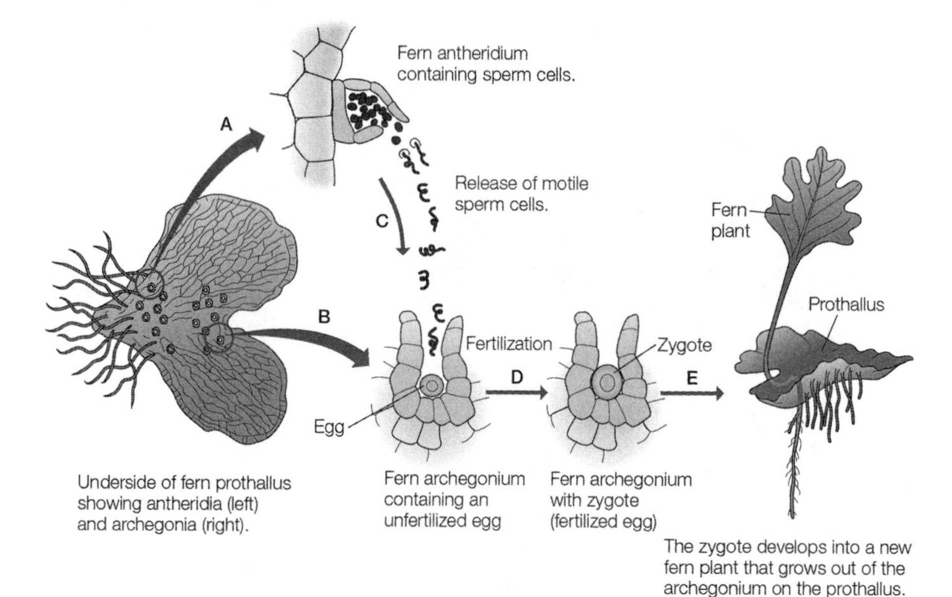

FIGURE 17.4 Stages of sexual reproduction in the life cycle of a typical fern. A. *Antheridia*, located among the rhizoids on the underside of the *prothallus*, produce *sperm cells*. B. *Archegonia*, located near the notch of the heart-shaped *prothallus*, produce *egg cells*. C. At maturity, the *sperm* are released from the *antheridia*. D. The sperm cells swim to the *archegonia* and fertilize the egg. E. The developing zygote forms roots and juvenile leaves and emerges from the *archegonium*. Subsequently, the *prothallus* undergoes senescence and withers away, leaving a fully independent fern plant.

SCHLEIDEN'S ASEXUALIST THEORY

Matthias Jacob Schleiden regarded the sexual theory of plants as an egregious example of scientific anthropomorphizing. As the co-founder and tireless champion of the cell theory, he argued in 1853 that the entire life cycle of seed plants is based on duplicative cell divisions, which we now call *mitosis*, which produces seeds by a vegetative (clonal) process:

> the individual vegetable is endowed with the power of forming new cells in its interior, and thus, as it were, of propagating itself. Now, the newly-formed cells have also this peculiarity, they grow and arrange themselves conformably to the cell in which they originate. This is the power given to all plants, to develop new plants out of any of their cells, when these come to be placed in favorable circumstances, and by this power is explained the facility with which almost all plants may be multiplied.[18]

Schleiden cited numerous examples of vegetative propagation, both natural and horticultural, in support of his theory. He repeated, without attribution, Goethe's interpretation of leaves, roots, and branches as separate individuals, consistent with the idea that an individual plant can be thought of as a clonal colony.

Although knowledge that plants are able to regenerate from cuttings can be traced at least as far back as Aristotle, Schleiden modernized our understanding of the phenomenon by connecting it to the mitotic divisions of individual cells. Unlike the somatic cells of animals, most diffentiated plant cells are potentially *totipotent*—that is, they can regenerate a complete organism under the appropriate conditions. In other words, the somatic cells of plants have nearly the same potential to generate a complete plant as the fertilized egg. Viewed in this context, it is easy to understand why Schleiden concluded that plant embryogenesis is a vegetative process. Like Goethe, Schleiden regarded ovules as specialized vegetative buds, referring to them as "seed-buds." Schleiden felt so confidant in his interpretation that he proposed that the word "sex" be banished from the botanical literature:

> It will of course be understood that the word "sex" means nothing beyond a mere indication, it being at any rate at present incorrect to attach to the term the meaning current with respect to animal life. It would be highly desirable wholly to banish the use of this equivocal term, as many misconceptions might be avoided. [19]

Schleiden thought it preposterous that anyone would think plants capable of sex: "I am surprised," he scoffed, "that no one has yet insisted upon the presence of the organs of sense, as eyes and ears, in plants, since they are possessed by animals."

Schleiden also rejected the "enigmatical" distinction between cryptogams and seed plants established by Linnaeus. Since he believed that cryptogam spores and the seeds of seed plants were equivalent structures, he referred to them both as "spores." Accordingly, he devised a new system of classification that left out sexuality. He divided the plant kingdom into two main groups: those with "covered spores" (Angiosporae) and those with naked spores (Gymnosporae). He placed algae and fungi in the Angiosporae. The cryptogams, on the other hand, he placed in the Gymnosporae under the heading *Plantae agamicae*—plants in which the "spore" develops into a new plant directly. Seed plants, including flowering plants, were all placed among the Gymnosporae as well, and were designated *Planta gamicae*. In this group, the "spore" (in this case referring to pollen), like the spores of the *Plantae agamicae*, develop asexually into a new plant but only by interacting with specialized cells in the ovule of the parent plant.

To understand Schleiden's asexual theory of embryogenesis, we will first need a brief introduction to the pollen tube.

AMICI'S DISCOVERY OF THE POLLEN TUBE

Following Nehemiah Grew's proposal of the sexual role of pollen in 1682, nearly 150 years were to elapse before the pollen tube was discovered. Thus, for more than a century, the process by which pollen grains landing on the stigma managed to fertilize the ovule contained in the ovary was a black box. Grew had speculated that the pollen grain spilled onto the outer surface of the ovary, "touching it with a prolific virtue." Koelreuter and Sprengel had shown that pollen must land on the stigma to effect fertilization. Samuel Morland in England, Geoffroy in France, and Malpighi in Italy advanced the theory that entire pollen grains slid down a central cavity in hollow styles to reach the ovule—but relatively few flowers have hollow styles. Several other observers, including Linnaeus, had suggested that pollen grains burst open on the stigma and the released contents diffused down the style into the ovule. Koelreuter, who opposed the idea that the pollen burst on the stigma, suggested that the pollen secreted an oily substance, the "essence" of the plant. He confessed ignorance as to what happened next, but he speculated that the male essence mixed with the female essence on the stigmatic surface, and the blended essences then diffused down the style to the ovule. Gaertner subscribed to a view similar to Kolreuter's in his prize-winning 1849 treatise.

The solution to the problem of how the pollen grain delivers the fertilizing material to the ovule had actually been discovered years earlier, in 1823, by Giovanni Battista Amici, a Florentine mathematician, astronomer, and accomplished microscopist.[20] Amici observed that the stigmas of *Portulaca oleracea* (common purslane) were covered with hairs containing swirling cytoplasm. Today we refer to such cytoplasmic swirling as "cytoplasmic streaming." While studying this phenomenon under the microscope, Amici happened to focus on a single pollen grain attached to the stigmatic hair he was observing. To his astonishment, a tubular "gut" suddenly emerged from the side of the pollen grain. For three hours he watched as the tube grew down the side of the hair and disappeared into the tissues of the stigma.

Amici's observations of pollen tubes were confirmed in 1827 by the French botanist Adolphe-Théodore Brongniart. Brongniart assumed that after penetrating the stigma and entering the style, the "spermatic tubules," as he called them, burst open, releasing "spermatic granules," which wriggled the rest of the way down to the ovule. However, Amici was skeptical of Brongiart's hypothesis. In 1830, after extending his own studies on *Portulaca* and *Hibiscus*, Amici framed the question succinctly:

> Is the prolific humor passed into the interstices of the transmitting tissue of the style . . . to be transported afterwards to the ovule, or is it that the pollen tubes elongate bit by bit and finally come in contact with the ovules, one tube for each ovule?

SCHLEIDEN'S ASEXUAL FEMALE PARTHENOGENIC POLLEN TUBES

Schleiden was among the many microscopists who were inspired to study pollen tube development after Amici's and Brongniart's reports. In 1837, he published a paper confirming Amici's observation that the pollen tube grew down through the style. He also reported that he observed the pollen tube entering the ovule through a pore in the ovule's surface.[21] Then, in 1838, Schleiden startled the scientific community with an astounding claim. He reported that upon entering the ovule, the tip of the pollen tube divided to form a separate cell, which then underwent repeated cell divisions inside the ovule to form the embryo.[22] In other words, the pollen tube is the sole source of the embryo that is present inside the seed. The ovule serves only as an incubator that houses and nourishes the embryonic plant.

According to Schleiden, this pattern of pollen tube growth paralleled the pattern observed during cryptogam spore development. For example, the germinating spores of ferns initially grow as linear protonema filaments. After reaching a certain length they grow laterally to form the flattened prothallus, which Schleiden regarded as the embryo precursor, or "proembryo" (see Figure 17.3). Schleiden claimed that pollen grains followed the same developmental program. Upon germinating on the stigma, they grow down the style to the ovule by means of a protonema-like pollen tube. Upon entering the ovule, the tip of the pollen tube then begins dividing laterally to form the embryo of the seed. The idea itself was not new with Schleiden. Several contemporary asexualists, including Schleiden's uncle, Johann Horkel, [23] were looking for parallels between the development of cryptogam spores and pollen grains. They believed that the main difference between the spores of cryptogams and the pollen grains of flowering plants was that pollen grains germinated on the stigma instead the soil. For example, in 1837, William Valentine, a British naturalist, stated that "it is a well-established fact that the embryo, or essential part of the seed, is derived from the pollen."[24] What set Schleiden apart from the other asexualists was his claim to have witnessed the process under the microscope. A. G. Morton has suggested that Schleiden simply confused the tip of the pollen tube with another type of cell in the embryo sac—yet another example of "believing is seeing."[25]

Whatever the source of the error, Schleiden's supreme confidence in the truth of his artifact led him to propose a radical revision of floral biology and sexual identity. Based on the Aristotelian doctrine that the female parent by definition provides the material substance of the embryo, Schleiden concluded that the pollen tube must therefore be a female structure that reproduces vegetatively. Since the ovule played only a nutritive role in housing the embryo, Schleiden inferred that it was a purely asexual structure.[26] In a bizarre twist of logic, Schleiden made the case for a unisexual, plants-as-female model based on his identification of the pollen tube as the female parent.

Because of his high standing in the field, Schleiden's pollen-parthenogenesis theory was accepted by many microscopists. In 1846, Amici presented compelling evidence from his studies with orchids that the role of the pollen tube in embryo formation was to activate a pre-existing egg cell within the ovule. Amici's theory also acquired adherents among microscopists, and a full-blown controversy ensued. To settle the matter, the Dutch Academy of Sciences offered yet another prize, which was awarded in 1850 to Hermann Schact. In fact, Schact's essay supported Schleiden's asexual pollination theory! Thus, the prize only succeeded in muddying the waters.

Even before its publication, the work was refuted by several eminent botanists, including Wilhelm Hofmeister. Hofmeister's paper, published in 1849, described observations on thirty-eight species belonging to nineteen genera and demonstrated that, in all cases, the egg cell that ultimately developed into the embryo was already present in the embryo sac prior to pollination. Indeed, one of Schleiden's own students confirmed the observations of Amici and Hofmeister,[27] and, six years later, Schleiden quietly retracted his asexual pollination theory.[28]

By 1856, sexual fusion of sperm and egg had been demonstrated in three different algae, *Fucus, Vaucheria*, and *Oedigonium*. Based on these and similar findings in animals, the German zoologist Oscar Hertwig formulated the general definition of *fertilization* as the union of the nuclei of the male and female gametes: sperm and egg.[29] Although evidence was still lacking in flowering plants, it was widely assumed that the fusion of egg and sperm was the basis of fertilization in angiosperm embryo sacs as well.

A KINGDOM DIVIDED CANNOT STAND

The discoveries of sex in the cryptogams and Schleiden's retraction of his theory of parthenogenic pollen tubes had removed the last barriers to the universal acceptance of sex in plants. But although sexuality was now deemed to be a feature of all plant life cycles, it was also indisputable that cryptogams had an additional stage based on the asexual production of spores. The problem was that neither angiosperms nor gymnosperms seemed to produce spores, suggesting that they lacked an asexual stage of the life cycle.

The history of the discovery of sex in plants had come full circle. Seed plants, once believed to be entirely asexual, were now thought to reproduce *exclusively* by sexual reproduction. Only the cryptogams seemed to have a true asexual stage in their life cycles. Both conceptually and biologically, the presence of an asexual, spore-producing stage in cryptogams suggested a profound taxonomic divide between the cryptogams and seed plants. Yet cryptogams and seed plants were supposed to belong to the same plant kingdom. The stage was now set for Wilhelm Hofmeister to unveil the asexual stage of the seed plant life cycle, revealing the unity of the life cycles of the cryptogams and seed plants.

NOTES

1. Charpa, U. (2003), Matthias Jakob Schleiden (1804–1881): The history of Jewish interest in science and the methodology of microscopic botany. *Aleph* 3:213–245; Farley, John (1982), *Gametes and Spores*. Johns Hopkins University Press, p. 48.

2. https://de.wikipedia.org/wiki/Johann_Horkel.

3. Charpa, Matthias Jakob Schleiden.

4. Schleiden, M. J. (1849/2013), *Principles of Scientific Botany, or Botany as an Inductive Science*. Forgotten Books, pp. 310–311.

5. Cited by Roberts, H. F. (1929), *Plant Hybridization Before Mendel*, Princeton University Press, p. 161.

6. Idem., p. 162

7. von Gaertner, C. F. (1849), *Experiments and Observations on the Production of Hybrids in the Plant Kingdom* (*Versuche und Beobachtungen über die Bastarderzeugung im Pflanzenreich*); see Roberts, *Plant Hybridization Before Mendel*, p. 168.

8. Mendel, G. (1865/66), *Experiments on Plant Hybrids*, trans. Eva R. Sherwood. In C. Stern, ed., *The Origin of Genetics: A Mendel Source Book*. W. H. Freeman and Company.

9. Fungi were grouped in the plant kingdom in most biology textbooks until 1969, when Robert Whittaker argued for placing them into a separate Kingdom. The new "Five Kingdom System" included the Monera, Fungi, Protista, Plantae, and Animalia.

10. *Online Oxford English Dictionary*.

11. Hedwig called the protonema a "cotyledon." The term *cotyledon* (from the Greek word *kotulēdōn* meaning cup-shaped cavity) was first applied to a patch of villi on the placenta of mammals in the mid sixteenth century. Linnaeus was the first to apply the term to the specialized leaves of embryos that act as absorptive organs, which take up nutrients from the endosperm like a placenta, rather than act as storage organs (*Philosophia Botanica*, 1751). The term later came to be applied to the primary leaves or "seed leaves" of the embryos of seeds plants.

12. Both archegonia and antheridia are present in algae, bryophytes, ferns, and fern allies, whereas archegonia (but not antheridia) are also found in most gymnosperms.

13. Morton, A. G. (1981), *History of Botanical Science*. Academic Press.

14. Hedwig, J. (1782), *Fundamentum Historiae Naturalis Muscorum Frondosorum*; (1784) *Theoria Generationis Fructificationis Plantarum Crptogamicarum*.

15. The authors thank Allan Rasmusson and Ian Max Möller for the Swedish expression!

16. Comte Leszczyc-Suminski (1849), Sur le développement des fougères. *Annales des Sciences Naturelles, Botanique*, III sér.:114–126.

17. Cited by Domański, C. W. (2004), Michał Hieronim Leszczyc-Sumiński (1820–1898): a biography and psychological portrait of the Polish naturalist and explorer. *Organon* 33:111–120.

18. Schleiden, M. J. (1853), *The Plant: A Biography*, second edition, trans. Arthur Henfrey, p. 65.

19. Schleiden, M. J. (1849/2013), *Principles of Scientific Botany, or Botany as an Inductive Science*. Forgotten Books, pp. 196–197.

20. Pollen tubes may actually have been observed eighteen years earlier by Austrian botanical artist and illustrator Ferdinand Lucas Bauer (1760–1826), who accompanied microscopist Robert Brown (1773–1858) on a journey of exploration to Australia from 1801 to 1805. Bauer and Brown had been assigned to the expedition led by the English navigator Matthew Flinders on the sloop *Investigator*. In 1805, Bauer made drawings illustrating the pollination mechanism of *Asclepias curassavica*. In the drawings, Bauer showed pollen tubes on the stigmatic surface and the entry of the tubes into the stylar canal. However, the results were not published and are only mentioned in a paper written in 1833 by Robert Brown. Thus, credit for the first published description of pollen tubes and a discussion of their significance still belongs to Giovanni Battista Amici. See Ducker, S. C., and R. B. Knox (1985), Pollen and pollination: a historical review, *Taxon* 34:401–419.

21. Schleiden, M. J. (1837), Einige Blicke auf die Entwicklungsgeschichte des vegetabilischen Organismus bei den Phanerogamen. *Archiv für Naturgeschichte* 3:289–320.

22. Schleiden, M. J. (1838), Some observations on the development of the organization in phanerogamous plants. *Philosophical Magazine* 12:244.

23. Zander, R., F. Encke, G. Buchheim, and S. Seybold (1984), *Handwörterbuch der Pflanzennamen*, 13th edition. Auflage. Ulmer Verlag. Cited in https://de.wikipedia.org/wiki/Johann_Horkel.

24. Valentine, W. (1837), Observations on the development of the theca, and on the sexes of mosses. *Transactions of Linneaus Society, London* 17:480 and 18:499.

25. Morton, History of Botanical Sciences, p. 396. According to A. G. Morton, Schleiden probably confused the suspensor, a large cell that attaches the embryo to the embryo sac, for an extension of the pollen tube pushing its way into the embryo sac.

26. Ibid.

27. von Sachs, Julius (1875), *History of Botany (1530–1860)*, Oxford University Press, p. 435.

28. In 1862, Schleiden gave up laboratory research at the University of Jena and moved to Dresden, where he briefly taught history and philosophy as a private tutor. In 1863, Schleiden accepted the position of Chair of Anthropology at the University of Dorpat (now the University of Tartu in Estonia). For the remainder of his life, he published voluminously and widely on topics in cultural history, as well as producing two books of poetry under the pseudonym "Ernst." Although he was not Jewish, he published two works on Jews and Judaism that were widely distributed. Among them was the article, "The Importance of Jews for the Preservation and Revival of Learning During the Middle Ages," in which he asserted that without Hebrew there would have been no Reformation. Charpa, Matthias Jakob Schleiden.

29. Maheshwari, P. (1950), *An Introduction to Embryology of Angiosperms*. McGraw-Hill.

30. The mechanism of spore dispersal in ferns is based on the tensile strength of water. As the sporangium wall dries, the mouth of the sporangium is pulled open by surface tension in the remaining water. At a critical point, cavitation bubbles form, breaking the surface tension. Since the sporangium wall is elastic, the spores are ejected from the sporangium at a velocity of 10 meters per second or more.

"In the garden!" he said, wondering at himself. "In the garden! But the door is locked and the key is buried deep."

—FRANCES HOGDSON BURNETT, *The Secret Garden* (1911)

18

Flora's Secret Gardens

UNBEKNOWNST TO WILHELM Hofmeister and his German colleagues, by 1838 Charles Darwin in England had already formulated his basic theory of evolution by natural selection:

> I happened to read for amusement Malthus on Population, and being well prepared to appreciate the struggle for existence which everywhere goes on from long-continued observation of the habits of animals and plants, it at once struck me that under these circumstances favourable variations would tend to be preserved, and unfavourable ones to be destroyed. The result of this would be the formation of new species. Here, then, I had at last got a theory by which to work.[1]

However, it wasn't until 1859 that Darwin, pressured by Alfred Russel Wallace, finally got around to publishing his ideas in *Origin of Species*. By this time, the apparent taxonomic divide based on their life cycles between the cryptogams and seed plants had at last been resolved by Hofmeister. Indeed, had it not been resolved, it would have been difficult for evolutionary morphologists to argue that seed plants had evolved from simpler, cryptogam-like ancestors. Julius von Sachs spoke for all botanists when he waxed rhapsodic in describing Hofmeister's stunning insights into the common features of the life cycles of cryptogams and seed plants:

> The results of the investigations published in the *Comparative Investigations on the Germination, Development and Fruit Formation of the Higher Cryptogams and Seed-formation in Conifers* in 1849 and 1851 were magnificent beyond all that has been achieved before or since in the domain of descriptive botany; the merit of the many

valuable particulars, shedding new light on the most diverse problems of the cell-theory and of morphology, was lost in the splendor of the total result.[2]

Sachs pointed out that Hofmeister's grand synthesis was incompatible with the prevailing idea of the fixity of species:

> The reader of Hofmeister's "Comparative Investigations" was presented with a picture of genetic affinity between Cryptogams and Phanerogams [seed plants], which could not be reconciled with the then reigning belief in the fixity of species. He was invited to recognize a connection of development which made the most different things appear to be closely united together, the simplest Moss with Palms, Conifers, and angiospermous trees, and which was incompatible with the theory of original types.[3]

After Hofmeister, Sachs declared, the so-called natural systems of classification had to be revised along evolutionary lines. Indeed, Hofmeister's research had prepared the ground for Darwin's *Origin of Species*:[4]

> The assumption that every natural group represents a [Divine] idea was here quite out of place; the notion entertained up to that time of what was really meant by the natural system had to be entirely altered. . . . When Darwin's theory was given to the world eight years after Hofmeister's investigations, the relations of affinity between the great divisions of the vegetable kingdom were so well established and so patent, that the theory of descent had only to accept what morphology had actually brought to view.[5]

HOFMEISTER'S CENTRAL INSIGHT

The concept of alternation of generations in plant life cycles has earned the reputation among plant biology instructors of being exceptionally hard to teach. And yet it is conceptually quite simple. The source of the pedagogical problem is fairly obvious: humans are animals, with typical animal life cycles. We naturally associate sexuality with the adult stage of the life cycle. This zoocentric idea of sexuality biases our expectations about sexuality in plants. When we look at, say, a fiddlehead fern, we feel that this large, elegant, mature individual must be the adult stage of the organism and therefore the sexual stage of the life cycle. But it isn't!

When we turn over a fern frond, we see clusters of sporangia filled with spores. They look something like seeds. Seeds, we know, contain the embryos of the next generation. We assume that if we place these spores on the ground, they will grow directly into another spore-producing fiddlehead fern plant. But they don't! As we have seen, fern spores germinate to form a totally different kind of plant called a *prothallus*, which is much smaller and simpler than the "parent" fern. Contrary to our expectations, it is the diminutive prothallus that bears the sexual organs: *archegonia* and *antheridia*. Counterintuitively, this tiny prothallus, which looks more like a humble liverwort than a venerable fern plant, is the *mature sexual stage of the fern*. The large, elegant organism we think of as the "fiddlehead fern" is actually the *asexual spore-producing generation* of the life cycle. It is the inconspicuous and unprepossessing prothallus—a completely separate individual—that constitutes the sexual generation of the fern. Such an alien life cycle violates all our expectations about how sex works, and that is why it is so hard to understand and accept.

We have the same false expectations of flowering plants. When we look at a glamorous rose bush, we assume it is the sexual stage of the life cycle. After all, aren't the stamens and pistils of the flowers the "sexual organs" of the rose, just as Nehemiah Grew, Camerarius, Sébastien Vaillant, Linnaeus, Erasmus Darwin, Joseph Koelreuter, and all the other sexualists said they were? In fact, they're not.

And when we turn over rose leaves and find no spores there, doesn't that demonstrate that roses do not make spores? No, it doesn't! The blossoming rose bush with its gorgeous hues and seductive fragrance is *not* the sexual stage of the rose life cycle. It is the *asexual spore-producing stage*, comparable to the mature fiddlehead fern plant.

The fundamental difference between ferns and rose bushes is that we can see the spores of a mature fern plant on the undersides of its leaves, while the spores of the rose bush are concealed within the anthers and ovaries of the flower, which are actually modified leaves. Otherwise, the fiddlehead fern and the rose share the same basic life cycle. The hidden spores *within* the anthers and ovaries develop into the true sexual stages of the rose, the *male* and *female gametophytes*, more commonly known as the *pollen tube* and the *embryo sac*.

This was Wilhelm Hofmeister's brilliant insight, which he elaborated in his *Theory of the Alternation of Generations*. Though hardly a household name, Hofmeister has been justifiably called "one of the true giants in the history of biology . . . [one who] belongs in the same pantheon as Darwin and Mendel."[6]

HOFMEISTER'S THEORY OF ALTERNATION OF GENERATIONS

Wilhelm Hofmeister was born in Leipzig, in 1824, the son of Friedrich Hofmeister, a music bookseller and occasional publisher, and his second wife Frederike. He had no formal schooling beyond that of a vocational high school. Originally destined to take over the family business, he served as an unpaid apprentice between the ages of fifteen and seventeen at a music store in Hamburg. This position apparently left him with sufficient free time to indulge his interest in natural history, sparked by his father who was an amateur botanist. Upon returning to Leipzig in 1841, he became increasingly fascinated by botany, inspired by Schleiden's new textbook. While working at his father's bookshop, he began a series of microscope investigations into the life histories of plants, usually carried out between four and six in the morning prior to opening the store.

Hofmeister was extremely nearsighted, but this handicap served him well as a morphologist. Severe myopics have the ability to focus on objects placed very close to their eyes, and the closer an object is to the eye the greater the magnification of the object displayed on the retina.[7] As pointed out by Kaplan and Cooke, [8] Hofmeister's myopia made it possible for him to observe the minutiae of living material with extraordinary clarity:

> Because Hofmeister had to bring living specimens so close to his unaided eyes, he became incredibly skilled at making hand sections through living material and making, for example, the most detailed observations on embryo development.[9]

By 1847, at the age of twenty-three, Hofmeister had begun the series of studies that was to become the foundation of his revolutionary theory of the alternation of generations. This research was published in book form by his father's firm in 1851, under the title: *Comparative Investigations on the Germination, Development and Fruit Formation of the Higher Cryptogams and Seed-formation in Conifers.*

Based mainly on freehand sections of fresh material, Hofmeister was the first to recognize that the life cycles of *all* the cryptogams, including the mosses, liverworts, hornworts, ferns, and related species, alternate between two distinct multicellular generations: a gamete-producing generation that begins with a spore, and a spore-producing generation that begins with a fertilized egg.[10] The gamete-producing generation, or *gametophyte* ("gamete-plant"), contains the sexual reproductive structures (antheridia and archegonia) that produce the sperm and egg cells, respectively. The spore-producing generation, or *sporophyte* ("spore-plant"), contains the asexual reproductive structures (sporangia) that produce the spores.

Now let's see how Hofmeister's general scheme for alternation of generations plays out in the cryptogams, starting with the mosses. As shown in Figure 18.1, asexual spores are released from the moss capsule, part of the sporophyte stage, which grows out of the top of the leafy gametophyte stage. The spores then germinate to form a filamentous *protonema*, which develops buds along its length. These buds grow to form the mature *leafy gametophyte* stage.

The sexual structures of mosses, the antheridia and archegonia, are located at the tips of the leafy gametophytes. The antheridia produce *motile sperm cells.* These motile sperm cells are picked up by splashing raindrops, and droplets containing sperm cells fall onto the archegonium. After being deposited near the mouth of the archegonium, they swim down the narrow canal and fertilize the egg cell situated at the base. The fertilized egg is termed the *zygote*. Finally, the zygote begins dividing and develops into the new mature sporophyte, and the cycle begins over again.

The sporophyte of mosses is called a "dependent" sporophyte because it remains attached to the leafy gametophyte its entire life. As shown in the diagram, spores are produced by a special type of cell division called meiosis, but since meiosis hadn't yet been discovered when Hofmeister published his work on alternation of generations, we'll defer our discussion of this topic until later in the chapter.

A similar, but somewhat different pattern is observed in ferns, which are vascular plants—that is, plants with specialized conducting tissues for water, minerals, and sugars. The vascular systems of fern sporophytes allows them to grow to a considerable size and become independent plants, in contrast to the dependent sporophytes of mosses. As shown in Figure 18.2, asexual spores are released from sporangia located on the undersides of fern leaves. Leaves that produce spores are called *sporophylls* (spore-leaves). The released spores germinate on the ground and divide to form the small flat prothallus, or mature gametophyte generation. Two types of gamete-forming structures (*gametangia*, or "gamete-vessels") are formed on the underside of the sexual stage, the prothallus: male antheridia and female archegonia. The antheridia produce motile sperm cells, and the archegonia contain single *egg cells* within their swollen bases. Since the antheridia and archegonia are located on the underside of the prothallus, they are in direct contact with a thin film of water at the soil surface. Upon being released, the motile sperm cells swim to the archegonia and fertilize the egg cells. Each fertilized egg becomes a zygote, and the zygote divides and develops into the young sporophyte, the asexual stage, within the

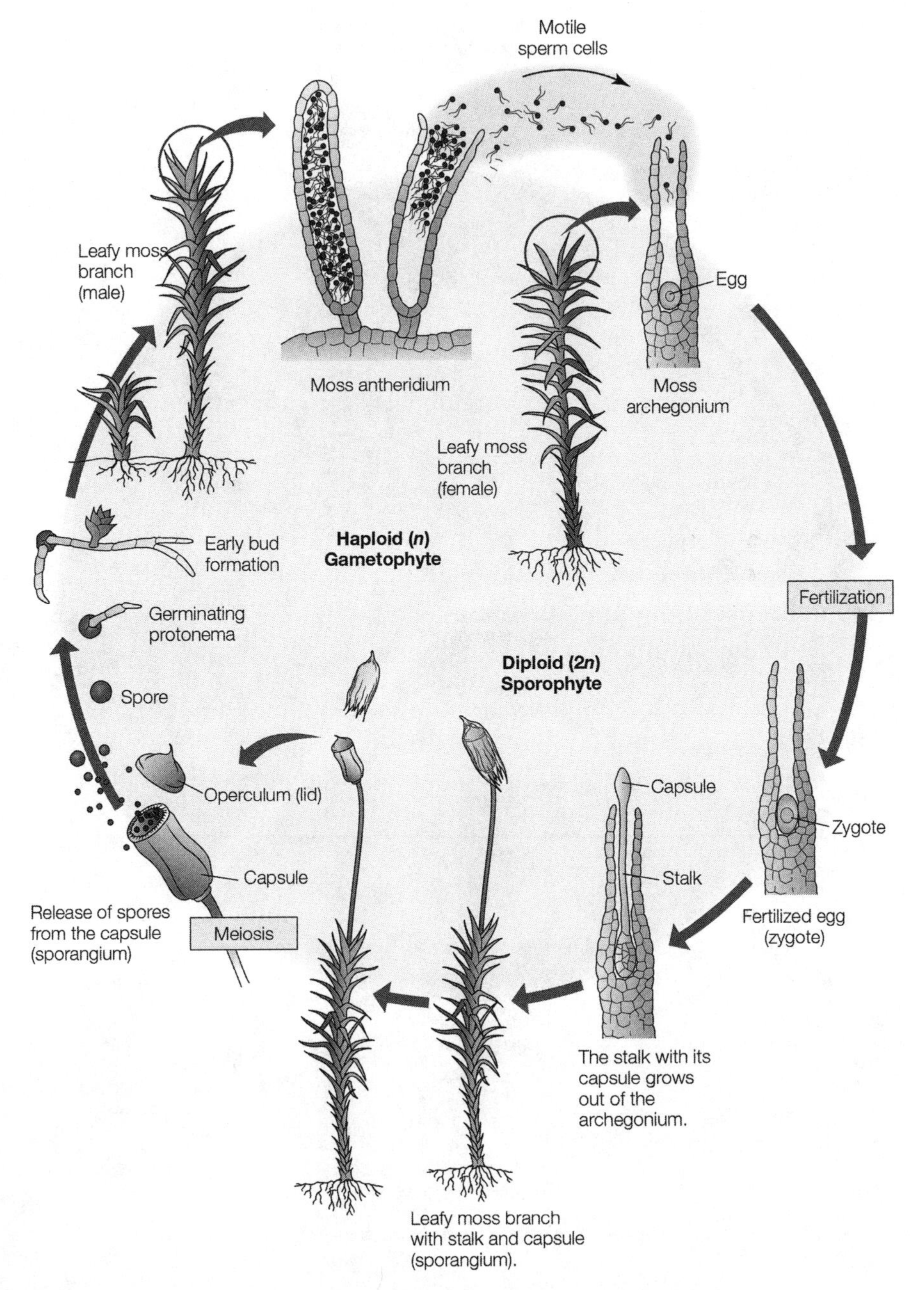

FIGURE 18.1 Diagram of life cycle of mosses illustrating the alternation of generations.

archegonium. Like the moss sporophyte, the fern sporophyte is initially dependent on the gametophyte (the prothallus) for its nutrition. However, unlike the moss sporophyte, which never becomes photosynthetic and therefore independent, the fern sporophyte soon outgrows the shorter-lived prothallus and forms roots, stems, and leaves. At maturity, the fern sporophyte produces sporangia on the undersides of its leaves, and the cycle repeats itself.

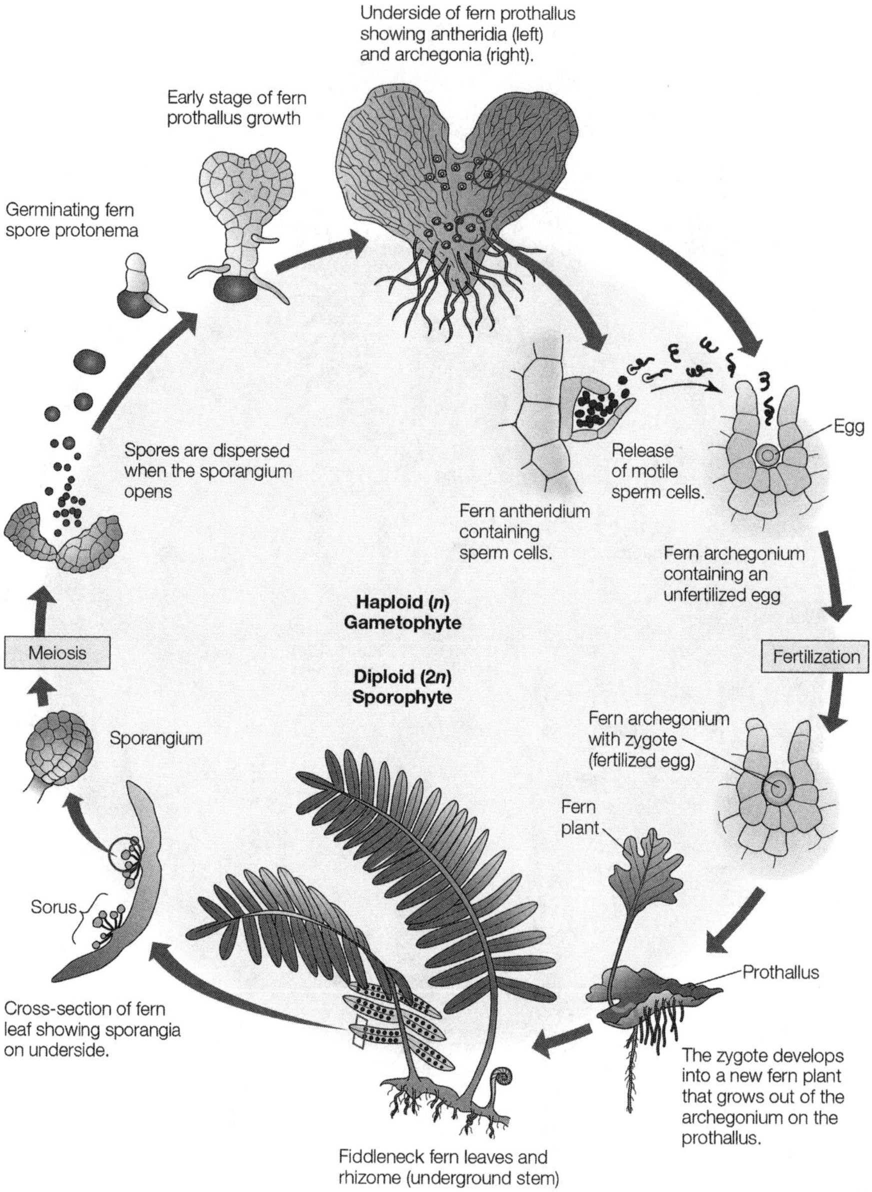

FIGURE 18.2 Diagram of the life cycle of ferns illustrating the alternation of generations.

HETEROSPORY: THE CLUE THAT UNIFIED A KINGDOM

Hofmeister noted that the gametophyte generations of various ferns and "fern allies"[11] could be either monoecious or dioecious. For example, the prothallus of the fiddlehead fern illustrated in Figure 18.2 is monoecious because both male and female sexual structures (antheridia and archegonia) are present on the same prothallus. In contrast, the gametophytes of the spikemoss *Selaginella* and the aquatic fern *Marselia* are dioecious because they possess two types of gametophytes: a male gametophyte containing antheridia, and a female gametophyte containing archegonia. Cryptogams with only one type of sporangium and one type of spore that produces one type of gametophyte containing both antheridia and archegonia (monoecious gametophytes), are termed *homosporous*. Cryptogams producing two types of sporangia, two types of spores, and separate male and female gametophytes (dioecious gametophytes), are termed *heterosporous*.

Heterospory in the spikemoss *Selaginella*,[12] a fern ally, is illustrated in Figure 18.3. The *Selaginella* sporophyte (left) produces leafy branches, some of which terminate with cone-like reproductive structures called *strobili*. These strobili consist of small, spore-bearing leaves called sporophylls, each one bearing either a female sporangium (*megasporangium*) or a male sporangium (*microsporangium*) (see Figure 18.3).[13]

The *megasporangium* of *Selaginella* produces female *megaspores*, and the *microsporangium* produces male *microspores*. These two types of spores then undergo cell divisions to form the mature female and male gametophytes, respectively. The megaporangium contains several archegonia, each with an egg at its base, whereas the microsporangium forms a single antheridium that gives rise to flagellated motile sperm cells.

The male and female gametophytes of *Selaginella* usually fall to the ground near one another. In damp or rainy weather, the motile sperm cells swim over to the female gametophytes, swim down the neck of the archegonia, and fertilize the egg cells, thus forming the zygote. The zygote then grows into another sporophytic *Selaginella* plant.

Hofmeister realized that heterospory was the crucial morphological link uniting the life cycles of the cryptogams and the seed plants. In flowering plants, the egg-containing embryo sacs and the sperm-containing pollen grains are produced by two separate structures, the ovules and anthers. Hofmeister reasoned that if the two sexual structures of angiosperms

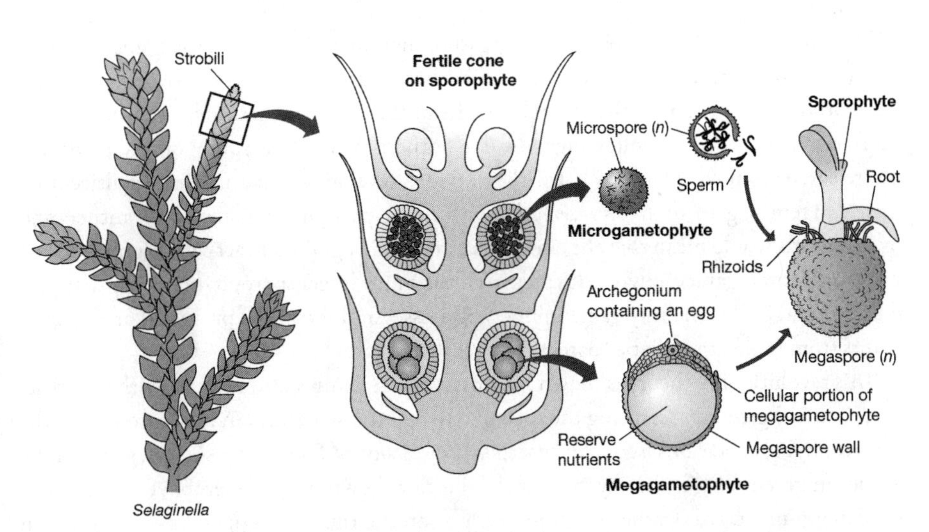

FIGURE 18.3 Reproduction in the spikemoss, *Selaginella*. (Left) Leafy branches and cone-like strobili of *Selaginella* sporophyte. (Middle) Strobili showing microsporangia containing microspores and megasporangia containing megaspores. (Right) Sperm released by the antheridia of the microgametophyte fertilize the egg in the archegonium of the megagametophyte, giving rise to a new sporophyte.

are, in fact, derived from two different types of spores, as in the heterosporous cryptogams, flowering plants would be considered heterosporous, like *Selaginella*.

HOFMEISTER UNIFIES THE LIFE CYCLES OF CRYPTOGAMS AND SEED PLANTS

There are two types of seed plants, the angiosperms, or flowering plants, and the gymnosperms, which include the conifers, cycads, *Ginkgo*, and the rarer group called the Gnetophyta (*Gnetum, Welwitschia*, and *Ephedra*). The term angiosperm literally means "vessel-seed," based on the fact that the ovules (immature seeds) are enclosed by the carpel. After fertilization, the ovule develops into the seed and the carpel forms the fruit. The term gymnosperm literally means "naked-seed." In Gymnosperms, the ovules are borne exposed on the surfaces of highly modified leaves, such as the cone scales of conifers. The ovules of both angiosperms and gymnosperms are fertilized by sperm cells delivered directly by pollen tubes. After fertilization, gymnosperm ovules develop into seeds, but since they are not enclosed by a carpel (fruit tissue), they are borne "naked." The shell surrounding a pine nut, for example, is a *seed coat* rather than a *fruit wall*.[14]

We have now arrived at the core of Hofmeister's theory, which is based on the following hypothesis: What if seed plants are like heterosporous ferns and fern allies, which produce two types of asexual spores: male and female? Could the ovule and the anther represent sporangia? If so, the ovule is actually a female sporangium or megasporangium, while the anther is a male sporangium, or microsporangium. If these conjectures are valid, the ovule should produce a megaspore, while the anther should produce a microspore, demonstrating the unity of the life cycles of the cryptogams and seed plants!

The supposition that the ovule is a megasporangium and the anther is a microsporangium is counterintuitive. Based on ferns and other cryptogams, we think of spores as asexual reproductive structures that are released from the plant and germinate on the ground. Hofmeister proposed the following radical hypothesis: what if the spores of seed plants are *not* released from the plant? What if they divide, develop, and mature into multicellular male and female gametophytes *internally*, within the sporophytic tissues of the anthers and ovaries? This would mean that the embryo sac and pollen grains are actually highly reduced male and female gametophytes, distinct from the much larger sporophytic parent that produced them, just as the tiny gametophytes of mosses and ferns (the prothalli) are separate plants from their sporophytic "parent."

This revolutionary theory is shown schematically in Figures 18.4 and 18.5, with the structures colored green representing the sporophyte and the structures in red representing the gametophyte generations (see color inserts). The left side of Figure 18.4 shows a sporophytic anther in green with a cutaway view of the interior showing the gametophyte generations (pollen grains) in red. In the two cross-sections on the right, the upper cross-section is an immature anther. At this early stage, the microsporangium is filled with "sporogenous cells" that will later give rise to spores. The lower cross-section shows a mature stage containing pollen grains. The pollen grains are colored red because they represent male gametophytes, while the surrounding tissues are green because they are sporophytic tissue.

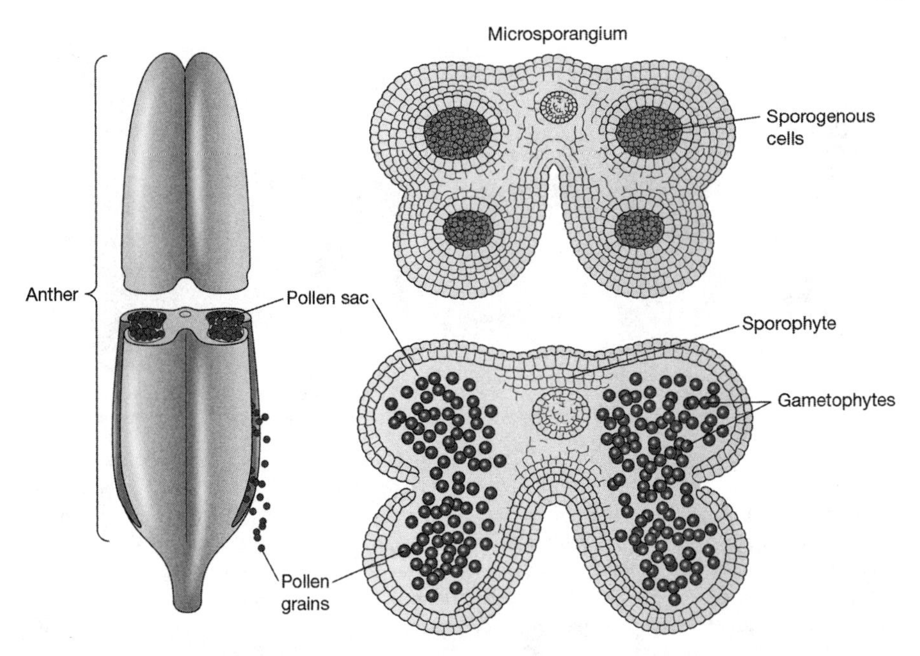

FIGURE 18.4 Anther, or *microsporangium*. The sporophyte generation is colored green and the gametophyte generation is colored red. A three-dimensional view of the mature anther is shown on the left, and two cross-sections representing the immature anther (top) and the mature anther (bottom) are shown on the right. The upper section is entirely green because the sporogenous cells in the four lobes have not yet undergone *meiosis* (discussed in the next section) to produce the haploid *microspores*, which represent the first stage of the gametophyte generation. The lower section shows a mature anther containing pollen grains, representing the male gametophyte. See color insert.

Figure 18.5 shows the pistil with ovules (megasporangia) inside the ovary, all of which are sporophytic tissue and therefore colored green. Four stages of ovule development are shown. The immature ovule is entirely green, including the megaspore mother cell, because it has not yet divided to produce megaspores. In the second stage, four megaspores have been produced, only one of which will survive. Since these represent the first stage of the gametophyte generation, they are colored reddish brown. In the third stage, the single functional megaspore has undergone two consecutive nuclear divisions to produce the four-nucleate stage of the embryo sac, or immature female gametophyte. During this time, the ovule gradually reorients itself 180 degrees. In the fourth stage, the nuclei of the embryo sac have undergone a final nuclear division to produce eight nuclei altogether. However, two of them subsequently fuse to form the single large nucleus of the central cell in the middle of the embryo sac. Cell membranes then form around each of the seven nuclei to produce the mature female gametophyte or embryo sac.

Hofmeister did not simply propose a hypothesis; he provided abundant evidence to support it. He showed that in the conifers (gymnosperms), a megaspore cell within the ovule divides to form a multicellular female gametophyte, which includes archegonia containing egg cells.[15]

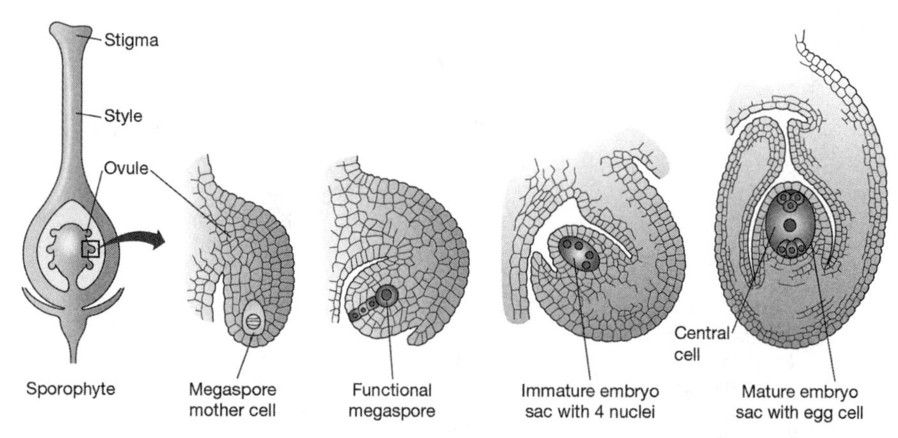

FIGURE 18.5 Ovule, or *megasporangium*. The diagram of a pistil on the left shows a longitudinal section through the ovary, with the ovules attached to a central structure called the *placenta*, a term borrowed from mammalian biology. The developmental sequence begins with the immature ovule colored entirely green because it is prior to the formation of the megaspore. Meiosis (discussed later) gives rise to the four haploid megaspores (colored reddish brown), only one of which survives. As the ovule undergoes further development, the functional megaspore undergoes three nuclear divisions to produce the mature female gametophyte, or *embryo sac*. The embryo sac contains the egg. See color insert.

In angiosperms, he demonstrated that the inner tissues of the ovule give rise to a single functional megaspore (see Figure 18.5), which develops into a highly reduced female gametophyte generation, or embryo sac. At maturity, the female gametophyte of angiosperms is called the embryo sac, consisting of an egg cell plus several other types of cells. As we shall see in the next section, this tiny gametophyte is genetically distinct from the ovule that produced it, so it truly represents a separate "generation."

Compared to the prothalli of heterosporous ferns and fern allies, this female gametophyte is so reduced in size that it doesn't even produce an archegonium. In the course of evolution, the female gametophyte generation of angiosperms has been reduced to its core function: egg production.

The microspores of flowering plants are formed in the anthers, which are the microsporangia of angiosperms. The highly reduced male gametophyte, the pollen grain, consists only of a vegetative, or tube cell that regulates the growth of the pollen tube, plus two sperm cells. As in the case of the female gametophyte, the male gametophyte of flowering plants has been reduced to its core functions, sperm production and transport. Unlike the embryo sac, however, the pollen grain escapes the confines of its microsporangium and is carried by wind or insects to the stigma (usually from another flower), where it promptly sends down a pollen tube that grows down through the style. Upon penetrating the embryo sac, the tip bursts, releasing the two sperm cells, one of which fuses with the egg cell. An overview of the angiosperm life cycle illustrating alternation of generations is shown in Figure 18.6.

Hofmeister correctly deduced the basic outline of alternation of generations, despite the fact that many of the details had not yet been elucidated. For example, Hofmeister had not actually observed the fusion of egg and sperm. In his studies with angiosperms, he observed the pollen tube tip reaching the embryo sac in the ovule, but he never actually saw

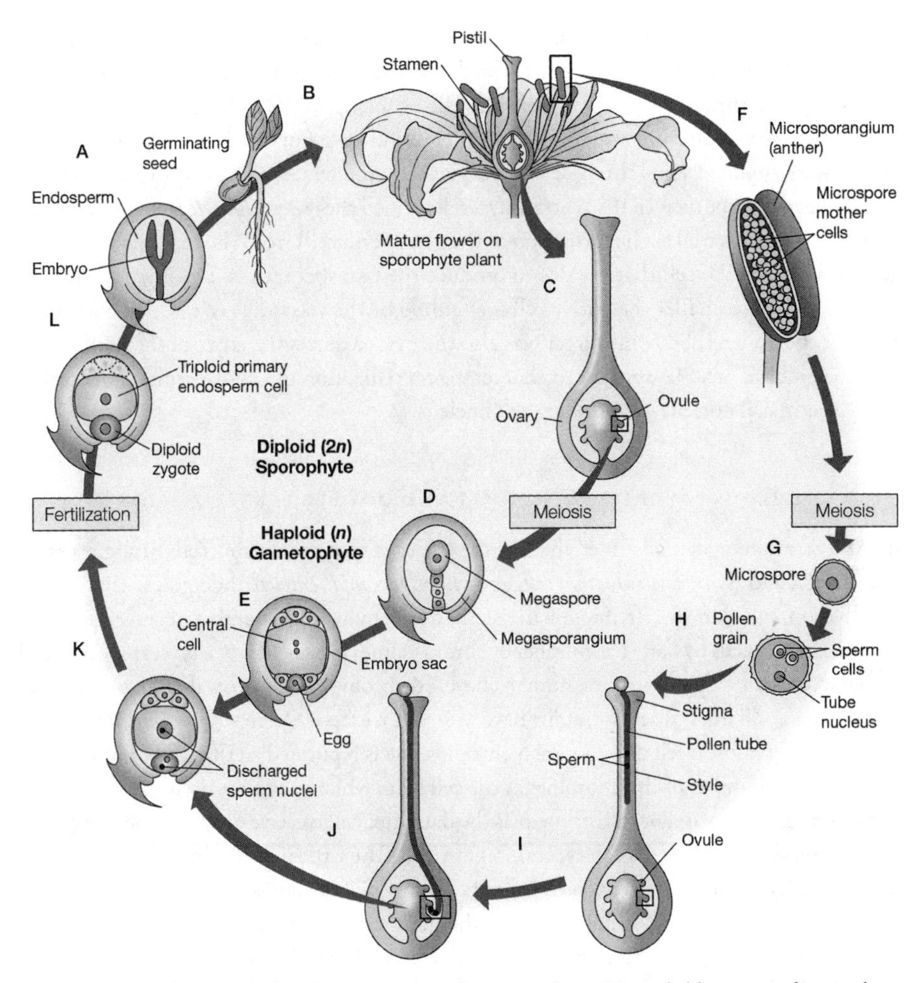

FIGURE 18.6 The alternation of generations in flowering plants. The pale blue area indicates the sporophytic stages and the light tan area indicates the gametophyte stages. A. The germinating seed gives rise to the young sporophyte generation. B. The mature vegetative sporophyte enters the reproductive phase and produces flowers. C. The ovary of the pistil gives rise to the ovules, or megasporangia. D. A specialized cell in the ovule undergoes meiosis to give rise to a single functional megaspore, the first stage of the female gametophyte generation. E. The megaspore divides mitotically to produce the embryo sac, or mature female gametophyte, containing an egg cell and several other types of cells. F. At the same time, specialized microspore mother cells in the anthers (microsporangia) give rise by meiosis to the microspores, the first stage of the male gametophyte generation. G. Inside the anther, the microspore divides twice to produce the pollen grain, or microgametophyte, consisting of a pair of sperm cells and a vegetative cell. H. The pollen grain is shed from the anther and lands on the stigma, where it germinates and grows down to the style. I. The tip of the pollen tube penetrates the embryo sac. J. The pollen tube tip bursts, releasing the two sperm cells in the embryo sac. K. One sperm cell fuses with the egg to create the zygote, the first cell of the sporophyte generation, and the other sperm fuses with the diploid central cell nucleus to produce the triploid primary endosperm cell. L. Completing the cycle, the zygote develops into an embryo (M), the seed matures and eventually germinates to produce a seedling (A). The seedling then continues to grow, eventually forming a new mature sporophyte, and a new cycle begins. See color insert.

it penetrate the embryo sac to the egg. He was also unaware of the existence of sperm cells inside the pollen tubes. He thus described the fertilization process as the diffusion of an activating fluid through the cell wall of the pollen tube tip to the embryo sac.

Hofmeister was also unaware of the specifics of pollen grain development. Many of the details were supplied some thirty-seven years later by the remarkable Polish-German botanist, Eduard Strasburger. In 1884, Strasburger described the *generative cell* and *vegetative cell* of pollen grains, as well as the formation of the two sperm cells from the generative cell. The generative cell is the cell that divides to produce the two sperm cells. During pollen grain development, the smaller generative cell is engulfed by the vegetative cell, which goes on to produce the pollen tube. Strasburger was also the first to detect the entry of the pollen tube into the embryo sac. He went on to characterize fertilization in angiosperms as the fusion of the sperm cell nucleus with the egg cell nucleus.

THE DISCOVERIES OF MEIOSIS AND DOUBLE FERTILIZATION

In 1888, Strasburger demonstrated that the two types of spore-producing cells of angiosperms (now referred to as *megaspore mother cells* and *microspore mother cells*) undergo a special type of cell division, called meiosis (reduction division), which reduces the number of chromosomes[16] of the daughter cells by half. The number of chromosomes per nucleus of a typical somatic cell is characteristic of a given species: humans have 46 chromosomes; Australian daisies have 2; roses have 14, alfalfa has 32, horsetails have 216; and the fern *Ophioglossum reticulatum* has 1,260. During mitotic cell division, each chromosome is replicated so that the daughter cells have the same number of chromosomes as the cell from which they were derived.

Strasburger showed that during meiosis the daughter cells receive only *half* the number of chromosomes in the nucleus of a typical somatic cell. Thus, the microspores and megaspores that result from meiosis receive only half the number of chromosomes as the microspore and megaspore mother cells that produced them. Since the male and female gametophytes develop from their respective spores by simple mitotic (replicative) division, they have the same number of chromosomes as the microspores and megaspores. If the number of chromosomes in the sporophyte is 2N, the number of chromosomes in the gametophyte is reduced during meiosis to 1N. ("N" in this case equals the number of chromosomes making up the complete set of genes of a given species.)

During the process of *syngamy*, the nuclei of the egg and sperm cells fuse, providing the zygote with two complete sets of chromosomes, one from the male gametophyte and the other from the female gametophyte. In this way, the 2N number of chromosomes is restored in the newly formed sporophyte. Cells with the 2N number of chromosomes are termed *diploid*, whereas cells with the 1N number of chromosomes are termed *haploid*. Thus, the sporophyte is made up of diploid cells, while the gametophyte is comprised of haploid cells.

Take, for example, a rose bush. The diploid number of chromosomes for a rose is $2N = 14$. All of the vegetative structures of a rose, including root, stem, leaves, and floral organs (sepals, petals, stamens, pistils), have the diploid number of chromosomes (14). Therefore, all of these structures comprise the sporophyte generation, which produces spores asexually. What you see when you look at a blossoming rose bush is thus an asexual, spore-producing plant, comparable to a mature fiddlehead fern.

The sexual phases of the rose (the male and female gametophytes) are highly reduced and hidden from view inside the sporophyte. Buried within the ovules and anthers of rose flowers are diploid megaspore mother cells and microspore mother cells, respectively. These spore-producing cells undergo meiotic ("reduction") division, yielding haploid megaspores and microspores containing only 7 chromosomes each. These two types of haploid spores are the first stages of the female and male gametophyte generations. The spores then divide mitotically to produce the mature gametophyte generations: the male pollen grains and the female embryo sacs, which produce the gametes, sperm and egg.

Pollination delivers the sperm to the egg via the pollen tube. During fertilization, the sperm nucleus fuses with the egg nucleus, thereby establishing the diploid number of chromosomes (2N = 14 in the case of roses) in the zygote. As a result, all of the following developmental stages of the sporophyte, from embryo to seedling to mature plant, are diploid. Strasburger was the first to demonstrate that morphological alternation of generations in plants is accompanied by a parallel alternation between the haploid (1N) and diploid (2N) generations. The haploid and diploid stages of the plant life cycle are indicated in Figure 18.6.

There is one more critical step, which, unlike alternations of generations, is unique to flowering plants. In 1902, the Russian botanist Sergius Nawaschin discovered the phenomenon of *double fertilization*, which occurs only in angiosperms. As described earlier, the pollen tube produces two sperm cells during development, but only one of them fertilizes the egg. The second sperm cell fuses with the *central cell*, a specialized binucleate cell (having two nuclei) within the embryo sac. Because fusion of three haploid nuclei is involved, the resulting cell has triple the number of chromosomes of the haploid nucleus and is therefore called *triploid* (3N). The triploid cell then undergoes repeated mitotic divisions to form the *endosperm* of the seed, the white tissue that provides nourishment to the developing embryo.[17] Cereal grains are especially rich in endosperm, which accounts for their importance as the primary staple crop of humans.

DARWIN AND MENDEL: SOLVING THE ANCIENT RIDDLES OF OUTCROSSING AND DEGENERATION

By the middle of the nineteenth century, the two-sex model of plants was at last firmly established. However, there remained several basic questions that would have to be answered before sexual reproduction in plants could be fully comprehended. Two of these questions, in particular, are of such long-standing and fundamental importance that they must be included in any full account of the history of the sexual theory: the questions of the biological significance of outcrossing and the cause of so-called "degeneration" in fruit trees. The scientific community could not even begin to address these two questions until two tectonic events shook the scientific community and changed the scientific landscape forever: the publication of Darwin's *Origin of Species* in 1859 and the rediscovery of Mendel's forgotten papers on inheritance in peas in 1900.

The phenomenon of outcrossing was especially contentious. Sprengel's comprehensive studies on the relationship between flower structure and insect pollination had been coldly received by Goethe and the nature philosophers, who believed that plants were by definition self-sufficient and needed no outside agents to complete their life cycles. They also objected to what they regarded as Sprengel's materialistic interpretation of insect-mediated pollination as nothing more than an elaborate game of deception and exchange of services. He had also referred to some insects as too "stupid" to navigate certain flowers and described insect "larceny," in which insects managed to steal nectar without performing pollination services (see Chapter 15). Where was God's wisdom in such an anarchic scenario?

Pollination by outside agents, such as wind or insects, was also bewildering because it implied that outcrossing was the norm and selfing the exception. As we saw in Chapter 15, the hereditary material was widely thought to be a uniform essence. It was assumed that selfing in hermaphroditic plants was nature's mechanism to preserve the purity—and thus the constancy—of the species by preventing the mixing of different species or varieties. Outcrossing, on the other hand, would actually *promote* such mixing, leading to the rapid dissolution of the species and universal chaos. Since species appeared to be constant in nature, selfing ought to be the primary pathway. Thus, Koelreuter and Sprengel were puzzled by the fact that most plants seemed to be outcrossers. As we have seen, Goethe had pointedly excluded outcrossing from his theory of metamorphosis.

The idea that eighteenth-century naturalists would assume selfing to be more beneficial than outcrossing may seem surprising today. Darwin observed the deleterious effects of selfing directly in his study of livestock breeding. Inbreeding, rather than contributing to the constancy of a particular breed, promoted the appearance of "sports" among the progeny, which differed markedly from the parental type. In contrast, the progeny of wild species that were free to interbreed were generally true to type. Darwin became convinced that, contrary to the conventional wisdom, cross-fertilization in nature actually *stabilized* species, whereas inbreeding promoted their dissolution.

Darwin was able to demonstrate the beneficial effects of outcrossing in plants by carrying out his own detailed cross- and self-pollination experiments in many different species and families, which he published in 1876. Darwin summarized his findings as follows:

> The first and most important of the conclusions which may be drawn from the observations given in this volume, is that cross-fertilisation is generally beneficial, and self-fertilisation injurious. This is shown by the difference in height, weight, constitutional vigour, and fertility of the offspring from crossed and self-fertilised flowers, and in the number of seeds produced by the parent-plants.[18]

Applying this principle to the evolution of flowers, Darwin proposed a thought experiment to illustrate how the nectar-secreting glands of flowers could have evolved through the process of natural selection:

> Now let us suppose that . . . nectar was excreted from the inside of the flowers of a certain number of plants of any species. Insects in seeking the nectar would get dusted with pollen, and would often transport it from one flower to another. The flowers of two distinct individuals of the same species would thus get crossed; and the act of crossing, as can be fully proved, gives rise to vigorous seedlings which consequently would have the best chance of flourishing and surviving. The plants which produced the flowers with the largest glands or nectaries, excreting most nectar, would oftenest be visited by insects, and would oftenest be crossed; and so in the long run would gain the upper hand and form a local variety.

In another section of *Origin of Species* titled "Utilitarian Doctrine, How Far True: Beauty, How Acquired," Darwin commented that "some naturalists" still objected to "the utilitarian doctrine that every detail of structure has been produced for the good of the possessor." Instead, they believe that the most attractive structures were "created for the sake of beauty, to delight man or the Creator." Darwin's argument against this idea is based on a comparison of the corollas of wind-pollinated versus insect-pollinated flowers:

> Flowers rank amongst the most beautiful productions of nature; but they have been rendered conspicuous in contrast with the green leaves, and in consequence at the same time beautiful, so that they can be easily observed by insects. I have come to this conclusion from finding it an invariable rule that when a flower is fertilized by the wind it never has a gaily colored corolla. . . . Hence we may conclude that, if insects had not been developed on the face of the earth, our plants would not have been decked with beautiful flowers, but would have produced only such poor flowers as we see on our fir, oak, nut and ash trees, on grasses, spinach, docks, and nettles, which are all fertilized through the agency of wind.[19]

One aspect of plant reproduction that Darwin was unable to explain, however, was the cause of variability among hybrid progeny. According to essentialist doctrine, the hereditary material was a uniform essence. It followed that during hybrid formation the two different essences mixed with one another like two miscible fluids. If such were the case, one would expect that all the progeny of a cross would be identical to each other and intermediate between the two parents. Furthermore, if the progeny of the hybrid cross (F_1 generation) were crossed with each other, the progeny of the second cross (F_2 generation) would be expected to be identical to the F_1 generation. Instead, the F_2 generation was far more variable than the F_1 generation. During the classical era, this phenomenon was referred to as "degeneration"—which, from the point of view of farmers wishing to perpetuate the desirable characteristics of cultivated fruits, it certainly was. Olive trees are, of course, the classic example.

Koelreuter, Gaertner, and others were puzzled by the fact that F_1 hybrids of different crosses sometimes exhibited slightly different phenotypes: some appeared to be intermediate between the two parents in all respects, some were mixtures of the features of both parents, whereas others tended to resemble one or the other of the parents. Koelreuter had tried to attribute these anomalies to incomplete mixing of the two parental essences during fertilization.

As for the extreme variability of the F_2 generation, Koelreuter was completely baffled. Darwin tried, but failed, to come up with an explanation. In the 1861 edition of *Origin,* he argued that the variability of the F_2 generation was caused by the physiological condition of the flower:

> For it bears on and corroborates the view which I have taken on the cause of ordinary variability; namely, that it is due to the reproductive system being eminently sensitive to any change in the conditions of life, being thus often rendered either impotent or at least incapable of its proper function of producing offspring identical with the parent-form. Now hybrids in the first generation are descended from species (excluding those long cultivated) which have not had their reproductive systems in any way affected, and they are not variable; but hybrids themselves have their reproductive systems seriously affected, and their descendants are highly variable.[20]

Darwin's belief that the variability in the F_2 generation is caused by the trauma of hybridization is reminiscent of Henschell's discredited view that Koelreuter's hybrids were actually monsters caused by emasculation of the flowers and root stress induced by growing the plants in pots. Even genius has its limits, and Darwin was no closer to an explanation of "degeneration" than any of his predecessors.

Meanwhile, the obscure monk Gregor Mendel was laboring away in his garden in the Augustinian monastery in Brünn, the provincial capital of what was then Moravia, Austria. Mendel had been carrying out crossing experiments with peas to try to answer the same questions about the causes of variability that Darwin was struggling with in his *Origin of Species*. But whereas Darwin thought of variability in physiological terms, Mendel, who had studied combination theory in mathematics—a method for determining all the possible arrangements of any group of objects—took a statistical approach to the problem. He had begun by studying albinism in mice, crossing wild-type mice with albinos and observing the colors of the progeny. However, after being reprimanded by the local bishop, Anton Ernst Schaffgotsch, he switched to plants. Bishop Schaffgotsch thought it unseemly and "unnecessarily titillating for a priest who had taken vows of chastity and celibacy" to be spending his time watching mice copulate.[21] "I turned from animal breeding to plant breeding," Mendel later explained with a sly smile, because "the bishop did not understand that plants also have sex."[22]

To apply his statistical methods to the study of heredity, Mendel divided heredity into discrete, physical units that could form different combinations with each other. Such a "particulate" theory of heredity represented a marked departure from the essentialist doctrine that envisioned crossing as the mixing of miscible fluids. For this approach to work, Mendel's hereditary units had to be distinct from each other and stable in their characteristics. After poring over the studies of Gaertner and other plant hybridizers, he chose peas as his experimental material because he was able to identify several phenotypic traits that behaved like his ideal stable hereditary units.

In a nutshell, Mendel's crossing experiments led him to discover the rules that govern the appearances of both the F_1 and F_2 generations. The hereditary material was not a uniform essence, but a collection of individual hereditary units. These units were later called *genes*. The pea genes Mendel studied existed in different forms, called *alleles*, which determined particular traits, such as smooth versus wrinkled peas. Crossing peas with different traits led to a combination of the hereditary units of the male and female parents, and the progeny (F_1 generation) all contained the same combination. However, when the F_1 generation was selfed, the hereditary units all came apart and formed new combinations. Because some of the alleles were *dominant* and some *recessive*,[23] the new combinations led to phenotypic variability. In modern terms, we say that Mendel discovered the two basic principles of the *segregation* and *independent assortment* of alleles during sexual reproduction.[24]

It is tantalizing to imagine what direction Darwin's work would have taken had he been aware of Mendel's results, which were originally published in 1866, only seven years after the *Origin of Species*. It is likely that he would have immediately realized that Mendel's new laws governing inheritance could explain the injurious effects of selfing as the accumulation and subsequent expression of deleterious recessive alleles that are not expressed in the progeny of outcrossed plants. This is thought to be one of the reasons why outcrossing was selected for in plants and why insect- and wind-mediated pollination became the norm. Without the existence of two sexes in plants there could be no outcrossing and thus no way to suppress the expression of mutated, deleterious alleles.

In addition, Darwin would no doubt have understood that an allele that is deleterious under one set of conditions may be advantageous under another set of conditions. Sexual reproduction is, in fact, a two-edged sword. On the one hand, it can suppress deleterious mutations that arise spontaneously during an organism's development. On the other hand, by creating new combinations of alleles, it can undo whatever advantages have been gained over many generations through the process of natural selection.[25] However, if the environmental conditions were to change suddenly (due, for example, to the presence of pathogens or shifts in sunlight, temperature, water availability, or soil chemistry), some of the so-called "deleterious" alleles might prove to be advantageous, thus allowing the species to adapt to the new conditions.

These two factors—the suppression of deleterious alleles and the increase in genetic diversity—provide us with a plausible explanation for the existence of two sexes in plants, although it is by no means the complete story. Although beyond the scope of this book, multiple factors have led to the evolution of sex, and different types of organisms have evolved sexual cycles for different reasons.[26] It is likely that additional evolutionary factors favoring the plant sexual cycle and alternation of generations will be discovered in the future.

SO WHO WON THE DEBATE, SEXUALISTS OR ASEXUALISTS?

Hofmeister's brilliant synthesis demonstrating that the alternation of generations is a universal feature of plant life cycles did not just unify the plant kingdom. It can also be viewed as the final resolution of the age-old quandary over plant sex.

Contrary to the usual triumphalist narratives of the history of science, both the sexualists and asexualists turned out to be correct, although for very different reasons. Aristotle and Theophrastus, along with their legions of asexualist disciples throughout the ages, arrived at their opinion via culturally biased intuition about the inner nature of plants, which led them to conclude, a priori, that plants were incapable of sex. On the other hand, the metaphors and analogies used by the asexualists to describe plant reproduction were often based on the female reproductive system, including eggs, uterus, and menstruation. Thus. the asexualists were in fact proposing that plants were functionally parthenogenic females.

Nehemiah Grew inaugurated the sexual theory by comparing stamens to penises, anthers to testicles, and pollen grains to sperm, while Camerarius provided the first experimental proof of the sexual theory in dioecious and monoecious species. However, unlike the testicles of male animals, anthers do not give rise to gametes (sperm cells) directly—they produce spores by meiosis instead. Since spores are not gametes—that is, they do not fuse with other spores to form a zygote—spore production is an asexual reproductive process. Thus the stamen itself is actually a spore-producing asexual structure. The microspores inside the anther divide mitotically to form a new haploid individual—the mature pollen grain with its elongated pollen tube containing two sperm cells.

Similarly, the ovary of the flower, unlike the ovaries of female mammals, does not give rise directly to the female gamete (egg) but rather gives rise by meiosis to the female spore, or megaspore. The megaspore then divides mitotically to form a new haploid plant—the embryo sac. Thus, the ovary, like the anther, is an asexual, spore-producing structure.

Although stamens and pistils are asexual structures, it is also true that they specifically give rise to the male and female gametophyte generations, respectively. It is thus convenient to assign these two asexual organs different genders: stamens are gendered male and pistils are gendered female. In adopting this convention, we are following the lead of Beukeboom and Perrin, who applied the term "gender" (normally defined as a culturally derived construction) to sporophytic structures that are specialized in "one or the other sexual function."[27] Although this practice risks a certain amount of confusion, which we are trying mightily to avoid, such a plant-specific definition of gender makes it easier to talk about the roles of stamens and pistils in the sexual cycle of angiosperms.

The two gametophytes, the pollen grain and embryo sac, are *genetically distinct* from the sporophyte that produced them, and not only because they are haploid instead of diploid. As noted earlier, since meiosis involves some exchange of alleles (gene variants) between chromosomes, new combinations and arrangements of maternal and paternal genes are generated. This makes the two gametophytes genetically distinct from the sporophyte, which is why they are referred to as the gametophyte *generation*. Despite their microscopic size, these two genetically distinct haploid individuals (male and female gametophytes) are the mature sexual stages of the plant life cycle.

Like the key to Frances Hogdson Burnett's *Secret Garden*, the key to the plant life cycle was "buried deep," not in the ground, but in the sporophytic tissues of the flower. Rather than germinate on soil as the spores of cryptogams do, angiosperm spores are retained within the two floral organs, stamens and pistils, where they develop into highly reduced male and female gematophytes. These two types of gametophytes growing inside the anther and ovary are, in a sense, Flora's "secret gardens." The pollen grain, or male gametophyte, became adapted over millions of years of evolution to hitch rides with wind, insects, or other animals and land with remarkable efficiency on the surface of the tiny stigma. There it sprouts a thread-like pollen tube that grows down through the style and forms a conduit for the accurate delivery of two sperm cells to the embryo sac, or female gametophyte. The fusion of sperm and egg marks the beginning of the new sporpophytic, or asexual stage, of the plant life cycle.

For centuries, Flora's sexuality was concealed under a veil of sporophytic tissue. Hofmeister's discovery that seed plants, like the cryptogams, consist of two "generations," an asexual diploid individual and two sexual haploid individuals, effectively lifted the veil of secrecy.

Because of the alternation of generations, the sexualists and the asexualists can both claim to have been right. However, the asexualists' position was based on ancient traditions and social constructions about gender rather than on science. It was the sexualists who successfully freed their minds from such cultural biases and glimpsed the true sexual nature of plants. The sexualists went on to demonstrate the presence of both the sexual and asexual generations in the life cycles of all plants, as revealed by Hofmeister's great synthesis, thus enabling the asexualists to share in the ultimate solution to the puzzle.

NOTES

1. Darwin, Charles (1958), *The Autobiography of Charles Darwin: 1809–1882*, N. Barlow, ed. Collins, p. 120.

2. von Sachs, J. (1906), *History of Botany (1530–1860)*, trans. H. E. F. Garnsey; revised by I. B. Balfour. Clarendon Press, p. 200. Originally published in German in 1860.

3. Ibid.

4. Ibid.

5. It is worth noting that Darwin himself never pursued the question of the evolution of the alternation of generations. Although the evolutionary relationship between the cryptogams and the seed plants was taken for granted by plant morphologists after Darwin, there was a long delay—several decades at least—before comparative morphologists began addressing the underlying selection mechanisms that gave rise to the plant life cycle.

6. Kaplan, D. R., and T. J. Cooke (1996), The genius of Wilhelm Hofmeister: the origin of causal-analytical research in plant development. *American Journal of Botany* 83:1647–1660.

7. The severity of nearsightedness is measured by the strength of the lens (in units of diopters) needed to correct it. For example, a −6 diopter myope can achieve a magnification of 1.5× with the naked eye by viewing an object at a distance of 6.2 inches. An even more severe −10 diopter myope (probably similar to Hofmeister's vision) can achieve a 2.5× magnification by viewing an object at only 4 inches from the eye. See Gorelick, Leonard, and A. John Gwinnett (1981), Close work without magnifying lenses: A hypothetical explanation for the ability of ancient craftsmen to effect minute detail. *Expedition* Winter: pp. 27–34.

8. Kaplan and Cooke, The genius of Wilhelm Hofmeister.

9. Ibid.

10. Hofmeister borrowed the term "Alternation of Generations" from zoology. The Danish zoologist Japetus Steenstrup had used it to describe the alternating sexual and asexual stages of the life cycles of certain invertebrates, such as tunicates, cnidarians, and trematodes. However, as was discovered later, the alternation of generations in plants differs from that in animals because the asexual sporophyte stage is *diploid* (having two sets of chromosomes), whereas the sexual gametophyte stage is *haploid* (having one set of chromosomes).

11. Fern allies are not true ferns but are similar in that they are nonseed plants with vascular systems for the transport of water and nutrients. Like ferns, they release asexual spores and exhibit the alternation of generations.

12. *Selaginella* is a heterosporous member of the Lycophyta, the oldest living Division of vascular plants, which first appeared in the fossil record around 410 million years ago, during the Silurian Period. There are about 700 species of *Selaginella*, most of which are small and delicate. They are found in a variety of climates, including tropical, arctic, temperate, and desert. One species, *S. lepidophylla*, is commonly known as the "resurrection plant." It grows in the Mexican desert and dries into a tight ball during droughts, but is able to resume growth upon wetting.

13. Gifford, E. M., and A. Foster (1987), *Morphology and Evolution of Vascular Plants*. W. H. Freeman, p. 136.

14. Juniper "berries" have fleshy walls and superficially resemble fruits, but they are actually *ovulate cones* comprised of a seed surrounded by small, scale-like leaves. Another apparent exception is the berry-like structures of members of the *Taxacaea*, such as yew trees. The seeds of yew

trees are also surrounded by a modified cone consisting of a cup-shaped fleshy leaf scale called an *aril*. When mature, the aril turns a bright red color.

15. For more detailed descriptions of gymnosperm and angiosperm morphology, see Willis, Kathy, and Jennifer McElwain (2014), *The Evolution of Plants*, second Edition. Oxford University Press.

16. Chromosomes are strands of DNA and protein that encode the complete set of genes, or *genome*, of an organism. In a typical nondividing cell, the chromosomes are loosely dispersed throughout the nucleus in the form of *chromatin*. Human cells contain forty-six chromosomes. If the unraveled DNA strands of the forty-six individual chromosomes were laid out end to end, the total length of the genomic DNA (all forty-six chromosomes) inside the nucleus would be about 6.5 feet. When 6.5 feet of dispersed DNA is packed into a nucleus that is only about 6 micrometers in diameter, it becomes impossible to see full-length strands of chromatin. Prior to nuclear division, however, the chromatin condenses into tightly packaged, highly organized bodies (*chromosomes*) that can be strongly stained by dyes for viewing under the microscope. The name "chromosomes" literally means "colored bodies." The aniline dye *basic fuchsin*, which stains chromosomes magenta red, is among the most effective for staining chromosomes.

17. The second sperm cell is not functioning as a true gamete because it doesn't fuse with the egg cell to form a zygote; it fuses with the central cell to form an endosperm cell. Therefore the term "double fertilization" is somewhat misleading. It is not yet known whether the two sperm cells are identical and interchangeable or whether one is specialized to serve as the gamete.

18. Darwin, C. R. (1876), *The Effects of Cross- and Self-Fertilisation in the Vegetable Kingdom*. John Murray.

19. Darwin, C. R. (1859), *On the Origin of Species by Means of Natural Selection*. A Mentor Book, New American Library, p. 185.

20. Darwin, C. R. (1859), *On the Origin of Species by Means of Natural Selection*, third edition. John Murray, p. 296.

21. Henig, R. M. (2000), *The Monk in the Garden*. A Mariner Book, Houghton Miflin.

22. Quote from Ibid.

23. Although members of a given species all have the same genes in their genome, individual members of a sexually reproducing population may, by random mutation, have different versions of the same gene, called *alleles*. Diploid organisms have two alleles for each gene in the genome, one from the male parent and one from the female parent. Different alleles of the same gene can give rise to different morphological or biochemical traits, or *phenotypes*. In some cases, one allele tends to mask the phenotype of another allele. The former is said to be *dominant* and the latter is said to be *recessive*.

24. In genetics, the *Law of Segregation* refers to the fact that, during meiosis, the two parental alleles of each gene *segregate* (i.e., separate from each other) so that the gamete (or in the case of plants, the gametophyte) receives only one of the two alleles. The *Law of Independent Assortment* states that when multiple traits (such as purple flowers, wrinkled seeds, and dwarfism) are inherited, the genes controlling these traits assort independently from one another during meiosis. Independent assortment allows traits to form new combinations with other traits, giving rise to novel phenotypes.

25. Beukeboom, L. W. and N. Perrin (2014), *The Evolution of Sex Determination*. Oxford University Press; Niklas, K. J., E. D. Cobb, and U. Kutschera (2014), Did meiosis evolve before sex and the evolution of eukaryotic life cycles? *Bioessays* 36:1091–1101.

26. The evolution of sex is a complex topic that is not yet fully understood. See Beukeboom and Perrin, *The Evolution of Sex Determination;* and Niklas et al., Did meiosis evolve before sex and the evolution of eukaryotic life cycles? for more detailed discussions of current theories.

27. Beukeboom and Perrin, *The Evolution of Sex Determination*, p. 10.

INDEX

Page references for figures are indicated by *f* and for tables by *t*.